ATOMIC PHYSICS 8

1973 – Atomic Physics 3

Proceedings of the Third International Conference on
Atomic Physics, August 7–11, 1972, Boulder, Colorado
S. J. Smith and G. K. Walters, Conference Chairmen and Editors

1975 – Atomic Physics 4

Proceedings of the Fourth International Conference on
Atomic Physics, July 22–26, 1974, Heidelberg, Germany
G. zu Putlitz, Conference Chairman, E. W. Weber and A. Winnacker, Editors

1977 – Atomic Physics 5

Proceedings of the Fifth International Conference on
Atomic Physics, July 26–30, 1976, Berkeley, California
Richard Marrus, Conference Chairman, Michael Prior and Howard Shugart, Editors

1979 – Atomic Physics 6

Proceedings of the Sixth International Conference on
Atomic Physics, August 17–22, 1978, Riga, USSR, A. M. Prokhorov, Conference Chairman
R. Damburg, Editor

1981 – Atomic Physics 7

Proceedings of the Seventh International Conference on
Atomic Physics, held August 4–8, 1980, at the Massachusetts
Institute of Technology, Cambridge, Massachusetts
Daniel Kleppner and Francis M. Pipkin, Conference Chairmen and Editors

1982 – Atomic Physics 8

Proceedings of the Eighth International Conference on
Atomic Physics, August 2–6, 1982, Göteborg, Sweden
Ingvar Lindgren, Conference Chairman
Arne Rosén and Sune Svanberg, Editors

A Continuation Order Plan is available for this series. A continuation order will bring delivery of each new volume immediately upon publication. Volumes are billed only upon actual shipment. For further information please contact the publisher.

ATOMIC PHYSICS 8

Editors

Ingvar Lindgren
Arne Rosén
Chalmers University of Technology
Göteborg, Sweden

and

Sune Svanberg
Lund Institute of Technology
Lund, Sweden

PLENUM PRESS • NEW YORK AND LONDON

The Library of Congress cataloged the first volume of this title as follows:

International Conference on Atomic Physics.

Atomic Physics; proceedings. 1st–
1968–
New York [etc.] Plenum Press.
v. illus. 26 cm. biennial.

1. Nuclear physics–Congresses. I. Title.
QC173.I 53 539.7 72-176581

Library of Congress Catalog Card Number 72-176581
ISBN 0-306-41361-2

Proceedings of
the Eighth International Conference on
Atomic Physics, held August 2–6, 1982,
at Chalmers University of Technology,
Göteborg, Sweden

A Division of Plenum Publishing Corporation
233 Spring Street, New York, N.Y. 10013

Printed in the United States of America

Organizing Committee

I. Lindgren Chairman	Chalmers University of Technology, Göteborg
S. Svanberg Vice Chairman	Lund Institute of Technology
A. Rosén Secretary	Chalmers University of Technology, Göteborg
V.V. Afrosimov	U.S.S.R. Academy of Sciences, Leningrad
E. Arimondo	Università di Pisa
B. Bederson	New York University
B. Cagnac	Université Pierre et Marie Curie, Paris
R.J. Damburg	Latvian Academy of Sciences, Salspils, Riga
G. Drake	University of Windsor
N. Fortson	University of Washington, Seattle
V.W. Hughes	Yale University, New Haven
D. Kleppner	M.I.T., Cambridge
J. Los	FOM-Institute, Amsterdam
R. Marrus	University of California, Berkeley
H. Narumi	Hiroshima University
G. zu Putlitz	Universität Heidelberg/GSI Darmstadt
F. Read	University of Manchester
A.L. Schawlow	Stanford University
G. Series	University of Reading
T. Skalinski	Institute of Physics PAN, Warsaw
S.J. Smith	J.I.L.A./N.B.S. and University of Colorado, Boulder

Programme Committee

In addition to the members of the Organizing Committee:

E.D. Commins	University of California, Berkeley
W. Johnson	University of Notre Dame
H. Kleinpoppen	University of Stirling
S. Lundqvist	Chalmers University of Technology, Göteborg
C. Nordling	Uppsala University
I. Sellin	Oak Ridge National Laboratory, Oak Ridge
H. Stroke	New York University

Sponsors

Chalmers University of Technology
Nobel Foundation
Swedish Natural Science Research Council
Department of Education
International Union of Pure and Applied Physics (IUPAP)
Nordic Institute of Theoretical Atomic Physics (NORDITA)
City of Göteborg
AB Volvo
Scandinavian Airlines System (SAS)

Host Institution

Chalmers University of Technology, Göteborg

PREFACE

The Eighth International Conference on Atomic Physics was held at Chalmers University of Technology, Göteborg, Sweden on August 2-6, 1982. Following the tradition established by earlier conferences in the series, it was attended by 280 participants from 24 countries. A total of 28 invited talks were delivered at the conference. These talks, which are presented in this volume, covered a wide range of topics in atomic physics in a broad sense. They extend from very basic problems (e.g., the interpretation of quantum mechanics in light of Bell's theorem and the feasibility of relativistic many-body calculations) to applied problems (e.g., laser detection of trace elements and spectroscopy of chemisorbed molecules).

Professor M.Ya. Amusia was unable to attend the conference but his invited paper is included here. Professor V.S. Letokhov presented a talk entitled "Prospects of Laser Detection of Very Rare Isotopes, but was unable to provide a manuscript. At the conference, 175 posters were presented. Abstracts have been published in a separate volume.

It is very much appreciated that all the 1981 Nobel laureates, Nicolaas Bloembergen, Arthur Schawlow and Kai Siegbahn, were able to attend and deliver their invited talks. Professor Schawlow summed up the conference and this too is presented here. The conference also benefited considerably from the presence of Professor I.I. Rabi, who gave a much appreciated talk at the conference dinner. As this talk was given without a manuscript, it could unfortunately not be included here.

Ingvar Lindgren
Arne Rosén
Sune Svanberg

CONTENTS

CONFERENCE OPENING

Kai Siegbahn

President of the International Union of
Pure and Applied Physics
Institute of Physics, University of Uppsala
Box 530, S-751 21 Uppsala, Sweden

It was only about two months ago that I had the pleasure to address myself as the president of the International Union of Pure and Applied Physics to an impressive number of physicists gathered here in Gothenburg on the occasion of the International Conference on Plasma Physics under the chairmanship of Hans Wilhelmsson. This time, Ingvar Lindgren together with Sune Svanberg and Arne Rosén have assembled here the 8th International Conference on Atomic Physics which also is sponsored by the International Union of Pure and Applied Physics. There are many connections between these two fields of research. Both at the generation of plasma and at the diagnostics of it a great number of atomic processes come into play. From that point of view plasma physics is partly atomic physics, although plasma physics is indeed living its own life, characterized among other things by the coupling between kinetic motion and magnetic field. We may recall in this connection the tremendous difficulties we have experienced in fusion research to achieve in the physics laboratory magnetic confinement and to reproduce the processes which occur under astrophysical conditions and which are basic in our understanding of the Universe.

It is not easy to define what we mean by atomic physics today when almost every branch of physics more or less is related to atomic concepts. The quantum theory is of course the foundation of it and for a long time much of the research in atomic physics was devoted to various precise ways to check the exact validity of quantum mechanics for simple systems. As a result of this the renormalization procedures were introduced in quantum electrodynamics which could remove the obvious remaining difficulties in quantum theory. In view of the outstanding success of this theory, in particular its ability to calculate physical quantities like the anomalous magnetic moment of the electron correctly to within an impressive number of decimals, it was surprising to hear recently at a meeting in Lindau that the grand old man of quantum theory, Paul Dirac, is still not quite happy. Instead of the established quantum electrodynamics he would rather prefer a theory which was basically free from renormalization procedures. His ideas how such a new quantum theory should be designed are still of a speculative nature, fascinating and of course controversial. For the time being we can safely use the quantum electrodynamics as it is since there does not yet seem to be any physical exception from it so far. It is interesting to note that during this conference other problems of a similar fundamental nature will be discussed under the title of "Experimental tests of Bell's inequalities in atomic physics".

With satisfaction we can also notice that the new field connected to the so-called "Quantized Hall effect" is on the program together with several other most important topics in atomic physics with bearing upon atomic constants and advanced physics metrology treated by pioneers in the fields.

Laser spectroscopy is rapidly developing and finds applications over wide fields in atomic physics. Apart from being applied to the more classical problems, for example to accurately measure hfs splittings, the precision and ultra-high resolution have enabled an increasing number of new problems to be studied. Also, continuous improvements in laser technology gradually offer new possibilities for research further out into the vuv part of the spectrum at higher power levels, increased pulse frequencies and with pulse lengths appropriate for different purposes. Again, we are glad to have the prominent leaders in the laser field among us here at the conference.

Some trends in atomic physics are obvious. Much of the interest has since long been concerned with the structural properties of free atoms. Experiments and quantum theory were applied to such systems mostly in atomic spectroscopy in different electromagnetic wave-length regions from hard X-rays to radio waves. By applying more and more refined and precise methods the atomic systems could be analyzed in terms of interactions between the atomic nucleus and the atomic electrons and between the electrons themselves. Present-day improvements in precision and also in sensitivity are remarkable: It is even possible to record the optical light being emitted from one single isolated atom, two atoms and three atoms, successively, trapped and confined in a special field configuration. As a complement to the study of the properties of single atoms in atomic physics, atomic and ionic interactions at collision events have also been studied. Such experiments belong to the traditional fields of atomic physics and give fundamental information about atoms when their electrons approach each other. A step further is when the atoms form bonds with each other in molecules. One can then distinguish between localized atomic or core electrons and more or less delocalized valence electrons, which are of particular interest for the chemists. Simple molecules, consisting of say 2 or 3 atoms, emit light extending into the vuv region, when the excited molecules are making electronic transitions in the valence region. A competitive mode is the emission of electrons at sufficiently high excitation energies through energy conversion which becomes quite dominating for most molecules. Such atomic systems are conveniently studied by means of electron spectroscopy. A rewarding branch is surface physics. Different forms of atomic aggregations, including clusters, solids and liquids, are subject to current research. A combination of laser excitation and electron spectroscopy is likely to develop in the future. Hopefully, the wave-length region for efficient laser excitation will soon be extended into the vuv at higher intensities by means of excimer lasers and other means.

A further glance at the conference program gives convincing evidence for the speed of progress in treating even large atomic and molecular systems by means of computer methods.

We have all been looking forward to meeting our colleagues on this occasion and as a member of the conference and as a representative of the International Union of Pure and Applied Physics I forward our thanks to Ingvar Lindgren and his associates here in Gothenburg for organizing this conference on Atomic Physics which no doubt will turn out to be most stimulating for us.

DYNAMICS OF RESONANT STATES

U. Fano

Department of Physics
University of Chicago
Chicago, Illinois 60637, U.S.A.

1. INTRODUCTION

The mechanics of atoms is often divided into "structure" and "collision" problems. I view here "collisions" as including any nonperturbative excitation process or chemical reaction, more generally any transformation of the structure of matter. The independent particle model has proved extremely successful for describing and interpreting structures; in fact it is commonly regarded as the theory of atomic systems. This model has also accounted for optical transitions and for fast collisions, which can be treated as weak perturbations of atoms. Its scope and power have been extended by configuration mixing and other procedures.

On the other hand, more elaborate transformations of matter, typically chemical - especially biochemical - reactions, involve coordinated motions of several electrons and nuclei, far beyond the scope of the independent particle model. Their microscopic description and interpretation require us to develop new theoretical concepts and tools. Initial stages of this development have been in progress for a couple of decades, centering on the study of N-particle wave functions in their configuration space with 3N dimensions.

Clues toward identifying and interpreting mechanisms of correlated motions have emerged repeatedly from the study of resonances, that is, of unstable states of two or more constituents that last long enough to appear as discrete structures in the energy spectra of collision processes. Resonance properties, typically their lifetimes, are generally expressed in terms of lumpparameters such as interaction matrix elements. Calculations of these parameters, often very accurate, utilize close-coupling or variational wavefunctions constructed from products of separate functions of single-particle coordinates. These descriptions serve their immediate purposes well but describe correlations only indirectly, through superposition coefficients. Mechanisms for the formation and decay of resonances remain thus rather obscure.

In this report I should like to deal with a paradoxical property of many resonances: The configuration of their particles looks totally unstable at first sight, in contrast to the modicum of stability required for a resonance to be observed. That is, the particle - or particles - appear to be localized in a region of configuration space free of obvious confining barriers. Very suggestive aspects of this phenomenon have become increasingly well documented in the last two years and seem worth reviewing even though the stability of the relevant resonances still lacks a firm theoretical basis.

2. PROTOTYPE PHENOMENA

Two seemingly different phenomena will be discussed, which I regard as essentially isomorphic from a physico-mathematical point of view. The first one concerns the famous resonance first discovered by Schulz, in e-He elastic scattering at 19.37 eV, and its analogs.[1] The Schulz resonance was attributed to the formation of He^- in a $1s(2s)^2$ configuration. Many analogous resonances have

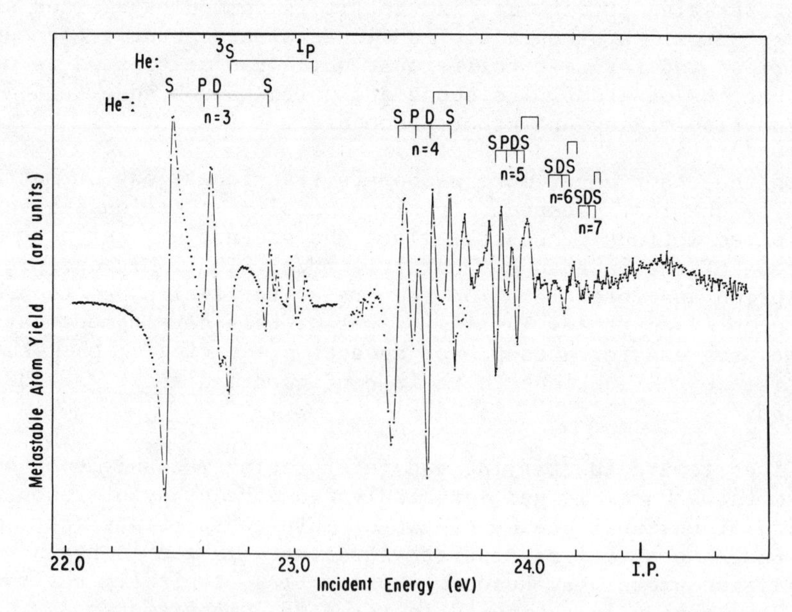

Fig. 1. Spectrum of $e+He \rightarrow e+He(\text{metast.})$ from ref. 2.

since been found, culminating in recent studies of the spectrum of e+He→e+He(metast.) which display quasi-Rydberg series of $He^{-}(1s\,n\ell^{2})$ resonances extending to n~10 (Fig. 1)[2]. Extensive work, including detailed calculations, has led to a consensus that the wave functions of the $n\ell$ pair (and of analogous configurations of doubly excited He) center in configuration space at points along the ridge of the potential barrier shown in Fig. 2[3]. The pair configuration shown in the inset is stabilized against bending away from $\theta_{12}=180^{o}$ by a potential rise in a direction θ_{12} orthogonal to the (r_1,r_2) plane but displayed in the alternative plot of Fig. 3. No comparable rise seems, however, to stabilize it against asymmetric stretch deformations which would shift the configuration away from $r_1 \sim r_2$ into the potential valleys on either side of the ridge. (A resonance with low quantum numbers might be stabilized by boundary conditions at $r_1=0$ and $r_2=0$ but the same would not hold in a semi-classical regime).

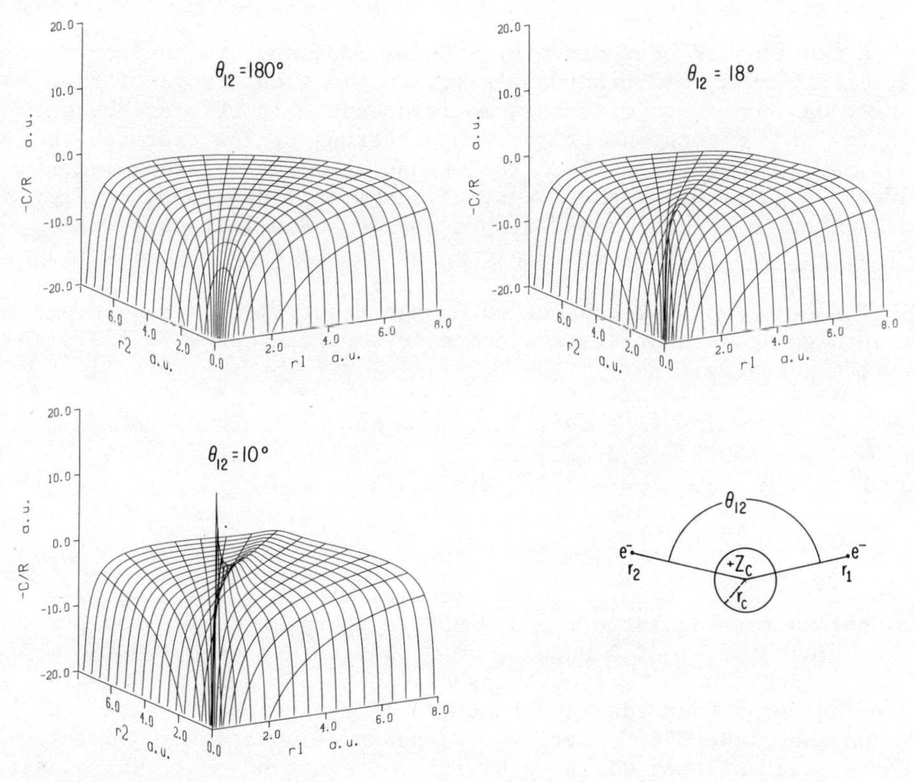

Fig. 2. Potential surface of a two-electron system over the (r_1,r_2) plane of configuration space. Lines are drawn on the surface at constant values of $R=(r_1^2+r_2^2)^{-\frac{1}{2}}$ and of $\alpha=\tan^{-1}(r_2/r_1)$; ridge line at $\alpha=45^{o}$. Surfaces are drawn for three different values of angle θ_{12} shown in inset. (Courtesy M. Cavagnero)

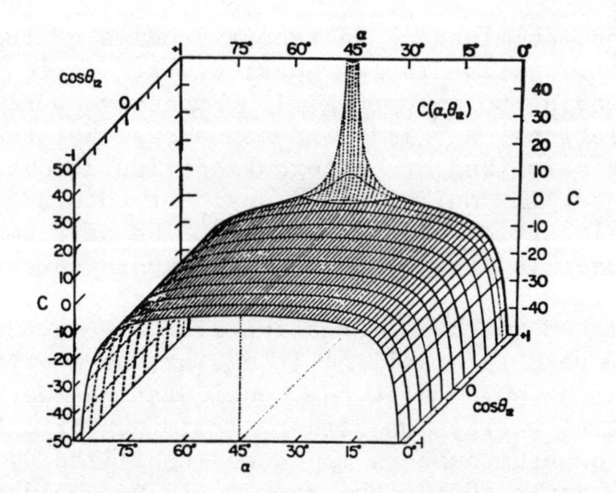

Fig. 3. Potential surface of a two-electron system over (α,θ_{12}) at constant R; ridge at $\alpha=45^{\circ}$ (ref. 8).

Attention had been drawn by Wannier long ago[4] to the peculiar instability of an electron pair's motion along the ridge of Fig. 2. Wannier was concerned with this motion because it affords the only path to joint detachment of the two electrons of low energies, but his remarks pioneered in fact the mechanics of correlated electron pairs.[3] In particular, the potential's symmetry displayed in Fig. 2 led Wannier to use and separate previously unfamiliar coordinates of the electron pair directed along and across the potential ridge.

The second phenomenon occurs in the spectrum of a single electron in high Rydberg states subjected to a strong magnetic field with the potential energy

$$V(r,\theta) = -\frac{e^2}{r} + \frac{e^2}{8mc^2} B^2 r^2 \sin^2\theta \ . \qquad (1)$$

As the diamagnetic term of this potential starts prevailing over the Coulomb term at large r and $\theta \sim 90^{\circ}$, Rydberg series of levels evolve into series of Landau resonances with spacings $\sim eB\hbar/mc$ which extend beyond the ionization threshold (Fig. 4)[5]. Here again evidence indicates that the wave function of the resonant states centers on the plane $\theta=90^{\circ}$ where $V(r,\theta)$ peaks at constant r. No influence is apparent that would stabilize the electron near this plane for low values of its angular momentum L_z. A plot of the potential (1) over the coordinate plane (r,θ) shows here too a ridge running along the line $\theta=90^{\circ}$. Studies of this phenomenon are reviewed in Sec. 3.

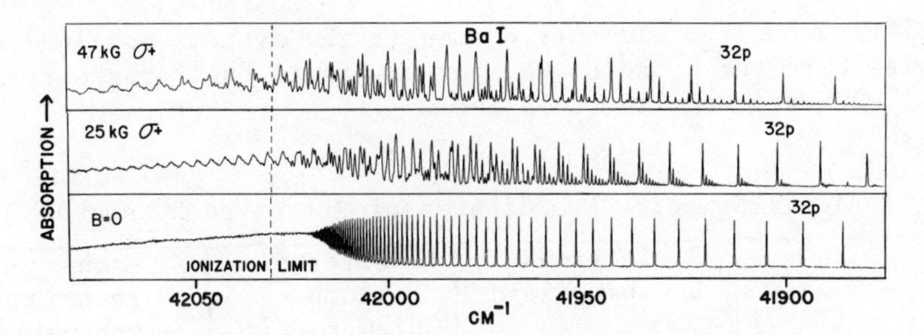

Fig. 4. Photoabsorption spectrum of barium near the ionization limit in magnetic fields B of different strengths (ref. 5).

The occurrence of resonances on potential ridges has obvious relevance for the "transition state" of chemical reactions.[6] This is the state of a transition complex as it traverses the activation threshold (i.e., the ridge) that separates the initial and final stable configurations of the reactants (Fig. 5). Indeed studies of the complex' wavefunction in coordinates appropriate to the potential symmetry on the ridge have attracted attention recently accounting, e.g., for a remarkable concentration of vibrational energy in the output of the reaction $H_2+F \rightarrow H+HF$.[7]

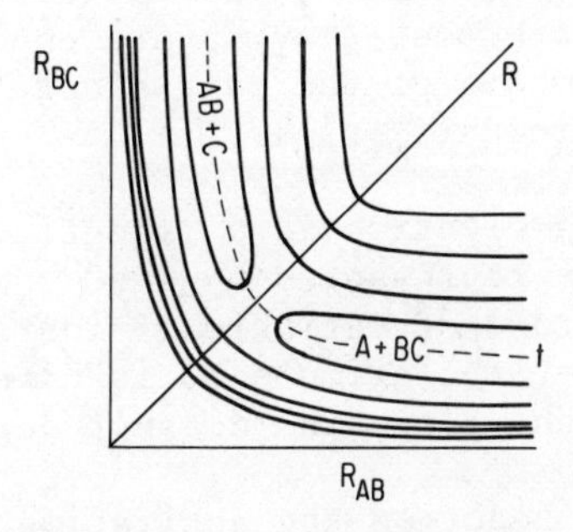

Fig. 5. Schematic illustration of a configuration space trajectory, t, for a chemical process $AB + C \rightarrow A + BC$. The solid lines depict equipotentials. The ridge line R represents here $R_{AB} = R_{BC}$. (Courtesy M. Cavagnero).

Sifting out the elements common to the systems outlined above, we review here the coordinates appropriate to their respective potential surfaces and listed in Table 1:

Table 1. Symmetry Coordinates of Prototype Phenomena

	2 electrons Coulomb Field	1 electron Coulomb + diamagnetic	exchange reaction (schematic)
Ridge coordinate	$R=(r_1^2+r_2^2)^{\frac{1}{2}}$	r	$R=(R_{AB}^2+R_{BC}^2)^{\frac{1}{2}}$
Unstable coord.	$\cos 2\alpha=\frac{r_1^2-r_2^2}{r_1^2+r_2^2}$	$\cos\theta$	t
Stable coord.	θ_{12}	–	–
Neutral coord.	–	ϕ	–
Ridge equation	$r_1=r_2$	$\theta=90^\circ$	$R_{AB}=R_{BC}$

a) For a pair of excited electrons the ridge line is represented by $r_1=r_2$ in Fig. 2, the potential being symmetric under reflection through it. In Fig. 3 the point ($\alpha=45^\circ$, $\theta_{12}=180^\circ$) marks the plot's intersection with the ridge line. The coordinate $\cos 2\alpha=(r_1^2-r_2^2)/(r_1^2+r_2^2)$, orthogonal to the ridge line, plays instead the role of "variable of instability", whereas displacements of the pair along the coordinate $\cos\theta_{12}=\hat{r}_1\cdot\hat{r}_2$ are opposed by a weak restoring force. Displacements of the pair along the ridge line are measured by the variable $R=(r_1^2+r_2^2)^{\frac{1}{2}}$.

b) For the Rydberg electron in a magnetic field the ridge line of the potential (1) is represented by $\theta=90^\circ$, whereas θ itself, at constant r, plays the role of variable of instability. The remaining electron coordinate, the azimuth ϕ, is here a variable of neutral equilibrium since the potential doesn't depend on it.

c) For a chemical reaction the activation threshold lies generally on a saddlepoint of a potential surface (Fig. 5). Distances from this point along a steepest descent trajectory represent the variable of instability. Motion along the ridge line, orthogonal to this trajectory, may be described in terms of a variable R analogous to that of the electron pair, which measures a "breathing deformation" of the complex.[8] The potential surface depends generally on a number of additional coordinates whose variations meet restoring forces or no reaction.

These diverse phenomena are viewed here as isomorphic in that all of them display the same feature involving a pair of independent variables: One of these variables runs along the ridge - i.e., an axis of symmetry - of the potential function; this coordinate is labelled R or r. The other variable runs orthogonally to the ridge and is called $\cos 2\alpha$ in example a), $\cos\theta$ in b) and represents the distance from the ridge along the trajectory of a chemical reaction in Fig. 5; the potential is a decreasing function of the square of this coordinate.

3. EVIDENCE ON THE DIAMAGNETISM OF RYDBERG STATES

The dominant influence of the diamagnetic term of the potential (1) on high Rydberg states was first pointed out by Schiff and Snyder in 1937.[9] Evidence of the wavefunction's concentration on the ridge emerged following the discovery of Landau resonances, from two remarks: i) The major phenomena are restricted to states of even parity under the reflection $\theta \rightarrow \pi - \theta$, which have an antinode at $\theta = 90^o$.[5] ii) The position of the Landau resonances is accounted for by a WKB treatment of one-dimensional motion along r at $\theta = 90^o$.[10]

a) Diagonalization of the Hamiltonian. More recently Rydberg levels of atomic hydrogen in a magnetic field have been calculated by diagonalizing the matrix of the diamagnetic term of (1) in a truncated base of zerofield H states.[11] As the field strength increases, removing the degeneracy of the manifold states with the same quantum number n, the levels with each n value fan out, and manifolds with $n' \neq n$ intersect as shown in Fig. 6. The expected

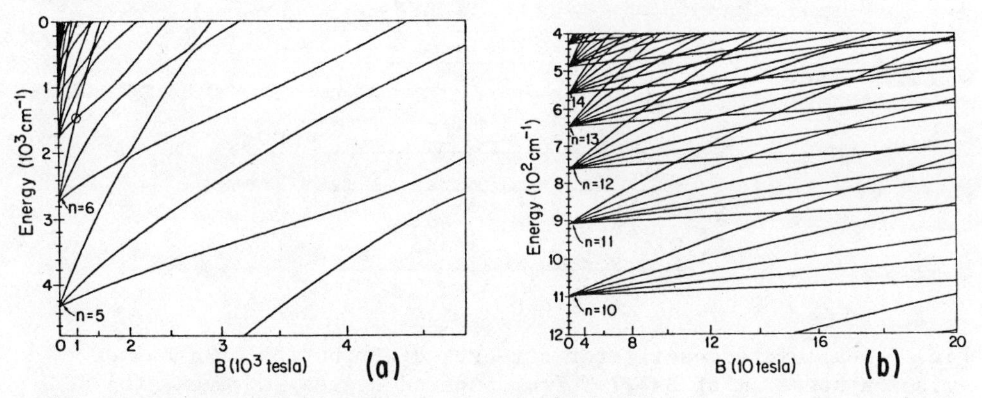

Fig. 6. Energy of hydrogen as a function of very strong magnetic field for (a) low-lying, and (b) highly excited states (ref. 11).

"avoided crossing" behavior at these intersections was observed only in the lower excitation diagram (Fig. 6a), suggesting that the crossings shown in Fig. 6b reflect an approximate symmetry which becomes relevant at higher excitations.

More extensive and revealing information was provided by a massive calculation which: a) diagonalized the matrix with the potential (1) in a very large Sturmian base, and b) evaluated the spectrum of photoabsorption from the ground state.[12] The spectrum of Fig. 7a shows, at $n\sim25$, how each Rydberg level is resolved into a manifold by the magnetic field but different manifolds still remain separate and are spaced according to Rydberg's law. At $n\sim35$, on the other hand, successive manifolds intermix but remain clearly recognizable. Within each manifold the intensity increases steadily with increasing excitation and the spacing of the highest lines of successive manifolds follows now a Landau pattern. Most important are the wave function plots in Fig. 8 showing direct evidence of electron localization in the range $50^o<\theta<130^o$, i.e., astride of the ridge line $\theta\sim90^o$, for the highest levels of each manifold.

The spectrum of Fig. 7a pertains to transitions with a dipole moment orthogonal to the magnetic field, leading to excited states with wavefunctions of even parity under the reflection $\theta\rightarrow\pi-\theta$ and hence with an antinode on the ridge. The spectrum of Fig. 7b pertains instead to transitions with a dipole parallel to B, leading

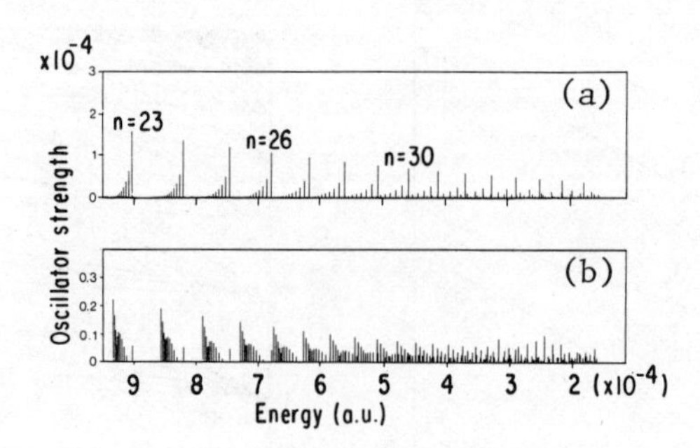

Fig. 7. Calculated oscillator strength distribution of a Balmer emission spectrum at B=4.7 T from excited states (a) even, (b) odd under $\theta\rightarrow\pi-\theta$ (ref. 12). Note in (b) the separation of each n-manifold into two bands.

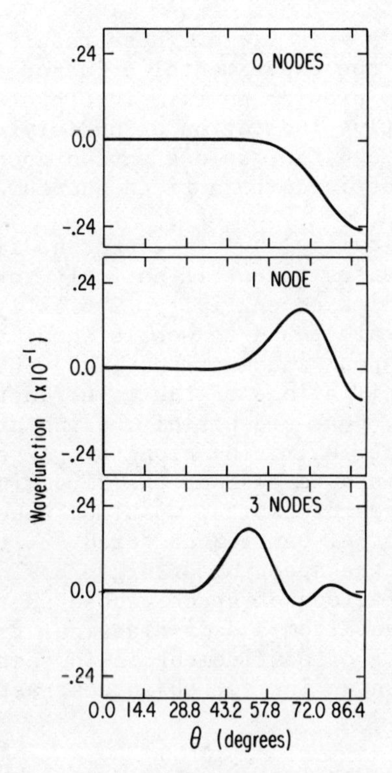

Fig. 8. Wavefunction vs. angle along the turning line in the region $0<\theta<90^o$ for states with 0,1, and 3 nodes. (1st, 2nd, and 4th highest line in a manifold of Fig. 7a) (ref. 12).

to odd states with a node on the ridge. The different distribution of intensities within corresponding manifolds of the two spectra is interpreted as follows: Excitation with a transverse dipole leads mainly to states of higher energy localized astride the ridge, excitation with a parallel dipole leads instead to states localized away from the ridge at $\theta\sim 0^\circ$ or $\sim 180^\circ$, i.e., where the diamagnetic potential is low. The free crossing of high levels of a manifold with the low levels of the next higher one, observed in Fig. 6b, would thus reflect the different localization of the corresponding wave functions rather than a hidden symmetry.

b) Stability in a weak field. These results of numerical calculations complement the experimental evidence of Fig. 4 in a very suggestive manner but provide no firm interpretation of the phenomena. A more substantive indication of underlying mechanisms arises from studies of the magnetic field's action upon semi-classical hydrogenic orbits restricted to a fixed quantum number n.

Goebel[13] and Soloviev[14] have treated the field $\vec{B}$ as a weak perturbation which causes the orbit not only to precess but also to nutate and to change its eccentricity. Orbits lying near the equatorial plane, $\theta=90^\circ$, are found to wobble about this plane reaching peak eccentricity at peak inclination. (The wobbling corresponds, of course, to the oscillations of the wave function over a limited range of θ in Fig. 8, peak eccentricity being attained at the classical turning point of the motion along θ.) A second set of orbits wobbles instead within a cone about the direction of the magnetic field. These two ranges of orbital wobbling appear to be separated by an effective potential barrier centered at $|\tan\theta|=\frac{1}{2}$ and manifested by a notch in the spectrum of Fig. 7b. The confinement of orbits in different regions of space, whose inertial-like origin is outlined below, emerges from a semi-classical treatment applicable at $n>>1$. The influence of confinement is in fact apparent only for the higher n values shown in Fig. 6b in contrast to those of Fig. 6a.

Analytically the wobbling is represented by oscillations of the Runge-Lenz vector $\vec{A}$ which points along the orbit's major axis. The components of $\vec{A}$ parallel and transverse to the field $\vec{B}$, which we label A_z and A_x respectively in a precessing frame, vary together with a transverse component of the orbital momentum, L_y. (The parallel component L_z keeps instead its fixed value m.) The orbit's diamagnetic energy is itself represented by a quadratic expression in $\vec{A}$ and $\vec{L}$ to lowest order in B^2,

$$E_d = \frac{1}{2}\,\frac{e^2B^2}{8mc^2}\left(\frac{\hbar^2}{e^2m}\right)^2 n^2(5A_x^2 + \vec{L}^2 + m^2), \qquad (2)$$

an expression that serves as the effective Hamiltonian of the orbital wobbling. These considerations of classical orbits can be translated into quantum mechanical WKB form. They also lead to two suggestive remarks: i) The three quantum operators $\{A_z,A_x,L_y\}$ form a subgroup of the hydrogenic symmetry, isomorphic to $\{L_x,L_y,L_z\}$.[15] ii) The effective Hamiltonian (2) is isomorphic to that of an asymmetric top.[13] Indeed the momentum space formulation of the Hamiltonian $p^2/2m-e^2/r+E_d$ is separable in the same ellipsoidal coordinates as the top's Hamiltonian.[14,16]

Soloviev's analysis of the stability of motion astride the ridge, or alternatively in the potential valleys parallel to $\vec{B}$[14], may be restated with reference to the limits imposed upon the direction of $\vec{A}$, i.e., upon the orbit's axis, by conservation of the energy E_d. Notice at the outset, from Eq. (2), that $\vec{A}$ must point across (along) the polar axis $\vec{B}$ for large (small) values of E_d. The wobbling of $\vec{A}$ may now be viewed as a rotation imposed by the orbital momentum component L_y through the commutation relation $[L_y,A_x]=-iA_z$. If $\vec{A}$ points initially across (along) the field $\vec{B}$, this action can proceed - at constant E_d and at constant $\vec{A}^2+\vec{L}^2=n^2-1$ - only by increasing $\vec{A}^2$ and correspondingly reducing $\vec{L}^2$. In either case the rotation comes to a halt whenever $\vec{L}^2$ has fallen to a level that entails $L_y=0$. The motion is thus confined by the interplay of the dynamical variables $\vec{A}$ and $\vec{L}$ rather than by a potential barrier.

A key element emerging from these results lies in the splitting of the levels of a fixed-n manifold into two separate bands, with the eigenfunctions of the upper band localized astride of the ridge in accordance with Fig. 8. However, this success is limited in two respects. First, it has been obtained by treating the magnetic field as a weak perturbation, an approximation that certainly does not hold for high Rydberg states whose diamagnetic energy exceeds the level spacing by far. (Specifically, the arguments of ref. 14 rest on the conservation of $\vec{L}^2+\vec{A}^2$, a parameter that has been used thus far only with reference to unperturbed H atoms.) Second, even though the splitting of a fixed-n manifold into two bands has been well established, and even though the analogous example of the asymmetric top has been worked out in full detail long ago[17], its physical interpretation has not yet emerged sufficiently clearly for us to anticipate its range of application. The asymmetric top example affords, however, the following clues which might have wide relevance: a) The wave function of the orientation of the top's angular momentum with respect to its axes of inertia is confined, in the highest energy eigenstates, to a cone of lowest inertia. b) The Schrödinger equation for these eigenstates displays a negative effective mass, analogous to that which is familiar for electron states near the top of a crystal band. (A negative mass occurs when a particle's velocity - i.e., group velocity - decreases with increasing wave number.)

c) Effects of stronger field action. The quantum mechanical calculations reported in a) above, by diagonalization of a truncated Hamiltonian matrix, extend into the range where the diamagnetic energy exceeds the spacing of Rydberg levels, thus complementing the analytical work based on a weak field approximation. In particular it emerges from ref. 12 that the main results of perturbation treatments reported in b) remain substantially valid far beyond the limit of their theoretical foundation. On the other hand the numerical calculations have themselves run into technical difficulties at still higher energies, a few Landau spacings below the ionization threshold for $B \sim 5$ T,[12] suggesting the onset of new physical circumstances. Numerical calculations with classical orbits have also shown departures from a simpler pattern of results in this range of energy and field strength.[13] However, no analogous manifestation has emerged from observations of the optical spectrum.

4. WAVE PROPAGATION ASTRIDE THE RIDGE

Since the theory of resonant Rydberg states in a magnetic field remains restricted to a perturbative treatment, we review here a complementary approach which derives from Wannier's treatment of an electron pair traveling along its potential ridge.[4] This approach aims at extending the one-dimensional integration of the Schrödinger equation along the ridge, developed in ref. 10 to account for the energy levels of Landau resonances, into a two-dimensional integration over a suitable range of two quasi-separable coordinates. Its ultimate goal is to show analytically whether and why the wave function would remain confined to a moderate range of distances on either side of the ridge in accordance with the computational results illustrated in Fig. 8.

a) Lowest order expansion away from the ridge. The initial step of this procedure was performed in 1970[18,19] to construct a wavemechanical WKB analog of Wannier's classical trajectories of an electron pair. That construction is transcribed here for wave propagation in the Coulomb+diamagnetic potential (1). The WKB integration of ref. 10 along the ridge assigns to each point of the ridge $(r=\rho, \theta=90^o)$ a phase

$$S(\rho) = (2m/\hbar^2)^{\frac{1}{2}} \int_0^\rho [E-V(r,90^o)]^{\frac{1}{2}} dr \ . \tag{3}$$

Its extension across the ridge requires us then to determine a constant phase line $S_\rho(r,\theta)=S(\rho)$ for values of $\theta \neq 90^o$.

The power expansion of this function

$$S_\rho(r,\theta) = S_{\rho 0}(r) + \tfrac{1}{2} S_{\rho 1}(r) \cos^2\theta + \cdots \tag{4}$$

is determined in an initial WKB approximation from the corresponding expansion of the equation

$$|\vec{\nabla} S_\rho|^2 = (2m/\hbar^2)[E-V(r,\theta)] = k^2(r,\theta)$$
$$= k_0^2(r) + \tfrac{1}{2} k_1^2(r) \cos^2\theta + \cdots . \tag{5}$$

The zeroth order term of Eq. (5) amounts to $dS_{\rho 0}(r)/dr=k_0(r)$, whereby $S_{\rho 0}$ coincides with the value (3) of $S(\rho)$. The first order term yields instead a quadratic equation in $S_{\rho 1}(r)$, with two alternative roots, whose origin can be traced as follows: The right-hand side of (5) is an increasing function of $\cos^2\theta$ in accordance with the decrease of $V(\underset{\sim}{r},\theta)$ as θ shifts away from the ridge. As the radial component of $\vec{\nabla}S$ remains constant in first order, the increase of $\cos^2\theta$ introduces a nonzero component of $\vec{\nabla}S_\rho$ transverse to the ridge, which can have two alternative signs since (5) determines only $|\vec{\nabla}S_\rho|^2$. Depending on whether the transverse component of $\vec{\nabla}S_\rho$ points away from (or toward) the ridge, the constant-phase line S_ρ=const. will be convex (or concave) toward large r (Fig. 9).

The normals to this line, which identify classical trajectories, will then diverge from (or converge to) the equatorial plane $\theta=90^o$ in the two cases. The occurrence of these two alternative

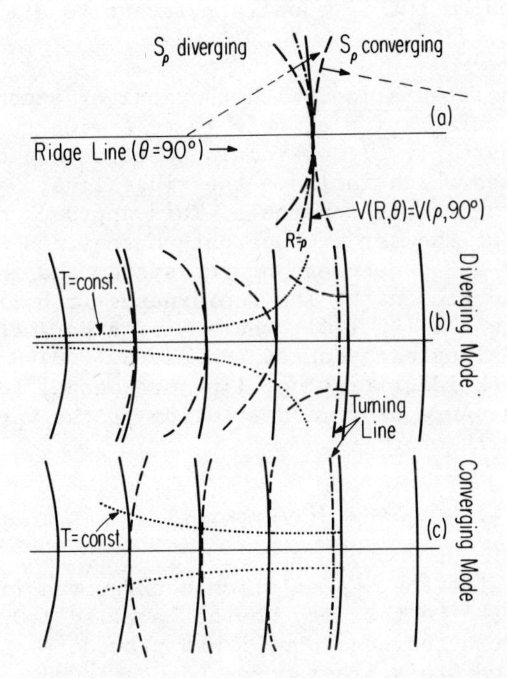

Fig. 9. Diagram of divergent and convergent modes of wave motion astride a ridge. a) Converging and diverging wavefronts (----) for a single value of ρ. The lines R=ρ (.—-.) and $V(r,\theta)=V(\rho,90^o)$ (——) are also shown. Normals to the wavefronts represent directions of electron trajectories. b) Sequence of diverging wavefronts, S_ρ=const (----). Orthogonal trajectories, T=const (····), equipotential lines (——), and the classical turning line (·-·-) are also shown. c) Same as b) for converging mode. (Courtesy M. Cavagnero).

types of solution of the wave equation parallels the splitting of the constant-n manifold of Sec. 3 into two bands. Divergence from $\theta=90^{o}$ would yield trajectories that bend away eventually into the potential valleys, convergence would yield trajectories that remain in the region of higher diamagnetic potential.

The curvature of each line S_{ρ}=const determines in turn the ρ-dependence of the wavefunction's amplitude A_{ρ} on the ridge, through the continuity equation $\vec{\nabla}\cdot A_{\rho}^{2}\vec{\nabla}S_{\rho}=0$,

$$A_{\rho} \propto \rho^{\zeta(\rho)} \quad . \tag{6}$$

The parameter ζ is the same root of the quadratic equation that determines $S_{\rho 1}$, its value is obviously smaller - indeed negative - for the divergent solution and positive for the convergent one. The variation of $\zeta(\rho)$ is assumed here to be slow, as appropriate to a WKB procedure. The treatment summarized here was developed for the two-electron problem but is equally relevant to all the examples described in Sec. 2.[20]

b) Extended integration. With the goal of extending the one-dimensional WKB integration of ref. 10 to a wide two-dimensional range of coordinates (r,θ), we should construct an orthogonal system of coordinates which includes the ridge line, $\theta=90^{o}$, and which might prove approximately separable. This approach has been initiated by choosing the set of constant-phase lines $S_{\rho}(r,\theta)=S(\rho)$ defined by Eq. (5) as one element of the system and their orthogonal set as the second element.[20] The coordinates (r,θ,ϕ) of a point are replaced by a new set (S_{ρ},T,ϕ), where T is a parameter that measures distances along each constant-phase $S_{\rho}(r,\theta)=S(\rho)$ and each value $T(r,\theta)$=const identifies one line orthogonal to all the constant-phase lines. An approximate solution of the wave equation is then sought with the form

$$\psi(r,\theta,\phi) = A_{\rho}F(S_{\rho})G(T)\exp(im\phi) \quad . \tag{7}$$

Here $F(S_{\rho})$ indicates the WKB solution constructed in ref. 10 along the ridge line, a line that may now be labelled T=0, while G(T) remains to be constructed, again by a WKB procedure. A preliminary question is whether this program can be developed consistently. The key question, for the purpose of assessing the stability of resonant states, is whether the wave function G(T) representing propagation across the ridge is bounded by turning points $\pm T_{t}$ in accordance with Fig. 8. (This bound on the range of T is required only for sufficiently large values of S_{ρ} since Landau resonances are observed only at the large radial distances.)

The following preliminary steps of this program have been taken in ref. 20:

i) The construction of successive terms of the expansion (4) of $S_\rho(r,\theta)$ by recursive use of Eq. (5) has been shown to proceed without apparent hindrance.

ii) The solution (6) was shown to hinge on the assumption (implicit in Refs. 18,19) that the factor G(T) of (7) be even under reflection through the ridge line T=0. Solutions with odd G(T) also exist but have an additional factor 3 in the exponent of (6); this factor intensifies the divergence or convergence of odd parity wavefunctions. Odd solutions are not expected to contribute to resonances but their superposition with even solutions is essential for representing reactive collisions.

iii) Substitution of the Ansatz (7) into the wave equation $\vec{\nabla}^2\psi+k^2(r,\theta)\psi=0$ reduces it to a separate equation for G(T) whose coefficients are independent of S to within variations of the scale factors of the curvilinear metric. Such variations are regarded as small in a WKB procedure.

An additional point, overlooked in ref. 20 but possibly decisive for the stability of resonances, is the following:

iv) The solution of Eq. (5) for the <u>convergent</u> waves yields

$$S_{\rho 1} = rk_0(\rho)\zeta(\rho) > k_1(\rho) \ , \tag{8}$$

meaning that the concavity of the constant-phase lines $S_\rho(r,\theta)$ = const, <u>exceeds</u> that of the constant potential line $k^2(r,\theta)$ = const, provided $k_0{}^2(r)>0$. [The diagram in Fig. 9c is drawn in accordance with Eq. (8).] The propagation of G(T) away from the ridge would thus meet a <u>rising potential</u> and hence eventually a turning point at $\pm T_t$. However, this remark affords only <u>prima facie</u> evidence of stability, pending actual construction of the coordinate lattice of Fig. 9 and of its metric.

c) <u>Localized breakdown of WKB solutions</u>. The WKB representation of the waves that propagate along the ridge, straddling it, fails partially at the approach of a turning point, $r=r_t$, where $k_0(r_t)=0$, owing to a divergence of the parameter $\zeta(\rho)$ in Eqs. (6) and (8).[21] (The mere vanishing of $k_0(r_t)$ could be bypassed by using Airy functions as WKB solutions.) Only <u>one</u> of the two alternative roots ζ diverges at each root r_t of $k_0(r_t)=0$, namely, the ζ corresponding to diverging trajectories when k_0 vanishes owing to the rise of a diamagnetic (or other) potential at large r, and the ζ corresponding to converging trajectories when k_0 vanishes at low r owing to the rise of a centrifugal barrier. The converging wavefunctions thus show themselves appropriate to the large-r region, where resonant excited states are usually localized, whereas the diverging wavefunctions seem appropriate to represent propagation away from the region of small radial distances whence excitations

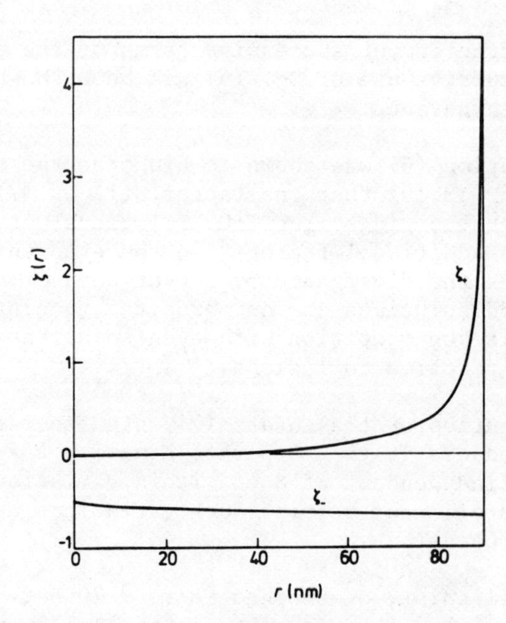

Fig. 10. Parameters $\zeta(\rho)$, Eq. (6), for diverging (+) and converging (-) modes (ref. 21).

originate. Figure 10 illustrates this behavior of the parameter ζ for wave propagation in the potential field (1). The resulting trajectories have been indicated in Fig. 9.

Failure of the WKB approximation for either the diverging or the converging wave implies presumably a local coupling of these two modes of propagation because the independent construction of each of them rests on that approximation. This coupling would in fact perform an essential role in the dynamics of resonances. Assuming that the converging mode will be shown to represent the main part of a resonant state's wavefunction astride the ridge at large r, its coupling to the diverging mode would provide the vehicle through which an electron (or electron pair, or molecular fragment) is excited into that state and decays out of it. Reference 21 only points to this likely mechanism without any detailed analysis.

A further essential aspect of the resonant state wave function, namely, its reflection at the large-r "turning line" where $k^2(r,\theta)$ vanishes, is only touched upon in ref. 21: The concave constant-phase lines represented by Eqs. (5) and (8) for $k^2(r,\theta) > 0$ are shown to coincide with the turning line itself to $O(\cos^4\theta)$ in the limit $k^2 \to 0$.

5. CONCLUDING REMARKS

The numerous contributions to the study of resonant states summarized in this report appear to provide a composite picture which is remarkably consistent - especially in the compatibility of the evidence of Secs. 3 and 4 - but still highly fragmented. This report aims at assembling the diverse evidence so as to provide a focus for further efforts. Its preparation, supported by the U.S. Department of Energy, Office of Basic Energy Sciences, owes much to discussions and to contributions of unpublished material by Charles W. Clark, Charles Goebel and several other colleagues.

REFERENCES

1. G. Schulz, Phys. Rev. Lett. 10, 104 (1963), Rev. Mod. Phys. 45, 378 (1973).
2. S.J. Buckman, P. Hammond, F.H. Read, and G.C. King, to be published.
3. F.H. Read, Atomic Physics 7, ed. D. Kleppner and F.M. Pipkin (Plenum, N.Y., 1981), p. 429; U. Fano, Rep. Prog. Phys. (in press).
4. G. Wannier, Phys. Rev. 93, 817 (1963).
5. W.R.S. Garton and F.S. Tomkins, Astrophys. J. 158, 839 and 1212 (1969); K.T. Lu, F.S. Tomkins, and W.R.S. Garton, Proc. R. Soc. London A362, 421 (1978).
6. J.O. Hirschfelder and E.P. Wigner, J. Chem. Phys. 7, 616 (1939).
7. J.M. Launay and M. Le Dourneuf, J. Phys. B 15, L455 (1982).
8. U. Fano, Phys. Rev. A 24, 2402 (1981).
9. L.I. Schiff and H. Snyder, Phys. Rev. 55, 59 (1939).
10. A.F. Starace, J. Phys. B 6, 585 (1973); see also A.R. Edmonds, J. Phys. (Paris) 31, Coll. C4-71 (1970).
11. M.L. Zimmerman, M.M. Kash, and D. Kleppner, Phys. Rev. Lett. 45, 1082 (1980).
12. C.W. Clark and K.T. Taylor, J. Phys. B 13, L737 (1980), Nature 292, 437 (1981), J. Phys. B 15, 1175 (1982).
13. C. Goebel, private communication. A.E.R. Edmonds and collaborators have also studied classical orbits in the potential (1). See the report by C.W. Clark, K.T. Lu, and A.F. Starace, Sec. 2.2, in Progress in Atomic Spectroscopy, Part C, eds. W. Hanle and H. Kleinpoppen (Plenum, N.Y., 1982) which complements the present report.
14. E.A. Soloviev, Pis'ma v Zh.E.T.F. 34, 278 (1981) [Transl.: Sov. Phys. JETP Lett. 34, 265 (1981)].
15. J.J. Labarthe, J. Phys. B. 14, L467 (1981).
16. D.R. Herrick, Phys. Rev. A 26, 323 (1981).
17. H.A. Kramers and G.P. Ittman, Z. Phys. 53, 553 (1929), 58, 217 (1929), 60, 663 (1930).

18. R. Peterkop, J. Phys. B 4, 513 (1971).
19. A.R.P. Rau, Phys. Rev. A 4, 207 (1971).
20. U. Fano, Phys. Rev. A 22, 2660 (1980).
21. U. Fano, J. Phys. B 13, L519 (1980).

IMPACT OF ATOMIC PHYSICS ON FUNDAMENTAL CONSTANTS

Richard D. Deslattes

Quantum Metrology Group
National Bureau of Standards
Washington, D.C. 20234

INTRODUCTION

This report offers one experimentalist's outlook on the constants area especially from the perspective of atomic physics. There are many recent reviews of selected areas of such work in addition to the well-known periodic overall studies including least squares adjustments. My intent is neither to offer the scientific depth of the specialized reviews nor to even approach the technical scope of an overall adjustment.

Instead I aim, after some brief indication of motivation, to present a rough classification of present-day efforts into three groupings. It will be evident that, although atomic physics spawned technologies intervene everywhere, one of the three sub-classes has an essential connection. The remainder of the report then aims first at a somewhat more detailed (though still very superficial) overview in this area and concludes with some outlook toward the future.

WHY ARE THE CONSTANTS INTERESTING?

There are, of course, valid and pragmatic reasons for wanting a consistent set of high quality values for the fundamental constants with documented uncertainties. This is the rationale for the

CODATA type efforts which aim, inter alia, to produce periodically revised "recommended" values for general use [1]. Thus, published work in any discipline area can conveniently indicate a source for numerical values and its output revised in the future.

A second viewpoint is often heard and was epitomized by Richtmyer's phrase "the romance of the next decimal place" [2]. I don't wish to trivialize this message but it has certain limitations. The positive content is that improved measurement precision and accuracy have often led to new discoveries and are likely to do so in the future. Examples are legion extending from early astrometry through the discovery of the rare gases and surely into our present decade. A problem with this perspective is that it offers little in the way of guidance as to what directions one should emphasize.

A third option, with which I tend to be more comfortable, takes note of the fact that a significant part of physics research aims at testing of our cherished beliefs. Work of this kind often appears as a test of some symmetry property or invariance. One is here looking for some small departure from a zero signal condition emerging, for example, from systematic flipping between presumably degenerate states. The history of such work is too well-known to need repeating [3].

There are yet other kinds of "cherished beliefs" which are occasionally interesting to call into question. An obvious case is at hand for leptonic atoms (especially positronium) where one's confidence in the calculability of its energy levels is deservedly very high but clearly worth questioning.

I tend to see the overall business of fundamental constants work first as generalizing this idea along the following line: In the particular case of calculable spectra, one might be asking if experiments on low and high Z electronic systems (or light and heavy leptonic atoms) are consistent. One way of phrasing such questions is to derive physical constants such as R_∞, m_e/m_p and α from low energy experiments and ask if the high energy results are well predicted.

To further generalize the above elementary example, one can look at the entire least squares exercise as testing consistency among not only physics of similar systems in different domains but also of inter-domain transfer of numerics. After all, as will be mentioned below, some of our available information for calculable spectra comes from classical measurements on macroscopic systems, some from macroscopic quantum mechanics and some from the domain of elementary particles and their aggregates. It is perhaps the highest art associated with a large-scale systematic study to extract from the tensions appearing in this over-determined system good clues about where difficulties are likely to lie [4].

CLASSES OF EXPERIMENTAL MEASUREMENTS

Several classifications have been introduced. The earliest seems to have been according to estimates of overall uncertainty. This is particularly useful in reducing complexity in the adjustment process if one group of experiments is significantly more accurate than a second group whose members are of approximately equal (and lesser) estimated accuracy [5]. A second type of classification partitioned the input measurements into a class whose data reduction required use of quantum electrodynamics in a significant way and the remainder which did not. This separation was particularly informative when the Josephson 2e/h measurement allowed for a choice between values for the fine-structure constant as derived from fine-structure and hyperfine structure in hydrogen [6].

As suggested in the introduction, I propose to examine a different grouping of experiments, specifically one which divides the totality of experiments into those which:

a) depend on dimensional measurements;
b) make use of macroscopic phase-coherence;
c) involve calculable spectra.

There are two aspects of constants type work which would make any or all of the topics (a-c) appropriate for discussion at this Conference. Firstly, one finds atomic physics type technology throughout, not least of all in the form of atomic and molecular primary and secondary frequency (wavelength) standards. Secondly, there is strong coupling among the effort groups a-c. For example, mechanical measurements (nowadays using laser wavelength standards) are used to fix coil geometries which are needed in establishing lepton and hadron magnetic moments. Josephson type processes, as mentioned above, are already important in estimates of calculable spectra for tests of basic theory.

Finally, of course, one must choose to make a particular emphasis for this conference. In this case it seems to me that the third category, namely that of calculable spectra is an almost inescapable choice. The organization of the remainder is thus a brief glimpse of categories a and b followed by a longer discussion of c. Finally an effort will be made to look to future problems and opportunities.

EXPERIMENTS WHICH DEPEND ON DIMENSIONAL MEASUREMENTS

The key feature emphasized in calling attention to this subgrouping is that extensive macroscopic quantities such as (object) length or mass present special problems. These problems are associated with the question of finding a rule for defining an object's boundary. As will be indicated below, this conundrum has been a serious limitation on several fundamental measurements. Difficulties are encountered not far below 1 ppm (or 0.02 nm) at the present time.

The situation of displacement determination stands in evident contrast. Given careful optical beam preparation, the interval between two positions of a retroreflecting mirror can be determined with considerable refinement. Simply stated, almost all questions about optical phase shift, surface location and certain classes of imperfection disappear when the same surface is illuminated at each terminus (and, if appropriate, during the "scan). There are important questions about illumination averaging but these can be controlled as has been already demonstrated below the 10^{-9} level, approaching even 10^{-10} [7].

As an aside, one notes that frequency measurements or, more specifically, frequency ratio measurements have no limitations of the sort emphasized above. Nor, for that matter, is our knowledge of any fundamental constant limited by frequency measurement technology. An often stated goal is that we should make serious efforts to reduce all measurement to the problem of frequency metrology, thereby eliminating all the above difficulties. Perhaps in that case we could get by with many fewer experimentalists, but that is another story.

A somewhat more modest goal is to see how far toward the precision available in displacements we could move with experiments which have been limited by boundary (or location) problems. Two approaches are evidently possible: the first seeks to avoid the problem by transforming the structure of experiments so that only displacements enter; the second seeks to face the interface problem and thereby better delineate object boundaries. It would appear that the first option offers the greater long term reward in the constants measurement business, provided we are clever enough to derive the required algorithms. The second approach is also interesting but difficult; it does, however, offer the potential reward of increased control of physical object morphology, not only for the area of work under discussion, but also suggesting applications outside this domain, perhaps even in the practical sphere.

The examples selected below are intended to illustrate these problems and strategies for coping with them. They are also chosen to represent instances in which improvement is currently possible and even interesting.

1. The Ampere and γ_p

The intent of this grouping is to include those electromagnetic measurements in which geometry (usually that of a coil) intervenes. Josephson type measurements are excluded, even though one of them, namely h/mc from a rotating superconducting ring, has a significant geometrical component. The remaining examples derive from earlier versions which either used a geometrically well-characterized coil to calculate a magnetic field strength, thereby permitting magnetic moment determination from Larmor precession frequency, or used a coil of known geometry in a calibrated field to transduce current into force.

By way of examples, consider the recent evolution of work on the proton's gyromagnetic ratio, γ_p and on the Ampere balance. Actually there are two separate experiments on γ_p, a low field version wherein a calculable field is produced (with subsequent measurement of the Larmor precession [8]) and a high field version which involves a current balance [9]. In fact, neither of these experiments uses free protons; but rather what is measured is γ_p' for those protons in a spherical water sample. As a further practical complication, electrical measurements are in terms of local units as maintained by your nearby national standardizing laboratory. For the most part, these are merely complications whose exploration is not particularly informative.

Where the real problems have been and where recent progress has been obtained is associated with experimental strategies. To begin with, consider the venerable calculable coil. At one time great effort went into dimensional metrology of the local details, with the results subsequently averaged, and an assumption made about the distribution of current within the wire. There were also detailed numerical calculations regarding finite length problems.

The list of progressive increments over the past few years is a catalog of increasingly clever moves which may soon be put together. For instance, coil pitch is now determined by magnetometry and laser interferometry with the proper averaging built into the method [8]. Variability of effective diameter is obtained by a differential induction scheme. The solenoid is "tricked" into thinking that it is over a kilometer long by trim coils. The list goes on but the spirit is evident. By injecting different currents into various (fixed) parts of a long coil or injecting the same current into successive regions, a proper characterization of the electrical geometry is possible and a rather ideal field produced.

Up to the present time, there has remained a required diameter measurement in those experiments from which numerical results have so far been obtained. This dependence has been a principal limitation, even though clever experimental design has reduced sensitivity to this parameter. Beginning with a proposal by Williams and Olsen for obtaining a line integral of $\underset{\sim}{B}\cdot d\underset{\sim}{\ell}$ (hence an average value of B), possible means for avoiding the diameter measurement altogether have been discussed [10]. The most recent of these was a proposal by Greene to use the separated oscillating field method of Ramsey directly on flowing water to obtain an essentially equivalent release from geometry [11]. In both these instances, laser interferometry is needed to reckon some key displacement (which enters to first order) but other length-type measurements are successfully evaded.

The high field γ_p measurements are derived from earlier schemes where the length of a current carrying conductor in a strong magnetic field directly established the force which was measured. In addition to the work of Kibble and Hunt mentioned above [9], there have been significant efforts elsewhere [12]. In the case of the work of Kibble and Hunt, the dimensional measurement problem was approached in two different ways. They used both an inductive method and a mechanical gauging technique to determine the key dimension of the suspended coil.

Future plans for work of the Ampere balance type call for total elimination of coil geometry. This permits use of the high field strengths from superconducting and permanent magnet structures because questions about the location of current flow no longer arise. An example of the next generation Ampere balance suggests the strategy involved [13]. A pair of coils carrying currents I_1 and I_2 experience a mutual force in the vertical (z) direction $F_z = I_1 I_2 M'$ where M' is the space derivative of M in the z direction. If this is integrated over a displacement from z_2 to z_1, one has it that $\int_1^2 F dz = I_1 I_2 \Delta M$. Similarly, if voltage is observed during the sweep, its magnitude is $V = I_2 \dot{M}$ where $\dot{M}$ is the time derivative of M. Thus on a sweep from t_2 to t_1 $\int_1^2 V dt = I_2 \Delta M$. Evidently one now has the clean result that

$$I_1 = \frac{\int F dz}{\int V dt} \qquad (1)$$

provided the space and time termini are associated with each other.

2. The Faraday and Avogadro Constants

While these constants have generally little interest in themselves, they do effectively enter as conversion factors between different scales of energy and of mass. Each can be thought of in terms of a hypothetical enumeration of atoms whereby one senses already their common sources of grief. No such atom by atom counting is practical, so measurement strategies avoid this in some way.

In the case of Faraday, one measures an integration over ionic (or electronic) charge and the corresponding mass transport, assuming known valency, stoichiometry and no electrical or material losses. In common with Avogadro constant determinations, Faraday measurement depends on absolute nuclidic abundances as obtained by mass spectroscopy and nuclidic mass measurements to obtain the mean molar mass, A. In this case an integration of F Coulombs deposits A (kilo) grams in a univalent system. Endpoint problems and material accounting seem unavoidable so that it is doubtful that this work progresses much beyond the recent studies, for example, those of Bower and Davis [14].

The Avogadro constant appears similarly plagued [15]. Atom counting is evaded by counting lattice planes (by X-ray interferometry nowadays but there are other approaches) in a "perfect" crystal. If mean molar mass is again A and one has a cubical sample (volume ℓ^3) of a cubic crystal with n atoms per unit cell (volume $a_o{}^3$), then the mass in a unit cell is equivalently nA atomic mass units or a^3M/ℓ^3, the ratio of these two numbers being N_A. Endpoint problems and assumptions regarding crystal perfection seem troubling here as well. These appear to be lessened by using density measurements, in which case one obtains the more familiar form:

$$N_A = \frac{nA}{\rho a_o{}^3} \quad (2)$$

Actually this result is more troubled than it appears, since determining ρ has (apparently) a direct dependence on the object size problem mentioned above. In principle this can be evaded by using two similar objects of different mass and different volumes, such that the "standard" is virtual. Such work may be carried out in the future. If so, and if abundance measurements and crystal perfection permit, the N_A problem may achieve accuracies beyond the limits set by object length considerations. How far is not clear, but with sufficient effort one might approach the 10^{-8} level.

MACROSCOPIC QUANTUM EFFECTS

The central point of all the work meant to be included here is that intrinsic properties of wave-functions (e.g., single-valuedness) apply constraints to pairs of macroscopic observables. Although superconductivity or superfluidity have historically close connections to many of the useful measurements completed to date, this connection is not intrinsic as will be evident in the following report by von Klitzing. It was at first surprising that effects occurring in the rough and tumble of a complex solid (or liquid) environment should be material and sample independent and expressible entirely in terms of fundamental constants. At first some of us were consoled by the association with superconductivity, but this has been shown to have the sole effect of replacing the charge e by 2e to acknowledge that the tunneling objects are Cooper pairs.

The first of the macroscopic quantum effects to be demonstrated was the ac Josephson effect, whereby 2e/h is established via the invariant ratio $n\nu/V_n$ for irradiation at frequency ν and noting the n'th step voltage, V_n. This has been a powerful tool in the service of electrical measurements. The history, which has been well reviewed by Petley, shows a rapid evolution from laboratory curiosity through a significant determination of α to the role of maintenance of the operational unit of voltage [16]. Incidentally, the early history of this area of work is an instance in which electrical metrology made a distinct impact on atomic physics. It will be recalled that a key issue of the early 1960's was the discrepancy between values of α obtained from fine-structure measurement and those deduced from hyperfine structure. This story is well told in the widely known summary, due to Taylor, Parker and Langenberg [6]. An important feature of the Josephson frequency-voltage measurements is their complete independence from geometry. It is precisely this fact which has permitted refinement to reach its present level near 0.01 ppm and there appears no obvious obstacle to further improvement.

The case of a second Josephson effect to which I now turn attention is less fortunate in the last mentioned regard. The experiment involves spinning a superconducting (maybe not required) ring at a range of angular frequencies [17]. One has again a procedure analogous to step counting, but in this case it is a matter of counting the number of magnetic nulls between an initial null at ω_n and a subsequent one at ω_m, n-m nulls away. In this case one obtains the invariant result:

$$\frac{h}{m} = 4\,\frac{\omega_n - \omega_m}{(n-m)}\,S \tag{3}$$

where S is the area of the superconducting ring. Thus, in addition to the formidable requirements of cryogenic technology and ultra low field magnetometry, one is faced with a serious object length problem which will ultimately place constraint on the limiting accuracy. Nonetheless, this experiment is moving forward, especially in Cabrerra's group at Stanford University [17] and the efforts of Gallop and collaborators at NPL [18]. It is conceivable that if other problems are solved, the remaining difficulty of dimensional metrology might be obviated either by mutual inductance or by using a big ring-little ring approach.

The last of the macroscopic quantum effect schemes that I shall mention is the one most recently discovered. Since the discoverer, von Klitzing, has a report following this one, my comments will be restricted to a few generalities. Essentially the Hall-effect resistance (i.e., the ratio of transverse voltage to longitudinal current) is quantized so that an V_T vs. I_ℓ curve exhibits steps in such a way that the ratio is a universal resistance [19]. If this is expressed in terms of fundamental constants, one obtains:

$$R_H = h/ne^2 \tag{4}$$

where n is a small integer. It is extraordinary that this already yields a competitive result for α^2 and the end is nowhere in sight. In particular, since there is no apparent material dependence and no dimensional metrology, the von Klitzing effect looks like a candidate to join that of Josephson as a new and thus far unimpeachable approach to quantum effects via purely macroscopic means. Clearly superconductivity has no role here. Of course, cryogenic experiments have other virtues, including lowered noise, clearer steps and the availability of sensitive comparators and magnetometers.

CALCULABLE SPECTRA

Although the treatment will remain fairly superficial, I wish to pay somewhat greater attention to the general area of calculable spectra than to the previously discussed groupings. The first, and most obvious, reason is that this is the principal domain of contact between atomic physics and fundamental constants. Secondly, it gives an opportunity to cite a few examples of technical evolution and to comment on issues which arise when linking work in different spectral regions. As will be clear in what follows, I am suggesting a very broad concept of calculable spectra. For instance, it seems to me that intervals between Landau levels in a magnetic field, the hyperfine splitting of the ground state of hydrogen and the Lyman α transitions in X-onium are all fair game. Actually when all the practicalities and sensitivities are reckoned carefully at the present time, it may well be that equivalent tests of QED are found in the inner vacancy energies of Fermium, the one-electron spectrum of 17 times ionized argon and atomic hydrogen.

1. Spectra of Hydrogen and Helium

Almost from the beginning, the calculable aspects of hydrogenic spectroscopy have played an important role in the fundamental atomic constants. It is perhaps surprising that this remains true today even though its role has changed significantly. Specifically, in the earlier times a readily measured visible transition, e.g., Balmer α, was used to fix the overall atomic energy scale by inversion of the Bohr formula or its successors to obtain a value for the Rydberg constant, R_∞. Also until the 1950's fine structure and hyperfine structure measurements, together with their theoretical estimations, offered alternative routes to a numerical value for the fine structure constant. As was mentioned above, a small but significant discrepancy finally arose between these two routes, which was ultimately resolved by the Josephson measurement of 2e/h.

It is inevitable that work of this kind faces limits as far as determination of numerical values for the constants and also as precision tests of fundamental theory. These limits may be seen to arise in two ways. First of all, the hydrogenic energy levels and their splittings are sensitive to some extent to the inner structure of proton. Secondly, since interesting electrodynamic effects scale with some power of $Z\alpha$, measurements of great precision are required. In the case of the former problem, one is attracted to the notion of the spectroscopy of purely leptonic systems, while the latter invites consideration of the single electron spectra of high Z ions.

Precision spectroscopy of helium has not, in general, been of importance for fundamental constants with the single conspicuous exception of optical spectroscopy of ionized helium. Taking advantage of the zero spin nucleus, E. Kessler was able to accomplish the last definitive Rydberg measurement before the advent of Doppler-free spectroscopy [20]. Beyond this, when two electrons are present, the physical constants and the QED effects are joined by the problem of how the two electrons learn to live with one another. This subject is rich and diverse and is treated to some extent in the contribution of Professor Fano to this Conference. It will not be discussed further in this report.

The advent of Doppler-free spectroscopy has reopened the discussion of atomic hydrogen at levels of precision and accuracy far beyond those traditionally available. The present-day laser spectroscopy concentrates on two issues: First is the fully resolved multiplet structure within Balmer α as pioneered by Hänsch and collaborators [21] and subsequently exploited by others; second is the hydrogen 1s-2s transition which is available by Doppler-free two-photon absorption [22]. It is interesting to note in passing that precision competitive with those obtained by non-linear spectroscopy has been recently reported by Amin, Caldwell and Lichten using linear spectroscopy and a well-prepared atomic beam [23]. The principal interest in 1s-2s is the ground state Lamb shift. The transition has been observed, but accurate numbers have not yet been reported although planned efforts are well underway.

I wish to make some specific comment regarding the reference wavelengths used in these studies. Traditionally, of course, krypton and mercury were the primary and secondary standards respectively against which hydrogenic and ionized helium spectra were measured. Subsequently, molecularly stabilized lasers have taken over this role, especially iodine stabilized HeNe lasers have taken over this role in the visible. Although these devices have serious known limitations and are not competitive in ultimate performance with alternative oscillator systems, they are fairly well behaved at the 10^{-9} level and, with care, somewhat below. These have for some time now been well connected with methane stabilized lasers in the infrared by high precision wavelength ratio measurements [7]. As is well known, the methane oscillator had been previously connected to cesium frequency standards via synthesis chains [24]. Very recently a synthesis chain has been extended into the visible and new ratio measurements completed, all of which exhibit a satisfactory degree of concordance [25]. As a result it has now been recommended that the practice of maintaining separate oscillators for the definitions of frequency and wavelength be discontinued and the current value for c taken as a convention [26].

A glance backward or a rather simple consideration of what has been published to date indicates a model of the situation which one encounters in calculable spectroscopy in general. Specifically the radiofrequency fine structure and hyperfine structure intervals were determined in terms of coherent microwave oscillators relative to the cesium frequency near 10 GHz. The red line, Balmer α, was in turn connected with another oscillator nearby in the visible. Integration of these results into an overall pattern required knowledge of the conversion factor connecting local standards in two different spectral regions. Although it has many other kinds of significance, the speed of light enters this discussion only as a conversion factor between local scales. As this discussion proceeds, it will be evident that other kinds of conversion factors or synthesis chains are required to fully exploit the information potentially available in calculable spectra.

2. High Z Systems

Interest in these more difficult species is associated with exploring Z dependence of radiative corrections. Also at very high Z there are possible precursing effects as nuclear charge approaches the region where spontaneous pair creation seems possible. Several approaches are possible and are easily seen to have close analogies to experiments on atomic hydrogen. (This report neglects the important studies by beam-foil technique of $\Delta n=0$ transitions in He-like species.) First of all, in the relatively low Z region, one finds at Z=17 that powerful quasi cw radiation from a CO_2 laser can be Doppler shifted into coincidence with the $2p_{1/2} - 2s_{1/2}$ (Lamb shift) transition in a fast beam experiment. Such a measurement has been carried out by Wood, et al [27] who report a value of 31.19 ± 0.22 THz for the shift in the atom's rest frame. When this is compared with available theoretical estimates of 31.35(2) THz from Mohr [28] and 31.93(13) THz from Erickson [29], there seems a favorable signal for the calculation of the first mentioned reference. The quality factor of the observed line is low, unfortunately ($\sim$ 4), while statistical uncertainty is high. Thus it is not easy to foresee big improvements in this type of experiment at Z=17. At higher Z's the Lamb shift increases, but so do power requirements for the laser. It may be that such classical Lamb shift exercises have thus already reached their asymptotic state. One notes, in passing, that this high Z measurement places almost negligible demands on our knowledge of the CO_2 oscillation frequencies.

Another approach is available for which there are now preliminary results and which offer some further promise for the future. This involves examination of $\Delta n \geq 1$ transitions with a final state n=1. The 1s level is perturbed to a greater extent than $n \neq 1$ levels resulting in a larger Lamb shift. Unfortunately the spectra of interest move into the X-ray region and require spectroscopy having both high resolution and a well-established connection to visible secondary standards which are, in turn, well tied to the spectrum of atomic hydrogen, i.e., the Rydberg constant. (Some aspects of the X-ray to visible connection will be discussed in the concluding section of this report.)

The spectral lines of immediate interest in the region $10 < Z < 25$ are Lyman α transitions. These occur in the spectral region from about 1 keV (1.2 nm) to 6 keV (0.2 nm). For the cases of $C\ell^{+16}$ and Fe^{+25}, the transitions are at 2.9 keV and 7.1 keV. The corresponding 1s Lamb shifts are about 1 eV and 4 eV, respectively. Line widths are 0.5 eV and 2 eV in these two cases. Thus the quality factor of the experiments is no better than in Ref. 27, and they are burdened by a hard calibration problem. So what is the point? First, these are manifestly independent measurements, the statistics are much better and the calibration problem is manageable. Secondly, they represent the beginning of a technology whose promise looks very great indeed. Consider that, by the region of Z=40, the Lamb shift amounts to more than 0.1% and that sub ppm spectroscopy is, in principle, available. Also, in this region the single photon emission 2s → 1s should be available with good intensity. This line, with a quality factor > 10^5, invites application of the full measurement potential already obtained in γ-ray spectra. Taken together these mean that the Lamb shift itself may be explored with a refinement in the range 10^{-3} - 10^{-4}.

Returning to some technical aspects of the spectroscopy itself, there are several ways to obtain n $\gtrsim$ 2 hydrogenic ions. They are, of course, well-known components of plasma sources, but this does not seem to offer a metrologically clean environment. They are also available with "heavy-ion" accelerators in two ways. First, by projectile stripping, i.e., beam-foil technique, one can obtain excited hydrogenic systems up to Z values limited by accelerator parameters. There are large Doppler shift and broadening problems, but these appear to be manageable in some cases of current interest. Alternatively, low energy recoil ions from a gas target are found in high charge states at low (∿ 1 eV) kinetic energies when a projectile of sufficient mass and energy is used.

Up to the present time at least two experiments have gone forward in this domain. First, Briand, et al. using the Bevalac at Berkeley recorded the Lyman α line in Fe^{25+} with a flat crystal and position sensitive proportional counter. Second, a group including the present author recorded Lyman α in $C\ell^{16+}$ produced at the Brookhaven tandem [30]. We used a bent crystal spectrometer and linear position sensitive detector. The normal X-ray spectrum ($K\alpha_{1,2}$) from gaseous argon excited by electron impact was used as a nearly coincident marker. In a separate exercise (see below), the argon Kα doublet was established in terms of R_∞ to sufficient accuracy that it was not limiting. The expected 1 eV shift was verified to approximately 1% in this exercise.

It might be mentioned in passing that much higher values of Zα are reached in the study of X-ray spectra of normal atoms than can be approached at this time by stripped ion spectroscopy. These spectra are, of course, objectionable since fundamental interaction problems are combined with those associated with many-body effects, hole-state calculations and so forth. On the other hand, recent systematic studies [31] show a pattern of general agreement within about 1 eV up to Z ∿ 92 where one expects a Lamb shift of 300 eV. We plan to extend this study into the trans-uranic region to see what pattern of discrepancy may emerge.

3. Exotic Atom Spectra

For present purposes, exotic atoms fall into two classes. First there are the purely leptonic systems especially positronium and muonium. Second are those atoms with a more or less normal nucleus and a single bound non-electronic particle. The first class offers some possibility for constants type measurements or clean tests of fundamental theory since, according to present understanding, no details of internal particle structure are known to be of significance. The second group has had various applications. For muonic atoms in sufficiently high angular momentum states, the spectra have primarily been used as a testing place for QED although one such transition was an early route to the muon mass. For low angular momentum states, muonic energy levels show effects due to details of the proton distribution in the nucleus.

In the case of other particle systems such as pions and kaons, there are, once again, important differences according to angular momentum. For relatively high values, the Coulomb energy levels permit Rydberg-type experiments which serve to determine particle mass values. For low values of angular momentum particles which couple to the nuclear force exhibit strong interaction shifts which constitute an area of long-standing and continuing activities.

The range of nuclei of interest in the types of studies mentioned above is extensive. Correspondingly, the range of spectroscopy is also rather broad. One has at the long wavelength end to deal with fine structure and hyperfine structure in, e.g., positronium and muonium. Also optical spectroscopy is possible for positronium, especially the recently demonstrated Doppler-free two photon absorption $1s \rightarrow 2s$ [32]. These spectral regions have been already mentioned above along with their problems in scale normalization.

As one proceeds to the heavier particle type atoms, the effective Rydberg scales by the ratio of particle mass to electron mass. This has the effect, for example, of moving low Z pionic spectra into the soft X-ray region of one to a few kilovolts. With increasing Z, spectra of heavy particle exotic atoms range through the region of atomic X-rays and are particularly interesting up to several hundred keV. Since interpretability of these spectra depends on having results well connected to the hydrogen spectrum, there arises a considerable problem in wavelength transfer which my colleagues and I have addressed over the past ten or more years. It is to a discussion of this process that the next section turns.

4. Marker Generation 2 eV - 2 MeV

Evidently there are requirements both for direct measurement algorithms where sufficiently intense sources are available and interesting and for a distribution of markers or secondary standards throughout this region. It is not surprising that no single technology is available for covering this entire range and that those available are of distinctly lower quality than one encounters in frequency synthesis to the visible. Nonetheless progress has been made and continues and there awaits at the high energy end a somewhat unexpected determination of an otherwise unavailable combination of physical constants.

The visible and near uv are well handled classically by interferometrically established secondary standards. These worked well to the limits of incoherent spectroscopy ($\sim$ 0.01 ppm). Nowadays laser wavemeter technology and non-linear optics permit covering this same range to improved accuracy, should this be required. Above 10 eV and up to a few hundred eV, interferometer techniques are not convenient. An alternative approach uses Ritz principle combinations of interferometrically determined components to obtain (slightly degraded) estimates of vacuum uv transition energies. An appreciation of this methodology may be gained from an earlier review by Edlén [33]. Accuracies available have been sufficient for applications encountered thus far, and it is reasonable to suppose that such work can be extended above the present range, should this be needed. Near 1 keV and above to the MeV region alternative technologies are needed to which this report now turns.

The entire thousandfold range just mentioned is accessible to crystal diffraction spectroscopy. Theoretically it was known for a long time that high resolving power could be obtained in principle by this means. It was also long understood that angle measurement (as required in application of a Bragg-Laue equation) could be done with great refinement. Unfortunately, since the uv procedure did not reach the X-ray region, there was a problem. This problem gave rise to a local scale (the so-called xu) based on variously agreed on values for certain prominent lines or, for a time, the lattice parameter of "the purest calcite". Approaches to handling that problem were through ruled grating measurement of long wavelength X-ray lines, calculable spectra or inversion of the X-ray determination of N_A. All of these approaches were and are seriously flawed. Main problems with the ruled grating work are poor resolution in the extreme grazing incidence regime and low quality factors for the X-ray lines. The alternative approaches via calculable spectra or physical constants have the effect of eliminating potential sources of information.

In the case γ-ray lines in the region $0.1 < E < 1$ MeV a similar situation arose. A single first-principles measurement using positron annihilation radiation (rather broad and ill-shaped) as a calculable marker normalized the convenient 411 keV line from the Hg daughter of ^{198}Au [34]. This excellent and difficult measurement used β-ray spectroscopy to measure the small difference between photo electrons promoted by 511 keV radiation from the K shell of uranium and photo electrons promoted from its L_{III} shell by the 411 keV radiation. Since this elaborate measurement was not easily duplicated or generalized, the resulting gamma-ray energy became a <u>de facto</u> local standard. At still higher energies, corresponding mass decrements become measurable to sufficiently high precision to yield still another type scale [35]. Unfortunately the use of a conversion factor from ΔM values in amu to energy (or wavelength) precludes access to another potentially observable grouping of constants.

Overall, then, one begins with at least three local scales and the corresponding interregion conversion factors. One way out of this is to use optical interferometry to determine the lattice period of a crystal and use it to measure Bragg-Laue angles. This has some practical as well as conceptual advantages. For instance, use of iodine stabilized HeNe lasers as the light source automatically connects any subsequent spectroscopy to atomic hydrogen. Second, although certain aspects of technique need to change as one goes from 1 keV to 1 MeV, the changes are not conceptually significant. The main overall technical problem is securing robust angle measurements for both the large angles that are encountered at low energies and the small ones encountered at γ-ray energies. Of significant nuisance value is the need to have several crystal types, orientations, and sizes to maintain optimum performance over the large range in energy. This is handled, though not yet as conveniently as one would like, by means of a quasi-null transfer method noted below.

The first step in current approaches to extending the optically based scale combines X-ray and optical interferometry to obtain the spatial period of a sample of single crystal silicon. This measurement has a particular advantage in that the X-ray interferometry is achromatic. Thus the great widths and peculiar shapes encountered in X-ray lines have no effect on the result. Also the crystals used have well defined lattice parameters to about the 0.01 ppm level which indicates the sort of limiting accuracy which one may hope for in the present era. Though inferior to accuracies mentioned above, it represents significant progress since no X-ray line has a feature that can be reliably located as well as 1 ppm and the best previous connection (via gratings) was surely worse than 10 ppm.

The next step is to distribute this initial calibration to other crystals whose type, orientation and shape are suited to various wavelength regions of interest. One wants to do this using X-ray sources (for convenience) but not being limited by their characteristics. Generically, it is easy to see how to proceed by arranging to measure relatively small differences between two crystals.

Given suitable crystals, refined versions of the techniques of classical Bragg-Laue spectroscopy are appropriate. Evidently one requires sensitive angle measurement over large angles or small angles. In the case of large angles, sub-ppm sensitivities are available using several well-known resolver or encoder techniques, or, for that matter, by a divided circle. In the case of small angles, adequate sensitivity requires use of optical interferometry with some degree of resolution enhancement depending on the particular range involved. In either the large angle case or the small angle one, the scale needs *in situ* calibration. This requires no external standard if the closure condition can be met. The large angle case has well known solutions, including one which uses an optical polygon and null pointing autocollimator. The small angle case can readily be treated also with a polygon and a more sensitive null-pointing autocollimator.

To date, this technology has been applied to obtain γ-ray markers, as required for muonic and pionic spectra, and to remeasure a number of mid to high Z K spectra. Results for the γ-rays and an overall summary of their applications to date have been summarized elsewhere [36]. New results on the X-ray spectra as a means of reaching higher values of $Z\alpha$ have also been communicated. Finally, a direct measurement of $AK\alpha_{1,2}$ has been used to establish the location of $C\ell^{+16}$ Lyman α and thereby a rough measure of the 1s Lamb shift as mentioned.

5. Summary and Outlook

A rather superficial overview has been presented of a few aspects of fundamental constants and of atomic physics and their intersection. Looking backward, one senses that some exercises are near their limits of quality and of interest. Toward the future, there are several directions where new results may be hoped for in the future. Clearly the new approaches in electrical measurement where object size and location need not be specified are very attractive. Also, as is suggested by von Klitzing's discovery, the end may not be in sight for macroscopic quantum phenomena occurring in real materials but depending only on fundamental constants. In the case of atomic hydrogen, the end is not yet in sight especially for 1s - 2s. When real accuracy is obtained, one may proceed to interpret this as a new R_∞ measurement or consider (ultimately) an overdetermined group of ratios among hydrogen transitions as an overall test of theory. The analogous transition in positronium is near to measurement at this time, although the exact significance is not yet obvious [37].

For higher Z one-electron systems, the future seems both fairly clear and somewhat promising. A landmark experiment, when it is obtained, will be accurate measurement of the 2s → 1s transition in otherwise bare krypton. Further advances in accelerator technology will make higher Z systems accessible but at the price of more difficult Doppler corrections, unless, of course, low energy recoils can be used. Perhaps at the next ICAP we shall see how these expectations have fared.

REFERENCES

1. The next major iteration of the continuing process is due to appear in the near future through the efforts of B.N. Taylor and E.R. Cohen under the aegis of the CODATA Task Group on Fundamental Constants. Its predecessor is to be found in E.R. Cohen and B.N. Taylor, Jour. Phys. Chem. Ref. Data 2, 663 (1973).
2. F.K. Richtmyer, Science 75, 1 (1932).
3. Last year's Nobel Lectures represent a convenient point of departure, V.L. Fitch, Science 212, 989 (1981); J.W. Cronin, ibid., 1221.
4. There are many alternative perspectives on this area among which the reader may wish to examine (titles are indicated for guidance):
   J.-M. Lévy-Leblond, "On the Conceptual Nature of the Physical Constants", Revista del Nuovo Cimento 7, Nr. 2, pp. 187-214 (1977).
   Kastler and P. Grivet, "The Measurement of Fundamental Constants (Metrology) and its Effect on Scientific and Technical Progress", in Atomic Masses and Fundamental Constants 5, J.H. Sanders and A.H. Wapstra, Eds., Plenum Press, N.Y. (1976), pp. 1-23.
   A.H. Cook, "Standards of Measurement and the Structure of Physical Knowledge", Contemporary Physics 18, No. 4, pp. 393-409 (1977).
   General guides to the field have been rare. The most recent is that due to B. Petley, "The Fundamental Physical Constants and the Frontier of Measurement," Adam Hilger (London), (1983).
5. See, e.g., E.R. Cohen and J.W.M. DuMond, Rev. Mod. Phys. 37, 537 (1965).
6. B.N. Taylor, W.H. Parker and D.N. Langenberg, Rev. Mod. Phys. 41, 375 (1969).

7. I refer here to independent exercises in optical interferometry designed to establish vacuum wavelength (frequency) ratios between CH_4 and I_2 stabilized HeNe lasers:
H.P. Layer, R.D. Deslattes and W.G. Schweitzer, Jr., Appl. Opt. 15, 734 (1976).
W.R.C. Rowley, BIPM document, CCDM/824.
G. Bönsch, PTB, Jahresbericht 1979, p. 135.
8. E.R. Williams and P.T. Olsen, Phys. Rev. Lett. 42, 1575 (1979); E.R. Williams, P.T. Olsen and W.D. Phillips, Proceedings of Precision Measurement and Fundamental Constants-II, B.N. Taylor and W.D. Phillips, Eds., U.S. G.P.O., in press.
9. B.P. Kibble and G.J. Hunt, Metrologia 15, 5-30 (1979).
10. E.R. Williams and P.T. Olsen, IEEE Trans. Instrum. Meas. IM-27, 467 (1978).
11. G.L. Greene, Metrologia 17, 83 (1981).
12. W. Chiao, R. Liu and P. Shen, IEEE Trans Instr. Meas. IM-29, 238 (1980).
13. P.T. Olsen, W.D. Phillips and E.R. Williams, J. Res. NBS 85, 257 (1980).
14. V.E. Bower and R.S. Davis, J. Res. NBS 85, 175 (1980). See also: V.E. Bower, R.S. Davis, T.J. Murphy, P.J. Paulsen, J.W. Gramlich and L.J. Powell, J. Res. NBS 87, 21 (1982).
15. Background in this area has been reviewed: R. Deslattes, "The Avogadro Constant" in Annual Reviews of Physical Chemistry 31, B.S. Rabinovitch, Ed., Annual Reviews, Inc., Palo Alto, USA (1980), p. 435.
16. B.W. Petley, "Electrical Metrology and the Fundamental Constant" in Metrology and Fundamental Constants Course 68 International School of Physics Enrico Fermi, A. Ferro Milone, P. Giacomo and S. Leschuitta, Ed., North Holland, Amsterdam, 1980, p. 358.
17. B. Cabrerra, S. Benjamin and J.T. Anderson, Physica 107B, 19 (1981).
18. J.C. Gallop, Jour. Phys. B 11, L93 (1978).
19. K. von Klitzing, G. Dorda and M. Pepper, Phys. Rev. Lett. 45, 494 (1980).
20. E.G. Kessler, Jr., Phys. Rev. A 7, 408 (1973).
21. T.W. Hänsch, M.H. Nayfeh, S.A. Lee, S.M. Curry and I.S. Shahin, Phys. Rev. Lett. 32, 1336 (1974).
22. C. Wieman and T.W. Hänsch, Phys. Rev. A 22, 192 (1980).
23. S.R. Amin, C.D. Caldwell and W. Lichten, Phys. Rev. Lett. 47, 1234 (1981).
24. K. Evenson, J.S. Wells, F.R. Petersen, B.L. Danielsen and G.W. Day, Appl. Phys. Lett. 22, 192 (1973).
25. C.R. Pollock, D.A. Jennings, F.R. Petersen, R.E. Drullinger, E.C. Beaty, J.S. Wells, J.L. Hall, H.P. Layer and K.M. Evenson, to be published.

26. The CCDM recommends that the metre be defined as follows: "The metre is the length of the path traveled by light in vacuum during the fraction 1/299 792 458 of a second."
27. O.R. Wood II, C.K.N. Patel, D.E. Murnick, E.T. Nelson, M. Leventhal, H.W. Kugel and Y. Niv, Phys. Rev. Lett. 46, 398 (1982).
28. P.J. Mohr, Phys. Rev. Lett. 34, 1050 (1975).
29. G.W. Erickson, J. Phys. Chem. Ref. Data 6, 831 (1977).
30. Participants were: P. Richard, M. Stockley and Redo Mann (Kansas State), R. Deslattes, P. Cowan and K.-H. Schartner (NBS), B. Johnson and K. Jones (BNL).
31. E.G. Kessler, Jr., R.D. Deslattes, D. Girard-Vernhet, W. Schwitz, L. Jacobs and O. Renner, Physical Review A (in press).
32. S. Chu and A.P. Mills, Jr., Phys. Rev. Lett. 48, 1333 (1982).
33. B. Edlén, Repts. Prog. Phys. XXVI, 181 (1963).
34. G. Murray, R.L. Graham and J.S. Geiger, Nucl. Phys. 63, 177 (1965).
35. For a discussion of use of mass differences, see for example, R.G. Helmer, P.H.M. van Assche and C. vander Leun, Atomic Data and Nuclear Data Tables 24, 39 (1979).
36. R.D. Deslattes, E.G. Kessler, Jr., W.C. Sauder and A. Henins, Annals of Phys. 129, 378 (1980).
37. S. Chu and A.P. Mills, Jr., Phys. Rev. Lett. 48, 1333 (1982).

THE QUANTUM HALL EFFECT

Klaus v. Klitzing
Physik-Department der Technischen Universität München
D-8046 Garching, FRG

INTRODUCTION

Basically, the quantum Hall effect (QHE) has nothing to do with atomic physics. Semiconductors are normally used to observe this quantum phenomenon, and the 10^{23} atoms per cubic centimeter of a semiconductor represent such a complicated system of interacting atoms that its electronic properties are normally described by phenomenological quantities and cannot be deduced from the properties of the isolated atoms. Nevertheless, our measurements on semiconductors demonstrate that the quantity h/e^2 (h = Planck constant, e = elementary charge) can be determined with an uncertainty of less than 10^{-6}. Since the fine-structure constant α is directly proportional to e^2/h (the proportional constant depends mainly on the well known velocity of light c), one can use the QHE for the determination of α with an uncertainty smaller than that resulting from high-precision measurements of the fine-structure and hyperfine-structure splitting of a hydrogen atom.

This article will give a survey about the different methods for the determination of the fine-structure constant including a simple introduction to the quantum Hall effect. For a more detailed discussion see Ref. 1.

METHODS FOR THE DETERMINATION OF THE FINE-STRUCTURE CONSTANT

Originally the fine-structure constant α was introduced by Sommerfeld in order to explain the optical spectrum of atomic hydrogen [2], where α was defined as the ratio of the electron velocity in the lowest Bohr orbit to the velocity of light. This ratio is given by the expression

$$\alpha = (e^2/h)(\mu_0 c/2) \qquad \text{(SI-units)}$$
$$\alpha = e^2/\hbar c \qquad \text{(cgs-units)} \qquad (1)$$

with $c = 299792458$ m/s and $\mu_0 \equiv 4\pi \times 10^{-7}$ H/m.

The officially recommended values for α are plotted in Fig. 1 as a function of time. The last least-squares adjustment of fundamental constants was in 1973 with a value [3]

$$\alpha^{-1} = 137.03604(11) \qquad (2)$$

The one standard deviation is given by ±11 for the last digits. A new adjustment is planned for 1983.

The small value of $\alpha \approx 1/137$ allows to use α as an expansion parameter in quantum electrodynamics (QED), and the corresponding theoretical equations are developed to a high accuracy. From a com-

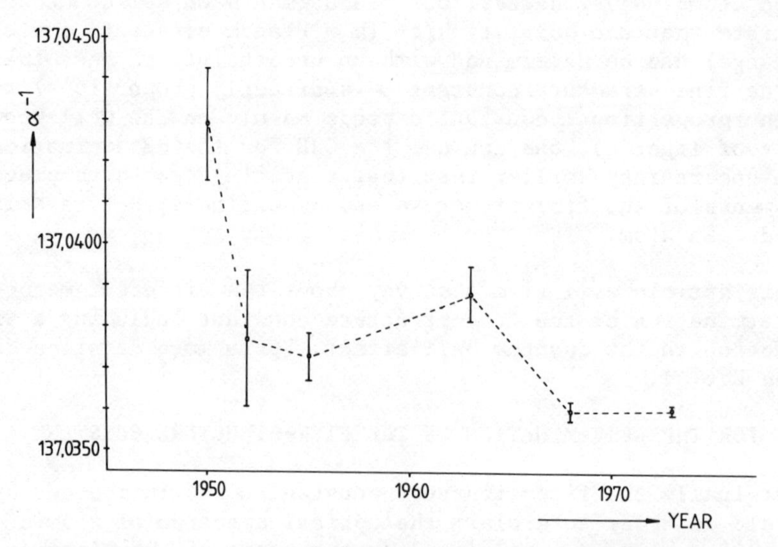

Fig. 1: Data for the officially recommended values of the fine-structure constant α (including one standard deviation) as a function of time.

parison between the result of the QED theory with the experimental data a value for α can be deduced. In order to check the QED theory (no proof exists that this theory is correct), an accurate value for α without using the quantum electrodynamic theory (WQED) is necessary. The quantum Hall effect and the γ_p-method seem to be the most reliable procedures for a WQED determination of the fine-structure constant. These methods will be discussed after the following summary of α- determinations using QED theory.

Determination of the fine-structure constant based on QED theory

The hydrogen spectrum is not the best one for a high-precision determination of the fine-structure constant. The fine-structure splitting $\nu_{fs} = 2P_{3/2} - 2P_{1/2}$ is a complicated function of α, electron mass m and proton mass M, anomalous magnetic moment of electrons a_e and Rydberg constant R and is given by the expression

$$\nu_{fs} = (1/16)\alpha^2 Rc \Big\{(1 + 5\alpha^2/8)(1 + m/M)^{-1} - (m/M)^2(1 + m/M)^{-3} + + 2a_e(1 + m/M)^{-2} - \alpha^3/\pi(\ln \alpha^{-2} + \delta)\Big\} \tag{3}$$

The term δ includes corrections from the QED theory.

The fine-structure splitting cannot be measured directly, it is the sum of the Lamb shift $S = 2S_{1/2} - 2P_{1/2}$ and the energy splitting $\Delta E = 2P_{3/2} - 2S_{1/2}$. The most accurate values are

S = 1057.862(20) MHz	(Ref. 4)
S = 1057.845(9) MHz	(Ref. 5)
S = 1057.8583(22) MHz	(Ref. 6)
ΔE = 9911.377(26) MHz	(Ref. 7)
ΔE = 9911.250(63) MHz	(Ref. 8)
ΔE = 9911.173(42) MHz	(Ref. 9)
ΔE = 9911.117(41) MHz	(Ref. 10)

The consistency among different measurements is not very good, and even if the theory (Eq.3) is correct, the uncertainty of α determined from the fine-structure splitting is at least ± 2ppm.

The hyperfine splitting of a hydrogen atom is known experimentally with an extremely small uncertainty [11]

$$\nu_{hfs} = 1420405.7517667(10) \text{ kHz}.$$

Unfortunately, the theory is limited to much less precision [12] by nucleon structure and radiative corrections with an estimated uncertainty of about 3 ppm. Similar calculations for muonium (no hadronic complication) are much more accurate, so that from a comparison between theory and experiment the following α- value is obtained [13]:

$$\alpha^{-1} = 137.035989(3)(47). \tag{4}$$

Here the error ± 3 originates from the experiment and the error ± 47 from the theory.

The best result for α using QED theory is obtained from measurements and calculations of the anomalous magnetic moment a_e of electrons or positrons. The experimental results are [14,15]

$$\begin{aligned} a_e &= 1\ 159\ 652\ 200\ (40) \times 10^{-12} \text{ (electron)} \\ a_e &= 1\ 159\ 652\ 222\ (50) \times 10^{-12} \text{ (positron)} \end{aligned} \tag{5}$$

The theory predicts a value [16]

$$\begin{aligned} a_e = {} & 0.5(\alpha/\pi) - 0.328478966\ (\alpha/\pi)^2 + 1.1765(13)(\alpha/\pi)^3 \\ & - 0.8(2.5)(\alpha/\pi)^4 + 4.4 \times 10^{-12}, \end{aligned} \tag{6}$$

where the coefficients for the α^3 and α^4 terms consist of 72 and 891 Feynman diagrams. Agreement between theory (Eq.6) and experiment (Eq.5) exists, if the following value for the fine-structure constant is assumed

$$\alpha^{-1} = 137.035993(5)(9). \tag{7}$$

The experimental and theoretical uncertainties lead in the last digit to a variation of ± 5 and ± 9, respectively. The total uncertainty is less than 10^{-7}, but systematic errors in the calculations cannot be excluded.

Determination of the fine-structure constant without using QED theory

The fine-structure constant (Eq.1, SI-units) can be rewritten in the following form

$$\alpha^{-1} = \left\{(c/4R) \cdot (\mu_p/\mu_B) \cdot (2e/h) \cdot (1/\gamma_p)\right\}^{1/2} \tag{8}$$

The Rydberg constant R, velocity of light c, magnetic moment of protons in units of Bohr magnetons, and the quantity e/h (from a.c. Josephson effect) are known with uncertainties of less than 0.03 ppm. The gyromagnetic ratio of the protons γ_p is normally not known with such a high accuracy,so that this quantity determines the uncertainty of α. Different groups measured α with a claimed uncertainty of less than 1 ppm using this γ_p-method, but the results disagree by about 4 ppm. Fig. 2 gives a summary of these data together with preliminary measurements of α using the quantum Hall effect (h/e^2-method). The α-values deduced from the QHE (measured by four different groups on two different materials) are more consistent than those obtained from γ_p-measurements. Therefore national laboratories in at least eight countries are investigating the application of the quantum Hall effect in the field of metrology.

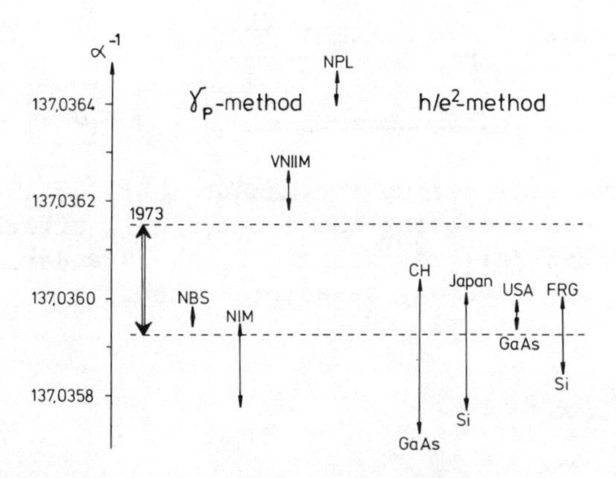

Fig. 2: Summary of values for the fine-structure constant without using quantum electrodynamic theory. The recommended value is characterized by the dotted lines, and the measured values of α (including one standard deviation) using the γ_p-method and the h/e^2-method (= quantum Hall effect) are plotted as vertical lines. The national laboratories are: NBS (National Bureau of Standards, U.S.), NIM (National Institute of Metrology, China), VNIIM (Mendeleev Institute, USSR), and NPL (National Physics Laboratory, U.K.)

QUANTUM HALL EFFECT

The discovery of the QHE in 1980 [17] is identical with the discovery of a new type of resistor: A resistor with a resistance value which depends only on the quantity $h/e^2 \approx 25813\ \Omega$, independent of the material or the geometry of the resistor. If such a resistor really exists, then the experimental arrangement for the measurement of the inverse fine-structure constant, which is directly proportional to h/e^2 (Eq.1), is extremely simple as shown in Fig. 3. The experimental set-up consists of a battery, a calibrated resistor R_N (resistors calibrated in SI-units with an uncertainty of about 10^{-7} are maintained at the national laboratories), the h/e^2-resistor R_H, and a resistor R_V which limits the current. The fine-structure constant can be determined directly from the ratio of the voltage

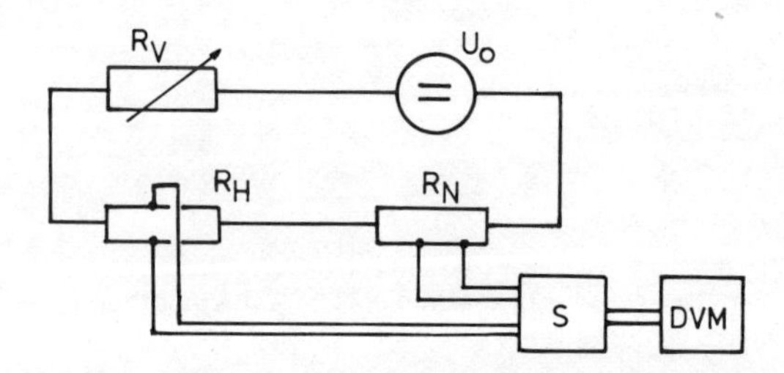

Fig. 3: Experimental set-up for measurements of the fine-structure constant using the quantum Hall effect. R_H is the quantized Hall resistance, R_N the standard resistance, and S is a two-way selector switch.

drops across the standard resistor and the h/e^2-resistor.

The existence of a quantum resistor with values $R_H = h/ie^2$ $(i = 1,2,3,...)$ is related to the Hall effect of a two-dimensional electron gas (2DEG). If a magnetic field $B = B_z$ is applied perpendicular to the plane of a 2DEG (x-y plane), an electric field E_y is built up, if a current I with a density j_x is flowing in x-direction (equilibrium between Lorentz force and force on an electron in the electric field E_y). With a two-dimensional carrier density n_s the Hall field E_y becomes

$$E_y = \frac{B}{n_s e} \cdot j_x \tag{9}$$

Since the ratio E_y/j_x is identical with the measurable Hall resistance $R_H = U_H/I$ (Hall voltage divided by the current through the sample), Eq.9 becomes

$$R_H = B/n_s e \tag{10}$$

A sketch of the sample is shown in Fig.4.

The pecularity of a 2DEG in a strong magnetic field is that the energy spectrum of the electrons is discrete due to orbital quantization of the cyclotron motion in a magnetic field. The energy spectrum can be written in the form [18]:

$$E = (n + \tfrac{1}{2})\hbar\omega_c \qquad n = 0,1,2,... \tag{11}$$

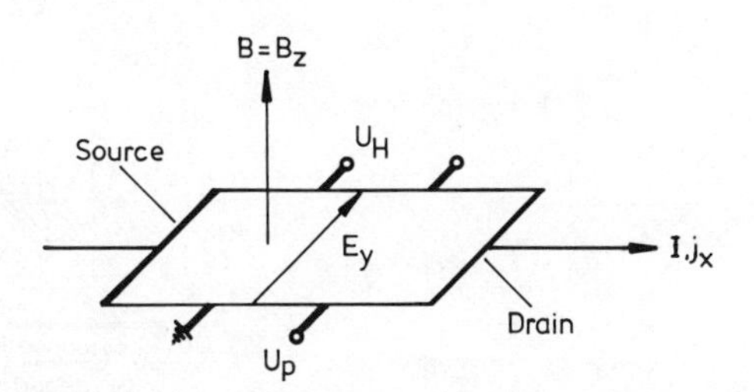

Fig. 4: For measurements of the QHE the Hall resistance $R_H = U_H/I = E_H/j_x$ and the resistivity $\rho_{xx} \sim R_x = U_x/I$ of a two-dimensional conductor are analyzed in a strong magnetic field. "Source" and "Drain" are the current contacts of the sample.

where the cyclotron energy $\hbar\omega_c$ is proportional to the magnetic field. Each of these energy levels (Landau levels) has a degeneracy per unit area of

$$N = eB/h \tag{12}$$

corresponding to the number of cyclotron orbits per unit area. This means that at a magnetic field of B = 10 T exactly 2 417 967 100 electrons/mm^2 occupy one energy level E, independent of the material.

If the two-dimensional carrier-density n_s and the magnetic field B are adjusted in such a way that exactly an integer number i of energy levels are occupied with electrons, this means

$$n_s = iN = ieB/h, \tag{13}$$

then the Hall resistance R_H (Eq.10) will adopt a value

$$R_H = h/ie^2. \tag{14}$$

This is the quantized Hall resistance necessary for a determination of α. The integer i is in practice not identical with the Landau quantum number n, because each energy level splits at least into two spin-split levels (not included in Eq.11). The situation is still more complicated for silicon where each Landau level consists of four energy levels ((100)-surface). However, the energy spectrum is unimportant for the QHE as long as the magnetic field is strong enough to separate the energy levels.

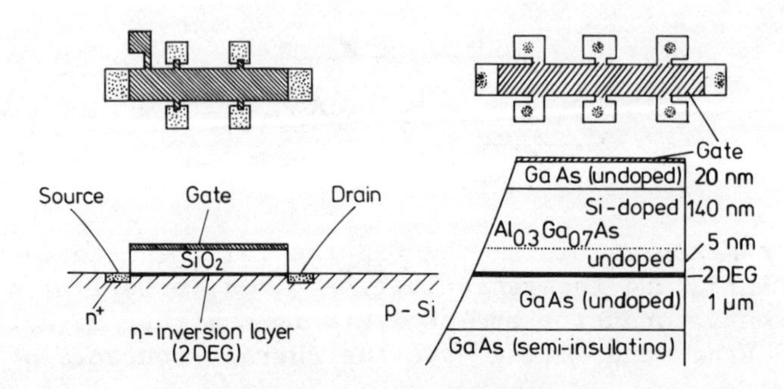

Fig. 5: Top view and cross section of typical devices used in our experiments. Left side: Silicon MOS field effect transistor. Right side: $GaAs/Al_xGa_{1-x}As$ heterostructure with Shottky gate.

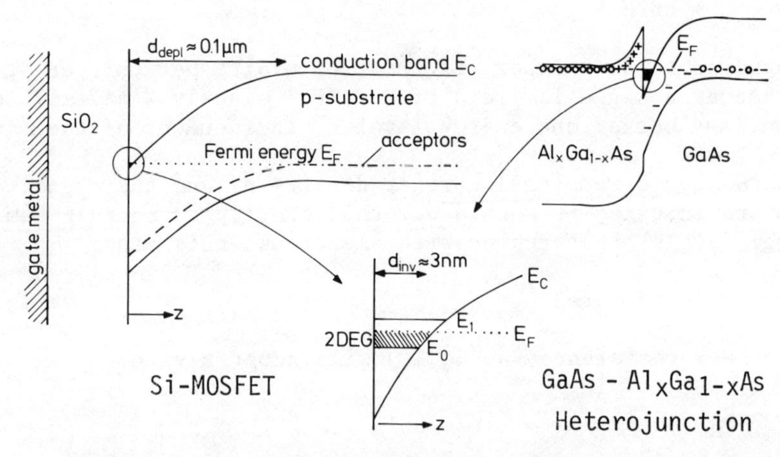

Fig. 6: Formation of a two-dimensional electron gas at the Si/SiO_2 interface of a MOSFET and at the $GaAs/Al_xGa_{1-x}As$ interface of a heterostructure. The confinement of the electrons in z-direction within a narrow potential well of about 3 nm leads to descrete energy eigenvalues E_0, E_1... (electric subbands). At low temperatures, only the lowest subband is occupied (electric quantum limit).

Theoretically, Hall effect measurements on all two-dimensional systems can be used for the h/e^2-determination, but only electrons at the semiconductor-insulator or semiconductor-semiconductor interface form a 2DEG useful for such experiments. A cross section and a top view of the two systems used in our measurements is shown in Fig.5.

Positive charges at the gate or in the dielectric between the gate and the semiconductor Si or GaAs induce electrons in the plane of the 2DEG. The surface carrier density n_s changes linearly with the gate voltage V_g. The restriction into two dimensions arises from a strong surface potential which confines the carriers within such a narrow channel close to the Si/SiO_2 or $GaAs/Al_xGa_{1-x}As$ interface that a motion of the electrons perpendicular to the surface is not possible. This is illustrated in Fig.6. The energy for the motion of the electrons perpendicular to the interface is quantized into

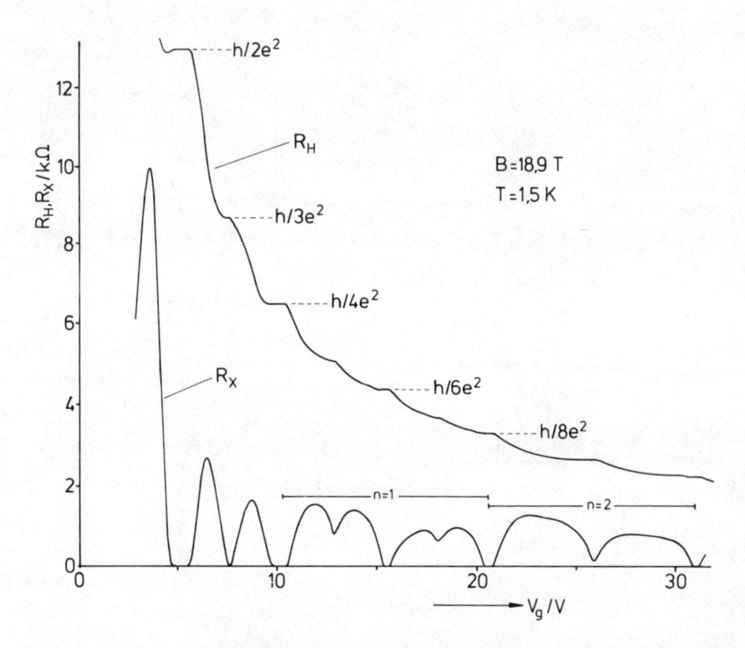

Fig. 7: Gate voltage dependence of the Hall resistance R_H and the resistivity R_x at B = 18.9 T for a silicon MOSFET.

discrete electric subbands E_0, E_1,... whereas the motion in the plane of the surface is not restricted. This means that with regard to the energy a real two-dimensional electron gas exists, if only the lowest electric subband E_0 is occupied. Higher subbands are normally not populated at helium temperatures.

A typical measurement of the Hall resistance $R_H = U_H/I$ and the resistivity $R_x = U_p/I$ on a silicon MOSFET (Metal-Oxide-Semiconductor Field-Effect-Transistor) is shown in Fig. 7. With increasing gate voltage $V_g \sim n_s$ different Landau levels are subsequently filled up with electrons resulting in oscillations for $R_x(V_g)$. Every time one energy level is completely filled (this means that the Fermi energy is located in the energy gap between two energy levels), R_x becomes zero because no elastic scattering from an occupied to an empty state and therefore no energy loss is possible. At these gate voltages V_g where R_x becomes zero, R_H adopts the value h/ie^2 as expected from Eq. 14.

Much more pronounced Hall steps are observed for the 2DEG of a GaAs-$Al_xGa_{1-x}As$ heterostructure, as shown in Fig. 8. In this case,

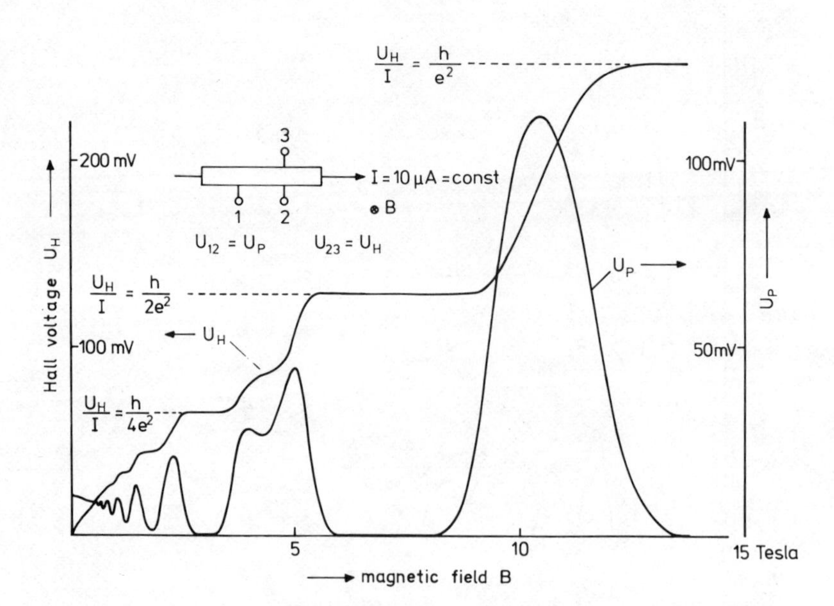

Fig. 8: Hall effect and resistivity measurements on a GaAs-$Al_{0.3}Ga_{0.7}As$ heterostructure as a function of the magnetic field B.

the Hall resistance $\rho_{xy} = R_H$ and the resistivity $\rho_{xx} \sim R_x$ are plotted as a function of the magnetic field at a fixed carrier density n_s. From an analysis of the Hall steps of such a system the following value for the fine-structure constant is found [19]:

$$\alpha^{-1} = 137.035968(23). \qquad (15)$$

This value is based on the assumption that Eq.14 is correct. The uncertainty arises mainly from instabilities in the value of the calibrated reference resistor necessary for the determination of α (see Fig.3). A large number of theoretical papers discuss the question whether microscopic details of the semiconductor may influence the accuracy of Eq.14. Up to now, no corrections to the value of the quantized Hall resistance are known, and the good agreement of the α-value deduced from the quantum Hall effect (Eq.15) with the recommended value (Eq.1) and data obtained from other experiments (Eq.4, Eq.7) demonstrates that corrections to the quantized Hall resistance (if any) should be smaller than the experimental uncertainty of about 10^{-6}. We believe that Eq.14 is correct even at a higher level of accuracy and that the QHE can be used on one hand as a standard resistor (if the value for h/e^2 is known or defined for metrological applications) and on the other hand for the determination of the fine-structure constant with an uncertainty corresponding to the uncertainty of the reference resistor.

ACKNOWLEDGEMENTS

The experiments were performed on a large number of devices, and I would like to thank K. Ploog (Max-Planck-Institut für Festkörperforschung, Stuttgart), M. Pepper (Cavendish Laboratory, Cambridge), and G. Dorda (Siemens-Forschungslaboratorien, München) for providing me with samples. The research work on the application of the quantum Hall effect in metrology is supported by the Stiftung Volkswagenwerk.

REFERENCES

[1] K.v.Klitzing, 1982, Two-dimensional systems: A method for the determination of the fine-structure constant, Surf.Sci. 113, 1.
[2] A. Sommerfeld, 1916, Annalen der Physik (Leipzig) 51, 1.
[3] E.R. Cohen and B.N. Taylor, 1973, The 1973 least-square adjustment of the fundamental constants, J.Phys.Chem.Ref. Data 2, 663.
[4] D.A. Andrews and G. Newton, 1976, Radio frequency atomic beam measurement of the Lamb-shift interval in hydrogen, Phys.Rev.Letters 37, 1254.
[5] S.R. Lundeen and F.M. Pipkin, 1981, Measurement of the Lamb-shift in hydrogen, n = 2, Phys.Rev.Letters 46, 232.

[6] Yu.L.Sokolov, 1982, Measurement of the Lamb shift in hydrogen, in Proc. of the Second Int.Conf. on Precision Measurements and Fundamental Constants, Eds B.V. Taylor and W.D. Phillips, Nat. Bur. Std. (US), Spec.Publ. 617

[7] S.L. Kaufman, W.E. Lamb, K.R. Lea, and M. Leventhal, 1971, Measurement of the $2^2S_{1/2}-2^2P_{3/2}$ Interval in Atomic Hydrogen, Phys.Rev. A 4, 2128.

[8] T.W. Shyn, T. Rebane, R.T. Robiscoe, and W.L. Williams, 1971, Measurement of the $2^2S_{1/2}-2^2P_{3/2}$ energy separation ($\Delta E - S$) in hydrogen (n = 2), Phys.Rev. A 3, 116.

[9] B.L. Cosens and T.V. Vorburger, 1970, Remeasurement of the $2^2S_{1/2}-2^2P_{3/2}$ splitting in atomic hydrogen, Phys.Rev. A 2, 16.

[10] K.A. Safinya, K.K. Chan, S.R. Lundeen, and F.M. Pipkin, 1980, Measurement of the $2^2P_{3/2}-2^2S_{1/2}$ fine-structure interval in atomic hydrogen, Phys.Rev.Letters 45, 1934.

[11] L. Essen, R.W. Donaldson, M.J. Bangham, and E.G. Hope, 1971, Frequency of the hydrogen maser, Nature 229, 110.

[12] E. de Rafael, 1971, The hydrogen hyperfine structure and inelastic electron-proton scattering experiments, Phys. Lett. 37 B, 201.

[13] V.W. Hughes, 1982, Precision exotic atom spectroscopy, in Proc. of the Second Int.Conf. on Precision Measurements and Fundamental Constants, Eds B.N. Taylor and W.D. Phillips, Natl. Bur. Std. (US), Spec. Publ. 617.

[14] R.S. van Dyck, P.B. Schwinberg, and H.G. Dehmelt, 1979, Progress in the electron spin anomaly experiment, Bull.Am. Phys.Soc. 24, 758 .

[15] P.B. Schwinberg, R.S. van Dyck, and H.G. Dehmelt, 1982, Comparison of the positron and electron spin anomalies, in Proc. of the Second Int.Conf. on Precision Measurements and Fundamental Constants, Eds B.N. Taylor and W.D. Phillips, Natl. Bur. Std. (US), Spec. Publ. 617.

[16] T. Kinoshita and W.B. Lindquist, 1981, Eight-order anomalous magnetic moment of the electron, Phys.Rev.Letters 47, 1573.

[17] K. v.Klitzing, G. Dorda, and M.Pepper, 1980, New method for high-accuracy determination of the fine-structure constant based on quantized Hall resistance, Phys.Rev.Letters 45, 494.

[18] for a review see: F. Stern, 1974, Quantum properties of surface space-charge layers, Crit.Rev.Solid State Sci. 4, 499.

[19] D.C. Tsui, A.C. Gossard, B.F. Field, M.E. Cage, and R.F.Dziuba, 1982, Determination of the fine-structure constant using GaAs-$Al_xGA_{1-x}As$ heterostructures, Phys.Rev.Letters 48, 3.

SUB-DOPPLER SPECTROSCOPY

T.W. Hänsch

Department of Physics
Stanford University
Stanford, California 94305, U.S.A.

The Doppler effect has long been one of the principal problems in high resolution spectroscopy of atoms and molecules. Atoms moving towards an observer appear to emit or absorb light at shorter wavelengths than atoms at rest, and atoms moving away appear to emit at longer wavelengths. In a gas, with atoms moving at random in different directions, the lines appear blurred, and important details in line shape and structure are often obscured.

But the advent of lasers and coherent light techniques opened a rich arsenal of spectroscopic tools which make it possible to study spectral lines in gas samples free of first order Doppler broadening.(1,2) Two conceptually different approaches have so far been used very successfully:
1) In saturation spectroscopy, a monochromatic laser beam "labels" a group of atoms within a narrow range of axial velocities through excitation or optical pumping, and a Doppler-free spectrum of these selected atoms is observed with a second, counterpropagating beam.
2) In two-photon spectroscopy it is possible to record Doppler-free spectra without any need for velocity selection by excitation with two counterpropagating laser beams whose first order Doppler shifts cancel.

A comprehensive review of all the work in this field is certainly beyond the scope of this presentation. In the first part, we will only attempt to gain a simplified overview of the wide variety of different techniques of Doppler-free saturation spectroscopy which have been used to date. Next we will discuss some recent advances in the development of tunable cw sources, which can extend Doppler-free laser spectroscopy into the important ultraviolet spectral region. Finally, we will briefly discuss Doppler-free two-photon spectroscopy, which offers some particularly intriguing challenges and opportunities for precision measurements in atomic hydrogen.

1. SATURATION SPECTROSCOPY

The possibility of Doppler-free saturation spectroscopy has been recognized since the early observations of Lamb dips in gas lasers (3-5). However, such studies were essentially limited to gas laser transitions themselves, or to those few molecular lines which happened to coincide with gas laser lines.

Around 1970, saturation spectroscopy became much more widely useful, when we learned that the linewidth of broadly tunable dye lasers could be reduced to a small fraction of a typical Doppler width.(6) At the same time, C. Bordé (7) and, independently, our group at Stanford (8-10) introduced a simple and versatile technique of saturation spectroscopy, which reaches high sensitivity even though the gas sample is placed conveniently outside the laser resonator.

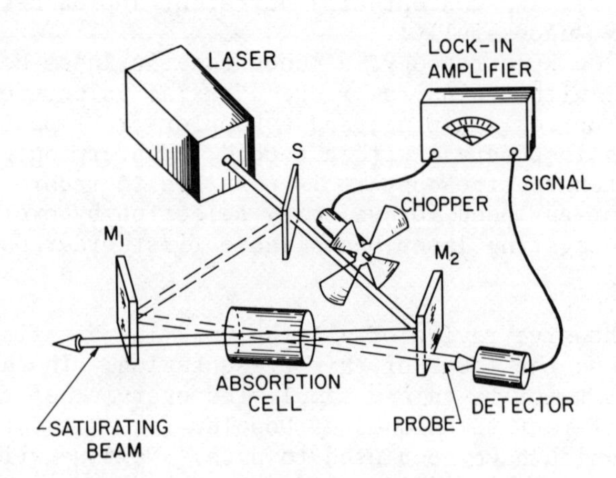

Fig. 1 Apparatus for Doppler-free saturated absorption spectroscopy in a gas sample.

In this technique, as illustrated in Fig. 1., the light from the laser is divided by a partial mirror into two beams which pass through the sample in nearly opposite directions. Each beam interacts only with those atoms which have the right axial velocity to be Doppler-shifted into resonance. The stronger saturating beam is chopped at an audio frequency. When it is on, it partly saturates the absorption of the gas by exciting the atoms and so removing them from the absorbing lower level. As a result, it can bleach a path for the probe beam so that the probe intensity becomes modulated at the chopper frequency. Such a signal is observed when the laser is tuned to the center of the Doppler-broadened line, where both beams are interacting with the same atoms, those with zero axial velocity.

The red Balmer-α line of atomic hydrogen (Fig. 2) was one of the first transitions to be studied with these powerful new tools. (11) The visible spectrum of this simplest of the stable atoms has, of course, played a key role in the development of atomic theory and quantum mechanics. (12) However, the important fine structure of its spectral lines had remained blurred in all classical observations by the particularly large Doppler broadening. We were, therefore, thrilled when even our first saturation spectra, recorded with a

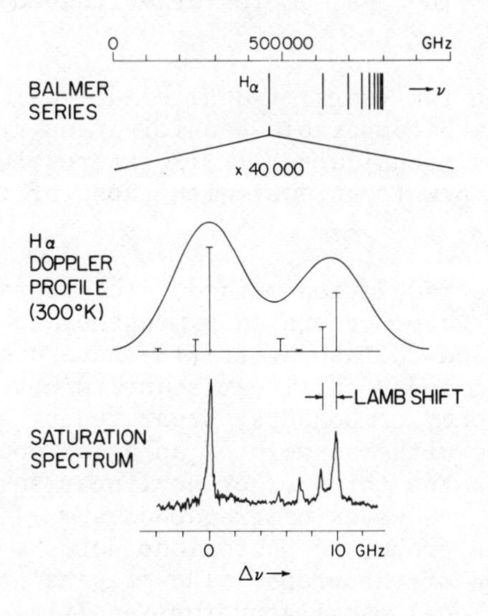

Fig. 2 Top: Balmer spectrum of atomic hydrogen. Center: Doppler profile of the Balmer-α line at room temperature and theoretical fine structure components. Bottom: Doppler-free spectrum of Balmer-α, recorded by saturated absorption spectroscopy with a pulsed dye laser.

pulsed dye laser in a Wood-type gas discharge, (11) showed cleanly resolved single fine structure components, and the n=2 Lamb shift could be observed directly in the optical spectrum. In 1974, we completed an absolute wavelength measurement, which yielded an almost tenfold improved value of the Rydberg constant. (13)

During these early measurements it became soon apparent that the chosen technique requires a sample of fairly high absorption and an intense laser source, resulting in problems such as pressure broadening, power broadening, and Stark broadening. Many alternative methods of saturation spectroscopy have since been developed, mostly motivated by the quest for higher sensitivity and higher resolution. All are based on the same phenomenon of spectral "hole burning", observed with two counterpropagating beams. They differ, however, in the strategy how the nonlinear interaction of the two light beams is detected.

Several researchers have demonstrated that Doppler-free signals can be obtained by observing light-induced changes in the refractive index or dispersion rather than the absorption, (14,15) although such saturated interference spectroscopy has so far found only limited use.

The well known method of polarization spectroscopy (16) gains a substantial increase in sensitivity by observing the induced dicroism and birefringence of the gas sample rather than changes in absorption.

During the past two years, two alternative methods have been introduced which reach comparable sensitivities via modulation and heterodyne detection techniques. It is interesting to compare the principles of these new techniques with those of the older polarization method.

In one of these modulation methods, the signal is detected via a new phase conjugate wave, generated by degenerate four wave mixing. (17,18) M. Ducloy and collaborators (18) have used a probe beam with a single frequency ω and a collinear counterpropagating saturating beam with two frequency components, separated by some small frequency interval δ, produced with the help of an acoustooptic modulator. Four wave mixing via the third order nonlinear susceptibility of the medium produces two new waves of frequencies $\omega \pm \delta$, travelling collinearly with the probe. A photodiode detects the resulting intensity modulation of the probe. The original probe beam serves thus as the local oscillator for sensitive heterodyne detection.

As first pointed out by C. Bordé, even the old chopper method can be described in similar terms. The chopper produces modulation sidebands for the saturating beam, and the resulting modulation of the probe beam implies that this beam is acquiring new frequency components via four wave mixing. Ducloy and collaborators have demonstrated, however, that a very substantial improvement in signal to noise ratio can be gained by chosing the modulation frequency δ so large that the signal is observed outside the noise spectrum of the fluctuating laser. For example, the amplitude fluctuations of a typical cw dye laser are confined to frequencies below 2 MHz. (19) By observing above this range, the noise amplitude drops by as much as 100 dB, and shot-noise limited detection becomes possible.

The second approach is based on FM sideband spectroscopy, a different modulation method, which is well known in microwave spectroscopy, but whose advantages in the optical region have only recently been demonstrated by J. L. Hall et al. (19), and, independently, by G. C. Bjorklund and collaborators (20,21). In this method, the probe beam is sent through an acoustooptic or electrooptic phase modulator which produces two (or more) FM sidebands of such amplitudes and phases that any constructive or destructive interference effects cancel completely. The intensity of the probe beam, before entering the sample, remains therefore exactly constant. If the sample now changes the amplitude or the phase of any of the sidebands or the phase of the carrier, this delicate balance is perturbed, and the light acquires an amplitude modulation, which can be readily observed with a fast photodiode, followed by rf heterodyne detection.

Spectral hole burning with a counterpropagating saturating beam results again in Doppler-free signals. Dependent on the phase of the detected signal, one can, in this way, observe two absorptive resonances or three dispersive resonances for a given transition. (Fig. 3) Despite this complication in line shape, the method has acquired enthusiastic followers, since the balancing FM sidebands provide a sensitive "null" method. Moreover, the signal can be detected at frequencies far above the noise spectrum of any lasers, so that the sensitivity is easily shot-noise limited.

It is interesting to note that the older polarization spectroscopy (16) owes its high sensitivity to a similar "null" approach: Here, the saturating beam is, for instance, made circularly polarized so that it depletes preferentially atoms with a particular orientation, leaving the remaining ones polarized. The resulting optical anisotropy is detected with a counterpropagating probe beam. This beam passes through a linear polarizer before entering the sample, and afterwards it encounters a second, crossed analyzer which prevents most of the light from reaching the detector. A vertically polarized probe beam can be described as a superposition of a right and a left hand circularly polarized wave, whose horizontal components cancel exactly through destructive interference. Any differential change in amplitude or phase will disturb this delicate balance, and the probe acquires a polarization

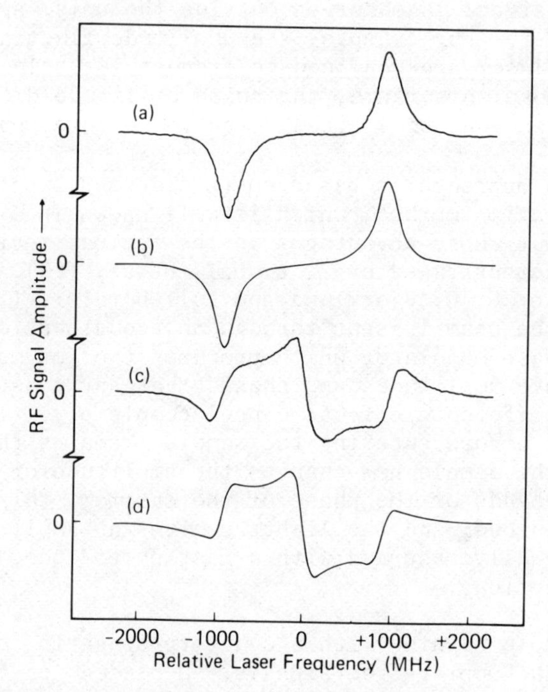

Fig. 3. Absorptive (top) and dispersive (bottom) line profiles, obtained by FM sideband spectroscopy.

component which can pass through the crossed polarizer. Again, this happens only at the line center, where both beams are interacting with the same atoms. The highest sensitivity is actually obtained with the analyzing polarizer slightly uncrossed, so that some probe light passes always into the detector. The new polarization components can then be detected through their interference with this "carrier," i.e. by degenerate optical heterodyne detection.

This polarization method was, for instance, used by J. E. M. Goldsmith et al. (22) for an improved wavelength measurement of the hydrogen Balmer-α line. A more than hundredfold increase in sensitivity, compared to the older chopper method, permitted measurements in a mild helium-hydrogen discharge with a low power cw dye laser, so that pressure broadening, power broadening and related problems were much reduced. These measurements lead to another threefold improvement in the accuracy of the Rydberg constant. (The precision of this measurement has only recently been surpassed by S. R. Amin et al. (23) observing laser-quenching of a well collimated beam of metastable hydrogen atoms. This approach is technically more complex but requires fewer systematic corrections than nonlinear saturation spectroscopy.)

Compared to the newer modulation techniques, polarization spectroscopy has the advantage of greater simplicity. And its reliance on atomic orientation or alignement can provide valuable information on angular momenta and disorienting collisions. However, the technique is limited to spectral regions where good polarizers are available, and considerable care is necessary to reach shot-noise limited sensitivity.

All techniques of saturation spectroscopy discussed so far monitor directly some change in the probe beam. For this reason, they work best with samples of non-negligible absorption. For very weakly absorbing samples, it is often advantageous to detect the absorption of light in the sample indirectly, for instance by observing the laser-induced fluorescence.

To selectively record Doppler-free signals in this way, it has become general practice to use an intermodulation method first introduced by M. S. Sorem and A. L. Schawlow.(24) In this method, the sample is irradiated with two counterpropagating laser beams which are chopped at two different audio frequencies f_1 and f_2. The signal is observed as a modulation of the total excitation rate at the sum or difference frequency $f_1 + f_2$. Such an intermodulation occurs at the center of a Doppler-broadened lines, where both beams are interacting with the same atoms so that they can saturate each other's absorption. Care is required to avoid spurious signals caused by nonlinear mixing in the detector or amplifiers.

Indirect detection can be convenient even if high sensitivity is not the primary issue. A particularly useful and simple detection method is provided by the optogalvanic effect: (25) the absorption of resonant laser light in a gas discharge can alter the discharge impedance, and hence the current or voltage. J. Lawler et al. (26) have been the first to record Doppler-free spectra by intermodulated optogalvanic spectroscopy. Optogalvanic detection has attracted much recent interest, because many excited species can be readily observed in gas discharges. Sputtering in hollow cathode discharges,(27,28) in particular, gives access to samples which would be difficult to vaporize by other means. And optogalvanic spectroscopy in electrodeless radiofrequency discharges (29) holds promise for studies of rare or corrosive species. On the other hand, such discharges with their unavoidable collision processes are clearly not the most ideal environment for very high resolution spectroscopy.

Recently, our group at Stanford has demonstrated the alternative intermodulation method of POLINEX (polarization intermodulated excitation) (30,31) which holds some interesting advantages, in particular for Doppler-free optogalvanic spectroscopy. Only seemingly minor changes are required to convert from the older approach to this new method: the chopper is simply replaced by two polarization modulators which modulate the polarizations of the two beams at two different frequencies, producing, for instance, alternating left and right hand circularly polarized light, while leaving the intensities unchanged. We still obtain an intermodulated signal at line center, because the combined absorption of the two beams depends on their relative polarization. If both fields have the same polarization, the two beams are predominantly interacting with atoms of the same orientation. If the polarizations are different, they tend to interact with different groups of atoms, and their mutual or cross saturation is reduced.

Like in polarization spectroscopy, the signal is associated with atomic orientation or alignment. However, the POLINEX signal due only to the dicroism, not the birefringence of the sample, so that the line asymmetries of the older polarization spectroscopy (16) are readily avoided.

We should also point out an important difference compared to the older amplitude intermodulation technique: neither beam alone is capable of modulating the total rate of excitation in an isotropic medium. Consequently, the signal does not have to be detected on a strongly modulated background, and high selectivity for the Doppler-free signals is maintained even if one of the polarization modulators is removed, i.e. if $f_2 = 0$.

Fig. 4 shows, as an example, a POLINEX spectrum of the yellow Cu laser line at 578.2 nm, recorded in a hollow cathode discharge. (31) Shown above, for comparison, is a spectrum of the same line recorded by the older intermodulated fluorescence method. The Doppler-free hyperfine components of the two stable isotopes in this spectrum appear on large Doppler-broadened pedestals. Similar pedestals have been observed for absorption lines starting from metastable levels in Mo and Ne.(28) They are ascribed to velocity changing elastic collisions which redistribute the atoms over the Maxwellian velocity distribution. (8) POLINEX spectra are free from these pedestals if there is a high probability that the atomic orientation is destroyed in any elastic collision. The method should provide, in fact, an interesting tool to study correlations between velocity changing and disorienting collisions in gases.

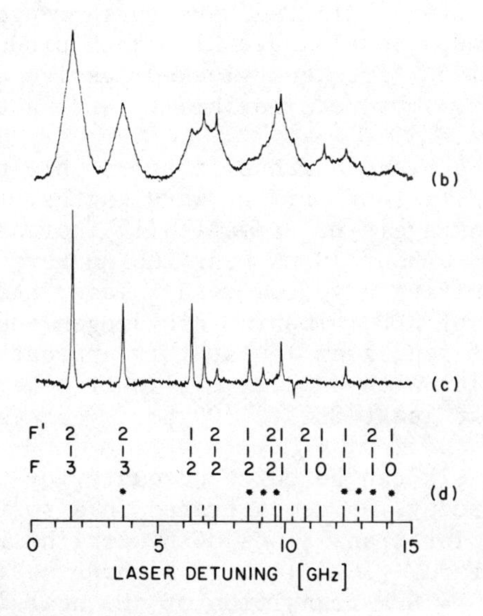

Fig. 4. Hyperfine spectrum of the 578.2 nm transition of Cu, observed in a hollow cathode discharge. (a) Intermodulated fluorescence spectrum. (b) POLINEX spectrum.

2. CONTINUOUS WAVE SATURATION SPECTROSCOPY IN THE ULTRAVIOLET

Today we have at our disposal a wide arsenal of different methods for Doppler-free saturation spectroscopy. It appears likely that future progress will come not so much from the development of still further techniques, but rather from an extension of the wavelength range of highly monochromatic tunable continuous wave laser sources, in particular towards the ultraviolet, where many interesting transitions remain unexplored.

Dye lasers and color center lasers provide very good sources at visible and near infrared wavelengths. But tunable radiation in the ultraviolet has until recently only been available from pulsed lasers whose linewidth is not easily reduced below typical atomic Doppler widths. During the past year, however, cw dye lasers with intracavity frequency doubler crystals have become commercially available which can produce about 30 mW of ultraviolet second harmonic radiation over a limited wavelength range from about 290 to 300 nm. Similar power levels have also been produced with a doubler crystal placed inside a frequency-locked passive enhancement cavity, (32,33) which offers improved flexibility and ease of adjustment.

A much wider wavelength range is accessible by sum frequency generation of two different primary wavelengths. (34) The efficiency for cw operation can again be dramatically increased with a passive enhancement cavity. About 10 mW near 265 nm have in this way been generated (33) by mixing a yellow cw dye laser and a 488 nm argon ion laser in a crystal of ADP (ammonium dihydrogen phosphate.) With a choice of different ion laser lines, this approach promises cw ultraviolet radiation with a multi-milliwatt power level over the entire range from at least 260 to 300 nm.

Such power levels can be quite adequate for Doppler-free saturation spectroscopy, as demonstrated in a series of recent studies of ultraviolet transitions of neutral helium in our laboratory at Stanford. (35-37) Fig. 5 shows as an example a spectrum of the 2^3S - 5^3P transition of ^3He near 294.5 nm, recorded by intermodulated fluorescence spectroscopy. The ultraviolet radiation was generated by a yellow cw ring cavity dye laser with cavity-enhanced external ADA (ammonium dihydrogen arsenate) frequency doubler. The absorbing metastable He atoms were produced by electron impact excitation of He gas at about 0.04 torr. The spectrum shows a cluster of resolved line components which could be assigned after the fine and hyperfine Hamiltonian had been diagonalized in an uncoupled representation. We were surprised to learn that the hyperfine structure of the 5^3P state of this simple 3-body system had been neither measured nor calculated before.

Although difficult in special cases, such measurements of fine level splittings could, in principle, be accomplished by alternative methods such as level crossing or optical-radiofrequency double resonance spectroscopy. Precise measurements of isotope shifts, on the other hand, would be difficult to accomplish by other means. H. Gerhardt and L. Bloomfield at Stanford (38) have recently determined the isotope shift of the 2^3S - 5^3P transition of ^3He and ^4He to within about ±5 MHz or 1 part in 10^4, about 20 times more accurate than the best available model calculations. Moreover, by using subsequent excitation with a cw color center laser operating near 2.5 um, they have also measured the isotopic shift of the infrared 5^3P - 13^3D transition. Since the upper Rydberg state has virtually no correlation-produced specific isotope shift, these measurements are yielding a first accurate experimental value for the specific isotope shift of the 5^3P level. Such measurements should provide an interesting testing ground for future refined model calculations.

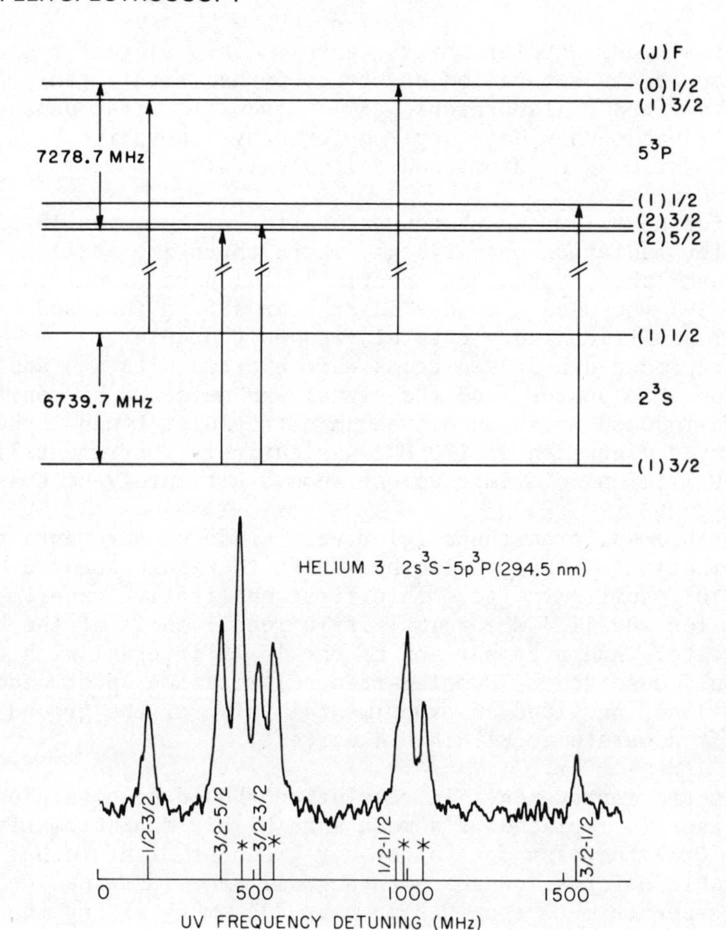

Fig. 5. Doppler-free intermodulated fluorescence spectrum of the 2^3S - 5^3P transition of ^3He near 296 nm.

3. DOPPLER-FREE TWO-PHOTON SPECTROSCOPY OF ATOMIC HYDROGEN 1S-2S

An extension of the wavelength range of tunable cw sources towards even shorter ultraviolet wavelengths promises particularly interesting opportunities for precision two-photon spectroscopy of atomic hydrogen. (39) The transition from the 1S ground state to the metastable 2S state is perhaps the most intriguing transition in this simplest of the stable atoms, because the 1/7 sec lifetime of the upper state implies an ultimate natural line width as narrow as 1 Hz, or an resolution better than 1 part in 10^{15}.

Doppler-broadening in two-photon spectroscopy can be conveniently eliminated without any need for velocity selection by excitation with two counterpropagating laser beams of equal frequency whose first order Doppler shifts cancel. This elegant technique which complements saturation spectroscopy had first been proposed by V. Chebotaev and collaborators. (40) Since its first demonstration in 1974 it has been widely applied for many innovative high resolution studies of atoms and molecules.(41)

Unfortunately, two-photon excitation of hydrogen 1S-2S requires ultraviolet radiation near 243 nm, where there are still no good cw sources available. The best spectra so far have been recorded by C. Wieman, (39) who used a cw dye laser near 486 nm followed by a pulsed dye laser amplifier and a crystal frequency doubler. Discharge-produced hydrogen atoms were excited with a standing wave field from this source, and the signal was detected by monitoring the collision-induced emission of vacuum ultraviolet Lyman-α photons. The observed linewidth of 120 MHz was entirely instrument limited, and a 100 million fold improvement should ultimately be possible.

Nonetheless, even these relatively crude measurements permitted a determination of the H-D isotope shift to within about 6 MHz or 1 part in 10^5, sufficient to obtain first qualitative experimental evidence for the 11.4 MHz relativistic recoil shift of the hydrogen ground state. And a comparison of the 1S-2S interval with the n=2 to 4 interval, observed by Doppler-free polarization spectroscopy of the Balmer-β line, provided an experimental value of the ground state Lamb shift accurate to within 0.4 percent.

Dramatic improvements in resolution should be possible if the pulsed laser is replaced by a much more highly monochromatic cw source. Over the past few years, our group at Stanford has devoted considerable efforts towards such a goal. For instance, we have been able to generate more than 0.7 mW near 243 nm by mixing the output of a yellow cw dye laser and a deep blue krypton ion laser in a crystal of ADP, cooled close to its Curie point near liquid nitrogen temperature. (33) However, the mixing crystal is quickly damaged under these conditions. Very recently, B. Couillaud and L. Bloomfield in our laboratory have obtained more encouraging results by mixing radiation from a deep red dye laser and an ultraviolet argon ion laser in an ADP crystal near room temperature, and we hope to generate up to several milliwatts with the help of an enhancement cavity.

Even if such high power levels cannot be sustained, it should still be possible to observe two-photon excitation with the help of more sensitive detection methods. Towards this end, U. Boesl and E. Hildum in our laboratory have recently completed construction of a hydrogen atomic beam apparatus, which permits the detection of 2S atoms via photoionization. The resulting charged particles are observed with a time-of flight mass spectrometer. Despite transit time broadening and uncompensated relativistic second order Doppler shifts, we hope to achieve line widths on the order of 1 Mhz in this way. Cooling of the hydrogen beam close to liquid helium temperature (42) could reduce these widths to a few kHz.

With the envisioned higher resolution, it should be possible to determine a better value of the electron/proton mass ratio from a precise measurement of the isotope shift. And a measurement of the absolute frequency or wavelength should provide a new value of the Rydberg constant with an accuracy up to 1 part in 10^{11}, as limited by uncertainties in the fine structure constant and the mean square radius of the proton charge distribution. A comparison with one of the Balmer transitions, or with a transition to or between Rydberg states could provide a value for the 1S Lamb shift that exceeds the accuracy of the best radiofrequency measurements of the n=2 Lamb shift. Such experiments can clearly provide very stringent tests of quantum electrodynamic calculations, and when pushed to their limits, they may well lead to some surprising fundamental discovery.

Ever since the 1S-2S two-photon excitation has been observed in hydrogen, it was obvious to us and many others that a similar experiment in positronium would be particularly interesting. This purely leptonic atom should provide an even cleaner testing ground for quantum electrodynamics, since it does not require large nuclear structure corrections. The energy levels of positronium are spaced roughly only half as far as those of hydrogen, and two-photon excitation of 1S-2S can be accomplished with visible laser light near 486 nm. However, even in the long living triplet 1S state, positronium annihilates within about 140 nsec, and the difficulties of producing sufficient numbers of such atoms for laser spectroscopy have long appeared quite formidable.

Using an ingenious trap to accumulate positrons from a radioactive source, S. Chu and A. Mills have finally succeeded in observing two-photon excitation of positronium. (43) The details of this pioneering experiment are presented in a subsequent paper.

Let me only mention that very recently a collaboration between Lawrence Livermore Laboratory, Stanford and Yale (44) has also begun to study two photon excitation of positronium 1S-2S. A scheme of the apparatus is shown in Fig. 6. Like in the experiment of Chu and Mills, the light is produced by a pulsed dye laser. The positronium atoms, however, are generated with the help of a pulsed electron accelerator: Pulses of several Amperere from the 100 MeV Electron Linear Accelerator at Livermore are sent into a solid tantalum target, where they produce intense bursts of positrons through repeated cycles of bremsstrahlung and pair production. A stack of tungsten foils serves as a moderator. The resulting slow positrons are sent through a tube with guiding magnetic solenoid onto a heated tungsten converter, which can produce up to 10^5 slow positronium atoms per pulse. Such a large number should make it readily possible to observe transitions to higher excited states, including Rydberg

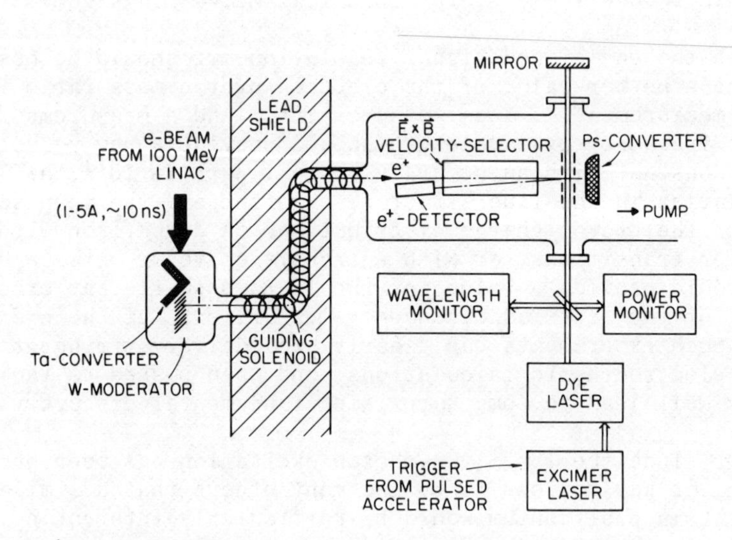

Fig. 6. Apparatus for Doppler-free two-photon spectroscopy of positronium, produced with a pulsed electron linear accelerator.

states of positronium, or to study rf transitions or level crossings between closely spaced excited levels of this interesting relativistic two-body system.

4. CONCLUSIONS

We have limited our discussions to Doppler-free laser spectroscopy of gaseous sample, and even here we had to ignore many interesting topics.

However, even our limited excursion may have shown that Doppler-free laser spectroscopy is still an exciting and lively field which continues to hold many interesting challenges and opportunities.

*Supported by the National Science Foundation under Grant NSF PHY-80-10689 and by the U.S. Office of Naval Research under Contract N00014-C-78-0403.

REFERENCES

1. W. Demtröder, "Laser Spectroscopy," Springer Series in Chemical Physics, Vol. 5, Springer Verlag, Berlin, Heidelberg, New York 1981.
2. M. D. Levenson, "Introduction to Nonlinear Laser Spectroscopy," Academic Press, New York 1982.
3. R. A. Macfarlane, W. R. Bennett, and W. E. Lamb, Jr., Appl. Phys. Letters 2, 189 (1963).
4. A. Szoke and A. Javan, Phys. Rev. Letters 10, 521 (1963).
5. P. H. Lee and M. L. Skolnick, Appl. Phys. Letters 10, 303, (1967).
6. T. W. Hänsch, Appl. Optics 11, 895 (1972).
7. C. Borde, Compt. Rend. 271, 371 (1970).
8. P. W.Smith and T. W. Hänsch, Phys. Rev. Letters 26, 740 (1971).
9. T. W. Hänsch, M. D. Levenson, and A. L. Schawlow, Phys. Rev. Letters 26, 946 (1971).
10. T. W. Hänsch, I. S. Shahin, and A. L. Schawlow, Phys. Rev. Letters 27, 707 (1971).
11. T. W. Hänsch, I. S. Shahin, and A. L. Schawlow, Nature 235, 63 (1972).
12. T. W. Hänsch, G. W. Series, and A. L. Schawlow, Sci. Am. 240, 94 (1979).
13. T. W. Hänsch, M. H. Nayfeh, S. A. Lee, S. M. Curry, and I. S. Shahin, Phys. Rev. Letters 32, 1336 (1974).
14. C. Borde, G. Camy, B. Decomps, and L. Pottier, Compt. Rend. 277, 381 (1973).
15. F. V. Kowalski, W. T. Hill, and A. L. Schawlow, Opt. Letters 2, 112 (1978).
16. C. Wieman and T. W. Hänsch, Phys. Rev. Letters 36, 1170 (1976).
17. P. F. Liao D. M. Bloom, and N. P. Economu, Appl. Phys. Letters 32, 813 (1978).
18. R. K. Raj, D. Bloch, J. J. Snyder, G. Camy, and M. Ducloy, Phys. Rev. Letters 19, 1251 (1980).
19. J. L. Hall, T. Baer, L. Hollberg, H. G. Robinson, in "Laser Spectroscopy V," A. R. W.. McKellar, T. Oka, B. P. Stoicheff, Eds., Springer Series In Optical Sciences, Vol. 30, Springer Verlag, Berlin, Heidelberg, New York (1981), p. 15.
20. G. C. Bjorklund, Opt. Letters 5, 15 (1980).
21. G. C. Bjorklund and M. D. Levenson, Phys. Rev. A24, 166 (1981).
22. J. E. M. Goldsmith, E. W. Weber, and T. W. Hänsch, Phys. Rev. Letters 41, 940 (1978).
23. S. R. Amin, C. D. Caldwell and W. Lichten, Phys. Rev. Letters 47, 1234 (1981).
24. M. S. Sorem and A. L. Schawlow, Opt. Commun. 5, 148 (1972).
25. J. E. M. Goldsmith and J. E. Lawler, Contemp. Phys. 22, 235 (1981).
26. J. E. Lawler, A. I. Ferguson, J. E. M. Goldsmith, D. J. Jackson, and A. L. Schawlow, Phys. Rev.

Letters 42, 1046 (1979).
27. J. E. Lawler, A. Siegel, B. Couillaud, and T. W. Hänsch, J. Appl. Phys. 52, 4375 (1981).
28. A. Siegel, J. E. Lawler, B. Couillaud, and T. W. Hänsch, Phys. Rev. A23, 2457 (1981).
29. D. R. Lyons, A. L Schawlow, and G.-Y. Yan, Opt. Commun. 38, 35 (1981).
30. T. W. Hänsch, D. R. Lyons, A. L. Schawlow, A. Siegel, Z-Y. Wang, and G.-Y. Yan, Opt. Commun. 37, 87 (1981).
31. Ph. Dabkiewicz and T. W. Hänsch, Opt. Commun. 38, 351 (1981).
32. M. Brieger, H. Busener, A. Hese, F. Moers, and A. Renn, Opt. Commun. 38, 423 (1981).
33. B. Couillaud, Ph. Dabkiewicz, L. A. Bloomfield, and T. W. Hänsch, Opt. Letters 7, 265 (1982).
34. S. Blit, G. Weaver, F. B. Dunning, and F. K. Tittel, Opt. Letters 1, 58 (1977).
35. L. A. Bloomfield, B. Couillaud, Ph. Dabkiewicz, H. Gerhardt, and T. W. Hänsch, Phys. Rev. A26, 713 (1982).
36. L. A. Bloomfield, H. Gerhardt, T. W. Hänsch, and S. C. Rand, Opt. Commun. accepted for publication (1982).
37. L. A. Bloomfield, H. Gerhardt, and T. W. Hänsch, Phys. Rev. A, accepted for publication (1982).
38. H. Gerhardt, L. A. Bloomfield, and T. W. Hänsch, to be published.
39. C. Wieman and T. W. Hänsch, Phys. Rev. A22, 192 (1980).
40. L. S. Vasilenko, V. P. Chebotaev, and A. V. Shishaev, JETP Letters 12, 113 (1970).
41. N. Bloembergen and M. D. Levenson, in "High Resolution Laser Spectroscopy," K. Shimida, ed., (Topics in Applied Physics, Vol. 13,) Springer Verlag, Berlin, Heidelberg, New York (1976), pp. 315.
42. S. B. Crampton, T. J. Greytag, D. Kleppner, W. D. Philips, D. A. Smith, and A. Weinrib, Phys. Rev. Letters 42, (1979).
43. S. Chu and A. P. Mills, Phys. Rev. Letters 48, 1333 (1982).
44. R. Alvarez, K. Danzmann, S. Dhawan, P. O. Egan, T. W. Hänsch, R. Howell, V. Hughes, M. Ritter, and K. Woodle, to be published.

DOPPLER NARROWING AND COLLISION-INDUCED ZEEMAN COHERENCE IN FOUR-WAVE LIGHT MIXING

N. Bloembergen, M. C. Downer and L. J. Rothberg

Division of Applied Sciences
Harvard University
Cambridge, MA 02138

1. REVIEW OF COLLISION-INDUCED COHERENCE

Consider a three-dimensional four-wave light mixing geometry as schematically shown in Fig. 1. The four beams all travel in the near-forward direction. Two beams have a frequency ω_1, which is offset by a fixed amount $\Delta = \omega_1 - \omega_{3S,3P}$ from a resonant line of the Na atom. Their wave vectors, $\underset{\sim}{k}_1$ and $\underset{\sim}{k}_1'$, respectively, lie in the vertical plane. A third beam, at frequency ω_2, has a wave vector $\underset{\sim}{k}_2$ in the horizontal plane. In four-wave mixing, the generation of a new beam in the horizontal plane is observed with wave vector $\underset{\sim}{k}_1 + \underset{\sim}{k}_1' - \underset{\sim}{k}_2$ and frequency $2\omega_1 - \omega_2$. As the frequency ω_2 is varied over an interval of about 40 cm^{-1} around the value of ω_1, seven resonances in the intensity of the new beam, $\mathscr{I}(2\omega_1 - \omega_2)$, have been observed by Prior et al.[1] in mixtures of Na vapor and helium. These are schematically indicated in Fig. 2. Four of these are familiar one-photon resonant enhanced four-wave mixing processes. The remaining three resonances, at $\omega_2 = \omega_1$, and at $\omega_2 = \omega_1 \pm 17$ cm^{-1}, are only observable in the presence of collisions, which can be controlled by the pressure of helium buffer gas.

The resonances at $\omega_2 = \omega_1 \pm 17$ cm^{-1} are due to a collision-induced coherence, $\rho_{bc}(\omega_1 - \omega_2)$ where $|b\rangle$ and $|c\rangle$ correspond to the $3^2P_{1/2}$ and $3^2P_{3/2}$ states of Na, respectively. The characteristics of these resonances have been reviewed by us at the 1981 Jasper conference on Laser Spectroscopy.[2] The width of this resonance varies linearly with helium pressure p_{He}; whereas its peak intensity approaches a constant value. The width is due to fine-structure changing collisions. The agreement between experiment[4] and theory,[2-4] which is well understood, is satisfactory.

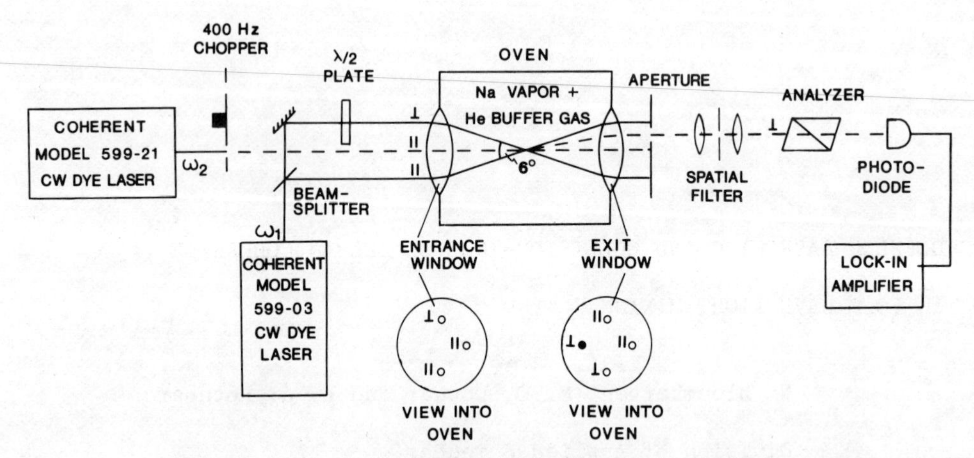

Fig. 1. Schematic diagram of experimental arrangement to detect collisionally induced resonances. The symbols "⊥" and "||" denote polarization directions of the electric field amplitudes with respect to an externally applied transverse magnetic field.

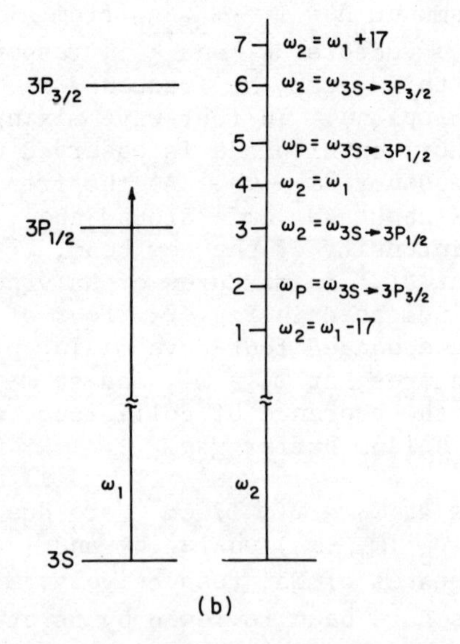

Fig. 2. Seven resonances in the four-wave mixing signal are observed as the frequency ω_2 is varied, while the frequency ω_1 is kept fixed at a detuning Δ from a Na resonance line. The resonances 1, 4 and 7 are collision induced.

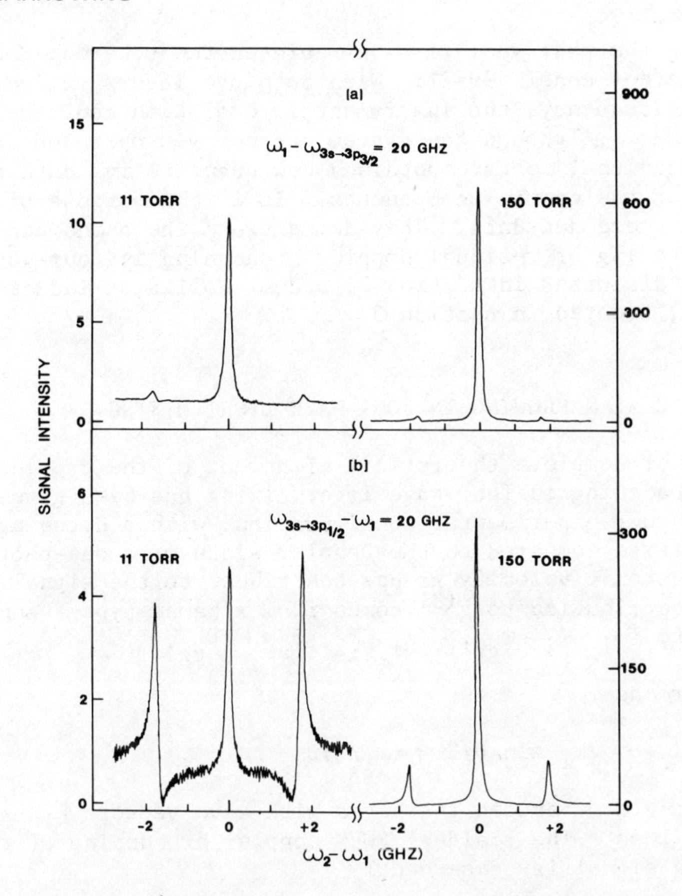

Fig. 3. The central resonance of Fig. 2 consists of three components in zero magnetic field, caused by the hyperfine splitting of $3^2S_{1/2}$ ground state of Na. The resonances are shown for two partial helium pressures and detunings near the $3^2P_{3/2}$ and $3^2P_{1/2}$ states, respectively. Note the change in vertical scale at the higher pressure. The widths are instrumentally limited.

The near-degenerate frequency resonance consists of several components, $\omega_2 = \omega_1 \pm 1.8$ GHz and $\omega_2 = \omega_1$. The data shown in Fig. 3 display a width of about 120 MHz independent of p_{He}, while the peak intensity is proportional to p_{He}^2. The width was instrumentation-limited and determined mostly by the frequency variations in the Coherent Model 599-03 dye laser. The data of Fig. 3 have also been discussed in our previous review.[2] The satellites are due to a collision-induced Raman coherence $\rho_{gg'}^{(2)}(\omega_1 - \omega_2)$, where $|g\rangle$ and $|g'\rangle$ represent the $3^2S_{1/2}$ ground state hyperfine components with F = 1 and 2, respectively.

During the past year we have replaced the Coherent Model 599-03 by the Coherent Model 599-21. With both dye lasers actively stabilized in frequency, the instrumental resolution could be increased to 1 or 2 MHz. Although the system has not yet operated according to specifications, we have obtained new quantitative data on the linewidth of the central components. It is the purpose of this paper to present these new data. They demonstrate the phenomena of collisional narrowing of residual Doppler broadening in four-wave light mixing, as discussed in section 2, and of collision-induced Zeeman coherences, treated in section 3.

2. COLLISIONAL NARROWING IN FOUR-WAVE LIGHT MIXING

A rather complete theoretical treatment of the problem of Doppler broadening in four-wave light mixing has been given by Druet et al.[5] In our experimental configuration, with a detuning $\Delta >> \omega_1(v_{at}/c)$ large compared to the Doppler width of a one-photon resonance, all atomic velocity groups contribute to the signal. The dominant contribution to $\chi^{(3)}$ comes from a Raman-type resonance proportional to $\{\omega_1 - \omega_2 - \omega_{gg'} - (\underset{\sim}{k}_1 - \underset{\sim}{k}_2)\cdot\underset{\sim}{v}_{at} + i\Gamma_{gg'}\}^{-1}$.

In our case,

$$|\underset{\sim}{k}_1 - \underset{\sim}{k}_2| \approx 2k_1 \sin \theta/2 \approx (\omega_1/c)\theta,$$

since the angle θ between the beams with wave vectors $\underset{\sim}{k}_1$ and $\underset{\sim}{k}_2$ is 6 degrees or less. The residual FWHM Doppler broadening of the four-wave mixing signal is consequently

$$\Delta\omega_D = 2(\ell n\ 2)^{1/2}\omega_1(v_{at}/c)\theta. \tag{1a}$$

If the Raman resonance is induced by the beam with $\underset{\sim}{k}_1'$ and $\underset{\sim}{k}_2$, the Doppler width would have a corresponding contribution with width

$$\Delta\omega_D' = 2(\ell n\ 2)^{1/2}\omega_1(v_{at}/c)\theta', \tag{1b}$$

where θ' is the angle between the waves with $\underset{\sim}{k}_1'$ and $\underset{\sim}{k}_2$. The data in Fig. 4 show that the linewidth indeed becomes narrower as the angles θ and θ' are decreased. Some data have been taken with angles as small as 1.5 degrees, but the separation of the new fourth beam, which is based principally on angular resolution, becomes experimentally more difficult. The linewidth is indeed proportional to θ at this relatively low value of p_{He}. For still lower values of p_{He}, the intensity of signals decreases, resulting in poor signal to noise. It should be kept in mind that the signals are induced by Na-He collisions.

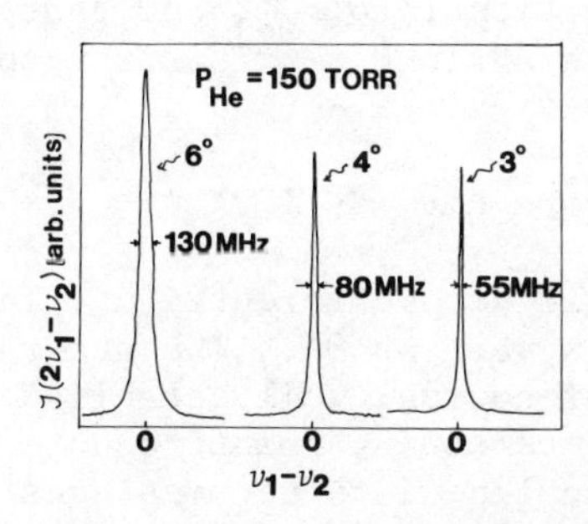

Fig. 4. The central component of the resonance in Fig. 3 is recorded with improved instrumental resolution. The width $\Delta\nu = \Delta\omega/2\pi$ is due to residual Doppler broadening and changes with the angle between the light waves at ω_1 and ω_2, respectively.

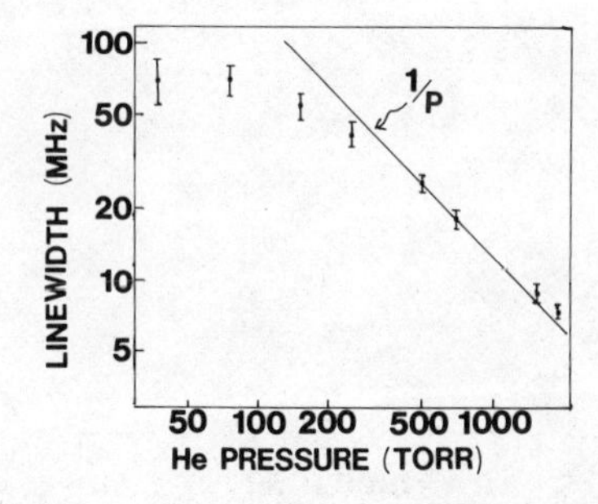

Fig. 5. Pressure narrowing of the residual Doppler broadening at a fixed angle $\theta = 3°$.

Fortunately, the residual Doppler broadening may be eliminated by collisions. The data in Fig. 5 show the variation in the width of the central components as a function of p_{He}. In contrast to the fine structure resonances at $\omega_2 = \omega_1 \pm 17\ \text{cm}^{-1}$, these linewidths tend to vary inversely proportional to p_{He} at sufficiently high values of p_{He}. This is a unique demonstration of motional narrowing in an atomic system at optical frequencies. When the time between collisions $\tau = \ell/v_{at}$, where ℓ is the mean free path, is smaller than the inverse inhomogeneous broadening, the latter is averaged out. The condition $\tau\Delta\omega_D \ll 1$ is equivalent to the condition that the mean free path is shorter than the grating constant, $|\underset{\sim}{k}_1 - \underset{\sim}{k}_2|\ell \ll 1$. Note that τ and ℓ are proportional to p_{He}^{-1}, and that the narrowing sets in sooner, the smaller the angle θ. Thus the observed width should become

$$\Delta\omega \approx (\Delta\omega_D)^2\tau \quad \text{for} \quad \tau\Delta\omega_D \ll 1 \tag{2}$$

and the lineshape should become Lorentzian. The linewidth at a fixed value of p_{He} should now vary as θ^2. The general phenomenon of motional narrowing is, of course, well established,[6,7] and excellent reviews exist.[8] The interesting point to note in this particular application is that the linewidth becomes considerably narrower than the natural linewidth of 20 MHz (FWHM) of the 3S-3P resonance, as shown by the recordings in Fig. 6. This necessitates a reinterpretation of the central resonance. Previously it was argued, on the basis of a two-level model,[2,9] that this component is due to the modulation at $\omega_1 - \omega_2$ of a population grating. The key terms in the

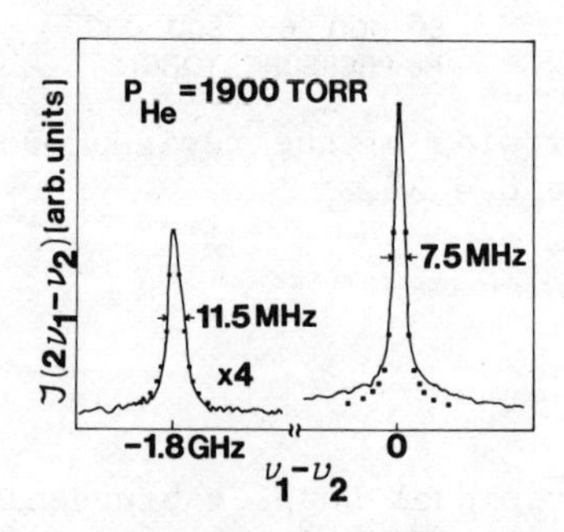

Fig. 6. Narrow resonances of collision-induced Zeeman and hyperfine structure coherences (FWHM is less than $2T_1^{-1}$). Power broadening was shown to be small at the beam intensities used in these recordings, taken with p_{Na} = 35 mtorr, p_{He} = 1900 torr, Δ = 30 GHz below $^2P_{1/2}$. The dots represent single Lorentzian fits to the observed lineshapes.

perturbation calculation were $\rho_{nn}^{(2)}(\omega_1 - \omega_2)$ and $\rho_{gg}^{(2)}(\omega_1 - \omega_2)$. The contribution from these terms has the natural width $\Gamma_{gg} \sim T_1^{-1}$. In the next section we shall discuss another contribution based on Zeeman coherences $\rho_{mm'}^{(2)}$ of the degenerate sublevels of the ground state $|g>$.

3. COLLISION-INDUCED ZEEMAN COHERENCE

It was realized previously[9] that the width of the satellite resonances at $\omega_2 = \omega_1 \pm 1.8$ GHz could be considerably narrower than T_1^{-1}. The signals correspond to Raman-type resonances between the two hyperfine levels of the ground state configuration. Although they have equal population, the collision-induced effects lead to a resonance with a width $\Gamma_{gg'}$, determined by hyperfine structure changing collisions. Presumably spin-exchange collisions between two Na atoms provide the dominant contribution to the homogeneous width $\Gamma_{gg'}$.

Since the central resonance $\omega_2 = \omega_1$ also displays a narrow component, it is natural to attribute this to a Raman-type resonance between two Zeeman sublevels $|m>$ and $|m'>$ of the degenerate ground state $|g>$. The density matrix element $\rho_{mm'}^{(2)}$, rather than the diagonal element $\rho_{gg}^{(2)}$, would be responsible for the narrow resonance. Its homogeneous width $\Gamma_{mm'}$ would be determined by collisions which change the orientation of the electron spin in each individual atom. Spin-exchange collisions are usually the dominant mechanism[10] establishing a Boltzmann distribution over the Zeeman sublevels in alkali vapor buffer gas cells. We would thus expect $\Gamma_{mm'} \approx \Gamma_{gg'}$, and the homogeneous widths should be between 1 and 3 MHz at the partial Na pressure (35 millitorr) used in the experiment.

The experimental widths in Fig. 6 are clearly still due to incomplete averaging of the Doppler broadening. The degenerate component can be fit to a superposition of a narrow Lorentzian and a wider one. The former is ascribed to a Raman-type resonance between two Zeeman sublevels with $m' = m \pm 1$. Such transitions can be induced if the linear polarization of the beam at ω_1 is orthogonal to that of the beam at ω_2. Thus the narrow component is ascribed a width $(\Delta\omega_D')^2\tau$, proportional to θ'^2. The Lorentzian fit to this component is indicated by the dashed line in Fig. 6. It is tempting to attribute the broader component to a process involving the diagonal density matrix element $\rho_{mm}^{(2)}$ with a width $\Gamma_{mm} \sim T_1^{-1}$. Such processes can be induced when the beams at ω_1 and ω_2 have parallel polarizations, as occurs in the experimental configuration of Fig. 1. The observed width of the broad component is, however, much larger than the natural linewidth, with a FWHM value of $2T_1^{-1} \approx 20$ MHz. At this time we have no explanation for this discrepancy.

The satellite line in Fig. 6 for $\omega_2 = \omega_1 - 1.8$ GHz has contributions to the hyperfine changing Raman transition for both parallel and orthogonal polarizations of the beams at ω_1 and ω_2, respectively. Thus its width has a contribution proportional to θ^2 as well as to θ'^2. Since the data in Fig. 6 were taken with $\theta > \theta'$, it may be understood why this component is somewhat wider than the narrow component for $\omega_1 = \omega_2$. It is also reassuring that there is no indication of a broad component in the hyperfine satellite line.

The application of an external magnetic field indeed causes a splitting of the narrow central component, confirming its nature as a collision-induced Zeeman coherence. Fig. 7 demonstrates this magnetic splitting of the four-wave mixing signal in a magnetic field B = 10 gauss, parallel to the electric field vectors of the waves with wave vectors $\underset{\sim}{k}_1$ and $\underset{\sim}{k}_2$, but orthogonal to the electric field of the wave with $\underset{\sim}{k}_1'$. Thus a Raman-type resonance between two Zeeman levels of the $3^2S_{1/2}$ ground state of the Na atom is induced with $\Delta m = \pm 1$. For a g-value ~ 2, the Zeeman splitting for a $\Delta m = 1$ transition should be 2.78 MHz/gauss. The observed separation of the $\Delta m = \pm 1$ components of about 60 MHz for a field of 10 gauss is consistent with this interpretation.

Zeeman coherences in degenerate four-wave light mixing were explicitly considered previously by Steel, Lam and McFarlane.[11] They used the standard phase-conjugate geometry with two pairs of beams going in opposite directions. In this geometry the phase mis-

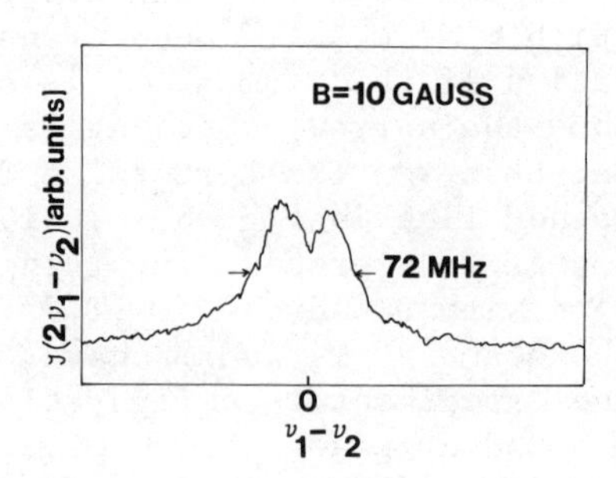

Fig. 7. Zeeman splitting, in an external magnetic field of 10 gauss, of the four-wave mixing resonance in the geometry of Fig. 1, caused by collision-induced Zeeman coherences with $\Delta m = \pm 1$.

matching is severe for detuning ω_2 away from ω_1. Furthermore, the observations are made exactly on resonance. Only one velocity packet with detuning $\Delta = 0$ contributes to the signal. Their resonance signal is not collision induced. Clearly the theory in both cases involves the selection of the significant terms in $\chi^{(3)}$, but the details are markedly different.

4. CONCLUDING REMARKS

The collision-induced resonances in four-wave mixing provide striking confirmation of very subtle features of damping in nonlinear situations, where simultaneously two, three, or four optical transitions are at or near resonance. In addition, they may become a versatile spectroscopic tool for gaining new information in situations such as those described below.

First of all, collisional narrowing of the collision-induced resonances with nearly degenerate frequencies can be studied up to very high buffer gas pressures without the inconvenience of simultaneous pressure broadening. Accurate tests in a simple atomic system of theoretical predictions regarding strong, weak, and intermediate velocity-changing collisions and the statistical dependence of velocity- and phase-changing collisions[8] may, therefore, be possible. With further improvements in instrumental resolution, the real homogeneous optical widths of the central Zeeman resonances and hyperfine satellites could be determined. However, information on various damping parameters is already known in many cases, or more readily obtainable, from radiofrequency spectroscopy[12] or from optical echo experiments.[13]

Secondly, the high peak intensity of the near-degenerate four-wave mixing, which scales as p_{He}^3 at high buffer gas pressures, may permit observation of the signal at frequency detunings Δ as large as 100 cm^{-1}. Observations at large detunings may, therefore, provide a new handle in the study of the transition from the impact regime to the quasi-static regime.

Thirdly, the collision-induced Raman resonances can expand the application of Raman spectroscopy to situations where the conventional Raman susceptibility vanishes. Andrews et al.,[14] for example, have demonstrated that sharp vibrational transitions in an initially unpopulated excited electronic state of a molecule can be observed as extra four-wave mixing resonances, even though they cannot be observed in absorption because of rapid dephasing of the electronic transition. In these experiments pentacene molecules were doped in a benzoic acid crystal, and phonon scattering rather than collisions provided the dephasing mechanism. CARS of equally-populated ground state rotational levels in molecules would also become possible by observing collision-induced resonances.

Finally, the degenerate frequency resonance permits straightforward lifetime measurements in the frequency domain of nonfluorescing excited states.[15]

Further refinements of our experiments in sodium vapor are planned, in addition to the above avenues of investigation. Improved discrimination of the signal beam from the three incident beams at smaller intersection angles and larger frequency detunings might be obtained by a double (or triple) modulation technique, in which beams at ω_1 and ω_2 are modulated at different frequencies. The signal is then synchronously detected at a linear combination of these modulation frequencies. Systematic variation of the partial pressures of Na and He, as well as the frequency detuning Δ, is also planned. Measurements can, of course, readily be extended to the investigation of other buffer gases and other alkali atoms.

Other combinations of the polarizations in the three beams would permit the Raman processes with $m' = m \pm 2$, $m' = m \pm 1$, and $m' = m$ to be investigated separately. With suitably applied magnetic fields, the influence of collisions on level crossing and anticrossing situations might be investigated.

This research was supported by the Joint Services Electronics Program of the United States Department of Defense under contract N00014-75-C-0648.

REFERENCES

1. Y. Prior, A. R. Bogdan, M. Dagenais and N. Bloembergen, Phys. Rev. Lett. 46:111 (1981).
2. N. Bloembergen, A. R. Bogdan and M. C. Downer, in: "Laser Spectroscopy V," A. R. W. McKellar, T. Oka and B. P. Stoicheff, eds., Springer, Heidelberg (1981), p. 157.
3. G. Grynberg, loc. cit. p. 174.
4. A. R. Bogdan, M. C. Downer and N. Bloembergen, Phys. Rev. A 24:623 (1981).
5. S. A. J. Druet, J. P. Taran and Ch. J. Bordé, J. de Phys. (Paris) 40:819 (1979).
6. N. Bloembergen, E. M. Purcell and R. V. Pound, Phys. Rev. 73: 679 (1948).
7. R. H. Dicke, Phys. Rev. 89:472 (1953).
8. S. G. Rautian and I. I. Sobel'man, Sov. Phys. Uspekhi 9:701 (1967).
9. A. R. Bogdan, M. C. Downer and N. Bloembergen, Opt. Lett. 6:348 (1981).
10. W. Happer, Rev. Mod. Phys. 44:169 (1972).
11. D. G. Steel, J. F. Lam and R. A. McFarlane, see ref. 2, p. 260.
12. W. Happer and A. C. Tam, Phys. Rev. A 16:1877 (1977).

13. T. W. Mossberg, F. Whittaker, R. Kachru and S. R. Hartmann, Phys. Rev. A 22:1962 (1980).
14. J. R. Andrews, R. M. Hochstrasser, R. M. Trommsdorff, Chem. Phys. 62:87 (1981).
15. T. Yajima, H. Souma and Y. Ishida, Phys. Rev. A 17:324 (1978).

EXCITATION OF THE POSITRONIUM $1^3S_1 - 2^3S_1$ TWO PHOTON TRANSITION

Allen P. Mills, Jr. and Steven Chu

Bell Laboratories
Murray Hill, NJ 07974 USA

INTRODUCTION

We report here the details of the first experiment to excite optically Ps from the n=1 triplet ground state to the 2^3S_1 state[1]. The accuracy of our measurement of the $2^3S_1 - 1^3S_1$ interval is approximately 0.8 ppm. Since the experiment was done with thermal positronium in vacuum, straightforward extension of the now proven technology should allow us to improve the accuracy of this measurement by several orders of magnitude.

Since the first observation of positronium (Ps) by M. Deutsch [2] in 1951, this atom has been recognized as one of the most fundamental bound-state systems available for study[3]. Being a purely leptonic atom consisting of an electron and its antiparticle, it provides a unique opportunity for studying a bound-state two body system and the quantum electrodynamic (QED) corrections to that system. These corrections to the energy levels of Ps are of particular interest because they contain virtual annihilation terms not found in hydrogen or muonium. Furthermore, unlike these atoms or hydrogen-like ions, the Dirac equation is not an adequate starting point for deriving the QED corrections in Ps, and the Bethe-Salpeter formalism[4] must be used. Unfortunately, the fully covariant Bethe-Salpeter equation has no known analytic solution and presents formidable calculational difficulties. Recent advances in treating the two-body problem have been reported by Caswell and Lepage[5,6] and Barbieri and Remiddi[7]. Nevertheless, Ps is one of the few bound-state systems where the exact Hamiltonian is believed to be known to great accuracy, and precision experiments provide tests of our ability to understand the two-body problem.

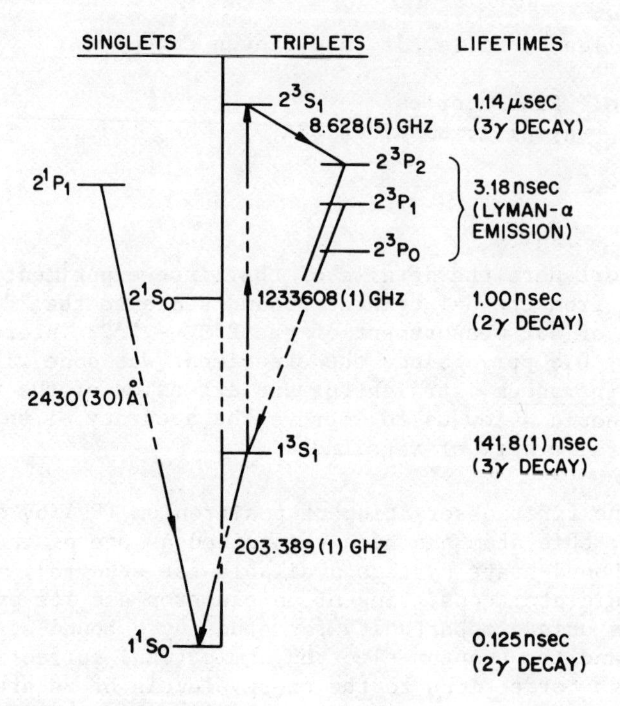

Fig. 1. Energy levels of the n=1 and n=2 states of positronium. The quantities with error bars in parentheses are measured values.

The lowest lying energy levels in Ps are shown in Fig. 1. Studies of the positronium atom have been previously limited to measurements of the ground state decay rates[8], and microwave studies of the n=1 "hyperfine"[9,10] $1^1S_0 - 1^3S_1$ interval, and a "Lamb shift" $2^3S_1 - 2^3P_2$ interval[11]. Nevertheless, the potential of precision optical spectroscopy on Ps has long been recognized, [12] especially since the pioneering work on the 1S → 2S transition in hydrogen done by Hänsch and his collaborators[13,14]. Since the natural linewidth of the $1^3S_1 - 2^3S_1$ interval in Ps is only 1 MHz, this energy splitting can in principle be measured to even greater precision than the 50 MHz wide $2^3S_1 - 2^3P_{0,1,2}$ microwave intervals. Our experiment is based on the slow positron beam technique[15] which resulted in the first observation of Ps in vacuum[16], the formation of n=2 Ps[17] and the measurement of the $2^3S_1 - 2^3P_2$ interval[11]. The present experiment was made possible by a million-fold increase in the instantaneous density of usable Ps since these earlier experiments. This advance was the result of the development of high intensity, slow positron (e^+) sources in vacuum[18,19], e^+ storage and bunching techniques[20] used to make a pulsed Ps source that matches the pulsed laser source, and the discovery of Ps emission from surfaces[16] that produce thermal Ps in vacuum[21,22]. Also crucial to our experiment were recent advances in the techniques of optical spectroscopy[13]. We combine the pulsed Ps source with a high power narrowband dye laser[23] capable of partially saturating the highly forbidden two photon 1S-2S transition over a sizable volume of space, the use of two-photon techniques[24] that allow us to avoid the considerable first order Doppler width and to excite all the Ps when the laser is tuned to the atomic resonance, and a single atom, resonant ionization detector [25] with a low background counting rate and 40% quantum efficiency.

APPARATUS

A schematic of the apparatus is shown in Fig. 2, and can be logically divided into a pulsed Ps source, a pulsed light source and frequency standard, and a single particle detector.

Positronium Source

The Ps source begins with a ^{50}Co β^+ source (150mCi) emitting positrons with an end point energy of 0.474 MeV. Some of the fast positrons enter a Cu(111) single crystal where they quickly thermalize (see Fig. 3). Although most of the positrons annihilate in the metal, a small fraction of them (0.2%) are stopped sufficiently close to the surface to allow them to diffuse back to the surface. Once at the surface, roughly ½ of the positrons are emitted into the vacuum with ∿0.7 eV kinetic energy (the negative work function of the e^+ in the metal)[26] and a spread in energy of ∿0.3 eV. The remainder escape as fast (∿3 eV) Ps or are trapped at the surface[27] in the "image" potential well. (Fig. 4) The overall conversion of high energy positrons to "moderated" low energy positrons with $\Delta E = 0.3$ eV is about 0.1%. The slow positrons (∿5×10^5 sec^{-1}) are guided through the 5m length of the apparatus by a ∿150G magnetic field produced by a 30 cm diam. solenoid.

In order to match [20] the e^+ source to the pulsed laser source, the positrons are stored in a magnetic bottle, the ends of which are closed by an electrostatic field, tuned to the cyclotron resonance A transverse rf electric field, tuned to the cyclotron resonance of the positrons, increases the transverse energy of the particles preventing them from escaping through the magnetic mirror. The positrons (1.6 eV initial longitudinal kinetic energy) are trapped with ∿50% efficiency and leak out of the bottle with a ∿100 μsec time constant (see Fig. 5).

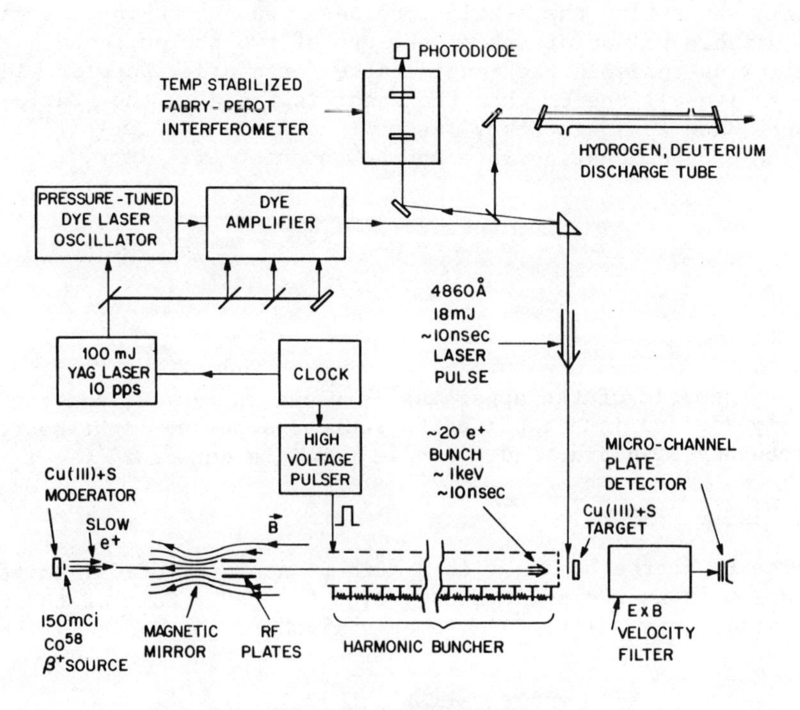

Fig. 2. Apparatus.

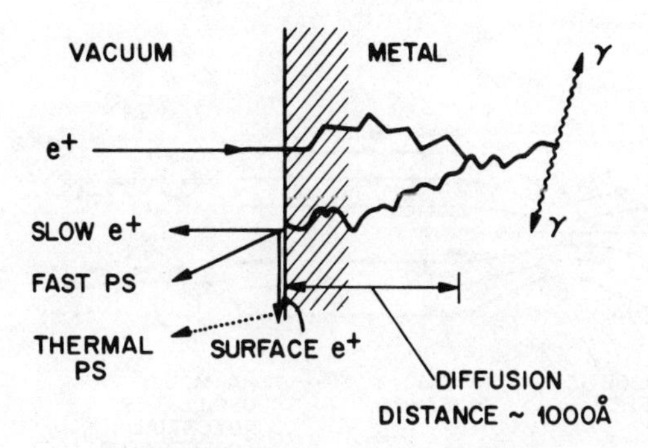

Fig. 3. Positron interactions with a surface.

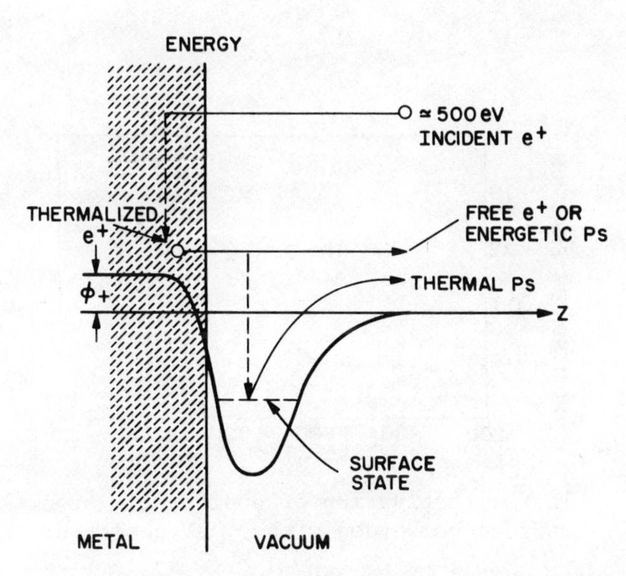

Fig. 4. Energetics of the positron surface interaction.

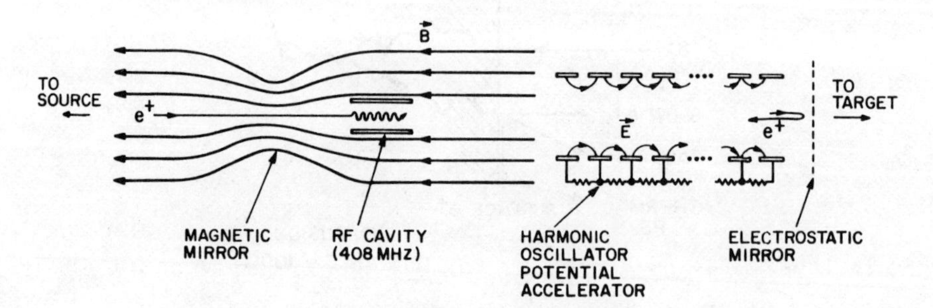

Fig. 5. Positron pulser (Ref. 20).

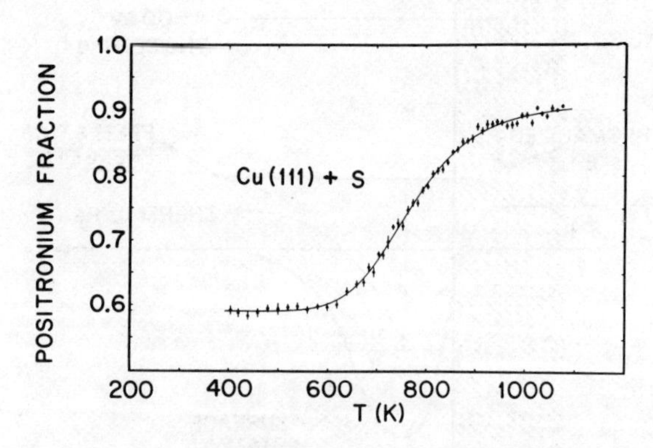

Fig. 6. Thermal activation of positronium from a Cu(111) surface bombarded with 30 eV positrons (Ref 21).

Positrons are extracted from the bottle by applying a 1.25 kV pulse to an accelerator consisting of a series of metal rings coupled by resistors. The resistor values are chosen so that the accelerator forms a 1.5m long harmonic potential well $V(z)=kz^2$, where z is the B-field direction and z=0 is the location of the positron target, another single crystal of copper. Since the stored positrons are essentially at rest, they will hit the target after a time corresponding to ¼ of the harmonic oscillator period. Roughly 20 positrons per high voltage pulse hit the target in a single bunch with a ∿10 nsec FWHM.

Once in the copper target, the positrons behave in the same way as in the copper moderator (Figs. 3,4) with two important differences. First, the average incident energy of the e^+ is ∿500 eV instead of 200 keV, so practically all the positrons are stopped in a short distance and diffuse back to the surface. Second, the target is heated so that the positrons trapped in the surface state are thermally activated to form thermal positronium in the ground state[21]. Fig. 6 shows the thermal activation of Ps from a Cu(111) surface. The process is analogous to thermionic emission of electrons or to thermal desorption of hydrogen, and can be described by the usual statistical mechanics arguments[28]. Fig. 7 shows the results of a time-of-flight measurement[22] that

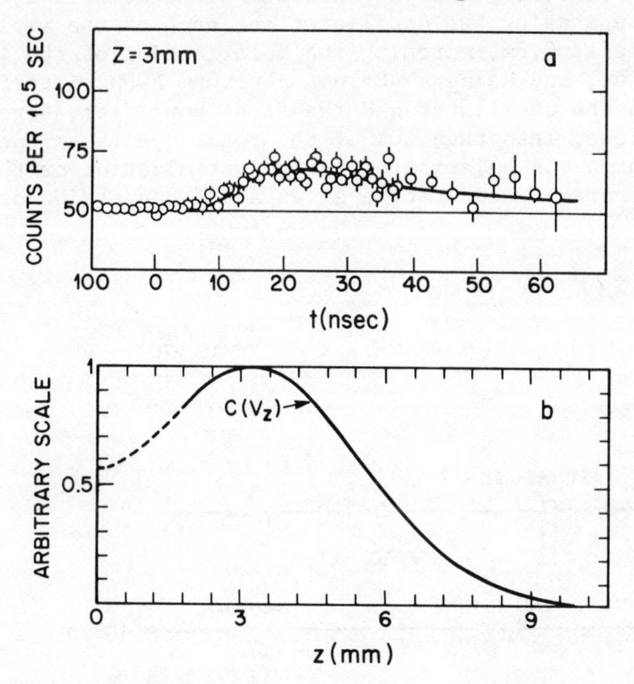

Fig. 7. (a) Time of flight spectrum of positronium from Cu(111). (b) Derived spatial distribution of positronium at t=30 nsec. (Ref. 22).

shows that the velocity distribution is a modified Maxwell Boltzman distribution. In particular, Fig. 7a shows how the distribution of thermal Ps atoms would appear a distance 3mm from the target. By delaying the laser pulse 20 to 30 nsec after the positrons hit the metal target we maximize the intersection of the light with the Ps if the light beam is a few millimeters from the target. We estimate that the laser beam intersects $\sim$1/3 of the roughly 5 thermal Ps atoms formed in each pulse.

Because the positron moderator and the Ps converter require well characterized metal surfaces, the entire positron apparatus is in an ultra high vacuum chamber ($\sim 2\times10^{-10}$ torr).

Light Source

The light source used to excite the Ps used a commercial, frequency tripled, Q-switched Nd-YAG laser (100 mj/pulse at 355nm, 12 nsec long FWHM, 10 pps) that pumps a homemade dye laser operating at 486 nm. The dye laser consists of a low power dye laser oscillator tuned by a diffraction grating used at grazing incidence [29] and an intercavity Fabry-Perot etalon, followed by three single-pass amplifier stages. (Fig. 8) Amplified spontaneous emission, a particularly annoying problem with coumarin dyes, was minimized by separating the oscillator and each of the amplifier stages by at least 0.6m, matching the polarization of the 486 nm and 355 nm light, and using a 486 nm, $\Delta\lambda$=1.0nm FWHM, interference filter between the oscillator and the first amplifier stage. With these precautions, less than 10% of the total dye laser output is amplified spontaneous emission. Careful optimization of the pump intensity delivered to each stage gives a 20% conversion of input

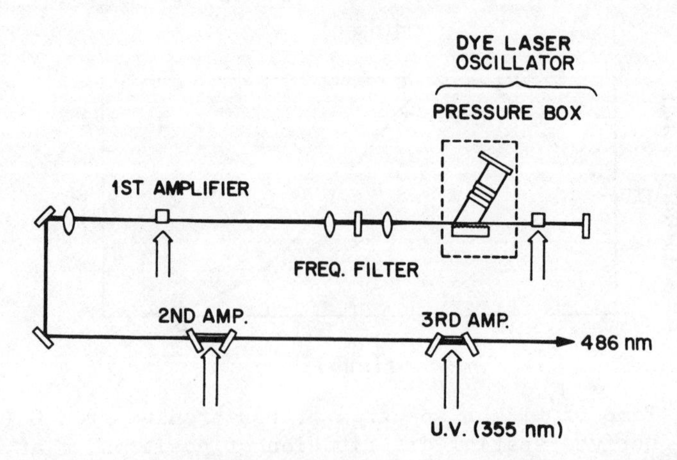

Fig. 8. Schematic of the tunable laser system.

pump energy to dye laser output energy. With fresh dye, the laser delivers over 20 mj/pulse, 10 nsec pulses, in a ~800 MHz bandwidth. The spatial structure of the beam as measured by burn patterns is smoothly varying and elliptically shaped ~3mmx2.5mm spot at the Ps source, ~4m after the last amplifier.

The laser can be linearly pressure tuned over the H_β, D_β and Ps 1S-2S resonances by passing nitrogen from a high pressure reservoir through a variable leak valve and into the vacuum/pressure chamber surrounding the diffraction grating and etalon.

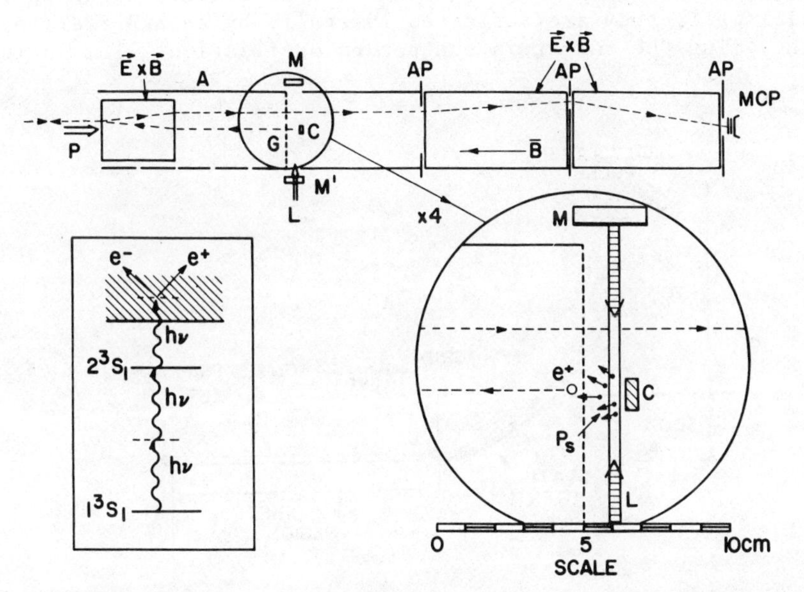

Fig. 9. Detail of the positronium-laser beam interaction region. CMA, channel multiplier array detector; AP, aperture; $\vec{E} \times \vec{B}$, $\vec{E} \times \vec{B}$ drift region; $\vec{B}$, 130-G magnetic induction; M, M', set of three mirrors that causes the laser beam to intersect the Ps in three places and produce counter propagating beams; A, accelerator electrode; C, Cu(111) + S e^+ target; G, Au grid; P, e^+ pulse incident on Cu target; L, laser pulse; Ps, thermal positronium emitted from the hot target. The dashed line shows path of the e^+ ionization fragment. The inset shows the resonant three-photon ionization process.

EXPERIMENT

The laser beam is made to intersect the positronium atoms ∿3mm away from the target. (See Fig. 9) Three mirrors inside the vacuum chamber are used to fold the beam into a counter propagating "z" pattern that increases the overlap between the Ps atoms and the light. The alignment is adjusted so that the laser beam simply returns back on itself.

We detect the Ps resonance excitation via a three photon process[30] (see inset Fig. 9). The photon flux is sufficient to induce the 1^3S_1 to 2^3S_1 transition with 20-40% probability. Once in the excited state, the Ps is ionized with greater than 95% probability, and we detect the resonance by collecting the ion fragment (e^+) as shown in Fig. 9. The ionized atoms (inset of Fig. 9) are swept away from the target by an electric field between the target and a grid 1.5 cm away. (See Fig. 10 for a schematic of the electrostatic potential.) As the ions are accelerated back up the potential well, they are deflected laterally by an ExB field so that they miss the target in the return oscillation. The positrons

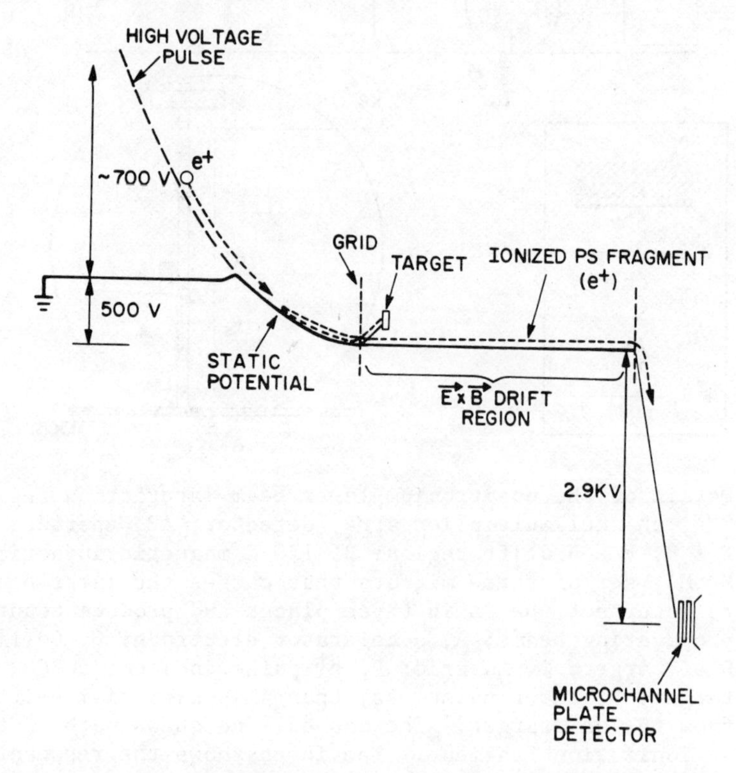

Fig. 10. Electrostatic potential diagram.

then enter two more ExB drift regions that velocity discriminate between the e^+s and heavy positive ions emitted from the hot target. The positrons are then accelerated to ∿2.9 kV and hit a 2 stage, 22mm dia., microchannel plate detector with a phosphor screen anode. The electric and magnetic fields are fine tuned on the ∿5 prompt slow positrons/pulse that are reemitted from the target. Each detected positron yields a 10 nsec wide anode signal and a green dot of light that can easily be seen through a window in the back of the vacuum chamber. The overall detection efficiency once the Ps is ionized is ∿40%.

A summary of the timing sequence is shown in Fig. 11. Fig. 11a shows the relative time delay between the positron pulse and the laser pulse measured using a NE 102 plastic scintillator and

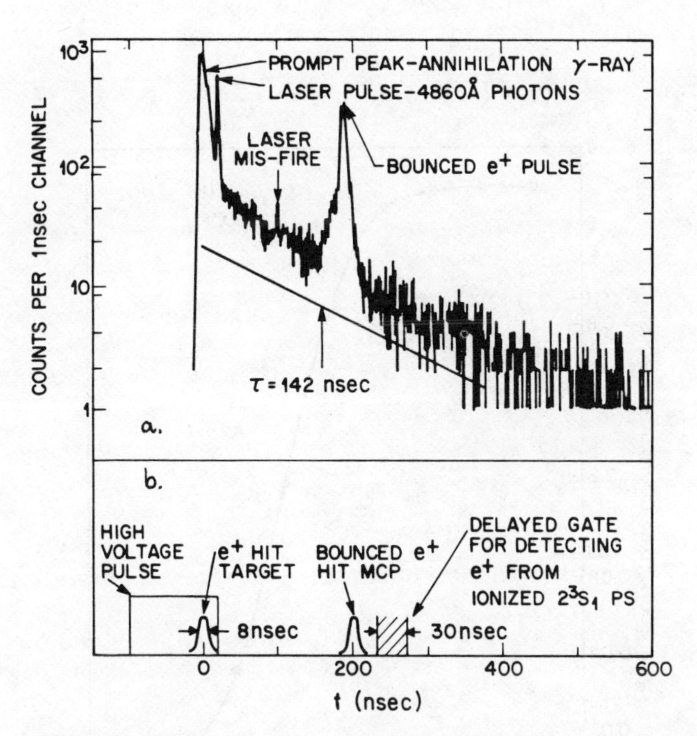

Fig. 11. Annihilation γ-ray and laser pulse time spectrum. The small secondary laser peak is associated with a timing jitter that was eliminated prior to obtaining the data of Fig. 14. The bounced e^+ pulse was made visible by turning off the E x B drift region so that the reemitted slow positrons hit the target again instead of travelling on to the microchannel plate detector (MCP).

a 9823 K photomultiplier tube that simultaneously detects the annihilation gamma rays and laser photons entering the light-tight scintillator wrapping through a pin hole. Fig. 11b shows the timing sequence of the buncher pulse, the e^+ hitting the target, the re-moderated e^+ that traverse the velocity selector, and the time window allowed for detecting the ionized Ps. By delaying the laser and gate by 30 nsec, we discriminate with 99% efficiency between the re-moderated e^+ and the e^+ resulting from the ionization of the Ps (see Fig. 12). As mentioned previously, we also maximize the overlap of the laser beam and the Ps atoms desorbed from the surface.

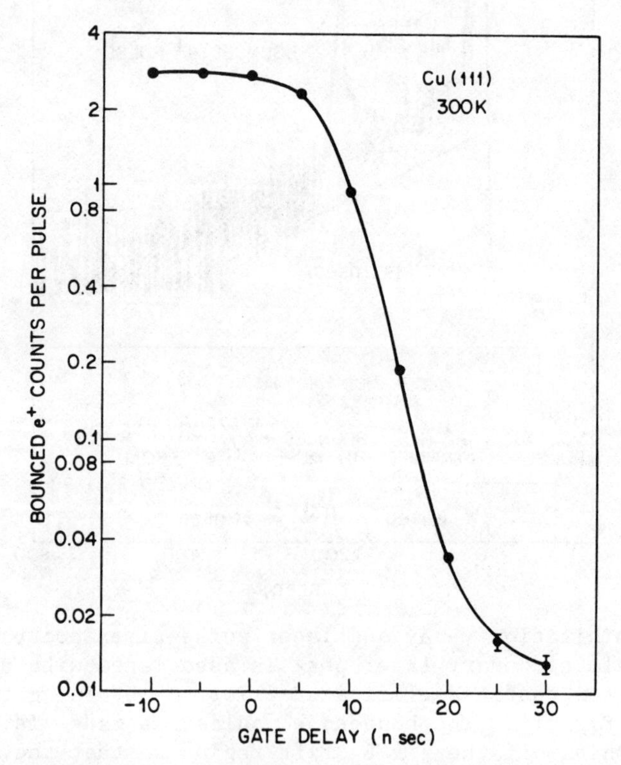

Fig. 12. Microchannel plate detector count rate vs. delay of coincidence gate (see delayed gate of Fig. 11).

The frequency of the laser is measured relative to the deuterium n=2→4 transitions. These excitations are detected optogalvanically[31] in a simple circuit consisting of a regulated power supply, 10 kΩ ballast resistor, and a flowing gas Woods discharge tube operating at 0.4 torr of hydrogen and deuterium, ∿2.0 torr of helium and ∿10 mA current, ∿30 V/cm electric field. When the pulsed dye laser is tuned to the hydrogen or deuterium resonances, the discharge conditions are momentarily perturbed, and damped voltage oscillation appears across the tube. The resonance signal is then capacitively coupled into a gated integrator. A stable discharge, essential for good signal/noise optogalvanic detection, is obtained most easily by simply running the tube for long periods of time. (>6 hrs.)

A temperature stabilized Fabry-Perot interferometer frequency marker (5.014 ± 0.016 GHz free spectral range) is calibrated on the hydrogen to deuterium n=2→4 isotope shift immediately before an experimental run. (See Fig. 13) The laser is then tuned first

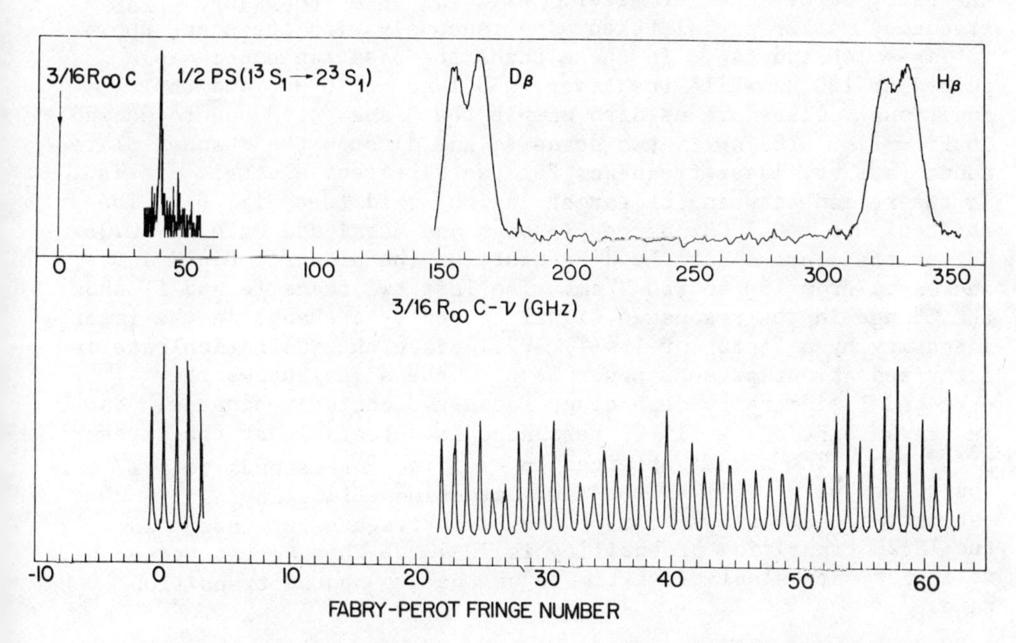

Fig. 13. Scan of the H_β and D_β optogalvanic signals and the positronium 1S→2S e^{+} count rate signal (see Fig. 14) plotted on the same frequency scale. The frequency shifts are measured using the Fabry-Perot signal shown in the lower half of the figure. The absolute frequency and scale factor are obtained from the known positions of the H_β and D_β resonances.

to the D_β lines, next to the vicinity of the Ps resonance, and then finally back to the D_β lines, so that each Ps scan is calibrated via the frequency marker relative to the D_β lines. Our calibration uses[32]

$$\nu_\beta(D) = \frac{3}{16} c\, R_\infty - 161.054 \text{ GHz and } \nu_\beta(H) = \frac{3}{16} c\, R_\infty - 328.812 \text{ GHz}.$$

We have made a 0.8 GHz correction to the frequency scale to account for the ac Stark shift of the calibration lines[14]. The correction is obtained by measuring the intensity of the laser in the discharge tube (the beam is expanded so that the uniform central portion fills the bore of the discharge tube) and applying the known ac Stark shift correction.

Results

Fig. 14 shows the results of 4 successive scans taken within a several hour period. The data were collected using a four-input multiscaler to record the gated microchannel plate detector signals (laser on and off), the optogalvanic signal and the signal from the Fabry-Perot interferometer. Fig. 14a shows the Fabry-Perot frequency marker signal taken simultaneously with the scans shown in Figs. 14b and 14c. In these scans the positron source is pulsed at 100 Hz while the laser is pulsed at 10 Hz, and the additional pulses are used to sample the "laser off" counts shown in Fig. 14b. The next two scans (c and d) show the channel plate count rate vs. laser frequency for two different electric fields in the region between the target and the grid (see Fig. 9). The shift of 0.8 ± 0.5 GHz agrees in sign and magnitude with the 0.36 GHz shift expected for the 2^3S_1 state as the electric field is increased from 160 to 280 V/cm. The last two scans (e and f) show the change in the resonance signal caused by a change in the laser intensity by a factor of $1.4(1) \approx \sqrt{2}$. Since the ionization rate is saturated at our present power levels, the signal shows the expected I^2 intensity dependence for a two photon resonance. No ac Stark shift of the 1S-2S resonance is detectable at the present precision. The signal observed in Fig. 14e corresponds to 0.17 counts per pulse. Given the ∿40% detection efficiency, and estimating that only ¼ to ½ of the Ps intersects the laser beam, the 1S→2S transition probability is 20-40%. This agrees reasonably well with our calculated estimate of the two-photon transition rate.

The linewidths observed in Fig. 14 are attributable to (i) a residual first order Doppler width due to the misalignment of the laser beam, (ii) the ∿800 MHz laser linewidth, (iii) the 200 MHz width due to the ionization rate, (iv) a ∿100 MHz second order Doppler width due to the thermal energy of the Ps, (v) the nonuniform Stark shift due to inhomogeneous electric fields (∿100 MHz). The field-free natural linewidth is only the ∿1 MHz

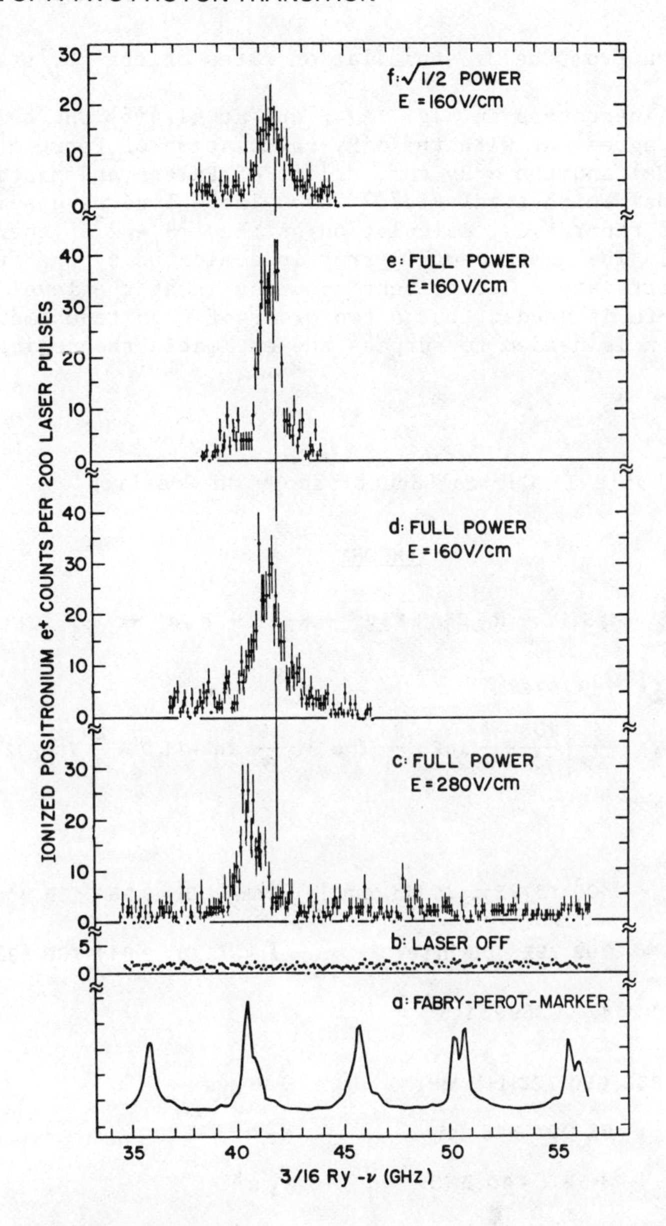

Fig. 14. Ionized positronium counts from the channel plate detector plotted vs. laser frequency. (a) Frequency marker signal. (b) Laser-off counting rate. (c) Resonance signal obtained with E=280 V/cm. (d), (e) Same as (c) except E=160 V/cm. (f) Same as (e) but with laser intensity decreased by $\sqrt{2}$. The vertical line shows the line center predicted by Ferrell (Ref. 33) and Fulton and Martin (Ref. 34). The uncertainty in the frequency scale is ±0.5 GHz.

contribution from the 3γ annihilation rates of the $^3S_1$ states.

The linecenters in Fig. 14d-f are at 41.4(5) GHz below 3/16Ry, in agreement with the α^2Ry calculation of Ferrell[33] (½82.006 GHz) and the α^3Ry corrections of Fulton and Martin[34] (½x1.527 GHz) which total 41.767 GHz. Table I gives a summary of the present theoretical calculation of the $1^3S_1 - 2^3S_1$ energy difference. The experimental error is dominated by the frequency marker uncertainty. Our present accuracy is at the level where the $\alpha^3 R_\infty$ term is needed, but a two order of magnitude improvement in precision is needed to surpass the estimated theoretical accuracy.

Table I Optical Spectroscopy on Positronium

<u>THEORY</u>

$$\Delta\nu(2^3S_1 - 1^3S_1) = \frac{3}{8} R_\infty c \left\{1 + K_2\alpha^2 + K_3\alpha^3 + K_4\alpha^4 + \cdots\right\}$$

$$K_2 = -719/576$$

$$K_3 = -\frac{2}{3\pi}\left[\frac{203}{72} + \frac{7}{12}\ln 2 + \frac{21}{4}\ln\alpha^{-1} - \frac{16}{3}\ln k_0(1,0) + \frac{2}{3}\ln k_0(2,0)\right]$$

$$K_4 = \ ?$$

$R_\infty =$ 109 737.315 21 (11) cm^{-1} $\ln k_0(1,0) = 2.984\ 128\ 555\ 8$

$c =$ 299 792 458 m sec^{-1} $\ln k_0(2,0) = 2.811\ 769\ 893\ 1$

$\alpha^{-1} =$ 137.035963 (15)

$\Delta\nu$ = 1 233 690 730 (1) MHz	$\frac{3}{8} R_\infty c$
− 82 005.616 (18)	$K_2\alpha^2$
− 1 527. 440 3 (5)	$K_3\alpha^3$
± 10. ?	$K_4\alpha^4$
1 233 607 197 (1) ± 10 ? MHz	TOTAL THEORY

<u>EXPERIMENT</u>

1 233 607 900 (1000)	MHz	
1	MHz	NATURAL LINEWIDTH

Discussion

An improved laser source and better laser metrology should enable us to obtain a measurement with a precision comparable to or better than the hydrogen 1S-2S measurements[14]. Fig. 15 outlines our immediate plans to improve our present observation. The laser being constructed should give an order of magnitude narrower linewidth with an order of magnitude higher average power. The improved counting rate should allow us to easily obtain a statistical uncertainty of less than 1 MHz in a matter of minutes. Other improvements include a flat Fabry-Perot interferometer inside the ultra high vacuum chamber that will filter the laser output to create a nearly Fourier transform limited pulse, and precision laser metrology techniques using well known iodine and tellurium reference lines[35].

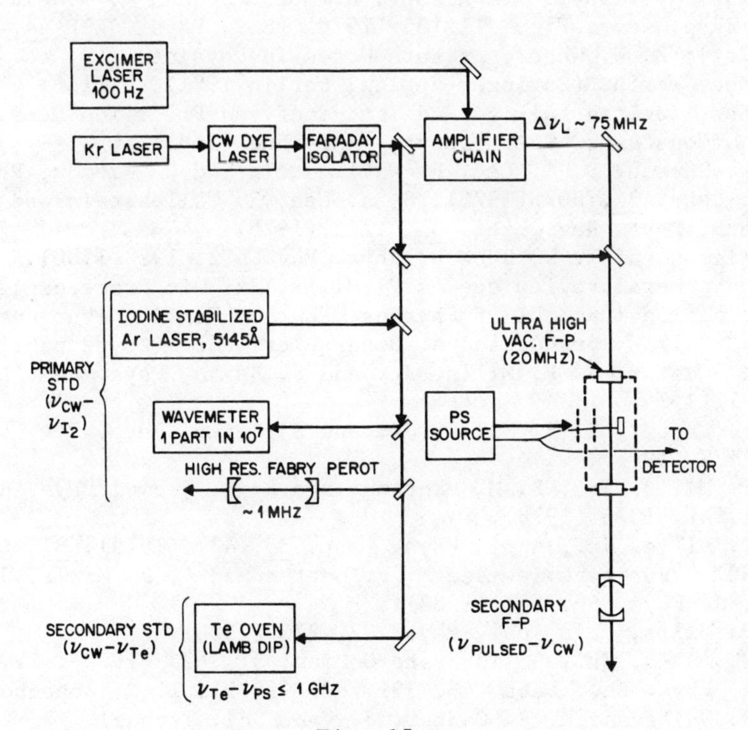

Fig. 15.

1. S. Chu and A. P. Mills, Jr., Phys. Rev. Lett. 48, 1333 (1982).
2. M. Deutsch, Phys. Rev. 82, 455 (1951). Se also Adventures in Experimental Physics, Vol. 4, Bogdan Maglich, ed. p. 63-127.
3. See reviews by S. Berko and H. N. Pendleton, Annu. Rev. Nucl. Part. Sci. 30, 543 (1980); A. Rich, Rev. Mod. Phys. 53, 127 (1981).
4. H. A. Bethe and E. E. Salpeter, Quantum Mechanics of One- and Two-Electron Systems in Handbuch der Physik, S. Flugge, ed. (Springer-Verlag, Berlin 1957) p. 281.
5. G. P. Lepage, Phys. Rev. A16, 863 (1977).
6. W. E. Caswell and G. P. Lepage, Phys. Rev. A18, 810 (1978); W. E. Caswell and G. P. Lepage, Phys. Rev. A20, 36 (1979).
7. R. Barbieri and E. Remiddi, Nucl. Phys. B141, 413 (1978).
8. R. H. Beers and V. W. Hughes, Bull. Am. Phys. Soc. 13, 633 (1968); D. W. Gidley and P. W. Zitzewitz, Phys. Lett. A69, 97 (1978).
9. A. P. Mills, Jr. and G. H. Bearman, Phys. Rev. Lett. 34, 246 (1975).
10. P. O. Egan, W. E. Frieze, V. W. Hughes and M. H. Yam, Phys. Lett. A54, 412 (1975).
11. A. P. Mills, Jr., S. Berko and K. F. Canter, Phys. Rev. Lett. 34, 1541 (1975).
12. H. W. Kendall, Ph.D., Thesis, Massachusetts Institute of Technology (1954). See also V. S. Letokhov and V. G. Minogin, Zh. Eksp. Teor. Fiz. 71, 135 (1976) [Sov. Phys. JETP 44, 70 (1976)]; E. W. Weber, Lecture Notes in Physics, 143, ed. Graff, Klempe, Werth, (Springer-Verlag, Berlin 1981), p. 146; V. W. Hughes, Invited talk at 2nd Int. Conf. on Precision Meas. and Fund. Constants, Gaithersburg, Maryland (1981).
13. T. W. Hänsch, S. A. Lee, R. Wallenstein and C. Wieman, Phys. Rev. Lett. 34, 807 (1975); S. A. Lee, R. Wallenstein and T. W. Hänsch, Phys. Rev. Lett. 35, 1262 (1975).
14. C. Wieman and T. W. Hänsch, Phys. Rev. A22, 192 (1980).
15. For a general review see A. P. Mills, Jr., in Proceedings of The International School of Physics, "Enrico Fermi", (Varenna, Italy July 1981) W. Brandt and A. Dupasquier, eds. (to be published).
16. K. F. Canter, A. P. Mills, Jr. and S. Berko, Phys. Rev. Lett. 33, 7 (1974).
17. K. F. Canter, A. P. Mills, Jr. and S. Berko, Phys. Rev. Lett. 34, 177 (1975).
18. A. P. Mills, Jr., P. M. Platzman and B. L. Brown, Phys. Rev. Lett. 41, 1076 (1978).
19. A. P. Mills, Jr., Appl. Phys. Lett. 35, 427 (1979); 37, 667 (1980). For a review see K. F. Canter and A. P. Mills, Jr., Can. J. Phys. 60, 551 (1982).
20. A. P. Mills, Jr., Appl. Phys. 22, 273 (1980).
21. A. P. Mills, Jr., Solid State Commun. 31, 623 (1979); K. G. Lynn, Phys. Rev. Lett. 43, 391, 803 (1979); I. J. Rosenberg, A. H. Weiss and K. F. Canter, J. Vac. Sci. Technol. 17, 253 (1980).

22. A. P. Mills, Jr. and L. N. Pfeiffer, Phys. Rev. Lett. 43, 1961 (1979).
23. R. Wallenstein and T. W. Hänsch, Opt. Comm. 14, 353 (1975); P. Drell and S. Chu, Opt. Comm. 28, 343 (1979).
24. L. S. Vasilenko, V. P. Chebotaev and A. V. Shishaev, Pis'ma Zh. Eksp. Teor. Fiz. 12, 161 (1970) [JETP Lett. 12, 113 (1970)]; M. Goppert-Mayer, Ann. der Physik 9, 273 (1973).
25. See for example, G. S. Hurst, M. M. Nayfeh, J. P. Young, M. G. Payne and L. W. Grossman, in Laser Spectroscopy III, ed., J. L. Hall and J. L. Carlsten, (Springer-Verlag, Berlin) p. 44 (1977).
26. C. A. Murray, A. P. Mills, Jr. and J. E. Rowe, Surf. Sci. 100, 647 (1980).
27. C. H. Hodges and M. J. Stott, Solid State Commun. 12, 1153 (1973).
28. J. B. Pendry, J. Phys. C. Solid State Phys. 13, 1159 (1980); H. J. Kreuzer, D. N. Lowy and Z. W. Gortel, Solid State Commun. 35, 781 (1980); S. Chu, C. A. Murray and A. P. Mills, Jr., Phys. Rev. B23, 2060 (1981).
29. M. Littman and H. Metcalf, Appl. Opt. 17, 224 (1978); I. Shoshan and U. P. Oppenheim, Optics Commun. 25, 375 (1978).
30. This type of detection was first used in hydrogen: G. C. Bjorkland, C. P. Ausschnitt, R. R. Freeman and R. Storz, Appl. Phys. Lett. 33, 54 (1978).
31. See for example, J. E. Lawler, Phys. Rev. A22, 1025 (1980).
32. J. D. Garcia and J. E. Mack, J. Opt. Soc. Am. 55, 654 (1965); G. W. Erickson, J. Phys. Chem. Ref. Data 6, 831 (1977).
33. R. A. Ferrell, Phys. Rev. 84, 858 (1951).
34. T. Fulton and P. C. Martin, Phys. Rev. 95, 811 (1954); and T. Fulton, Johns Hopkins Univ. preprint JHU-HET 8206, to be published, and private communication.
35. J. Hall, private communication.

EXPERIMENTAL TESTS OF BELL'S INEQUALITIES IN ATOMIC PHYSICS

Alain Aspect

Institut d'Optique Théorique et Appliquée
Bâtiment 503 - Centre Universitaire d'Orsay - BP 43
91406 ORSAY CEDEX - FRANCE

1 - INTRODUCTION

Bell's Inequalities provide a quantitative criterion to test some reasonable Supplementary Parameters Theories versus Quantum Mechanics. Thanks to Bell[1], the debate about the possibility of completing Quantum Mechanics by an underlying substructure has been brought into the experimental domain.

The motivations for considering supplementary parameters will be found in the analysis of the famous Einstein-Podolsky-Rosen Gedankenexperiment[2]. Introducing a reasonable Locality Condition, one can derive Bell's theorem, which states

(i) that Local Supplementary Parameters Theories are constrained by Bell's Inequalities :

(ii) that certain predictions of Quantum Mechanics sometimes violate Bell's Inequalities.

We will point out that a fundamental assumption for the conflict is the Locality assumption. We will show that in a more sophisticated version of the E.P.R. thought experiment ("timing experiment"), the Locality Condition may be considered as a consequence of Einstein's Causality, preventing faster-than-light interactions.

The purpose of this discussion is to convince the reader that the formalism leading to Bell's Inequalities is very general and reasonable. What is surprising is that it conflicts with Quantum Mechanics.

As a matter of fact, situations exhibiting such a conflict are very rare, and it was necessary to design special experiments for getting a sensitive test. Atomic physics is the field where the experiments that follow most closely the ideal scheme of the Gedankenexperiment have been carried out. We will review these experiments, and their results.

2 - WHY SUPPLEMENTARY PARAMETERS ? THE EINSTEIN-PODOLSKY-ROSEN-BOHM GEDANKENEXPERIMENT

Experimental scheme

Let us consider the optical variant of the E.P.R. Gedankenexperiment modified by Bohm[3]. A source S emits a pair of photons with different energies, ν_1 and ν_2 , counterpropagating along $\pm\vec{Oz}$ (Fig. 1). Suppose that the polarization part of their state vector is :

$$|\Psi(\nu_1,\nu_2)\rangle = (1/\sqrt{2})\left[|x,x\rangle + |y,y\rangle\right] \tag{1}$$

where $|x\rangle$ and $|y\rangle$ are linear polarizations states.

We perform on these photons linear polarization measurements. The analyzer I in orientation $\vec{a}$, followed by two detectors, gives + or - result, corresponding to a linear polarization found parallel or perpendicular to $\vec{a}$. Analyzer II, in orientation $\vec{b}$, acts similarly.*

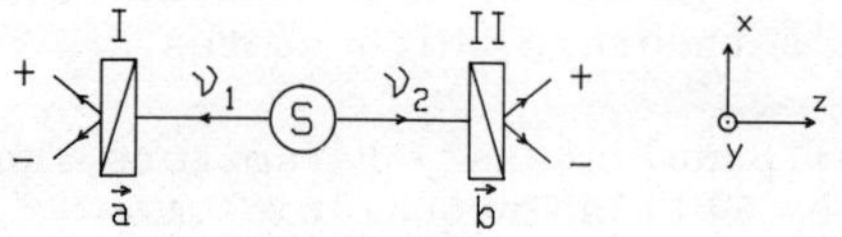

Fig. 1. Einstein-Podolsky-Rosen-Bohm Gedankenexperiment with photons. The two photons ν_1 and ν_2, emitted in the state (1), are analyzed by linear polarizers in orientations $\vec{a}$ and $\vec{b}$. One can measure the probabilities of single or joint detections after the polarizers.

* There is a one-to-one correspondence with the Gedankenexperiment dealing with a pair of 1/2 spin particles, in a singlet state, and analyzed by two Stern-Gerlach filters[3].

It is easy to derive the Quantum Mechanical predictions for these measurements, single or in coincidence.

Let $P_{\pm}(\vec{a})$ be the probability of getting the result ± for ν_1 ; similarly $P_{\pm}(\vec{b})$ is related to ν_2. Quantum Mechanics predicts :

$$P_{+}(\vec{a}) = P_{-}(\vec{a}) = 1/2$$
$$P_{+}(\vec{b}) = P_{-}(\vec{b}) = 1/2 \qquad (2)$$

Let $P_{\pm\pm}(\vec{a},\vec{b})$ be the probability of joint detection of ν_1 in channel ± of I (in orientation $\vec{a}$), and of ν_2 in channel ± of II ($\vec{b}$). Quantum Mechanics predicts :

$$P_{++}(\vec{a},\vec{b}) = P_{--}(\vec{a},\vec{b}) = \frac{1}{2}\cos^2(\vec{a},\vec{b})$$
$$P_{+-}(\vec{a},\vec{b}) = P_{-+}(\vec{a},\vec{b}) = \frac{1}{2}\sin^2(\vec{a},\vec{b}) \qquad (3)$$

Correlations

In the special situation $(\vec{a},\vec{b}) = 0$ one finds

$$P_{++}(\vec{a},\vec{b}) = P_{--}(\vec{a},\vec{b}) = 1/2$$

while

$$P_{+-}(\vec{a},\vec{b}) = P_{-+}(\vec{a},\vec{b}) = 0$$

So, if ν_1 is found in the + channel of I (the probability of which is 50 %), we are sure to find ν_2 in the + channel of II (and similarly for the - channels). There is a strong correlation between the results of measurements on ν_1 and ν_2.

A convenient way of displaying these correlations is the polarization correlation coefficient :

$$E(\vec{a},\vec{b}) = P_{++}(\vec{a},\vec{b}) + P_{--}(\vec{a},\vec{b}) - P_{+-}(\vec{a},\vec{b}) - P_{-+}(\vec{a},\vec{b}) \qquad (4)$$

The prediction of Quantum Mechanics is

$$E_{MQ}(\vec{a},\vec{b}) = \cos 2(\vec{a},\vec{b}) \qquad (5)$$

For $(\vec{a},\vec{b}) = 0$, we find $E_{MQ}(0) = 1$, i.e. a complete correlation.

Supplementary parameters

Correlations between distant measurements on two systems that have separated may be easily understood in terms of some common properties of the two systems. Let us consider again the correlations of polarization measurements in the case $(\vec{a},\vec{b}) = 0$. When we find +

for ν_1, we are sure to find + for ν_2. We are thus led to admit that there is some property (Einstein said "an element of physical reality") pertaining to this particular pair, and determining the result ++. For another pair, the results will be -- ; the invoked property is different.

Such properties, differing from one pair to another one, are not taken into account by the Quantum Mechanical state vector $|\Psi(1,2)\rangle$ which is the same for all pairs. This is why Einstein concluded that Quantum Mechanics is not complete. And this is why such properties are referred to as "supplementary parameters" (sometimes called "hidden-variables").

As a conclusion, one can hope to "understand" the E.P.R. correlations by such a classical-looking picture, involving supplementary parameters differing from one pair to another one. It can be hoped to recover the Quantum Mechanical predictions when averaging over the supplementary parameters. It seems that so was Einstein's position[4]. At this stage, a commitment to this view point is just a matter of taste.

Remark. Since Einstein spoke of "an element of the physical reality", some authors call these theories invoking supplementary parameters "Realistic Theories"[5].

3 - BELL'S INEQUALITIES

Formalism

Bell tried to translate into mathematics the preceding discussion, by introducing explicitly supplementary parameters, denoted λ. Their distribution on an ensemble of emitted pairs is specified by a probability distribution $\rho(\lambda)$, such that

$$\rho(\lambda) \geqq 0 \text{ and } \int d\lambda\, \rho(\lambda) = 1 \tag{6}$$

For a given pair, characterized by a given λ, the results of measurement will be

$$A(\lambda,\vec{a}) = \begin{cases} +1 \\ \text{or} \\ -1 \end{cases} \quad \text{at analyzer I (orientation } \vec{a})$$

$$B(\lambda,\vec{b}) = \begin{cases} +1 \\ \text{or} \\ -1 \end{cases} \quad \text{at analyzer II (orientation } \vec{b}) \tag{6'}$$

A particular theory must be able to supply explicitly the functions $\rho(\lambda)$, $A(\lambda,\vec{a})$ and $B(\lambda,\vec{b})$.

It is then easy to express the probabilities of various results.

For instance $P_+(\vec{a}) = \frac{1}{2}\int d\lambda\ \rho(\lambda)\left[A(\lambda,\vec{a}) + 1\right]$ etc...

In particular, we will use the correlation function :

$$E(a,b) = \int d\lambda\ \rho(\lambda)\ A(\lambda,\vec{a})\ B(\lambda,\vec{b}) \tag{7}$$

A (naive) exemple

Let us suppose that the two photons of a pair are emitted with the same linear polarization, defined by its angle λ with $\vec{Ox}$ (Fig. 2).

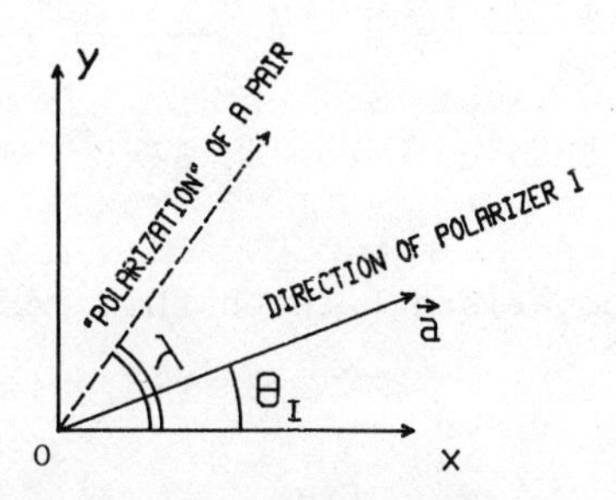

Fig. 2. Our example. Each pair has a "direction of polarization," defined by λ.

The probability distribution is taken to be isotropic :

$$\rho(\lambda) = 1/2\pi$$

As a simple model for the polarizer I we assume that we get the result + 1 if

$$|\theta_I - \lambda| \leq \frac{\pi}{4} \quad \text{or} \quad |\theta_I - \lambda| \geq 3\,\frac{\pi}{4}$$

The result − 1 is obtained for

$$\frac{\pi}{4} < |\theta_I - \lambda| < 3\pi/4$$

The response can thus be written

$$A(\lambda,\vec{a}) = \frac{\cos 2(\theta_I - \lambda)}{|\cos 2(\theta_I - \lambda)|}$$

Similarly

$$B(\lambda,\vec{b}) = \frac{\cos 2(\Theta_{II}-\lambda)}{|\cos 2(\Theta_{II}-\lambda)|}$$

With this model, we find

$P_+(\vec{a}) = P_-(\vec{a}) = P_+(\vec{b}) = P_-(\vec{b}) = 1/2$ which is identical to the Quantum Mechanical result.

As correlation function, we find :

$$E(\vec{a},\vec{b}) = 1 - 4\,\frac{|\Theta_I-\Theta_{II}|}{\pi} = 1 - 4\,\frac{|(\vec{a},\vec{b})|}{\pi}$$

Like the Quantum Mechanical result, $E(\vec{a},\vec{b})$ depends only on the relative angle $(\vec{a},\vec{b})$.

Fig. 3 shows a comparison between this result and the Quantum Mechanical prediction.

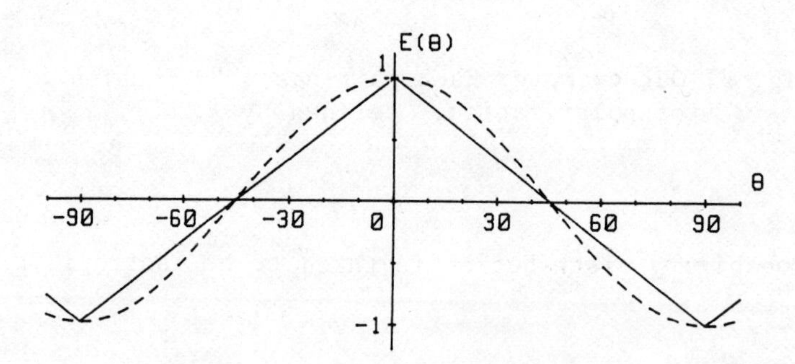

Fig. 3. Polarization correlation coefficient, as a function of the relative orientation of the polarizers.

---- : Calculated by Quantum Mechanics ;

—— : Given by our simple model.

The agreement is not too bad. It might be hoped that some more complicated model will be able to reproduce exactly the Quantum Mechanical predictions.

Bell's discovery is the fact that this search is hopeless.

Inequalities

Let us consider the quantity

$$s = A(\lambda,\vec{a}).B(\lambda,\vec{b})-A(\lambda,\vec{a}).B(\lambda,\vec{b}') + A(\lambda,\vec{a}').B(\lambda,\vec{b})+A(\lambda,\vec{a}').B(\lambda,\vec{b}')$$

$$= A(\lambda,\vec{a})\left[B(\lambda,\vec{b})-B(\lambda,\vec{b}')\right] + A(\lambda,\vec{a}')\left[B(\lambda,\vec{b}) + B(\lambda,\vec{b}')\right] \quad (8)$$

Remembering that the four numbers A, B take only the values ± 1, we find that

$$s(\lambda,\vec{a},\vec{a}',\vec{b},\vec{b}') = \pm 2$$

The average over λ is therefore included between + 2 and - 2, i.e.

$$-2 \leq \int d\lambda \; \rho(\lambda).s(\lambda,\vec{a},\vec{a}',\vec{b},\vec{b}') \leq 2$$

According to (7), we rewrite this

$$-2 \leq S(\vec{a},\vec{a}',\vec{b},\vec{b}') \leq 2 \quad (9)$$

with

$$S = E(\vec{a},\vec{b})-E(\vec{a},\vec{b}')+E(\vec{a}',\vec{b})+E(\vec{a}',\vec{b}')$$

These are B.C.H.S.H. inequalities, i.e. Bell's inequalities generalized by Clauser, Horne, Shimony, Holt [6]. They bear upon a combination of four polarization correlation coefficients, measured in four orientations of the polarizers. S is thus a measurable quantity.

4 - CONFLICT WITH QUANTUM MECHANICS

Evidence

Let us take the particular set of orientations of Fig. 4a. Replacing the E's by their Quantum Mechanical values (5) for pairs in state (1), we obtain :

$$S_{MQ} = 2\sqrt{2}$$

This Quantum Mechanical prediction strongly violates the upper limit of inequalities (9). We thus find it impossible to reconcile the formalism defined in (6) and (6') with the predictions of Quantum Mechanics for the particular (E.P.R.-type) state (1).

General study

We look for the greatest conflict, and we derivate S with

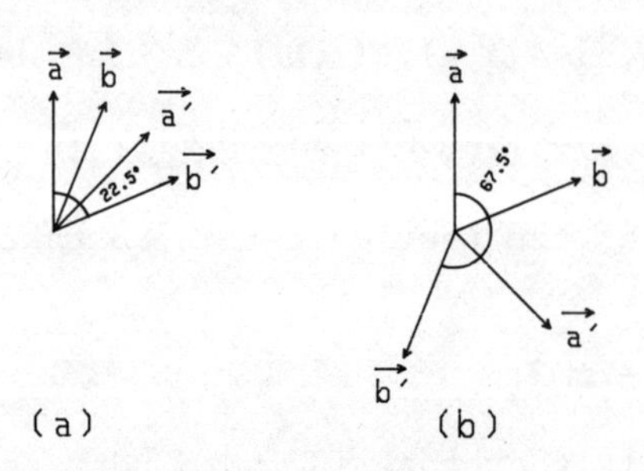

Fig. 4. Orientations yielding the largest conflict between Bell's Inequalities and Quantum Mechanics.

respect to the three angles $(\vec{a},\vec{b})$, $(\vec{b},\vec{a}')$ and $(\vec{a}',\vec{b}')$ (which are independent). S_{MQ} is extremum if

$$(\vec{a},\vec{b}) = (\vec{b},\vec{a}') = (\vec{a}',\vec{b}') = \Theta$$

and it takes the value

$$S_{MQ}(\Theta) = 3 \cos 2\Theta - \cos 6\Theta$$

Derivating now with respect to Θ, we obtain the maximum and minimum values of S_{MQ}.

$$S_{MQ}^{Max} = 2\sqrt{2} \qquad \text{for } \Theta = \pi/8$$

$$S_{MQ}^{Min} = -2\sqrt{2} \qquad \text{for } \Theta = 3\pi/8 \tag{10}$$

The corresponding orientations are displayed in Fig. 4.

Figure 5 displays the variations of $S_{MQ}(\Theta)$, and the limits given by B.C.H.S.H. inequalities. One sees that the conflict is serious.

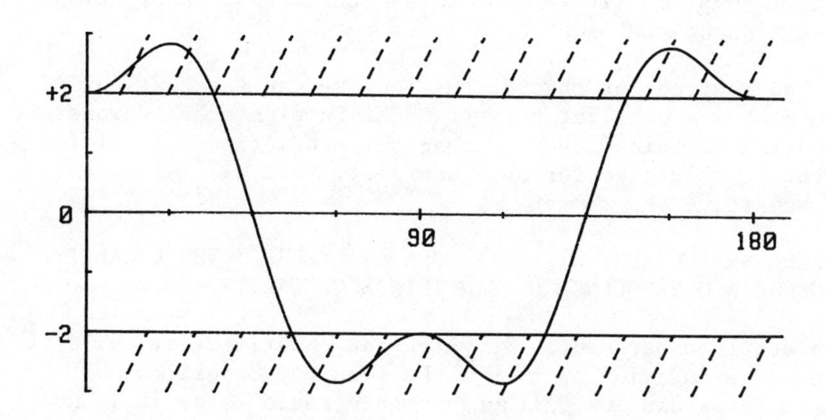

Fig. 5. $S(\Theta)$ as predicted by Quantum Mechanics for pairs in state (1). The conflict arises in the /// zone.

5 - DISCUSSION OF THE HYPOTHESES

To try to understand which part of the formalism causes this conflict, let us point out the hypotheses implied by formalism (6) and (6'). The supplementary parameters λ have been introduced for explaining the E.P.R. correlations by some common properties of the two photons. This point has already been discussed.

The used formalism is deterministic. When λ is fixed, then the results of measurements $A(\lambda,\vec{a})$ and $B(\lambda,\vec{b})$ are certain, i.e. λ determines the result. It might be thought that it is the reason for the conflict with Quantum Mechanics. But Bell[1], and Clauser and Horne[7] have exhibited Stochastic Supplementary Parameters Theories that are not deterministic, and which nevertheless lead to Bell's Inequalities. The deterministic character does not seem sufficient to lead to a conflict*.

As stressed by Bell, the formalism follows a Locality Condition. The result of measurement at I, $A(\lambda,\vec{a})$, does not depend on the orien-

* This conclusion is not shared by all authors. For instance, A. FINE [8] argues that the stochastic theories of Bell or of Clauser and Horne achieve no further generality, since they can be mimicked by a deterministic theory.

tation $\vec{b}$ of the remote polarizer II, and vice-versa, nor does $\rho(\lambda)$ (i.e. the way in which pairs are emitted) depend on the orientations $\vec{a}$ and $\vec{b}$. Bell's Inequalities no longer hold if we don't make the locality assumption (It is easy to see that the demonstration of § 3 fails with quantities such as $A(\lambda,\vec{a},\vec{b})$ or $\rho(\lambda,\vec{a},\vec{b})$).

As an abstract of this discussion, we can say that Bell's theorem states a conflict between Local Supplementary Parameters Theories and certain Quantum Mechanical predictions. It yields a quantitative criterion for this conflict, that will allow us to design sensitive experiments.

6 - GEDANKENEXPERIMENT WITH VARIABLE ANALYZERS : THE LOCALITY CONDITION AS A CONSEQUENCE OF EINSTEIN'S CAUSALITY.

In static experiments, in which the polarizers are held fixed for the whole duration of a run, the Locality Condition must be stated as an assumption. Although highly reasonable, it is not prescribed by any fundamental physical law. To quote J. Bell "the settings of the instruments are made sufficiently in advance to allow them to reach some mutual rapport by exchange of signals with velocity less than or equal to that of light". If such interactions existed, the Locality Condition would no longer hold for static experiments, nor would Bell's Inequalities.

Bell thus insisted upon the importance of "experiments of the type proposed by Bohm and Aharonov [4], in which the settings are changed during the flight of the particles"*. In such a timing-experiment, the locality condition would become a consequence of Einstein's Causality that prevents any faster-than-light influence.

As shown in our 1975 proposal[9], it is sufficient to switch each polarizer's orientation between two particular settings ($\vec{a}$ and $\vec{a}'$ for I, $\vec{b}$ and $\vec{b}'$ for II). It then becomes possible to test experimentally a larger class of Supplementary Parameters Theories: those obeying Einstein's Causality. In such theories, the response of polarizer I at time t is allowed to depend on the orientation $\vec{b}$ (or $\vec{b}'$) of II at time $t - L/C$ (L being the distance between the polarizers). A similar retarded dependence is considered for the way in which pairs are emitted at the source (characterized by the supplementary parameters distribution). For random switching times, with both sides uncorrelated, the predictions of these more general theories are constrained by generalized Bell's Inequalities[9].

On the other hand, it is easy to show that the polarization correlations predicted by Quantum Mechanics depend only on the

* The idea was already expressed in Bohm's book[3].

orientations $\vec{a}$ or $\vec{a}'$ and $\vec{b}$ or $\vec{b}'$ at the very time of the measurements, and do not involve any retardations terms such as L/C. For a suitable choice of the set of orientations $(\vec{a},\vec{a}',\vec{b},\vec{b}')$ - for instance the sets displayed in Fig. 4 - the Quantum Mechanical predictions still conflict with generalized Bell's Inequalities.

Such a timing-experiment with variable analyzers would thus provide a test of Supplementary-Parameters-Theories, obeying Einstein's Causality, versus Quantum Mechanics.

7 - GENERAL CONSIDERATIONS FOR A REAL SENSITIVE EXPERIMENT

Sensitive situations are seldom

Quantum Mechanics has been so much upheld in a great variety of experiments that Bell's Theorem might appear as an impossibility proof of supplementary parameters. However, situations in which this conflict arises (sensitive situations) are rare ; in 1965 none had been realized.

Bell's Inequalities obviously constrain the whole classical physics, i.e. Classical Mechanics and Classical Electrodynamics, which can be expressed according to the formalism (6) and (6'). (For instance, in Classical Mechanics, we can take as λ the initial positions and velocities...). Moreover, in a situation involving two correlated measurements onto two separated subsystems, Quantum Mechanics will very seldom predict a violation of Bell's Inequalities. Without being exhaustive, we can point out to important necessary conditions for a sensitive experiment (according to Quantum Mechanics) :

(i) : the two subsystems must be in a non-factorizing state, such as a singlet state for two spin 1/2 particles, or the similar state (1) for two photons ;

(ii) : for each subsystem, it must be possible to choose the measured quantity among at least two non-commuting observables (such as polarization measurements along directions $\vec{a}$ and $\vec{a}'$ neither parallel nor perpendicular).

As a matter of fact, these are stringent conditions.

Time conditions

As we have seen, the Locality Condition may be derived from Einstein's Causality, if the experiment fulfils some requirements, that can be split in two conditions :

(i) : the measurements onto the 2 subsystems are space-like separated ;

(ii) : the choices of the quantities measured on each subsystem are made at random, and are space-like separated from the measurement on the opposite side. It is obviously much more difficult to fulfil the second condition.

Production of pairs of photons correlated in polarization

As pointed out by C.H.S.H.[6], pairs of photons emitted in suitable atomic radiative cascades are good candidate for a sensitive test. Consider for instance a $(J = 0)\rightarrow(J = 1)\rightarrow(J = 0)$ cascade, in the singlet states of an alkaline earth (Fig. 6). Suppose that we select, with the use of wavelengths filters and collimators, two plane waves of frequencies ν_1 and ν_2 propagating along $-\vec{Oz}$ and $+\vec{Oz}$ (Fig. 7.)

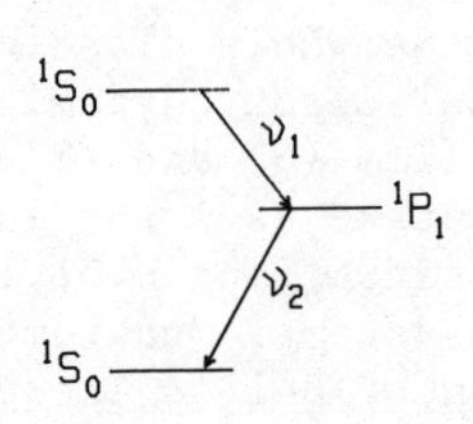

Fig. 6. Radiative cascade emitting pairs of photons correlated in polarization. that only even isotopes can be used[10]).

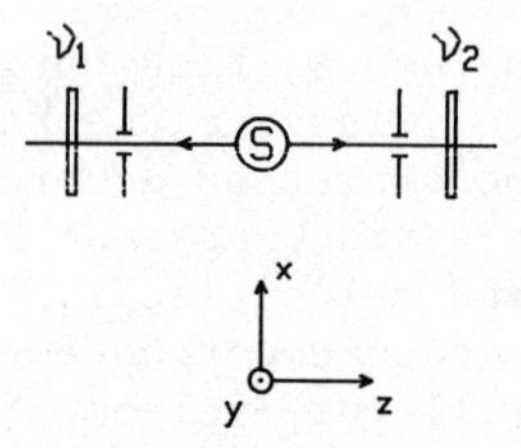

Fig. 7. Ideal configuration (infinitely small solid angles).

It is easy to show, by invoking parity and angular momentum conservation, that the polarization part of the state vector describing the pair (ν_1,ν_2) can be written :

$$(1/\sqrt{2})\left[|R,R\rangle + |L,L\rangle\right] \tag{11}$$

where R and L are circularly polarized states. By expressing $|R\rangle$ and $|L\rangle$ on a linear polarization basis, we obtain the state (1)

$$|\Psi(\nu_1,\nu_2)\rangle = (1/\sqrt{2})\left[|x,x\rangle + |y,y\rangle\right]$$

We know that such a pair is a good candidate for a sensitive experiment, since corresponding Quantum Mechanical predictions violate Bell's Inequalities.

Real experiment

A real experiment differs from the ideal one in several respects. For instance, the light should be collected in finite solid angles, as large as possible (Fig. 8). One can show[10] that the contrast of the correlation function then decreases, since (5) is replaced by :

$$E_{MQ}(\vec{a},\vec{b}) = F(\vec{u}).\cos 2(\vec{a},\vec{b}) \qquad (12)$$

where $F(\vec{u}) \leqq 1$.

Fig. (9) displays F(u) for a $0 \rightarrow 1 \rightarrow 0$ alkaline-earth cascade (with no hyperfine structure). Fortunately, one can use large angles without great harm. For u = 32° (our experiments), F(u) = 0.984.

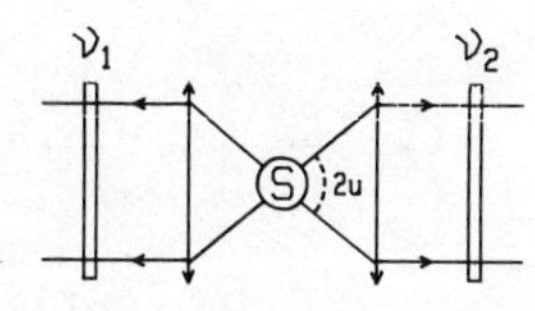

Fig. 8. Realistic configuration, with finite solid angles.

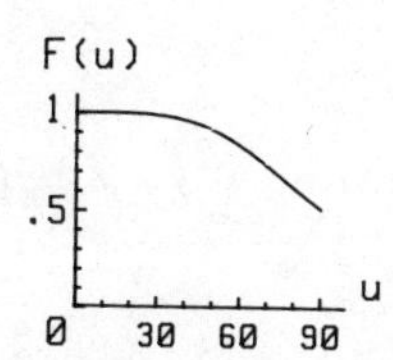

Fig. 9. F(u) for a 0 -1 - 0 cascade.

All other inefficiencies - polarizers defects, accidental birefringences etc... - will similarly lead to a decrease of the correlation function $E(\vec{a},\vec{b})$. The function $S_{MQ}(\Theta)$ (Fig. 5) is then multiplied by a factor less than 1, and the conflict with Bell's Inequalities decreases, or even vanishes.

Therefore, an actual experiment must be carefully designed and every auxiliary effect must be evaluated. Everything must be perfectly controlled since one can assume that a forgotten effect would similarly lead to a decrease of the conflict (one knows for instance that hyperfine structure dramatically decreases F(u), so that only even isotopes can be used [10]).

8 - PREVIOUS EXPERIMENTS (1970-1976)[5,11]

The C.H.S.H. paper[6] in 1969 had shown the possibility of sensitive experiments in atomic physics. Two groups began to build an experiment. Following the C.H.S.H. proposition, they used a simpler experimental scheme, involving one-channel polarizers.

Experiments with one channel polarizer

In this simplified experimental scheme, one uses polarizers that transmit light polarized parallel to $\vec{a}$ (or $\vec{b}$), and blocks the orthogonal one. One thus only detects the + results, and the coincidence measurements only yield $N_{++}(\vec{a},\vec{b})$.

Auxilliary runs are performed with one or both polarizers removed (we denote ∞ the "orientation" of a removed polarizers). We can write relations such as :

$$N(\infty,\infty) = N_{++}(\vec{a},\vec{b}) + N_{+-}(\vec{a},\vec{b}) + N_{-+}(\vec{a},\vec{b}) + N_{--}(\vec{a},\vec{b})$$

$$N_{++}(a,\infty) = N_{++}(\vec{a},\vec{b}) + N_{+-}(\vec{a},\vec{b})$$

etc...

By substitution into inequalities (9), one gets new B.C.H.S.H. inequalities

$$-1 \leq S' \leq 0 \tag{13}$$

with

$$S' = (1/N(\infty,\infty)) \left[N(\vec{a},\vec{b})-N(\vec{a},\vec{b}')+N(\vec{a}',\vec{b})+N(\vec{a}',\vec{b}')-N(\vec{a}',\infty)-N(\infty,\vec{b})\right]$$

(we omitted the subscripts ++)

For the same orientation sets as previously (Fig. 4), the Quantum Mechanical predictions violate ineq. (13) :

$$S'^{Max}_{MQ} = \frac{\sqrt{2}-1}{2} \qquad \text{for} \qquad \Theta = \pi/8$$

$$\tag{14}$$

$$S'^{Min}_{MQ} = \frac{-\sqrt{2}-1}{2} \qquad \text{for} \qquad \Theta = 3\pi/8$$

The derivation of ineq. (13) requires a supplementary assumption. Since the detection efficiencies are low (due to small angular acceptance and low photomultipliers efficiencies), the probabilities involved in the $E(\vec{a},\vec{b})$ (Eq. (4)) must be redefined on the ensemble of pairs that would be detected with polarizers removed. This procedure is valid only if one assumes a reasonable hypothesis about the detectors. The C.H.S.H. assumption states that, "given that a pair of photons emerges from the polarizers, the probability of their joint detection is independent of the polarizer orientations" (or of their removal) [6]. Clauser and Horne have exhibited another assumption[7], leading to the same inequalities.*

Results

In the Berkeley experiment (Clauser and Freedman[12]), the $4p^2\ ^1S_0$ - $4s4p\ ^1P_1$ - $4s^2\ ^1S_0$ cascade of Calcium was excited by ultraviolet absorption towards a $^1P_1$ upper state. Since the signal was weak, and spurious cascades occurred, it took more than 200 hours of measurement for a significant result. The experiment upheld Quantum Mechanics, and violated inequalities (13) by several standard deviations.

At the same time, in Harvard, Holt and Pipkin[11] found a result in disagreement with Quantum Mechanical predictions, and in agreement with Bell's Inequalities. They excited the $9^1P_1 \rightarrow 7^3P_1 \rightarrow 6^3P_0$ cascade in Mercury 200 by an electron beam. The data accumulation lasted 150 hours.

Clauser[13] repeated their experiment in Mercury 202. He found an agreement with Quantum Mechanics, and a violation of Bell's Inequalities.

In 1976, in Houston, Fry and Thompson[14] used the $7^3S_1 \rightarrow 6^3P_1 \rightarrow 6^3S_0$ cascade in Mercury 200. Their selective excitation involved a C.W. single-line-laser. The signal was several order of magnitude larger than in previous experiments, allowing them to collect the data in a period of 80 minutes. Their result was in excellent agreement with Quantum Mechanics and violated generalized Bell's inequalities by 4 standard deviations.

9 - ORSAY EXPERIMENTS (1980-1982)

The source

Since our aim was to use more sophisticated experimental schemes, we had first to build a high-efficiency and very stable and well controlled source. This was carried out (Fig. 10) by a two-

* Although these assumptions are reasonable, let us mention that there exist supplementary-parameters theories that do not obey them. From the view point of supplementary-parameters theories, there is no way for experimentally testing these assumptions[5].

photon-excitation of the $4p^2\ ^1S_o$ - $4s4p\ ^1P_1$ - $4s^2\ ^1S_o$ cascade of calcium[15]. This cascade is very well suited to coincidence counting experiments since the lifetime τ_r of the intermediate level is rather short (5ns). If one can reach an excitation rate of about $1/\tau_r$, then an optimum signal-to-noise ratio for this cascade is attained.

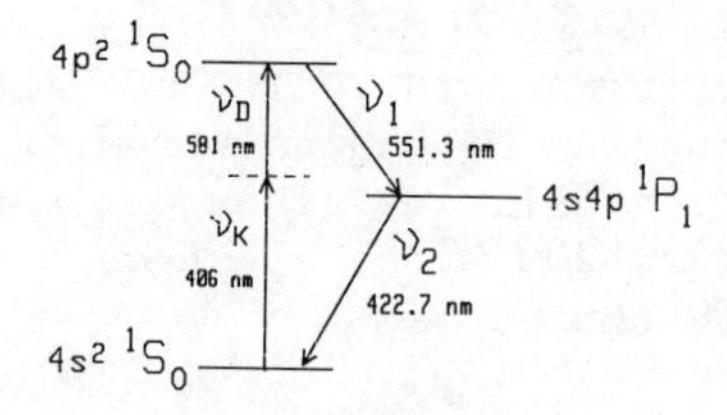

Fig. 10. Two-photon excitation of the chosen cascade in Calcium.

We have achieved this optimum rate with the use of a Krypton laser (λ_K = 406.7 nm) and a dye laser (λ_D = 581 nm) tuned to resonance for the two-photon process. Both lasers are single-mode operated. They have parallel polarizations.

They are focused onto a Calcium atomic beam (laser beam waists about 50 μm).

Two feedback loops provide the required stability of the source (better than 0.5 % for several hours) : the first loop controls the wavelength of the tunable laser to ensure the maximum fluorescence signal ; a second loop controls the power of one laser and compensates all the fluctuations.

With a few tens of milliwatts from each laser, the cascade rate is about $N = 4 \times 10^7\ s^{-1}$. An increase beyond this rate would not significantly improve the signal-to-noise ratio for coincidence counting, since the accidental coincidence rate increases as N^2, while the true coincidence rate increases as N.

Detection - Coincidence counting

The fluorescent light is collected by large-aperture aspherical lenses, followed by a set of lenses and the polarizers.

The photomultipliers feed the coincidence-counting electronics, that includes a time-to-amplitude converter and a multichannel analyzer, yielding the time-delay spectrum of the two-photon detections (Fig. 11). This spectrum involves a flat background due to accidental coincidences (i.e. between photons emitted by different atoms). True coincidences yield a peak around the null-delay, with an exponential decrease (time constant τ_r).

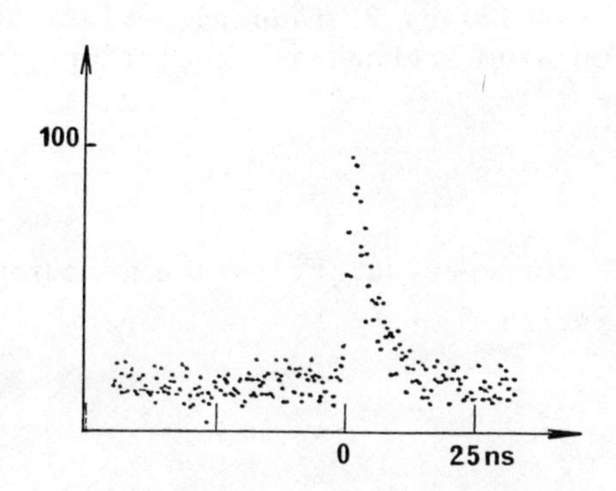

Fig. 11. Time-delay spectrum. Number of detected pairs as a function of the delay between the detections of two photons.

The true-coincidence signal is thus taken as the signal in the peak.

Additionally, a standard coincidence circuit with a 19 ns coincidence window monitors the rate of coincidences around null delay, while a delayed-coincidence channel monitors the accidental rate. It is then possible to check that the true coincidence rate obtained by substraction is equal to the signal in the peak of the time-delay spectrum.

In the second and third experiments, we have used a fourfold coincidence system, involving a fourfold multichannel analyzer and four double-coincidence circuits. The data were automatically gathered and processed by a computer.

Experiment with one-channel polarizers[16]

Our first experiment was carried out using one-channel-pile--of-plates polarizers, made of ten glass-plates at Brewster angle.

Thanks to our high-efficiency source, the statistical accuracy was better than 2 % in a 100 s run (with polarizers removed). This allowed us to perform various checks.

The test of Bell's inequalities has yielded

$$S'_{exp} = 0.126 \pm 0.014 \tag{15}$$

violating inequalities (13) by 9 standard deviations, and in good agreement with the Quantum Mechanical predictions (for our polarizers and solid angles) :

$$S_{MQ} = 0.118 \pm 0.005$$

(this error accounts for uncertainty in the measurements of the polarizers efficiencies).

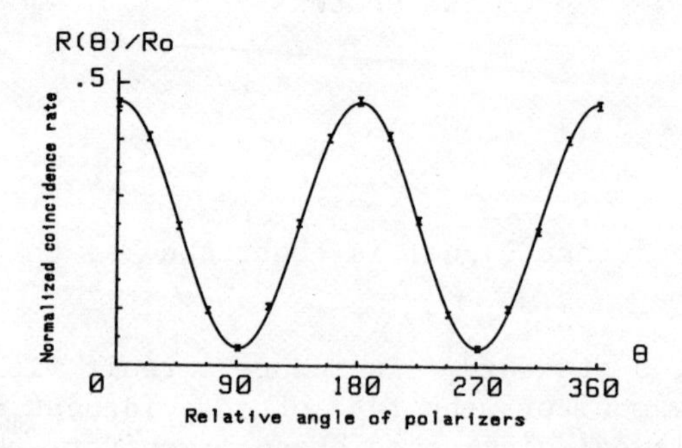

Fig. 12. Experiment with one channel polarizers : Normalized coincidence rate as a function of the relative polarizers orientation. Indicated errors are ± 1 standard deviation. The solid curve is not a fit to the data but the prediction by Quantum Mechanics.

The agreement between the experimental data and the Quantum Mechanical predictions has been checked in a full 360° range of orientations (Fig. 12).

In order to fulfil the first time-condition (§7) we have repeated these measurements with the polarizers at 6.5 m from the source. At such a distance (four coherence-lengths of the wave packet associated with the lifetime τ_r) the detection events are space-like separated. No modification of the experimental results was observed.

Experiment with two-channel analyzers[17]

With single-channel polarizers, the measurements of polarization are inherently incomplete. When a pair has been emitted, if no count is obtained at one of the photomultipliers, there is no way to know if "it has been missed" by the detector or if it has been blocked by the polarizer (only the later case corresponds to a result $-$ for the measurement). This is why one had to resort to auxiliary experiments, and indirect reasoning, in order to test Bell's inequalities.

With the use of two-channel polarizers, we have performed an experiment following much more closely the ideal scheme of Fig. 1.* Our polarizers were polarizing cubes transmitting one polarization (parallel to $\vec{a}$, or respectively to $\vec{b}$) and reflecting the orthogonal one. Such a polarization splitter, and the two corresponding photomultipliers, are mounted in a rotatable mechanism. This device (polarimeter) yields + and - results for linear polarization measurements along $\vec{a}$ (respectively $\vec{b}$). It is an optical analog of a Stern-Gerlach filter for spin 1/2 particles.

With polarimeters I and II in orientations $\vec{a}$ and $\vec{b}$, and the fourfold coincidence counting system, we are able to measure in a single run the four coincidence rates $R_{\pm\pm}(\vec{a},\vec{b})$. We then get directly the correlation coefficient for the measurement along $\vec{a}$ and $\vec{b}$:

$$E(\vec{a},\vec{b}) = \frac{R_{++}(\vec{a},\vec{b}) + R_{--}(\vec{a},\vec{b}) - R_{+-}(\vec{a},\vec{b}) - R_{-+}(\vec{a},\vec{b})}{R_{++}(\vec{a},\vec{b}) + R_{--}(\vec{a},\vec{b}) + R_{+-}(\vec{a},\vec{b}) + R_{--}(\vec{a},\vec{b})} \qquad (16)$$

It is then sufficient to repeat the same measurement for three other orientations, and the B.C.H.S.H. inequality (9) can directly be tested.

* A similar experiment, using calcite polarizers, has been undertaken at the University of Catania, Italy.

This procedure is sound if the measured values (16) of the correlation coefficients can be taken equal to the definition (4), i.e. if we assume that the ensemble of actually detected pairs is a faithful sample of all emitted pairs. This assumption is very reasonable with our very symmetrical scheme, where the two measurements +1 and −1 are treated in the same way (the detection efficiencies in both channels of a polarimeter are equal). Moreover, we have checked that the sum of the four coincidence rates $R_{\pm\pm}(\vec{a},\vec{b})$ is constant when changing the orientations, although each rate strongly varies. The size of the selected sample of pairs is thus found constant.

The experiment has been done at the set of orientations of Fig. 4, for which the greatest conflict is predicted. We have found

$$S_{exp} = 2.697 \pm 0.015 \qquad (17)$$

violating the inequalities (9) ($|S| \leq 2$) by more than 40 standard deviations ! This result is in excellent agreement with the predictions by Quantum Mechanics (for our polarizers and solid angles) :

$$S_{MQ} = 2.70 \pm 0.05$$

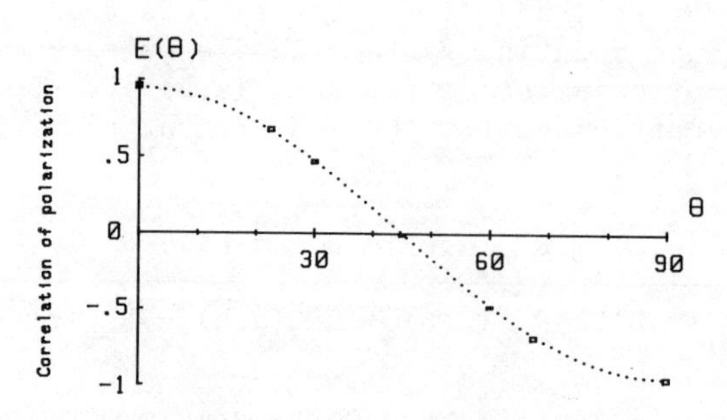

Fig. 13. Experiment with two-channels polarizers : Correlation of polarizations as a function of the relative angle of the polarimeters. The indicated errors are ± 2 standard deviations. The dashed curve is not a fit to the data, but Quantum Mechanical predictions for the actual experiment. For ideal polarizers, the curve would reach the values ± 1.

The uncertainty of S_{MQ} accounts for a slight lack of symmetry of both channels of a polarizer (± 1%). The effect of these dissymmetries has been computed and cannot create a variation of S_{MQ} greater than 2%.

We have also performed measurements of $E(\vec{a},\vec{b})$ in various orientations, for a direct comparison with the predictions of Quantum Mechanics (Fig. 13). The agreement is clearly excellent.

Timing experiment[18]

As stressed in § 6, an ideal E.P.R. type experiment would involve the possibility of switching a random times the orientation of each polarizer. We have done a step towards such an ideal experiment by using the modified scheme displayed in Fig. 14.

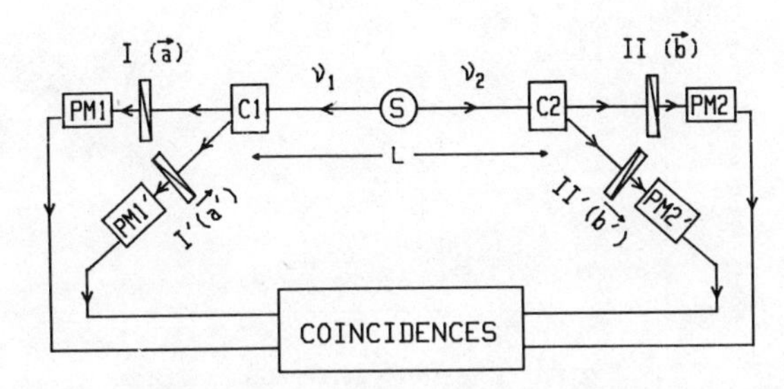

Fig. 14. Timing-experiment with optical switches. (C_1 and C_2). A switching occurs each 10 ns. The two switches are independently driven.

Each (single-channel) polarizer is replaced by a setup involving a switching device followed by two polarizers in two different orientations : $\vec{a}$ and $\vec{a}'$ on side I, $\vec{b}$ and $\vec{b}'$ on side II. The optical switch is able to rapidly redirect the incident light from one polarizer to the other one. Each setup is thus equivalent to a variable polarizer switched between two orientations. The distance L between the two switches is 12 m.

The switching of the light is effected by acousto-optical interaction of the light with an ultrasonic standing wave in water. The incidence angle (Bragg angle) and the acoustic power are adjusted for a complete switching between the 0th and 1st order of diffraction. At an acoustical frequency of 25 MHz, the

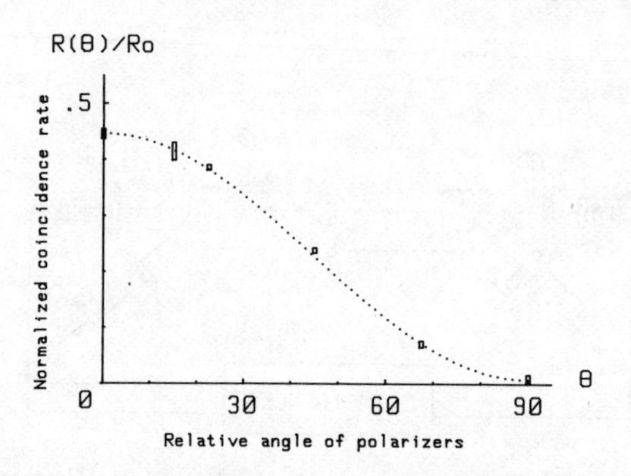

Fig. 15. Timing experiment : average normalized coincidence rate as a function of the relative orientation of the polarizers. Indicated errors are ± 1 standard deviation. The dashed curve is not a fit to the data but the predictions by Quantum Mechanics for the actual experiment.

switching frequency is 50 MHz. A change of orientation of the equivalent variable polarizer then occurs each 10 ns. Since this period (10 ns) as well as the lifetime τ_r (5 ns) are small compared to L/C (40 ns), a detection event on one side and the corresponding change of orientation on the other side are separated by a space-like interval. The second time-condition is thus partially fulfilled.

With the large beams used in the experiment, the commutation was not complete, since the incidence angle was not exactly the Bragg angle. Instead of being 0, the minimum of transmitted light in each channel was 20 %.

Since we had to reduce the divergence of the beams, the detected coincidence rates were weaker by an order of magnitude than in our previous experiments. Accordingly, the duration of data accumulation was longer.

The test of Bell's Inequalities (13) involves a total of 8 000 s of data accumulation with the 4 polarizers in the orientations of Fig. 4. A total of 16 000 s was devoted to auxilliary calibration measurements with half or all polarizers removed.

In order to compensate the effects of systematic drifts, the data accumulation was alternated between the various configurations each 400 s. The average yields

$$S'_{exp} = 0.101 \pm 0.020 \qquad (18)$$

violating inequ. (13) by 5 standard deviations, and in good agreement with the Quantum Mechanics predictions

$$S_{MQ} = 0.113 \pm 0.005$$

Another run has been carried out for a direct comparison with Quantum Mechanics. Fig. 15 exhibits an excellent agreement.

According to these results, Supplementary-Parameters Theories obeying Einstein's Causality seem to be untenable. To escape this conclusion, one might argue that the switching was not complete. However, a large fraction of the pairs undergoes forced switching. If Bell's Inequalities were obeyed by these pairs, it is hard to believe that we would not have observed a significant discrepancy between our results and the Quantum Mechanical predictions.

Our experiments differs from the ideal scheme in another respect : the switching are not truly at random, since the acousto-optical switches are driven by quasi-periodic generators. Nevertheless, the two generators on the two sides function in a completely uncorrelated way, especially considering their frequency drifts.

10 - CONCLUSION

Our last experiment (timing-experiment), as well as the previous ones, has some technical imperfections. Some loopholes thus remain open for the advocates of Supplementary-Parameters Theories obeying Einstein's Causality. Improved experiments will probably become feasible in the future[19], but we already have an impressive agreement with Quantum Mechanics. Supplementary Parameters Theories obeying Einstein's Causality and compatible with our results appear somewhat artificial, since the experimental results would have to change dramatically (disagreement with Quantum Mechanics) with certain technical improvements(such as an increase of the efficiencies of the photomultipliers)*.

According to Bell[20], we are thus forced to admit :

(i) either that there are, at the level of the supplementary parameters, faster-than-light influences**;

(ii)or to renounce an explanation in terms of supplementary parameters.

The second position seems a priori more confortable. But, to quote Mermin[22], "I challenge the reader to suggest any... other way to account for what happens" (i.e. the observed strong correlations).

I hope that even those who are not committed to such discussions will be convinced that Einstein has pointed out one of the most extraordinary property of Quantum Mechanics. We must thank J. Bell to have provided us with the possibility of experimentally evidencing this property.

---

* Another far-fetched issue is to admit that the two switches, although they look randomly driven (in an ideal experiment) are in fact correlated with each other, and also with the pairs. We then have to admit that the whole world is completely entangled, and that there is no possibility of a free choice of what we decide to measure !

**We must emphasize that in a timing experiment - even ideal - these hypothetical faster-than-light influences cannot be controlled for practical telegraphy[21].

REFERENCES

1. J.S. Bell, On the Einstein-Podolsky-Rosen Paradox, Physics 1 : 195 (1964).
J.S. Bell, Introduction to the Hidden-Variable Question, in : "Foundations of Quantum Mechanics", B. d'Espagnat ed., Academic, N.Y. (1972).

2. A. Einstein, B. Podolsky and N. Rosen, Can Quantum-Mechanical description of physical reality be considered complete ?, Phys. Review 47 : 777 (1935). See also Bohr's answer :
N. Bohr, Can Quantum-Mechanical description of physical reality be considered complete ?, Phys. Rev. 48 : 696 (1935).

3. D. Bohm, "Quantum Theory", Prentice-Hall, Englewoods Cliffs, N.J. (1951).

4. D. Bohm and Y. Aharonov, Discussion of Experimental Proof for the paradox of Einstein, Rosen and Podolsky, Phys. Rev. 108 : 1070 (1957).

5. J.F. Clauser and A. Shimony, Bell's Theorem : Experimental Tests and Implications, Rep. Progr. Phys. 41 : 1881 (1978).

6. J.F. Clauser, M.A. Horne, A. Shimony and R.A. Holt, Proposed experiment to test local hidden-variable theories, Phys. Rev. Lett. 23 : 880 (1969).

7. J.F. Clauser, and M.A. Horne, Experimental consequences of objective local theories, Phys. Rev. D 10 : 526 (1974).

8. A. Fine, Hidden Variables, Joint Probability, and the Bell Inequalities, Phys. Rev. Lett. 48 : 291 (1982).

9. A. Aspect, Proposed Experiment to Test Separable Hidden-Variable Theories, Phys. Lett. 54 A : 117 (1975).
A. Aspect, Proposed Experiment to test the nonseparability of Quantum Mechanics, Phys. Rev. D 14 : 1944 (1976).

10. E.S. Fry, Two-Photon Correlations in Atomic Transitions, Phys. Rev. A 8 : 1219 (1973).

11. F.M. Pipkin, Atomic Physics Tests of the Basic Concepts in Quantum Mechanics, in : "Advances in Atomic and Molecular Physics", D.R. Bates and B. Bederson, ed., Academic (1978).

12. S.J. Freedman and J.F. Clauser, Experimental test of local hidden-variable theories, Phys. Rev. Lett. 28 : 938 (1972).

13. J.F. Clauser, Experimental Investigation of a Polarization Correlation Anomaly, Phys. Rev. Lett. 36 : 1223 (1976).

14. E.S. Fry, and R.C. Thompson, Experimental Test of Local Hidden-Variable Theories, Phys. Rev. Lett. 37 : 465 (1976).

15. A. Aspect, C. Imbert, and G. Roger, Absolute Measurement of an Atomic Cascade Rate Using a Two Photon Coincidence Technique. Application to the $4p^{2}\,{}^{1}S_{0}$ - $4s4p\ {}^{1}P_{1}$ - $4s^{2}\,{}^{1}S_{0}$ Cascade of Calcium excited by a Two Photon Absorption, Opt. Comm. 34 : 46 (1980).

16. A. Aspect, P. Grangier and G. Roger, Experimental Tests of Realistic Local Theories via Bell's Theorem, Phys. Rev. Lett. 47 : 460 (1981).

17. A. Aspect, P. Grangier and G. Roger, Experimental Realization of Einstein-Podolsky-Rosen-Bohm Gedankenexperiment : A New Violation of Bell's Inequalities, Phys. Rev. Lett. 49 : 91 (1982).

18. A. Aspect, J. Dalibard and G. Roger, Experimental Test of Bell's Inequalities Using Variable Analyzers, submitted to Phys. Rev. Lett.

19. T.K. Lo, and A. Shimony, Proposed Molecular Test of Local Hidden-Variables Theories, Phys. Rev. A 23 : 3003 (1981)

20. J.S. Bell, Atomic-cascade Photons and Quantum-Mechanical Nonlocality, Comments on Atom. Mol. Phys. 9 : 121 (1980).

21. A. Aspect, Expériences basées sur les Inégalités de Bell, J. Physique Colloque C2 : 63 (1981).

22. N.D. Mermin, Bringing home the atomic work : Quantum mysteries for anybody, Am. J. Phys. 49 : 940 (1981).

RELATIVISTIC EFFECTS IN MANY-BODY SYSTEMS

Lloyd Armstrong, Jr.
Physics Division
National Science Foundation
Washington, DC 20550
and
Department of Physics
Johns Hopkins University
Baltimore, Maryland 21218

INTRODUCTION

The introduction of relativity into many-body systems produces a number of new effects, and a number of new problems. First among the latter might be the difficulty in defining what a "relativistic effect" might be. As we shall discuss below, relativity in a many-body system necessarily implies the use of QED. Thus, any attempt to differentiate between "relativistic effects" and "QED effects" must be rather arbitrary, and not necessarily desirable. At the opposite extreme, many "relativistic effects" can be approximately incorporated into nonrelativistic calculations through introduction of familiar operators such as spin-orbit, spin-other-orbit, etc., into the Schrodinger equation. Thus the wisest procedure seems to be to not specifically define "relativistic effects" at the beginning, but rather to plunge right in and let the discussion itself provide a rough definition.

I should begin by giving some reasons why "relativistic effects in many-body systems" is a timely subject for discussion. The basic physics that I will dicuss has been known for many years, of course. There have been changes recently, however, in our ability to investigate computationally the implications of these theories, and in our ability to study experimentally systems where relativistic effects are prominent. On the computational side, it is only relatively recently that computer codes for relativistic atomic structure calculations[1,2] have become widely available. These codes tend to be rather complicated and time consuming to run, and it is again only recently that a significant number of researchers have had access to computers of sufficient power to enable them to carry out systematic studies. The fusion program, with its great need for parameters for highly ionized atoms, has added a great stimulus to these studies, particularly in the United States. On the experimental side, high temparature plasmas, x-ray spectroscopy, and beam foil spectroscopy have been provided us with some relatively high precision data on structures of highly ionized atoms, and inner shell energies of atoms.[3] On the horizon is the new and very exciting field of ion trapping and cooling, which should lead to very high precision spectroscopic studies of ions. The resulting comparison of theory and experiment has already lead to numerous fairly difficult tests of our abilities to carry out relativistic and QED calculations in many-body systems, and even more exciting tests will be produced in the near future.

Let us turn now to our discussion of relativistic systems. In general, we shall assume that relativistic calculations are based on the Dirac equation.[4] Therefore, before discussing the many-body situation, it is worthwhile to briefly review the relativistic effects contained in the one electron Dirac equation for an electron moving in a central field:

$$(\vec{\alpha}\vec{p} + \beta m - eU(r))\,\Psi_{nljm} = E\,\Psi_{nljm} \qquad (1)$$

($\hbar=c=1$) The quantities $\vec{\alpha}$ and β are 4 x 4 matrices, and Ψ is a column matrix of four rows, which can be given more explicity as:

$$\Psi_{nljm} = \begin{pmatrix}\psi_{nljm}\\ \phi_{nl'jm}\end{pmatrix} \qquad (2)$$

with ψ and ϕ being column matrices of two rows each. Because $U(r)$ is central, both ψ and ϕ can be written as products of spin (χ), angular (Y), and radial (F) functions:

$$\psi_{nljm} = \Sigma C(\tfrac{1}{2}lj|m_s m_l m_j)\,\chi_{\frac{1}{2}m_s} Y_{lm_l}(\theta,\phi)\,F_{nlj}(r) \qquad (3)$$

where $\chi_{\frac{1}{2}}$ is a two component spinor. ϕ_{nljm} has a similar description, except that the angular part is associated with an orbital momentum $l' = l \pm 1 = j \pm \frac{1}{2}$. That is, j, not l, is a good quantum number relativistically. In addition, ϕ contains a second radial function $G_{nlj}(r)$, and the two radial functions F(r) and G(r) satisfy two coupled first order differential equations. In the nonrelativistic limit, F_{nlj} becomes the radial function for the corresponding Schrodinger equation. The function G_{nlj} is generally the smaller of the two functions, being roughly of order $(Z\alpha)F_{nlg}$.

One problem is that (1) has negative energy solutions (particles of energy $\backsim$ -m) in addition to the perfectly reasonable positive energy solutions which correspond to particles of energy of the order of m. Dirac, of course, had to interpret all of these negative energy states as being filled under "normal conditions" in order to prevent all of the positive energy electrons from falling into them.[4] A hole in this filled negative energy sea is now thought of as a positron. This interpretation makes the predictions of Dirac's equation agree with the world as we see it, but the fact that this information is not contained in the mathematical expression, but rather must be added as a written "codicil" can lead to some confusion when one tries to define a two (or more) electron Dirac equation.

Before moving on to this more complicated two body problem, let us look at a few more properties of the simple one-electron equation. If we expand the central field Dirac equation roughly in powers of $Z\alpha$, we find a number of familiar interactions. In order $(Z\alpha)^2$, one finds a p^4/m^3 mass correction term, the Darwin term that shifts s states, and the spin-orbit interaction $-(e/2m^2r)(dU/dr)\vec{s}\cdot\vec{l}$. If we look at the special hydrogenic case, U=Ze/r, we can get a rough idea of what these lower order relativisic terms do to the wavefunction of an atom or ion. We find that the spin-orbit interaction causes j-dependent shifts in the one-electron energy, and that the mass correction term causes an l-dependent shift. Adding the Darwin term to these two corrections, one finds that the energy is given approximately by

$$E = m - \frac{mZ^2e^4}{2n^2}\left[1 + \frac{Z^2e^4}{n^2}\left(\frac{n}{j+\frac{1}{2}} - 3/4\right)\right] \qquad (4)$$

That is, the energy is lowest for the $j = 1-\frac{1}{2}$ state, but it is lower than the nonrelativistic energy for both states of the doublet. The expectation value of r can be used to give us some feeling for changes in the spatial extent of the electron produced by relativity. One finds that the expectation value of r predicted using the hydrogenic Dirac equation[5] is less than that predicted using the hydrogenic Schrodinger equation, with the $j = 1-\frac{1}{2}$ state showing a larger contraction than the $j = 1+\frac{1}{2}$ state, i.e., relativity causes orbital contraction.

THE TWO-ELECTRON DIRAC EQUATION

Now let us move on to the case where two Dirac electrons interact with each other and with the nucleus. If we take the interaction between the two electrons to be Coulombic, we can write down a two electron Dirac equation:

$$\left[\sum_{i=1}^{2}\left(\vec{\alpha}_i\cdot\vec{p}_i + \beta_i m - \frac{Ze^2}{r_i}\right) + \frac{e^2}{r_{12}}\right]\Psi(1,2) = E\,\Psi(1,2) \qquad (5).$$

Brown and Ravenhall[5] pointed out that this equation, if solved, probably would have no stable bound states; in addition, the equation ignores something Dirac told us, and is consequently not good physics. Both of these statements can be understood if we imagine writing (5) as

$$\left[H_0 + \frac{e^2}{r_{12}}\right]\Psi(1,2) = E\ \Psi(1,2) \qquad (6)$$

and use perturbation theory based on the use of H_0 as the unperturbed Hamiltonian. The zeroth order wavefunction describing two positive energy electrons will obviously be a product of two positive energy hydrogenic wavefunctions corresponding to some total energy E_0. Exactly degenerate in total energy with this pair of positive energy electrons will be an infinite set of pairs composed of one bound or free positive energy electron, and one negative energy electron in a continuum state. As a consequence, the original pair of positive energy electrons can "autoionize" into a positive energy electron and a negative energy electron through the action of the Coulomb term e^2/r_{12}. We know, of course, that atoms with two electrons do exist, and are stable, at least with respect to this kind of decay, so this picture must not be correct. Indeed, to have gotten this picture we had to forget what Dirac said about negative energy states: this picture involves a positive energy electron falling into a negative energy state (pushed, actually, by e^2/r_{12}), which should not happen since the negative energy states are all filled. We can patch this up by rewritting a more nearly correct two electron Dirac equation as

$$\left[\sum_{i=1}^{2} (\vec{\alpha}_i\cdot\vec{p}_i + \beta_i m - \frac{Ze^2}{r_i}) + \Lambda_+ \frac{e^2}{r_{12}} \Lambda_+\right] \Psi(1,2) = \Psi(1,2) \qquad (7)$$

where Λ_+ is a positive energy projection operator, whose effect is to prevent either or both of the two positive energy electrons in our example from being turned into negative energy electrons. Equation (7) can be obtained in a rigorous way starting from QED.[7]

Having modified our Dirac equation, we should also require that Ψ contain only electrons, i.e.,

$$\Lambda_+\ \Psi = \Psi \qquad (8)$$

The question now arises as to how to define this positive energy projection operator. Obviously the easiest way is to write

$$\Lambda_+ = \sum_{n+} \prod_{i=1}^{2} |\psi_{n+}(r_i))\ (\psi_{n+}(r_i)|$$

where n+ is a positive energy solution to some one-electron Dirac equation. However, since the positive energy space resulting from the use of one potential is different from that resulting from another potential, the ultimate "correctness" of the two electron Dirac equation will be a function of the choice of one electron wavefunctions for use in Λ_+. We shall discuss below the choice of Λ_+ in the Hartree-Fock case.

Let us now consider what physics is left out of the two electron Dirac equation of eq(7). To really describe the two body interaction, we should go to QED. QED starts with the one vertex interaction between a field and a current; the corresponding Fock space Hamiltonian is[8]

$$H = \int j^{\mu} A_{\mu} d^4x \tag{9}$$

where*

$$j^{\mu} = -e : \bar{\psi}\gamma^{\mu}\psi : \tag{10}$$

The ψ are Fock space operators which can be defined by

$$\psi = \sum_{n+} b_n u_n (x,t) + \sum_{n-} c_n v_n (x,t) \tag{11}$$

where $u_n(x)$ and $v_n(x)$ are, respectively, positive and negative energy solutions to some one-electron Dirac equation, and b^+_n and c^+_n are creation operators for the corresponding electron and positron states, etc. Again, we see that the form the theory takes depends on the solution to the problem.

The simplest way to turn (9) into a two body equation is to assume that A more-or-less satisfies the classical equation relating a field to a current:

$$\Box A_{\mu} = -4\pi j_{\mu}$$

To do this we rewrite (10) as

$$j^{\mu} = \Sigma\, j^{\mu}_{nm}(\vec{x},t) : a^+_n a_m : \tag{12}$$

where

$$a_n = c^+_n \quad \text{if } n=n-$$

$$a_n = b_n \quad \text{if } n=n+$$

*Actually, a better form would be $\frac{-e}{2} \left[\bar{\psi}\gamma^{\mu}, \psi\right]$, but this seems slightly more cumbersome to work with, and would not change the discussion which follows.

and assume that the desired current can be written as

$$A_\mu = \Sigma A_\mu^{nm} (\vec{x},t): a^+_n a_m: \tag{13}$$

Then, we require that

$$\Box A_\mu^{nm} = -4\pi j_\mu^{nm} \tag{14}$$

i.e., that A_μ be a field produced by the current j_μ. To proceed, we must make some choice of gauge. I will consider two different gauges, Coulomb and Feynman, since the gauge dependence of our results will be rather informative.

Solution of (14) in the Feynman gauge is particularly simple. In this gauge, all four components of A_μ are treated equivalently. Then, one immediately obtains

$$A_\mu^{nm} (\vec{x},t) = \int \frac{j_\mu^{nm}(\vec{y}, t \pm |\vec{x}-\vec{y}|)}{|\vec{x}-\vec{y}|} d^3y \tag{15}$$

where the retarded solution (outgoing wave) is used if the transition $m \to n$ corresponds to the emission of a photon ($E_m > E_n$), with the advanced solution (incoming wave) being used for the opposite situation.[8] Combining (9), (12), (13), and (15), one finds

$$H = \int \mathcal{H} \, d^3x \, d^3y \tag{16}$$

where

$$\mathcal{H} = \tfrac{1}{2} \Sigma H^F(n,m;p,q):a^+_n a_p: :a^+_m a_q: \tag{17}$$

Here,

$$H^F(n,m;p,q) = e^2 \phi^*_n(x) \, \phi^*_m(y) \, \frac{(1-\vec{\alpha}_1 \cdot \vec{\alpha}_2)}{R} e^{iwR} \, \phi_p(x)\phi_q(y)$$

$$\equiv \phi^*_n(x)\phi^*_m(y) h^F \, \phi_p(x) \, \phi_q(y) \tag{18}$$

with $R = |\vec{x}-\vec{y}|$, and $w = |E_m - E_q|$; $\phi_n(x)$ is the configuration space wavefunction corresponding to a_n, etc.

In the Coulomb gauge, we have $\vec{\nabla} \cdot \vec{A} = 0$, and

$$\nabla^2 A_0 = -4\pi j_0$$

leading to the "instantaneous" Coulomb interaction

$$A_0^{nm}(\vec{x},t) = \int \frac{j_0^{nm}(\vec{y},t)}{R} d^3y = -e \int \frac{\phi^*_n(x)\phi_m(y)}{R} d^3y$$

The longitudinal component of A is, of course, zero in this gauge, and the transverse components are given by

$$A_t^{nm}(\vec{x},t) = \int \frac{j_t^{nm}(\vec{y},t \pm R)}{R} d^3y$$

Again, combining all the pieces, we find that

$$= e^2\phi^*_n(x)\phi^*_m(y) \left[\frac{1}{R} - \frac{\vec{\alpha}_1\cdot\vec{\alpha}_2}{R} e^{iwR} + \frac{(\vec{\alpha}_1\cdot\vec{\nabla}_1)(\vec{\alpha}_2\cdot\vec{\nabla}_2)(e^{iRw}-1)}{RW^2} \right] \phi_p(x)\, \phi_q(y)$$

$$\equiv \phi^*_n(x)\, \phi^*_m(y)\, h^C\, \phi_p(x)\, \phi_q(y) \qquad (19)$$

QED is a gauge invariant theory, so both eqs (18) and (19) should lead to equivalent Hamiltonians. However, if one must truncate the space of functions ϕ_n in which the operators are expanded, this will obviously destroy the gauge invariance of the theory. Since such a truncation is always necessary in practice, we are faced with the fact that our final results are going to depend to some degree on the particular gauge chosen.

There are many different "interactions" contained in $\mathcal{H}$, depending on the mix of electrons and positrons represented by the quantum numbers n, m, p, and q. There is even a part of $\mathcal{H}$ which is a one body operator; this part is obtained when the product of two normal ordered pairs of operators appearing in (17) is turned into a single normal ordered product of four operators (which is a two-body operator).

For our purposes, the most important interaction contained in $\mathcal{H}$ corresponds to the case in which all of the particles involved are electrons. In this case, the one-body part of $\mathcal{H}$ corresponds to part of the self energy of the electron and to part of the Lamb shift (which may or may not be the same thing, depending on your nomenclature). I shall define the Lamb shift as being almost outside the scope of this talk, and so shall give no details concerning its form. I shall, however, consider the magnitude of this effect later on.

Returning to the form of the two-body part of $\mathcal{H}$ in the "all electron" case, we see that generally $wR \ll 1$, thus allowing us to expand the exponential in (19) to a good approximation. "Retardation terms" are those which disappear when we take the lowest approximation, $e^{iwR} = 1$. In that limit, one finds that $h^C = h^F$:

$$h^C = h^F = e^2(1 - \vec{\alpha}_1 \cdot \vec{\alpha}_2)/R \qquad (20)$$

This is called the Gaunt interaction. Obviously, if n=p and m=q in (19) and (20), w=0 and the retardation terms are identicially equal to zero. That is, for diagonal terms, $h^C = h^F$, and the Gaunt interaction is the correct two electron interaction. Keeping the next term in the expansion of e^{iwR}, we find that the next contribution to h is

$$\frac{e^2}{2R}\left(\vec{\alpha}_1 \cdot \vec{\alpha}_2 - \frac{(\vec{\alpha}_1 \cdot \vec{R})(\vec{\alpha}_2 \cdot \vec{R})}{R^2}\right) \qquad \text{Coulomb gauge} \qquad (21)$$

$$-e^2 w^2 R/2 \qquad \text{Feynman gauge}$$

Obviously it is the Coulomb gauge which leads to the familiar Breit interaction, obtained by adding the second term in (20) to the expression obtained in (21):

$$H_B = -\frac{e^2}{2R}\left(\vec{\alpha}_1 \cdot \vec{\alpha}_2 + \frac{\vec{\alpha}_1 \cdot \vec{R}\, \vec{\alpha}_2 \cdot \vec{R}}{R^2}\right) \qquad (22)$$

In the limit of eqs (21), h^C has the benefit of being an operator, whereas h^F is not purely an operator since it depends on the wavefunctions through w.

We can make the connection between the QED Hamiltonian given above and the two electron Dirac equation given previously by noting that restricting the creation and annihilation operators of $\mathcal{H}$ (eq(17)) to refer only to electrons is the Fock space equivalent of the configuration space operators Λ_+. Thus, we see that keeping only the $\Lambda_+(e^2/R)\Lambda_+$ term in the two electron Dirac equation is equivalent in the Coulomb gauge (eq(19)) to dropping the contribution from the exchange of a transverse photon between the two electrons; in the Feynman gauge (eq(18) it is equivalent to dropping effects of both retardation and exchange of vector photons. We also see that all of these effects could be introduced into our two electron Dirac equation by the simple expedient of replacing $\Lambda_+(e^2/R)\Lambda_+$ by the two-body parts of either $\Lambda_+ h^C \Lambda_+$ or $\Lambda_+ h^F \Lambda_+$. The resulting equation would have the defect that it would no longer be Hamiltonian-like due to the retardation terms. However, because of the projection operators Λ_+, one could expand h^C as above, obtaining the Breit interaction and

Table I

Mercury $K_{\alpha 1}$ x-ray energy (eV)[9]

	E(1s)	E(2p)	E(1s)-E(2p)
Dirac Fock, point nucleus	-451193.9	-532479.0	71285.2
Finite nucleus	+77.8	+131.2	-53.4
Breit interaction	+313.5	+581.3	-267.8
Transverse correction	-7.5	-10.9	+3.4
Self energy	+282.6	+481.1	-198.5
Vacuum polarization	-64.7	-108.7	+44.1
	-450592.2	-521405.1	70812.9
experiment			70819.0

thus a Hamiltonian form for the resulting two electron Dirac equation. That is, it is perfectly correct to put the Breit interaction into our two electron Hamiltonian so long as it is bracketed by Λ_+ projection operators.

Suppose that we have put $\Lambda_+ H_B \Lambda_+$ into the two electron Hamiltonian. What is then left out in terms of interactions? What is primarily left out is any effect which involves creation or destruction of a positron-electron pair, or pairs. For example, any effect described by a second order S matrix with an extra positron-electron pair existing in the intermediate state is not included in this Hamiltonian. Specifically excluded (perhaps for the best from a computational standpoint) are all of the familiar "divergent" terms such as electron self energy, vacuum polarization, etc. All of these effects can, however, be treated perturbatively using the two electron Dirac equation as a starting point.

Before discussing some approaches to relativistic many-body calculations, I would like to quote a few results so that we can get some idea concerning the magnitudes of the various terms discussed in this section. Table I shows results of a calculation by Grant and McKenzie[9] of energies of two states of Hg having inner shell holes. The "Breit" contribution is obtained by evaluating H_B perturbatively. The "transverse" contribution is the change produced in this quantity when the corresponding operator from eq(19) is evaluated, that is, when the transverse contribution is evaluated keeping the full effects of retardation. "Self energy" and "vacuum polarization" are the equivalent of the Lamb shift, with the former being obtained by using an effective charge with Mohr's hydrogenic results,[10] the latter coming from an approximation of Fullerton and Rinker.[11] We see that the Breit interaction provides about 0.1% of the total binding energy; the retardation correction is only about 5% of the Breit term. Note that this result is for energetic inner shell states, where one might expect the retardation correction to be greatest. Note also the size of the Lamb shift relative to that of the Breit interaction. This is consistent with results of Desclaux et al.[12] shown in Table II, where some energy levels of FeXXI are shown. One sees that the Lamb shift contributes about ½ angstrom to a typical transition wavelength in this case. This is clearly a significant number when high precision studies of such levels are carried out. Cheng and Johnson[13] have pointed out that the self energy (Lamb shift) grows faster than $(Z\alpha)^4$ in the region Z >100, and that it is already about 0.3% of the binding energy of the 1s electron at Z=100. In addition, they also demonstrated the importance in a heavy atom of calculating the self energy using Dirac-Fock rather than Coulomb wavefunctions, and of including finite nuclear size effects in the wavefunctions. Thus, overall, it is clear that QED effects (such as the self energy) are for many quite "ordinary" situations, as important as the smaller relativistic effects (such as the Breit interaction).

MANY-BODY EFFECTS

Having found a two electron Hamiltonian, we can easily extend it to the n-electron case by replacing $\sum_{i=1}^{2}$ by $\sum_{i=1}^{n}$, and $1/r_{12}$ by $\sum_{i<j} 1/r_{ij}$. As pointed out by several authors, there will also be 3-, 4- etc., body operators introduced by our forcing the Hamiltonian to act only in the space of electrons.[14] These more-than-two-body operators lead, however, to very small corrections and we shall not consider them further.

The simplest relativistic many-body calculation one can do is probably a Hartree or Hartree-Fock type calculation using the many electron Hamiltonian obtained by extending eq. (7) as described above. Many such calculations have been done, but without including the projection operators Λ_+. Why is it that these calculations seem to give reasonable results where the Brown-Ravenhall analysis would seem to tell us they should not? The answer lies in the interative way in which these ostensibly many body equations are solved. One really solves the one electron Dirac equation for each individual electron moving in a field produced by the electron distribution found in the previous iteration. The solutions to the one electron Dirac equation are, of course, well behaved, and only the positive energy solutions are kept in preparing the next iteration. When convergence is finally obtained, one has, in effect, solved the many body equation using projection operators constructed with the solutions of the many body equation. Not only does this intuitively seem to be a good way to define the Λ_+, but Mittleman[15] has shown that using projection operators obtained from a Hartree-Fock

Table II
Selected energy levels of Fe XXI (in cm^{-1}).[12]

Configuration	State	MCDF	Breit	Lamb shift	Total
$2s^2 2p^2$	$^3P_0$	0	0	0	0
	$^3P_1$	76789	-2834	22	73977
	$^1D_2$	253596	-6981	-121	246494
$2s2p^3$	$^5S_2$	481080	-3405	-4646	473029
	$^3D_1$	784303	-1192	-4896	778215
	$^3D_3$	815957	-6562	-4835	804560
	$^3P_2$	953210	-4994	-4820	943396
	$^1P_1$	1283066	-4841	-5252	1272975
$2p^4$	$^3P_0$	1756182	-691	-8276	1737215
	$^3P_1$	1741262	-2335	-8252	1750675
	$^1D_2$	1846959	-4035	-8257	1834667

equation has another benefit: Use of such projection operators produces an energy eigenvalue of the many body equation which is stable with respect to small changes in the projection operators.

One of the interesting relativistic effects which has been seen in this simple type of calculation has to do with the average value of the radius of the different orbitals in a many electron atom compared to their nonrelativistic values. Generally, one finds that the "inner" electrons in a many electron atom--most often s or p orbitals--contract as a result of relativity just as in the hydrogenic atom (see Introduction). This contraction increases the shielding of the nucleus as seen by the outer electrons, resulting in an expansion of the orbitals of outer electrons. Thus, the simple hydrogenic picture of the effects of relativity on orbital size is not really descriptive of what happens in a many-electron atom.

This change in orbital size has implications in a number of areas. For example, Manson[16] has shown that in the photoionization of the 6p subshell in atoms, one finds that the branching ratios and the β parameters which determine the photo-electron angular distribution are influenced strongly by this effect. What happens is that the $6p_{\frac{1}{2}}$ is "inner" and contracts significantly compared to a nonrelativisitic 6p orbital, while the $6p_{3/2}$ is almost "outer" and does not change much. The Cooper-Seaton minima corresponding to the $p_{1/2} \rightarrow \varepsilon d$ and $p_{3/2} \rightarrow \varepsilon d$ channels are correspondingly pulled apart, with the $p_{1/2} \rightarrow \varepsilon d$ minimum being moved to much higher energy. This separation alters the β parameters and branching ratios.

A relativistic Dirac-Hartree-Fock calculation is somewhat more complicated than the corresponding nonrelativistic calculation due to the fact that each wavefunction has a large and a small component. Thus for the n electron problem there are 2n coupled equations in the relativistic calculation rather than n as in the nonrelativistic calculation. There is an even more severe complication however, produced by the fact that each nonrelativistic (nl) orbital corresponds to two relativistic orbitals $(n,l,j = l+\frac{1}{2})$ and $(n,l,j = l-\frac{1}{2})$ (except of course, if $l = 0$). Consequently, what is a one configuration Hartree-Fock (HF) calculation nonrelativistically usually corresponds to a multi-configuration Hartree-Fock (MCHF) relativistically. What this implies is that a single configuration Hartree-Fock calculation is usually less likely to give accurate results in the relativistic case than in the nonrelativistic case.

One can, of course, use MCHF techniques to study correlation effects both relativistically and nonrelativistically. Again, the large number of relativistic states which "correspond" to a particular nonrelativistic state may limit the amount of real correlation that can be included in a relativisitc MCHF. This problem is particularly severe for open shell atoms.

Another difficulty which has recently surfaced with respect to the relativistic MCHF has to do with the relationship between the two relativistic orbitals corresponding to a given nl.[17,18] In principal, as $c \to \infty$, these two orbitals should become the same; however, tests show that they do not when obtained with commonly used codes. This energy error introduced in the $c = \infty$ limit by the appearance of these two independent nl orbitals seems to remain unchanged when c is returned to a more seemingly value, thus providing a receipe for correcting the relativistic results. In this way Huang et al.[18] have recently been able to correctly evaluate a number of doublet separations which had previously been quite poorly predicted by multiconfiguration relativistic codes.

Despite these drawbacks, the relativistic MCHF has been extensively used in relativistic studies. An example of an area in which numerous such calculations have been done is the study of transition energies and oscillator strengths along isoelectronic sequences. A relatively old but rather typical result[19] is shown in Fig. 1, showing for the Be isoelectronic sequence oscillator strengths for both the allowed $^1P_1 - {}^1S_0$ transition and the $^3P_1 - {}^1S_0$ intercombination line. The "allowed" oscillator strength starts high, then drops rapidly for increasing Z, finally turning around and increasing again. This behavior is produced by two effects. First, the line strength is a steadily decreasing function of Z, primarily produced by the contraction of the atom with increasing Z. In fact, in this case, the nonrelativistic calculation shows exactly the same decrease, so this is not really a relativistic effect. What is relativistic is the huge increase in the energy of the $^1P_1 - {}^1S_0$ transition, which begins about Z = 32. This is primarily a spin-orbit type of effect on the energy and consequently on the oscillator strength. The spin orbit also greatly affects the line strength for the intercombination line, since it mixes in a small amount of the allowed transition. At around Z=50, one finally begins to see a relativistic change in the line strengths, as the relativistic orbitals begin to contract faster than the nonrelativistic calculation would show. This effect is strong enough to cause a flattening in the intercombination line oscillator strength. Generally, what one finds is that for $\Delta n=0$, transitions, the transition energy is what really shows the relativistic effects. For $\Delta n=1$ transitions, on the other hand, relativity plays a much smaller role in the determination of the transition energy, and as a result the effect of relativity on the line strength and on the transition energy is more nearly the same.

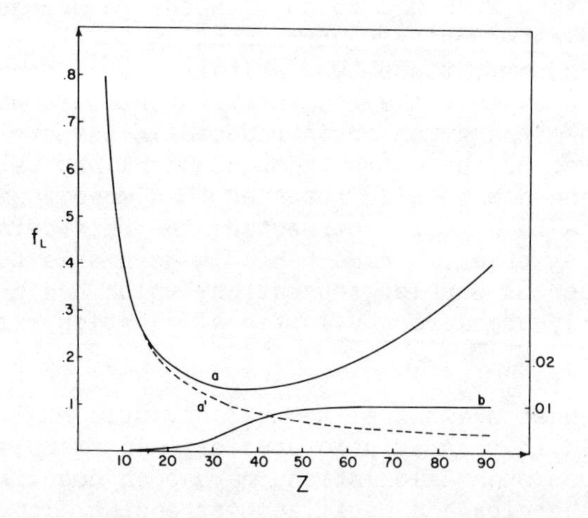

Fig. 1. Oscillator strengths in the Be isoelectronic sequence. The curve marked a is the result of a relativistic "$^1S_0$ - $^1P_1$" calculation, with a' being the result of a nonrelativistic calculation. The curve marked b is the result of a relativistic "$^1S_0$ - $^3P_1$" calculation.

What is shown in Fig. 1 is the oscillator strength in the "length" gauge. As has been described in detail by Grant,[20] the relativistic transition operators which correspond in the nonrelativistic limit to the length and velocity operators are related to each other by a gauge transformation of the photon operator. The MCHF proceedure is not gauge invariant, basically because a truncation has been carried out in the sum over creation and annihilation operators in the definition of the many-electron

Hamiltonian. Consequently, the different gauges will lead to different results for oscillator strengths. As a practical matter, the length gauge usually seems to give more nearly correct results.

In general, most of the methods which have been used to study many body systems nonrelativistically have been extended so that they may be used for relativistic studies. Diagramatic many-body perturbation theory, for example, has been used for several relativistic studies. Although the methods are essentially unchanged by addition of relativity, again the calculations are more complicated due to the two component nature of the wavefunction, the doubling of the number of orbitals that need to be used in a calculation due to the fact that j rather than l is a good quantum number, and the need (at least in principal) to allow for pair creation in intermediate states. This approach has been used extensively by Das and coworkers[21] in studies of hyperfine structure, and by several authors in studies of weak neutral current effects in atoms.[22] These latter calculations have graphically demonstrated to a wide audience how difficult it is to do relativistic many-body calculations.

Another nonrelativistic many-body theory which has a relativistic extension is the random phase approximation. This theory is discussed in detail by W. Johnson in this volume, so we will not consider it further, other than to remark that one possible attraction of this technique is that oscillator strengths it produces are gauge invariant.

As we remarked at the beginning, one can in many cases do quite reasonable semi-relativistic calculations by starting with the Schrodinger equation and adding the various fine structure interactions as perturbations. This approach usually works so long as the primary relativistic effect is to cause intermediate coupling or a shift in the energy, but is not so useful when the wavefunctions of the individual electrons begin to be significantly altered by the effects of relativity. Very roughly, this seems to become important for electrons moving in a field corresponding to a Z greater than about 30.

HOW GOOD ARE THE CALCULATIONS?

In many cases, one can make very accurate relativistic calculations of atomic parameters. One area in which rather extensive comparisons have been made is that of inner shell transitions. Figure 2 was taken from a recent report by Mohr.[23] It shows a comparison of theoretical[13, 24] and experimental[25] $K\alpha_1$ x-ray wavelenghts. The circles and squares show $(T_0-E)/T_0$, where T_0 is the theoretical energy not including self energy, and E is the experimental energy. The remaining curves show

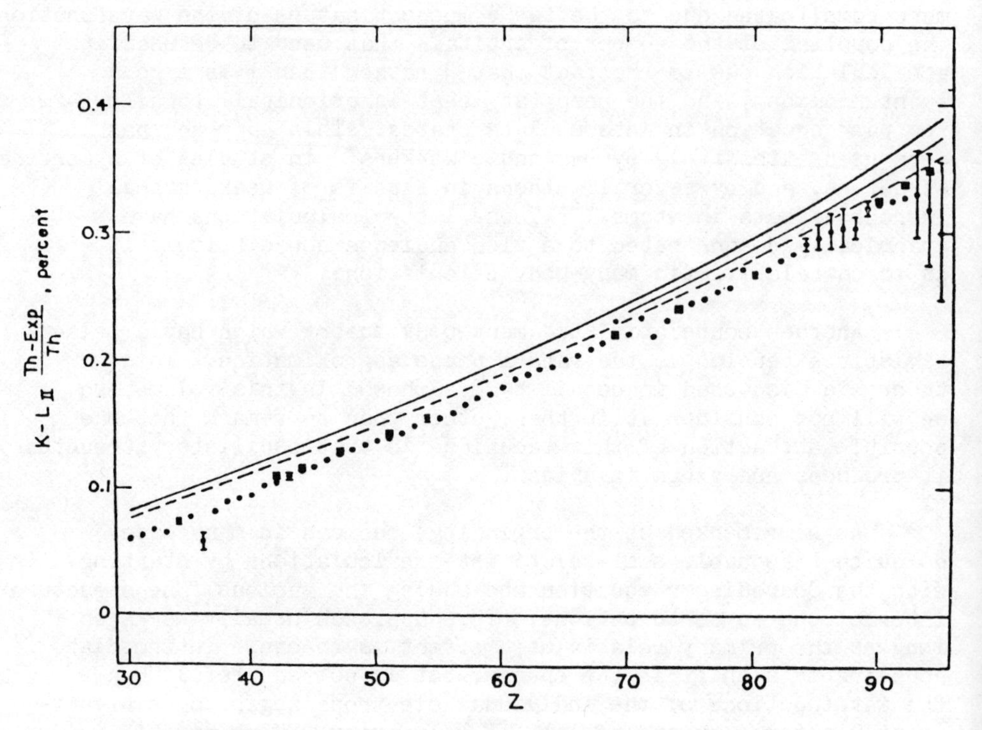

Fig. 2. Comparison of theory and experiment for the energy of the $K-L_{II}$ x-ray energy in heavy atoms.

$-T_s/T_0$, where T_s is the self energy contribution in several approximations.[5] Since T, the total calculated energy, is equal to T_0+T_s, in the limit of complete agreement between theory and experiment (T=E) the curve of $(T_0-E)/T_0$ would coincide with that of $-T_s/T_0$. The solid line shows T_s in the hydrogenic limit,[10] the dashed curve used the hydrogenic T_s with an effective Z,[24] and the second solid line from Z=70 to 95 shows the Hartree Fock self energy of Cheng and Johnson.[13] Overall, agreement is obviously excellent, with less than about .01% error. Other x-rays, involving higher shells, are not generally predicted so accuratly, however. Hypersatellite lines are also generally not predicted quite so well as the main line. For example for the $K\alpha_1$ line of Hg (Table I), the calculated energy is only six eV off for a line of energy 70819 eV; for the hypersatellite, the calculated value differs from the experimental 71964 eV by +26 eV.

Another area where several good comparisons between theory and experiment have now been made is in highly stripped ions. For example, for the eighteen levels of $(2s+2p)^4$ in FeXXI (Table II), the difference between theory and experiment is 1% or less.[12] Figure 3, taken from a paper by Reader and Luther,[26] shows the difference between observed and calculated[27] wavelenghts of the $4s^2\ ^1S_0-4s4p\ ^1P_1$ line in the Zn isoelectronic sequence. Again, the agreement is at the 1% or better level.

Not all results are, of course, as good as the ones shown here. These are, however, examples of what careful calculations can produce under favorable conditions. "Favorable" conditions generally implies that correlation effects are produced by a fairly small number of states, e.g., states within the complex for small n. "Unfavorable" conditions would have to include situations where there is a significant amount of correlation arising from a large number of configurations, or where the complex contains too large a number of states to be considered as a whole. In other words, at this point it appears that the major difficulty in relativistic many body calculations lies not in the "relativistic" aspect, but rather in the "many-body" part.

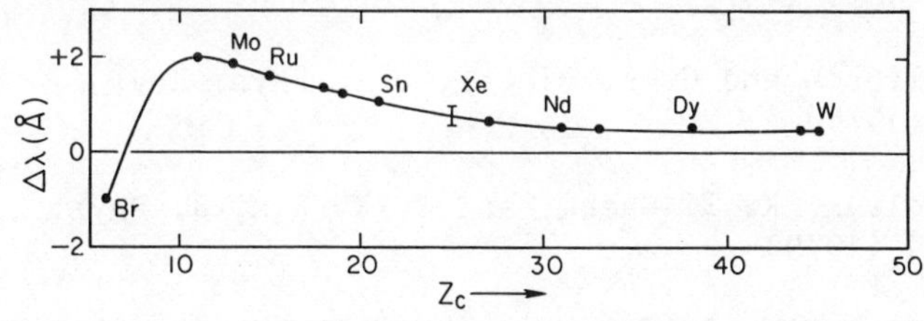

Fig. 3. Difference between observed and calculated wavelengths of the $4s^2\ ^1S_0$ - $4s4p\ ^1P_1$ resonance line in the Zn isoelectronic sequence.[26]

REFERENCES

1. J. P. Desclaux, Comp. Phys. Comm. 9, 31 (1975)

2. I. P. Grant, B. J. McKenzie, P. H. Norrington, D. F. Mayers, and N. C. Pyper, Comp. Phys. Comm. 21, 207 (1980); B. J. McKenzie, I. P. Grant, and P. H. Norrington, Comp. Phys. Comm. 21, 233 (1980)

3. Recent reviews of the experimental situation have been given, for highly ionized atoms by W. Martin, and for inner shell energies, by R. D. Deslattes and E. G. Kessler. Both reviews can be found in the Proceedings of the Workshop on the Foundations of the Relativistic Theory of Atomic Structure, ed by H. G. Berry, K. T. Cheng, W. R. Johnson, and Y.-K. Kim, Argonne National Lab. Report ANL--80-126 (1980)

4. P. A. M. Dirac Quantum Mechanics (Oxford University Press, London) 4th ed 1958.

5. R. H. Garstang and D. F. Mayers, Proc. Camb. Phil. Soc. 62, 777 (1966); V. M. Burke and I. P. Grant, Proc. Phys. Soc. 90, 297 (1967)

6. G. E. Brown and D. G. Ravenhall, Proc. Roy Soc. London A 208, 532 (1951)

7. G. E. Brown, Philos. Mag. 43, 467 (1952) J. Sucher, Phys. Rev. A 22, 348 (1980) M. H. Mittleman, Phys. Rev. A 24, 1167 (1981)

8. A. I. Akheizer and V. B. Berestetskii, Quantum Electrodynamics (Interscience, New York) 1965

9. I. P. Grant and B. J. McKenzie, J. Phys B 13, 2671 (1980)

10. P. J. Mohr, Ann. Phys. (NY) 88, 52 (1974); Phys. Rev. Lett. 34, 1050 (1975)

11. L. N. Fullerton and G. A. Rinker, Jr., Phys. Rev. A 13, 1283 (1976)

12. J. P. Desclaux, K. T. Cheng, and Y.-K. Kim, J. Phys. B 12, 3819 (1979)

13. K. T. Cheng and W. R. Johnson, Phys. Rev. A 14, 1943 (1976)

14. H. Promakoff and T. Holstein, Phys. Rev. 55, 218 (1938); C. Chanmugam and S. S. Schweber, Phys. Rev. 1, 1369 (1970) M. H. Mittleman, Phys. Rev. A 4, 893 (1971)

15. M. H. Mittleman, as reported at the Workshop on the Foundation of the Relativistic Theory of Atomic Structure, Argonne National Lab.; Dec. 4-5, 1980

16. S. Manson, Private communication

17. C. P. Wood and N. C. Pyper, Mol. Phys. 41, 149 (1980); N. C. Pyper, S. J. Rose, and I. P. Grant, J. Phys. B 15, 1319 (1982)

18. K.-N. Huang, Y.-K. Kim, K. T. Cheng, and J. P. Desclaux, Phys. Rev. Lett. 48, 1245 (1982)

19. L. Armstrong, Jr., W. R. Fielder and D. L. Lin, Phys. Rev. A 14, 1114 (1976)

20. I. P. Grant, J. Phys. B 7, 1458 (1974)

21. See e.g., M. Vajed-Samii, S. N. Ray, T. P. Das, and J. Andriessen, Phys. Rev. A 20, 1787 (1979)

22. S. P. Carter and H. P. Kelly, Phys. Rev. Lett. 42, 966 (1979); A. M. Martensson, E. M. Henley, and L. Wilets, Phys. Rev. A 24, 308 (1981); M. J. Harris, L. E. Loving, and P. G. H. Sandars, J. Phys. B 11, L749 (1978)

23. P. J. Mohr, given at the NATO Advanced Institute on Relativistic Effects in Atoms, Molecules, and Solids, Vancouver, BC., August 1981. To be published in the Proceedings thereof (Plenum)

24. K.-N. Huang, M. Aoyagi, M. H. Chen, B. Crasemann and H. Mark, Atomic Data and Nuclear Data Tables 18, 243 (1976)

25. J. A. Bearden and A. F. Burr, Rev. Mod. Phys. 39, 125 (1967); R. D. Deslattes, E. G. Kessler, Jr., L. Jacobs, and W. Schitz, Phys. Lett. 71A, 411 (1979)

26. J. Reader and G. Luther, Phys. Rev. Lett. 45, 609 (1980)

27. P. Shorer and A. Dalgarno, Phys. Rev. A 16, 1502 (1977); P. Shorer, Phys. Rev. A 18, 1060 (1978)

RELATIVISTIC MANY-BODY CALCULATIONS

Walter R. Johnson

Department of Physics
Notre Dame University
Notre Dame, IN 46556 USA

I. INTRODUCTION

Inner electrons in atoms and ions of high nuclear charge, Z, move with velocities comparable to the velocity of light. Consequently, many-body calculations of structures and transitions for high Z atomic systems must account for the effects of relativity. Treating relativistic effects perturbatively is not satisfactory for heavy systems since the relevant expansion parameter, αZ ($\alpha = e^2/4\pi\hbar c \sim 1/137$), may be close to one. For such systems approaches based on the approximate Dirac-Coulomb hamiltonian[1]

$$H = \sum_i h_i + \sum_{i>j} \frac{e^2}{r_{ij}} \tag{1}$$

where

$$h = c\vec{\alpha}\cdot\vec{p} + (\beta - 1)mc^2 - \frac{e^2 Z}{r}$$

is a one-electron Dirac hamiltonian, have proved to be satisfactory.

There are formal difficulties which must be borne in mind when working with the hamiltonian of Eq. (1):

1. This hamiltonian does not follow from QED[1-4], even to lowest order in α.
2. The hamiltonian may not possess any normalizable eigenstates[5].

Despite these difficulties the use of H in connection with relativistic Hartree-Fock (HF) calculations [referred to below as Dirac-Fock (DF) calculations] have been fruitful. Basing DF calculations on H has been justified[6] on the grounds that the DF wave function provides an approximation to the no-pair wave function of QED. Two questions then arise concerning such DF calculations:

1. What are the correlation corrections required to give the exact no-pair wave function.
2. What are the leading QED corrections required to go beyond the no-pair approximations.

Progress in obtaining the answers to these questions has been significant during the past decade, but completely satisfactory answers to these questions have yet to be given.

In the paragraphs below we review some of the recent progress on relativistic many-body calculations which provide partial answers to the first of these questions and we also describe work on the Breit interaction and QED corrections which addresses the second question. We begin in Section II with a review of applications of the DF approximation to treat inner-shell problems, where correlation corrections are insignificant, but where the Breit interaction and QED corrections are important. Next, we discuss, in Section III, the multiconfiguration Dirac-Fock (MCDF) approximation which is a many-body technique appropriate for treating correlation effects in outer shells. Finally, in Section IV, we turn to applications of the relativistic random-phase approximation (RRPA) to treat correlation effects, especially in systems involving continuum states.

We will not discuss methods based on 1/Z expansion techniques[7-9], nor shall we discuss those employing parametric potentials[10], even though these approaches have led to interesting and important advances in the understanding of relativistic atomic systems.

II. DIRAC-FOCK METHOD

We construct a trial N-electron wave function as an antisymmetric produce of orthonormal one-electron Dirac orbitals $u_i(\vec{r})$:

$$\Psi_{DF} = A(u_1(1)u_2(2)\ldots u_N(N)) \tag{2}$$

The expectation value of H is determined using the trial wave function and is optimized with respect to the one-electron orbitals to give the DF equations

$$(h+V)u_i=\varepsilon_i u_i \qquad i=1,\ldots,N \tag{3}$$

where the DF potential V is given by

$$Vu = \sum_{i=1}^{N} e^2 \int \frac{d^3r'}{|\vec{r}-\vec{r}'|} \; [(u_i^\dagger u_i)'u-(u_i^\dagger u)'u_i] \tag{4}$$

For a closed-shell atom Eq. (3) has central-field, one-electron orbitals as solutions. If we consider an orbital with quantum numbers n,κ,m then the central field decomposition takes the form:

$$u_{n\kappa m}(\vec{r}) = \frac{1}{r}\begin{bmatrix} P_{n\kappa}(r)\ \Omega_{\kappa m}(\hat{r}) \\ iQ_{n\kappa}(r)\ \Omega_{-\kappa m}(\hat{r}) \end{bmatrix} \tag{5}$$

where $\Omega_{\kappa m}(\hat{r})$ is a spherical spinor. The large and small component radial functions $P_{n\kappa}(r)$ and $Q_{n\kappa}(r)$ satisfy coupled radial equations:

$$\begin{bmatrix} \varepsilon_{n\kappa} + \frac{e^2 Z}{r} - V_D & c\left(\frac{d}{dr} - \frac{\kappa}{r}\right) \\ -c\left(\frac{d}{dr} + \frac{\kappa}{r}\right) & \varepsilon_{n\kappa} + 2mc^2 + \frac{e^2 Z}{r} - V_D \end{bmatrix}\begin{bmatrix} P_{n\kappa} \\ Q_{n\kappa} \end{bmatrix} = \begin{bmatrix} X_{n\kappa} \\ Y_{n\kappa} \end{bmatrix}, \tag{6}$$

where V_D is the direct contribution to the DF potential and $X_{n\kappa}$, $Y_{n\kappa}$ are the exchange contributions. The DF eigenvalue $\varepsilon_{n\kappa}$ is a frozen-orbital approximation to the binding energy of an electron in

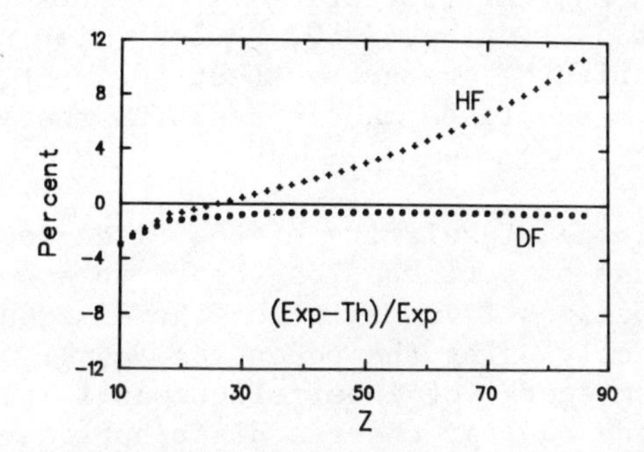

Fig. 1 Relative difference between frozen orbital calculations of K-shell binding energies and the measurements of Ref. 12. HF calculations are from Ref. 14.

subshell n,κ. Variational collapse of the radial DF equations is prevented by requiring that the radial functions $P_{n\kappa}$ and $Q_{n\kappa}$ be exponentially damped at large r. Equations (6) were originally considered by B. Swirles[11] in 1935; they are readily solved using numerical techniques similar to those employed in non-relativistic HF calculations.

To illustrate the importance of relativity in inner-shell physics, we plot in Fig. 1 the percentage difference between experimental[12] K-electron binding energies and frozen-orbital binding energies, determined from both HF[13] and DF[14] calculations, against nuclear charge. The non-relativistic and relativistic theoretical results agree with each other for low Z, but both calculations differ from the experimental K-binding energies, principally because of relaxation and correlation effects. As nuclear charge increases the error in the HF calculation changes sign and increases quadratically to about 12% at Z=90, due to the neglect of relativistic orbital contraction effects. The error in the DF calculation, on the other hand, decreases to an approximately uniform -0.5% for large Z. This residual small error in the DF calculation is due to the interplay of a number of factors. A careful analysis of these factors has led to a clearer understanding of the virtues and limitations of the DF method, so we will outline below the corrections to the DF eigenvalues which account for the discrepancy.

1. Finite Nuclear Size Effects

Nuclear size corrections are introduced into DF calculations by replacing the nuclear Coulomb potential in Eq. (6) by a potential $V_N(r)$ derived from a realistic nuclear charge distribution $\rho_N(r)$. The parameters defining the charge distribution are in turn determined from electron-nuclear scattering experiments[15] or from studies of muonic atoms[16]. For Hg, Z=80, nuclear size corrections reduce the K-shell binding energy by about 50eV out of a total of 83keV. Two comments should be made concerning these small corrections:

i) Because of the singularity of the Dirac-Coulomb orbitals at r=0, the use of perturbation theory to account for finite size corrections is unreliable; the DF equations must be solved exactly using the potential $V_N(r)$.

ii) The binding energy of inner-electrons is sensitive to the shape of the nuclear charges distribution. This later point was emphasized recently by Chen, et al.[17] who showed that the difference in K-binding energies calculated using a Fermi-distribution and a uniform distribution with the same RMS radius varied from 2 to 10 eV as Z varied from 80 to 100.

2. Relaxation Effects

As mentioned previously the DF eigenvalue is only a frozen-orbital approximation to the binding energy. Within the DF framework the binding energy is the difference between the total energy of the ion formed when an electron is removed from the atom and the total energy of the atom. If separate DF calculations are performed for the ion and atom one finds

$$E_{ion} - E_{atom} = |\varepsilon_k| - \Delta_k$$

where ε_k is a DF eigenvalue and Δ_k is a correction due to relaxation of the ionic electrons around the hole. This relaxation energy is about 90eV for the K-shell of Hg.

3. Breit Interaction

The correction to the instantaneous Coulomb interaction between two electrons due to the exchange of a transverse photon with wave number k can be described by the configuration space interaction[1,18]

$$b_{12}(k) = -e^2\left[\vec{\alpha}_1\cdot\vec{\alpha}_2\,\frac{\cos kr_{12}}{r_{12}} + \vec{\alpha}_1\cdot\vec{\nabla}_1\vec{\alpha}_2\cdot\vec{\nabla}_2\,\frac{(1-\cos kr_{12})}{\kappa^2 r_{12}}\right] \tag{7}$$

in calculations of direct two-electron matrix elements k=0 and $b_{12}(k)$ reduces to the familiar form given by Breit[1].

$$b_{12}(0) = -e^2\left[\frac{\vec{\alpha}_1\cdot\vec{\alpha}_2}{r_{12}} - \frac{(\vec{\alpha}_1\cdot\vec{\alpha}_2 - \vec{\alpha}_1\cdot\hat{r}_{12}\vec{\alpha}_2\cdot\hat{r}_{12})}{2\,r_{12}}\right] \tag{8}$$

$$= -\frac{e^2}{2r_{12}}\,(\vec{\alpha}_1\cdot\vec{\alpha}_2 + \vec{\alpha}_1\cdot\hat{r}_{12}\vec{\alpha}_2\cdot\hat{r}_{12}) \tag{9}$$

The first term in Eqns. (7) and (8) is referred to as the magnetic interaction while the second is called the retardation interaction. In calculations of exchange matrix elements $k=(\varepsilon_1-\varepsilon_2)/\hbar c$ where ε_1 and ε_2 are the orbital energies of the two interacting electrons. In multiconfiguration calculations where off-diagonal two-electron matrix elements are required there are two different wave numbers k_1 and k_2 which can be constructed from the four orbital energies. In this case one should make the replacement[3,6]

$$b_{12}(k) \rightarrow \frac{1}{2}\,(b_{12}(k_1)+b_{12}(k_2))\ . \tag{10}$$

Since the difference between $b_{12}(k)$ and the Breit interaction, $b_{12}(0)$, is of order $\alpha^2 Z^2$, which is a small number in many applications, it often suffices to replace $b_{12}(k)$ by $b_{12}(0)$ when computing exchange or off-diagonal matrix elements.

The Breit interaction treated as a first order perturbation decreases the K-binding energy of Hg by 303eV. Most of the contribution (328eV) is from the magnetic part of Eq.(9); the remainder (-25eV) is from the retardation interaction. Including the dependence of b_{12} on k increases the K-binding in Hg by another 6eV.

The Breit interaction can be included along with the Coulomb interaction in SCF calculation at the expense of a large increase in computational time and effort. For the specific case of K-binding in Hg the difference in the magnetic contribution computed as a perturbation and computed self-consistently is only 0.25 eV, a value well below the level of other uncertainties in DF calculations. Therefore, treating b_{12} as a perturbation appears to be well justified.

The relative size of the magnetic and retardation contributions to total energies along the periodic system is illustrated in Fig. 2 where the magnitude of the total DF energy of neutral atoms is compared to that of the magnetic and retardation energies. The DF and retardation energies are negative while the magnetic energy is positive.

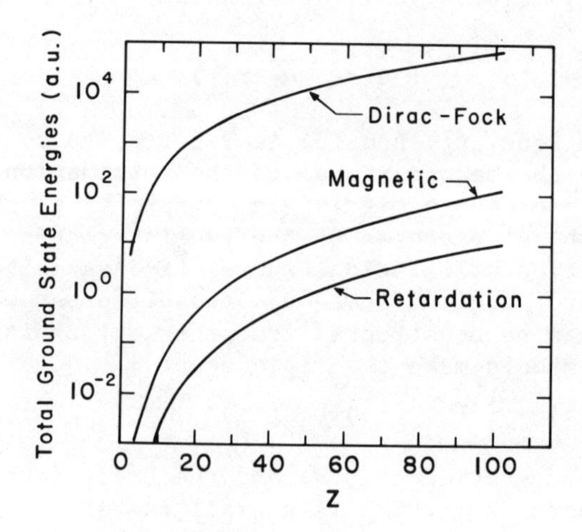

Fig. 2 Magnetic and retardation corrections to the Dirac-Fock ground state energies of neutral atoms.

4. QED Effects

a) Self-Energy

To lowest order in perturbation theory each electron may be considered as moving independently in a screened Coulomb field. In this approximation the atomic self-energy is the sum of one-electron contributions. The only available one-electron calculations of self-energy are those of Desiderio and Johnson[19], and those of Cheng and Johnson[20]; both of these calculations are restricted to 1s electrons in heavy atoms. More extensive self-energy calculations are available for the unscreened Coulomb field case from the work of Mohr[21], who considered 1s, 2s, $2p_{1/2}$ and $2p_{3/2}$ electrons for nuclear charges, Z, ranging from 10 to 110. In practical atomic calculations[22,23], the self-energy contributions for electrons with principal quantum numbers n=1 and 2 are often included by replacing Z in Mohr's results by an effective charge Z^*, which is chosen to account for electron screening. The prescription for choosing Z^* can be checked for 1s electrons by comparing with the available screened calculations.

The self-energy contribution for 1s electrons in heavy atoms is comparable to the Breit contribution, reducing the K-binding energy for Hg by 196±2eV. The uncertainty here reflects the various opinions on how best to treat screening corrections.

b) Vacuum Polarization

Including to lowest order the effects of vacuum polarization in SCF calculations is a simpler matter since one may express these corrections to lowest order in α and αZ as the expectation value of the Uehling potential[24], and the Uehling potential may be written in terms of the charge density of the screening field. Higher order corrections to the Uehling potential have been worked out by Wichmann and Kroll[25] and by Källén and Sabry[26]. While the Uehling contribution is commonly included as a correction in DF calculations, the relatively small higher order effects have been included in only a few tabulations[17,23]. The vacuum-polarization correction is of opposite sign to that of the self-energy, and about 20% as large. For the K-shell of Hg the vacuum-polarization contribution increases the binding by about 44 eV.

5. Summary of small corrections

In Table 1 we list the various contributions discussed above as determined in three different calculations. The electric contributions are from relaxed DF calculations and include both relaxation and finite nuclear size corrections. The differences in the electric contributions in the three calculations shown is mainly due to dif-

Table 1. Hg (Z = 80) K-Shell Binding Energy (eV)

Term	Mann & Johnson[18]	Desclaux[22]	Chen, et al.[17]
Electric	83562.0	83559.1	83560.0
Magnetic	-317.4	-327.9	-317.6
Retardation	20.2	24.4	20.2
$k \neq 0$	-	6.0	-
Self-Energy	-199.3	-197.0	-198.7
Vac.-Pol.	44.8	44.2	43.2
Corr. Vac.-Pol.	-	-	-1.7
Total	83110.3	83102.9	83105.6

ferent assumptions concerning the nuclear charge distribution parameters. The large differences in the magnetic and retardation terms occur because $b_{12}(0)$ rather than $b_{12}(k)$ was used in Ref. 22. The effects of higher order terms in kr_{12} are included separately in this calculation so that all three calculations agree well on the net effect of the Breit interaction. Major differences show up in the self-energy calculations because of differences in the treatment of screening. The final binding energies agree to three significant figures and can be compared with the experimental value of Bearden and Burr[12] 83102 eV which is subject to unknown chemical and surface corrections that probably increase the value by 6 or 8 eV.

Comparisons less sensitive to experimental uncertainties can be made with inner-shell x-rays, since one expects chemical or solid state effects to produce approximately equal shifts on all deap inner levels. Comparisons of this type of Chen, et al.[17], by Desclaux[27] and by Kessler, et al.[28] show agreement between theoretical and experimental x-ray energies to a level of 1-2 eV. This agreement is perhaps fortuitous considering the uncertainty in such calculations with regard to nuclear size corrections, to self-energy screening corrections, and to higher order radiative corrections. Nevertheless, such results are remarkable in view of the enormous complexity of the systems being considered.

III. MULTICONFIGURATION DIRAC-FOCK CALCULATIONS

The DF equations written down in Sec. II give an independent particle picture of atomic structure. One of the important schemes developed to go beyond the independent particle picture and to include correlations is the MCDF method[22,29]. In MCDF calculations the wave function describing an atomic state is taken to be a linear superposition of configurational wave functions $\Phi_n(\pi JM)$ coupled to give a definite parity π and angular momentum J,M

$$\Psi(\pi JM) = \sum_n c_n \Phi_n(\pi JM) \tag{11}$$

The expectations value of H is calculated using this wave function and the resulting expression for the energy is optimized with respect to the orbitals defining Φ_n and with respect to the weight coefficients c_n. One is led to a set of coupled differential equations for the orbitals which involve the weight coefficients, and to a set of coupled algebraic equations for c_n (an algebraic eigenvalue problem) which involves the orbitals. These equations are solved self-consistently using iterative schemes similar to those used in non-relativistic multiconfiguration Hartree-Fock calculations[13].

A number of large scale computer programs are available to perform MCDF calculations. Three of these programs which have been used extensively are the Desclaux code[30], the Grant code[31], and the Bogdanovich code[32]. These programs have provisions for including the Breit interaction and QED corrections. There have been a large number of applications of MCDF techniques to calculations of atomic levels and transitions in recent years and several excellent review papers[22,29] are available so we will restrict our attention here to a few elementary examples which illustrate important features of MCDF calculations.

As a first example we consider the MCDF calculation of transition energies and oscillator strengths for low lying, J=1, odd parity, levels of Be-like ions[33-35]. Because of the near degeneracy of 2s and 2p states this is a problem involving strong correlation effects, requiring one to go beyond the DF approximation to obtain meaningful results.

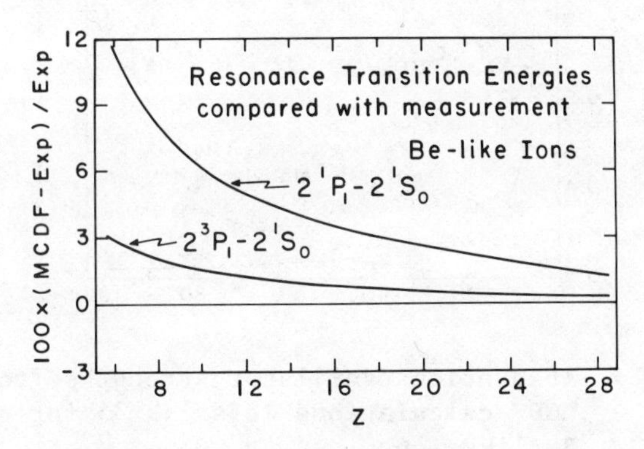

Fig. 3 Comparison of MCDF calculations with measurements of Ref. 37 for Be-like ions.

The ground state of a Be-like ion is taken to be a sum of three configurations $2s^2$, $2p_{1/2}^2$, and $2p_{3/2}^2$. Addition to the $2p^2$ configurations to the ground state wave function accounts for the effects of correlation most important at high Z[8,9,36]. Since there are three configurations, the MCDF equations have three solutions corresponding to low-lying J=0, even-parity states; the lowest energy solution represents the ground state.

Excited state wave functions are constructed in a similar way. The low-lying, J=1, odd parity state is described approximately by two configurations: $2s2p_{1/2}$ and $2s2p_{3/2}$. In this case, there are two solutions to the MCDF equations corresponding to $2\,^1P_1$ and $2\,^3P_1$ states. For low Z ions the two configuration weights, c_1 and c_2, are those appropriate to LS coupling. As Z increases along the isoelectronic sequence the coefficients gradually change reflecting the change from LS to jj coupling of the MCDF results.

Comparisons with precise spectroscopic values of the transition energies[37] for the $2\,^3P_1$ and $2\,^1P_1$ states are given in Fig. 3 where the difference between theory and measurement is shown as a function of Z. The differences shown in Fig. 3 are due to the limited number of configurations included in the trial wave functions. As Z increases, the omitted configurations play a less and less important role.

In Fig. 4 we show the results of calculations of absorbtion oscillator strengths for Be-like ions determined with the MCDF wave

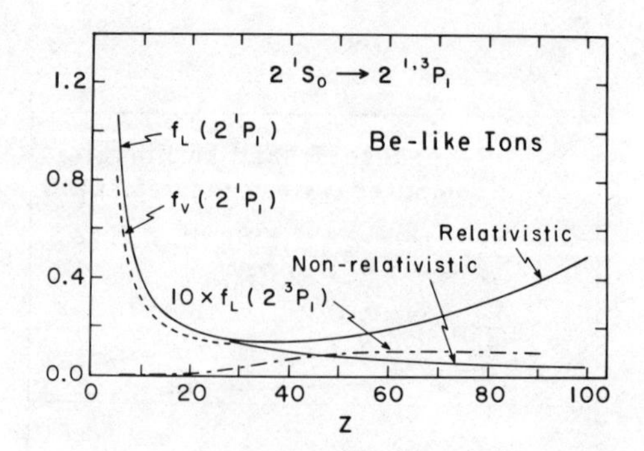

Fig. 4 Absorbtion oscillator strengths from MCDF calculations Refs. 33-35 for Be-like ions.

functions described above. Several important features of MCDF transition probabilities are illustrated in this figure.

Firstly, the curves labeled $f_L(2\,^1P_1)$ and $f_V(2\,^1P_1)$ give resonance oscillator strengths calculated using two different gauges for the electromagnetic transition operator. Describing the radiation field in the Coulomb gauge leads to a dipole transition operator which takes on the velocity form, V, non-relativistically; while a different gauge is required to give a length, L, form dipole operator in the non-relativistic limit. The fact that the L and V forms are different illustrates the lack of gauge invariance in MCDF calculations. For high Z ions the two results become essentially identical, while for low Z, the f_L results are expected to be more reliable[38]. Secondly, the substantial difference at high Z between the non-relativistic and relativistic predictions shown in Fig. 4 is a result of sizable relativistic corrections to the transition energy at high Z. Such relativistic effects on transition energies are especially pronounced for $\Delta n=0$ transitions. Thirdly, the curve representing the $^3P_1$ oscillator strength illustrates how forbidden transitions become progressively more allowed as Z increases, due to the transition from LS to jj coupling.

The practical importance of MCDF calculations of transition energies and probabilities along isoelectronic sequences such as those illustrated in the example above is obvious. By comparing such calculations with known transitions, systematic differences may be established which in turn can be used to help identify unknown transitions.

An example illustrating some of the problems arising in MCDF calculations is the study of the ground state fine structure. The ground state of a Be-like ion can be represented by a single $2s^2 2p$ configuration[40]. Calculations carried out rising $2p_{1/2}$ and $2p_{3/2}$ orbitals give slightly different energies; the difference in energies represents the ground state fine structure. In Cols. 2 and 3 of Table 2 we compare single-configuration DF calculations of the fine structure with measurements for B-like ions.

While the agreement between theory and measurement is impressive it is still of interest to examine the changes which occur in the predicted fine-structure interval when one improves the ground state wave functions. Since the $(2p)^3$ configuration is expected to give the most important correlation corrections one is led to consider the configurations:

$(2p_{3/2})^2\ 2p_{1/2}$, for J=1/2; and

$(2p_{1/2})^2\ 2p_{1/2}$, $(2p_{3/2})^2\ 2p_{1/2}$, $(2p_{3/2})^3$, for J=3/2.

Table 2. Ground State Fine-Structure (cm^{-1}) for the Boron Isoelectronic Sequences[a].

Z	Exp.	DF[b]	MCDF	Corr.
5	16	15.7	435.3	15.7
6	63.4	64.9	271.5	62.7
7	179.5	179	335.8	172.4
10	1310	1346	1472	1298
11	2139	2199	2308	2124
14	6990	7194	7183	6968

a) extracted from Huang, Kim, Cheng, and Desclaux, Ref. 39
b) Ref. 40

There is only one possible $(2p)^3$ configuration for the J=1/2 state, while there are three for the J=3/2 state. Because of this unbalance in the number of configurations, there is a corresponding unbalance in the correlation corrections to the energies of the two fine-structure components, and the "true" fine-structure is masked by correlation effects. This difficulty is illustrated in Col. 4 of Table 2 where the results of MCDF calculations are shown. These MCDF results are seen to be in worse agreement with measurement than the single configuration DF values.

The remedy which was proposed by Huang, et al. in Ref. 39 is to separate the contributions to the MCDF fine-structure interval ΔE_{MCDF} into two parts.

$$\Delta E_{MCDF} = \Delta u + \alpha \Delta v. \qquad (12)$$

The contribution Δu arises from the unbalanced treatment of correlation, and would remain even in the non-relativistic limit. The term $\alpha\Delta v$, proportional to the fine-structure constant α, represents the "true" fine-structure and would vanish non-relativistically. The term Δu in Eq. (12) can be evaluated by setting $\alpha=0$ in the MCDF computer code and then calculating the resulting energy difference. This term can be subtrated from ΔE_{MCDF} to give the true fine-structure $\alpha\Delta v$. The resulting values are shown in Col. 5 of Table 2 and are indeed in better agreement with measured values than either the MCDF or the DF results.

This example and related studies of the fine-structure[41 42] illustrate the care required in applying the MCDF method to studies of atomic systems.

IV. RELATIVISTIC RANDOM-PHASE APPROXIMATION

An entirely different approach to the correlation problem for relativistic atomic systems, which has been used to study both bound states[41,44] and the continuum[45], is the RRPA. The point of departure in RRPA calculations is the hamiltonian of Eq. (1). One considers a closed shell atom or ion in its ground state to be described by a DF wave function. An applied external field excites particle-hole pairs from the DF ground state. In the sense of many-body perturbation theory[46], the lowest order correlation corrections to the particle-hole excitation amplitude are those due to one-particle, one-hole final state processes and those due to two-particle, two-hole ground-state processes. As in non-relativistic RPAE calculations[46] these lowest order correlation corrections can be iterated to all orders of perturbation theory, giving an integral equation for the excitation amplitude which includes the lowest order correlation corrections and many, but not all, of the higher order corrections. The integral equation for the excitation amplitude can be converted into a family of coupled linear integro-differential equations for the excited DF orbitals. These excited orbital equations are time-dependent Dirac-Fock equations; the effects of correlation being included in a modification of the ground state DF potential. In the applications described below the orbital equations are treated numerically using methods pioneered by Dalgarno and Victor[48].

Table 3. Transition Wavelengths (Å) in He-like Ions.

	$1^1S_0 \to 2^1P_1$		$1^1S_0 \to 2^3P_1$		$1^1S_0 \to 2^3P_2$		$1^1S_0 \to 2^3S_1$	
Z	obs[a]	RRPA[b]	obs	RRPA	obs	RRPA	obs	RRPA
12	9.168	9.163	9.231	9.230	9.228	9.227	9.313	9.315
14	6.647	6.644	6.688	6.687	6.685	6.884	6.739	6.740
16	5.038	5.036	5.066	5.065	5.063	5.082	5.101	5.101
18	3.948	3.947	3.969	3.968	3.965	3.965	3.993	3.993
20	3.178	3.176	3.192	3.192	3.189	3.188	3.210	3.210
26	1.850	1.849	1.859	1.859	1.855	1.854	1.868	1.867
28	1.588	1.587	1.596	1.596	1.592	1.591	1.603	1.602
29	1.477	1.477	1.485	1.484	1.481	1.480	1.491	1.491

a. Gabriel, Ref. 51
b. Johnson, Lin, Dalgarno, Ref. 43

Applications of the RRPA to bound state problems have been reviewed previously[49,50] and we will give just one example here, to illustrate certain features of the RRPA approach. In Table 3 we show a comparison of the wavelengths of low lying excited states of ions of the He isoelectronic-sequence determined by solving the RRPA equations[43] and values determined from the spectra of solar flares[51]. Corrections from the Breit interaction are included in the RRPA results. One sees that the predicted values of the transition wavelengths are in excellent agreement with the observed values; however, several comments are in order concerning these results:

1. The transition wavelengths predicted by RRPA for lower members of the isoelectronic sequence where correlation corrections are relatively more important are not as accurate.
2. While the transition probabilities for the allowed transitions shown in Table 3 are in excellent agreement with the most sophisticated calculations, those for forbidden transitions (which are very correlation dependent) are not accurate at low Z.
3. In contrast to MCDF calculations the RRPA calculations are independent of the gauge of the applied field[43], so length-form and velocity-form calculations of oscillator strengths give identical results.

Extensive applications of the RRPA have also been made to study relativistic effects in low energy photoionization of atoms and ions of high nuclear charge[52]. A typical result is shown in Fig. 5 where we plot the ratio of partial photoionization cross-section for $5p_{3/2}$ and $5p_{1/2}$ subshells of Xe (Z=54) against photon energy. Nonrelativistically this ratio is expected to be 2, the ratio ground state occupation numbers. The ratio can depart from 2 relativistically because of spin-orbit effects. In the figure we plot experimental values of the ratio[53,54] together with the results of several relativistic calculations. The curves labeled DS are from uncorrelated Dirac-Slater calculations[54] and those labeled DF from uncorrelated DF calculations[55], while the solid curves are from RRPA calculations[52]. In those labeled (5s+5p) only excitation of the outer 5s and 5p subshells were considered, while in that labeled (5s+5p+4d) excitations of the 4d subshell were also included. In this later calculation there are 13 coupled channels associated with dipole excitation. The relatively good agreement between (5s+5p+4d) RRPA calculation and experiment in this case illustrates the importance of including many-body effects in relativistic calculations. All of the relativistic calculations predict a deviation of the branching ratio from 2 with the correct order of magnitude, but only the correlated RRPA calculation leads to a quantitatively correct result.

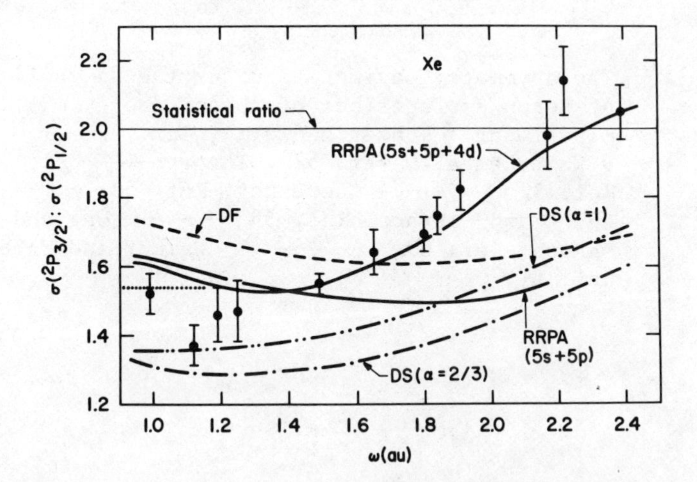

Fig.5 $^2P_{3/2}$: $^2P_{1/2}$ branching ratio for xenon as a function of photon energy ω. Exp: ....Samson et al., Ref. 53; Φ Wuilleumeir et al., Ref. 54. Theory: ——— Johnson and Cheng, Ref. 52; - - - Ong and Manson, Ref. 55; ——.—— and ——..——. Wuilleumeir et al., Ref. 54.

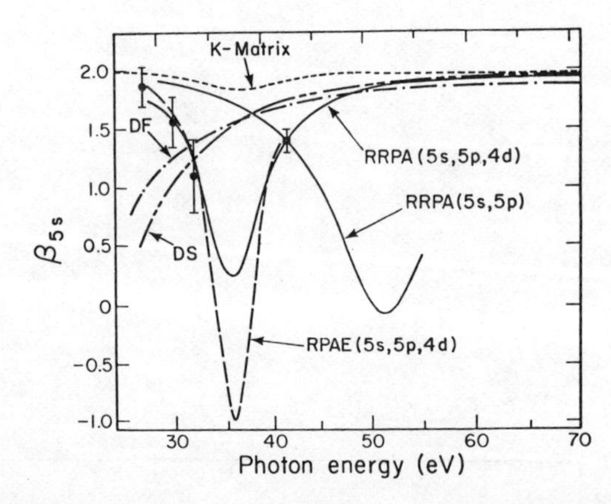

Fig. 6 The asymmetry parameters β for the 5s shell of xenon as functions of photon energy ω. Experiment: Dehmer and Dill, Ref. 56; White et al., Ref. 57. Theory: ——— RRPA, Ref. 52; — — — Cherepkov, Ref. 58; ---- Huang and Starace, Ref. 59; —.— Ong and Manson, Ref. 60; —— — — —— Walker and Waber, Ref. 61.

A second example, which illustrates the interplay between relativistic and correleation effects, is the energy dependence of the angular distribution asymmetry parameter β for the 5s subshell in xenon shown in Fig. 6. Non-relativistically β has the value 2, independent of energy, for an ns subshell. As the experimental points on the figure indicate there can be large relativistic deviations of β from 2 at certain energies. These deviations occur near the Cooper-minimum of the cross section where the non-relativistic photoionization amplitude is small. In the neighbourhood of this minimum relativistic effects play an important role. The DF[60] and DS[62] results shown in Fig. 6 predict large deviations from β=2; however, since the location of the Cooper-minimum is sensitive to correlation effects, these calculations are quantitatively incorrect. The RRPA calculations including 2-shell and 3-shell correlations are also shown illustrating the sensitivity of β to correlations. The 3-shell calculation is in good agreement with measurement, as is the RPAE[58] curve. This later result is from a non-relativistic calculation including 3-shell correlations to which an empirical relativistic (spin-changing) amplitude was added. The results of a non-relativistic K-matrix calculation[59] in which the spin-orbit interaction was treated perturbatively is also included for comparison.

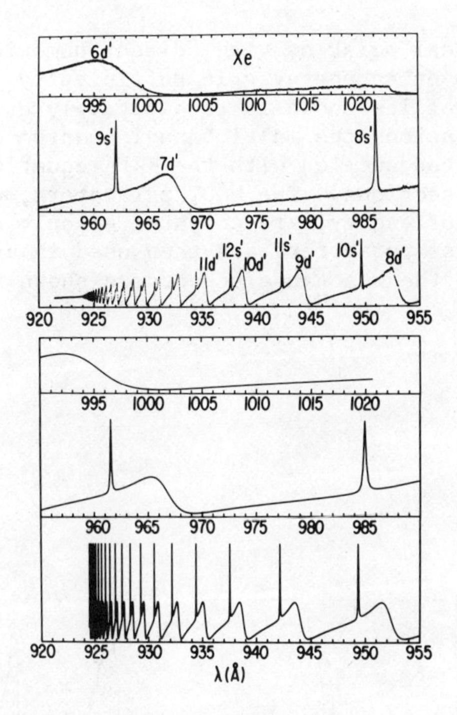

Fig. 7 Beutler-Fano resonances in Xe. Top panels: Measurements of Eland, Ref. 62. Bottom panels: RRPA calculations, Ref. 63.

A quite different application of the RRPA equations is illustrated in Fig. 7 where we compare experimental[62] and theoretical[63] Beutler-Fano resonances in the xenon photoabsorbtion cross section. These resonances occur for photon energies just above the $5p_{3/2}$ threshold and are a result of the coupling between nd and ns states converging to the $5p_{1/2}$ threshold and the continuum. Since the RRPA automatically provides for the $5p_{3/2}$- $5p_{1/2}$ threshold separation, and includes couplings between the relevant open and closed channels, it is a theory ideally suited to study such resonances. The comparison in Fig. 7 illustrates how well the theory works in such applications.

From a practical point of view, direct numerical solutions of the RRPA equations on an energy grid sufficiently fine to produce the theoretical profiles shown is prohibitively expensive. To circumvent this limitation, the multichannel quantum-defect theory (MQDT) is used in conjunction with the RRPA equations[64] to give the theoretical cross-sections. The MQDT parameters, which are slowly varying functions of energy, are computed at only a few points and interpolated. These parameters are then used in a standard MQDT analysis to determine the resonance structures shown in the bottom panels of Fig. 7.

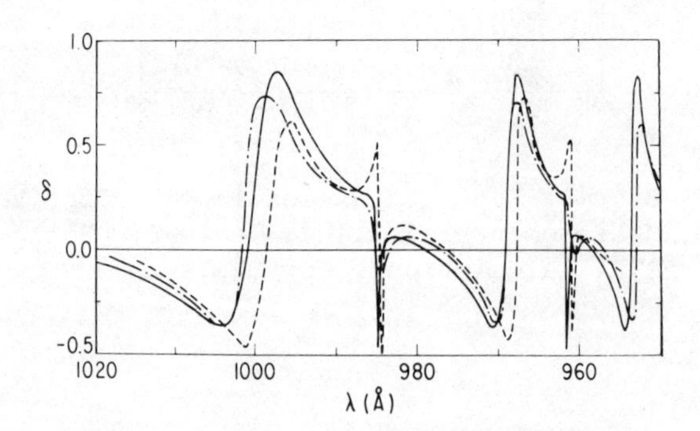

Fig. 8 Total spin polarization parameter δ in the resonance region for Xe. The dash-dot curve is the experimental result of Ref. 65, the solid curve is the RRPA result of Ref. 63, and the dashed curve is the MQDT result of Lee, Ref. 66.

The RRPA-MQDT analysis leads to detailed photoionization amplitudes and phases which can be used to predict the various measurable quantities related to the ionization process. To illustrate, we show in Fig. 8 the spin polarization, δ,as a function of wavelength in the Beutler-Fano resonance region. The solid curve gives the theoretical values[63], which can be compared with the experimental measurements[65], and with semi-empirical MQDT predictions[66]. Suprisingly, the ab-initio RRPA calculations describe the experimental better than the semi-empirical analysis; this is perhaps because of the large number of parameters entering into the semi-empirical determination.

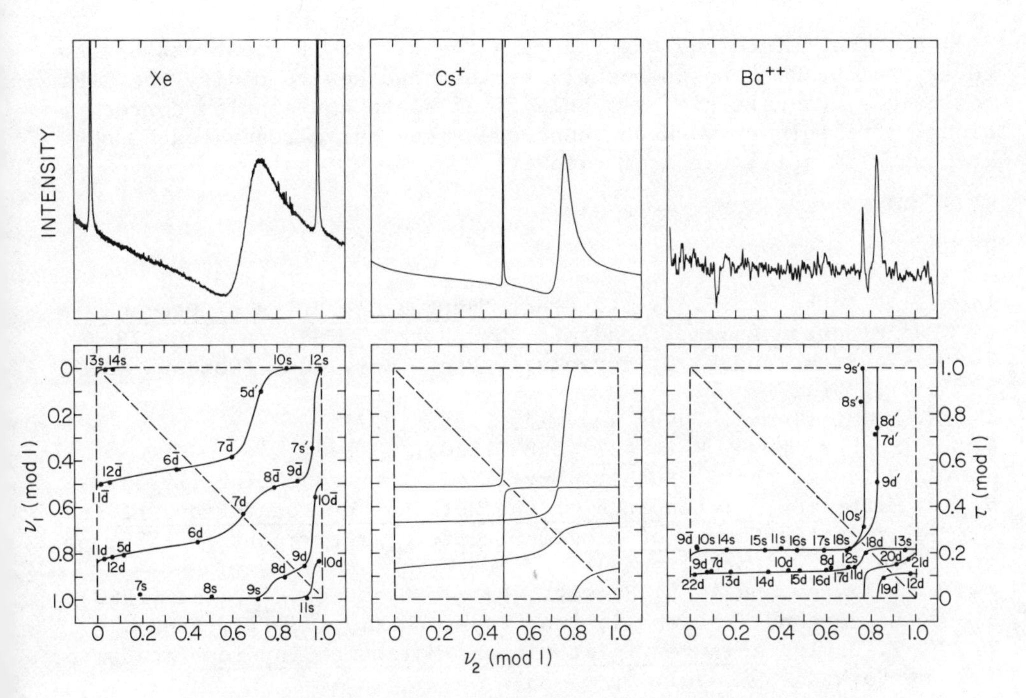

Fig. 9 Beutler-Fano resonances for Xe-like ions. Upper panels: Intensities vs. effective quantum number ν_2. Lower panels: Open-chanel phase-shifts $\delta/\pi = \tau(\text{mod } 1)$ vs. ν_2. Results for Xe and Ba^{++} are from measurements of Refs. 62 and 68. Results for Cs^+ are from an RRPA calculation.

Resonance studies using RRPA-MQDT can of course be extended from neutral species along isoelectronic sequences. In the upper three panels of Fig. 9 we show the Beutler-Fano resonances in Xe, Cs^+, and[67] Ba^{++}, resp. as functions of the effective quantum numbers ν_2. For Xe and Ba^{++} the curves are taken from experimental studies[62,68], while for Cs^+ the curve is from an RRPA calculation. From studies such as this, the gradual transition of shapes and locations can be analyzed. In the lower panels we show, in a Lu-Fano plot, the phases-shifts δ/π for the three open channels as functions of ν_2. Again, these curves are constructed from spectrocopic data for Xe and Ba^{++}, but from theory for Cs^+. The widths of the resonances shown in the upper panels are determined by the rate at which the phase-shifts shown in the lower panels increase through π radians. By inverting the procedure used to construct the Lu-Fano plot for Xe, and Ba^{++} from spectral data, one can determine line positions from the phase-shifts. Thus from studies such as these, detailed predictions of positions (and strengths) along entire spectral sequences can be obtained.

For the future one must address the difficult questions of how to extend the RRPA to open-shell systems and how to modify the RRPA to include two-particle, two-hole final slate correlation corrections. Preliminary work on these questions has already begun, but such work is still far from complete.

REFERENCES

1. H. A. Bethe and E. E. Salpeter, Quantum Mechanics of One-and Two-Electron Atoms, (Academic, New York, 1978), pp. 170-178.
2. G. E. Brown and D. G. Ravenhall, Proc. Roy. Soc. (London) A208, 552 (1951).
3. M. H. Mittleman, Phys. Rev. A5, 2395 (1972).
4. J. Sucher, Phys. Rev. A22, 348 (1980).
5. G. Feinberg and J. Sucher, Phys. Rev. Lett. 35, 1740 (1975).
6. M. Mittleman, Proceedings of the Workshop on Foundations of the Relativistic Theory of Atomic Structure, ANL report #80-126, pp. 27-36 (1980, unpublished).
7. A. Dalgarno and A. L. Stewart, Proc. Phys. Soc. 75, 441 (1960).
8. D. Layzer and J. Bachall, Ann. Phys. (N.Y.) 17, 177 (1962).
9. H. T. Doyle, Advances in Atomic and Molecular Physics (Academic, New York, 1969), Vol. 5, p. 337.
10. L. N. Ivanov, E. P. Ivanova, and U. I. Safronova, J. Quant Spectrosc, Radiat. Transfer, 15, 553 (1975).
11. R. Swirles, Proc. Roy. Soc. (London) A152, 625 (1935).
12. J. A. Bearden and A. F. Burr, Rev. Mod. Phys. 39, 125 (1967).
13. C. F. Fischer, The Hartree-Fock Method for Atoms (Wiley, New York, 1977).
14. J. B. Mann, private communication.

15. H. B. Collard, L. R. B. Elton, and R. Hofstadter, Nuclear Radii, ed. by H. Schopper (Springer, Berlin, 1967), Landolf-Börnstein, New Series, Group I, Vol. 2.
16. R. Engfer, H. Schneuwly, J. L. Vuillenmier, H. K. Walter, and A. Zehnder, At. Data Nucl. Data Tables 14, 509 (1974).
17. M. H. Chen, B. Crasemann, M. Aoyagi, K.-N. Huang, and H. Mark, At. Data Nucl. Data Tables 26, 563 (1981).
18. J. B. Mann and W. R. Johnson, Phys. Rev. A4, 41 (1971).
19. A. M. Desiderio and W. R. Johnson, Phys. Rev. A3, 1267 (1971).
20. K. T. Cheng and W. R. Johnson, Phys. Rev. A14, 1943 (1976).
21. P. J. Mohr, Phys. Rev. Lett. 34, 1050 (1975); Ann. Phys. (N.Y.) 88, 26, 52 (1974); private communication.
22. J. P. Desclaux, Physica Scripta 21, 436 (1980).
23. K.-N. Huang, M. Aoyagi, M. H. Chen, B. Crasemann, and H. Mark, At. Data Nucl. Data Tables 18, 243 (1976).
24. E. A. Uehling, Phys. Rev. 48, 55 (1935).
25. E. H. Wichmann and N. M. Kroll, Phys. Rev. 101, 843 (1956).
26. G. Källén and A. Sabry, Dan. Mat. Fys. Medd. 29, 17 (1955).
27. J. P. Desclaux, Proceedings of the Workshop on Foundations of the Relativistic Theory of Atomic Structure, ANL report #80-126, p. 86.
28. R. Deslattes, private communication.
29. N. Beatham, I. P. Grant, B. J. McKinzie, and S. J. Rose, Physica Scripta 21, 423 (1980).
30. J. P. Desclaux, Comp. Phys. Comm. 9, 31 (1975).
31. I. P. Grant, B. J. McKenzie, and P. H. Norrington, Comp. Phys. Comm. 21, 207 (1980).
32. P. O. Bogdanovich, private communication.
33. Y. K. Kim and J. P. Desclaux, Phys. Rev. Lett. 36, 139 (1976).
34. L. Armstrong, Jr., W. R. Fielder, and D. L. Lin, Phys. Rev. A19, 1114 (1976).
35. K. T. Cheng and W. R. Johnson, Phys. Rev. A15, 1326 (1977).
36. D. Layzer, Ann. Phys. (N.Y.) 8, 271 (1959).
37. B. Edlén, Physica Scripta 20, 129 (1979).
38. A. F. Starace, Phys. Rev. A3, 1242 (1971); 8, 1141 (1973).
39. K.-N. Huang, Y.-K. Kim, K. T. Cheng, and J. P. Desclaux, Phys. Rev. Lett. 48, 1245 (1982).
40. K. T. Cheng, Y.-K. Kim, and J. P. Desclaux, At. Data Nucl. Data Tables 24, 111 (1979).
41. C. P. Wood and N. C. Pyper, Mol. Phys. 41, 149 (1980).
42. J. Bauche and M. Klapisch, J. Phys. B5, 29 (1972).
43. W. R. Johnson, C. D. Lin, and A. Dalgarno, J. Phys. B9, L303 (1976).
44. P. Shorer, C. D. Lin, and W. R. Johnson, Phys. Rev. A16, 1109 (1977).
45. W. R. Johnson and C. D. Lin, Phys. Rev. A20, 964 (1979).
46. A. L. Fetter and J. D. Walecka, Quantum Theory of Many-Particle Systems, (McGraw-Hill, N.Y., 1971).
47. M. Ya. Amusia and N. A. Cherepkov, Case Studies in Atomic

Physics 5, 47 (1975).
48. A. Dalgarno and G. A. Victor, Proc. Roy. Soc. A291, 291 (1966).
49. W. R. Johnson, C. D. Lin, K. T. Cheng, and C. M. Lee, Physica Scripta 21, 409 (1980).
50. K. T. Cheng and W. R. Johnson, Proceedings of the Workshop on Foundations of the Relativistic Theory of Atomic Structure, report #80-126, pp. 115-132 (1980, unpublished).
51. A. H. Gabriel, Mon. Not. R. Astron. Soc. 160, 99 (1972).
52. W. R. Johnson and K. T. Cheng, Phys. Rev. Lett. 40, 1167 (1978); Phys. Rev. A20, 978 (1979).
53. J. A. R. Samson, J. L. Gardner, and A. F. Starace, Phys. Rev. A12, 1459 (1975).
54. F. Wuilleumeir, M. Y. Adam, N. Sandner, V. Schmidt, and W. Mehlhorn, Phys. Rev. A16, 646 (1977).
55. W. Ong and S. T. Manson, J. Phys. B11, 163 (1978).
56. J. L. Dehmer and D. Dill, Phys. Rev. Lett. 39, 1049 (1976).
57. M. G. White, S. H. Southworth, P. Kobrin, E. D. Poliakoff, R. A. Rosenberg, and D. A. Shirley, Phys. Rev. Lett. 43, 1661 (1979).
58. N. A. Cherepkov, Phys. Lett. 66A, 204 (1978).
59. K.-N. Huang and A. F. Starace, Phys. Rev. A21, 697 (1980).
60. W. Ong and S. T. Manson, J. Phys. B11, L65 (1978).
61. T. E. H. Walker and W. T. Waber, J. Phys. B7, 674 (1974).
62. See J. Berkowitz, Photoabsorbtion, Photoionization and Photoelectron Spectroscopy (Academic, N.Y. 1979), p. 181.
63. W. R. Johnson, K. T. Cheng, K.-N. Huang, and M. LeDourneuf, Phys. Rev. A22, 989 (1980).
64. C. M. Lee and W. R. Johnson, Phys. Rev. A22, 979 (1980).
65. U. Heinzmann, F. Schäfers, K. Thimm, A. Wolcke, and J. Kessler, J. Phys. B12, L679 (1979).
66. C. M. Lee, Phys. Rev. A10, 1598 (1974).
67. U. Fano, Nuovo Cim. 12, 156 (1935); K. T. Lu, Phys. Rev. A4, 579 (1971); C. M. Lee and K. T. Lu, Phys. Rev. A8, 1241 (1973).
68. W. T. Hill, T. B. Lucatorto, T. J. McIlrath, and J. Sugar, private communication.

ONE- AND TWO-ELECTRON SYSTEMS

G. W. F. Drake*

Research and Engineering Staff
Ford Motor Company
Dearborn, Michigan 48121

I. INTRODUCTION

One- and two-electron atoms provide the traditional testing ground for the techniques of atomic physics. Since highly accurate electronic wave functions are obtainable, high precision comparisons between theory and experiment can be made. Much of the current research interest centers on relativistic and quantum electrodynamic (QED) effects in highly ionized systems. The purpose of the present paper is to summarize recent progress in this area.

Interest in highly ionized systems stems in part from the recent availability of high precision measurements of transition frequencies in these systems, and in part from potential applications in plasma diagnostics. The theory of relativistic and QED effects in two-electron systems requires further development before full agreement with experiment can be achieved. For example, the fine structure splittings of F^{7+} are now known about a factor of 10 more accurately than what has been achieved theoretically.

Section II.A of this paper reviews the current state of theory and experiment for Lamb shifts in hydrogenic ions. Section II.B discusses a number of recent experiments to detect a variety of asymmetries in the quenching of metastable $2s_{\frac{1}{2}}$ hydrogen and He^+. This work provides fundamental tests of radiation theory and damping. Finally, Section III discusses relativistic and QED effects in two-electron ions.

---

*Permanent address: Department of Physics, University of Windsor, Windsor, Ontario, Canada

II. ONE-ELECTRON SYSTEMS

II.A Lamb Shifts

The discovery of the Lamb shift by Lamb and Retherford[1] initiated the development of modern QED. According to one-electron Dirac theory, the $2s_{\frac{1}{2}}$ and $2p_{\frac{1}{2}}$ states should be exactly degenerate. However, electron self-energy and vacuum polarization corrections (see Fig.1) raise the energy of $2s_{\frac{1}{2}}$ relative to $2p_{\frac{1}{2}}$ by about 0.03 cm^{-1} (∿1000 MHz) (see Fig.2). Lamb and Retherford's observation of a microwave resonance at this frequency was rapidly followed by Bethe's[2] first estimate of the electron self-energy, with a result in rough agreement with experiment. This work has initiated a long sequence of Lamb shift measurements of ever increasing precision in hydrogen and the hydrogenic ions. The experimental progress has

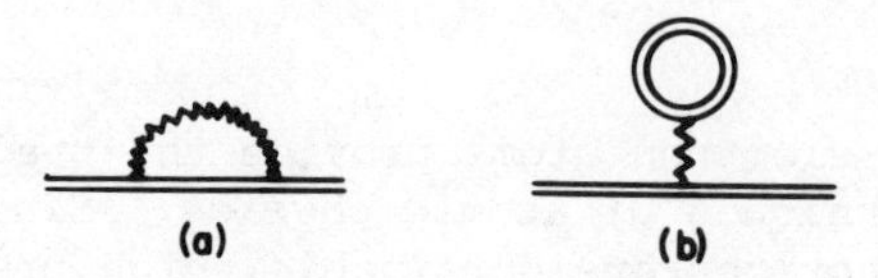

Fig.1. Feynman diagrams for (a) the electron self-energy and (b) the vacuum polarization. Double solid lines represent a bound electron in the Coulomb field of the nucleus.

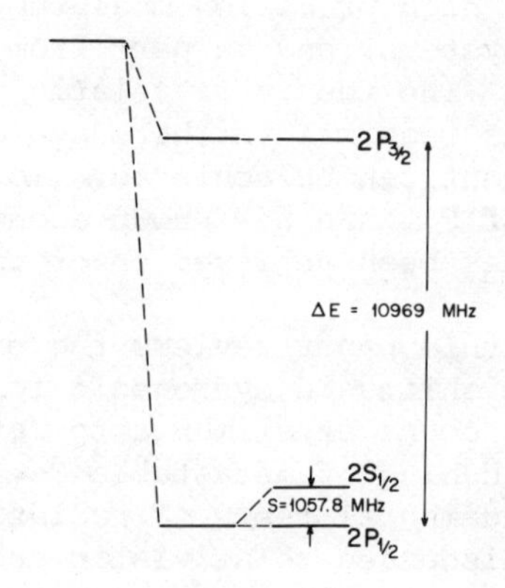

Fig.2. Energy level diagram for the n=2 states of hydrogen showing the progressive splittings produced first by relativistic corrections in one-electron Dirac theory and then by QED corrections. The diagram is drawn roughly to scale.

been paralleled by a corresponding sequence of increasingly accurate theoretical calculations. The generally good agreement between theory and experiment has provided an important check on the computational techniques of QED in the presence of strong fields (i.e. the Coulomb field of the nucleus). Both theory and experiment have now reached a high state of refinement, as reviewed by Lautrup et al.[3], Kugel and Murnick[4], Brodsky and Mohr[5] and Drake[6]. Recent progress and the current issues are discussed below.

The traditional method of calculating the Lamb shift is to expand in powers of $Z\alpha$. For an electron with quantum numbers $n\ell j$, the expansion has the form[7,8]

$$\Delta E = \frac{8mc^2\alpha(Z\alpha)^4}{6\pi n^3}\{A_{40} + A_{41}\ln(Z\alpha)^{-2} + A_{50}(Z\alpha)$$

$$+ (Z\alpha)^2[A_{62}\ln^2(Z\alpha)^{-2} + A_{61}\ln(Z\alpha)^{-2} + G(Z\alpha)]$$

$$+ (\alpha/\pi)[B_{40} + O(Z\alpha)] + O(\alpha^2/\pi^2)\} + \Delta E_M + \Delta E_R \tag{1}$$

where $G(Z\alpha) = A_{60} + O(Z\alpha)$ (2)

and ΔE_M and ΔE_R are finite nuclear mass and radius corrections. Each of the constants A_{mn} can be decomposed into sums of the form

$$A_{mn} = A_{mn}^{SE} + A_{mn}^{VP} + A_{mn}^{MM} \tag{3}$$

arising from electron self-energy, vacuum polarization and anomalous magnetic moment parts respectively. The lowest order terms are

$$A_{40}^{SE} = \ln[(Z\alpha)^2 R_\infty/\varepsilon_{n\ell}] + \frac{11}{24}\delta_{\ell,0} \tag{4}$$

$$A_{40}^{VP} = \frac{-1}{5}\delta_{\ell,0} \tag{5}$$

$$A_{40}^{MM} = \frac{3c_{\ell,j}}{8(2\ell+1)} \tag{6}$$

and $c_{\ell,j} = \begin{cases} 1/(\ell+1) & \text{for } j = \ell+\frac{1}{2} \\ -1/\ell & \text{for } j = \ell-\frac{1}{2} \end{cases}$

All the other A_{mn}'s displayed in (1) and (2) are known exactly[3-6] with the exception of A_{60}^{SE}. The quantity $\varepsilon_{n\ell}$ is the Bethe average excitation energy defined by

$$\ln\varepsilon_{n\ell} = \frac{\sum_{n'}(E_n - E_{n'})^3|\langle n|\vec{r}|n'\rangle|^2\ln|E_n - E_{n'}|}{\sum_{n'}(E_n - E_{n'})^3|\langle n|\vec{r}|n'\rangle|^2} \tag{7}$$

Highly accurate values of $\ln \varepsilon_{n\ell}$ have been calculated by Klarsfeld and Maquet[9] for several low lying states, and asymptotic formulas have been obtained by Poquerusse[10].

Since different calculations of the Lamb shift differ in lowest order by the value adopted for A_{60}^{SE}, the percentage differences in the final results increase roughly in proportion to Z^2. The experimental precision required for a significant test of theory therefore correspondingly decreases with increasing Z. Both Erickson[11] and Mohr[12] have extracted values for A_{60}^{SE} from relativistic calculations of the complete electron self-energy diagram. After subtracting the known lower order terms in (1), what remains is $G^{SE}(Z\alpha)$, the self-energy part of (2). Fitting the results as a function of Z to the form

$$G^{SE}(Z\alpha) = A_{60}^{SE} + bZ\alpha \ln(Z\alpha)^{-2} + cZ\alpha \tag{8}$$

results in

$$A_{60}^{SE}(\text{Erickson}) = -17.246 \pm 0.5$$

$$A_{60}^{SE}(\text{Mohr}) = -24.064 \pm 1.2$$

for the $2s_{\frac{1}{2}} - 2p_{\frac{1}{2}}$ Lamb shift. The difference corresponds to a Lamb shift difference of 0.049 MHz in hydrogen, which is much greater than the accuracy of recent experimental measurements. A recent recalculation by Sapirstein[13], which treats certain small terms in the equation for the Dirac-Coulomb Green's function, yields

$$A_{60}^{SE}(\text{Sapirstein}) = -24.9 \pm 0.9$$

in agreement with Mohr's value. Although Sapirstein's calculation is for the $1s_{\frac{1}{2}}$ ground state, the n dependence is expected to be small. His result therefore indicates that Mohr's values for $G^{SE}(Z\alpha)$ are probably substantially correct. An experimental precision of approximately $\pm$ $10Z^2$ ppm (parts per million) is required for a significant test of the $G^{SE}(Z\alpha)$ term.

A further recent contribution to the theory of the Lamb shift is an argument by Borie[14] that finite nuclear size effects should be included directly in the calculation of the lowest order level shift of $ns_{\frac{1}{2}}$ states. Eq.(1) contains an overall multiplying factor of $4\pi e\langle\rho(\vec{r})\rangle$, where $\rho(\vec{r})$ is the nuclear charge density. The numerical factor in (1) corresponds to choosing $\rho(\vec{r}) = Ze\delta(\vec{r})$ for a point nucleus. Borie's correction results from using instead a finite nuclear distribution with the result

$$\langle\rho(\vec{r})\rangle_{ns} = Ze|\psi_{ns}(0)|^2[1 - \frac{2Z}{a_0}\langle r\rangle_{(2)}] \tag{9}$$

where a_0 is the Bohr radius and

$$\langle r\rangle_{(2)} = \iint r\rho(|\vec{r} - \vec{u}|)\rho(u)d^3u d^3r \tag{10}$$

Assuming an exponential form for the charge distribution of mean radius R, then

$$\langle r\rangle_{(2)} = \frac{35R}{16\sqrt{3}} \tag{11}$$

to give a correction of

$$\Delta E_B = \frac{8mc^2\alpha(Z\alpha)^4}{6\pi n^3}\left(\frac{-35ZR}{8\sqrt{3}\, a_0}\right)[A_{40} + A_{41}\ln(Z\alpha)^{-2}]\delta_{\ell,0} \tag{12}$$

For H and D, the additional shifts are - 0.042 and - 0.104 MHz respectively, assuming R_H = 0.86 fm and R_D = 2.10 fm. However, the validity of the Borie correction has been questioned by Page et al.[15] on the grounds that it may be cancelled by higher order relativistic effects. The comparison with experimental data is discussed below.

Experimental methods of measuring the Lamb shift can be broadly classified into two basic types. The first is the microwave resonance technique originally used by Lamb and Retherford[1], in which the $2s_{\frac{1}{2}}$ - $2p_{\frac{1}{2}}$ transition is observed directly in a microwave cavity. The prime factor limiting the accuracy is the precision with which the resonance line center can be located. Since the width Γ of the $2p_{\frac{1}{2}}$ state is about one tenth of the Lamb shift, the line center must be located to a precision of 10x ppm relative to Γ for a precision of x ppm in the Lamb shift. Line narrowing can be achieved by use of Ramsey's[16] separated oscillatory fields technique, but at the expense of a lower signal-to-noise ratio. Since the Lamb shift increases in proportion to Z^4, the frequency lies outside the microwave region for ions heavier than Li^{++}. A tunable laser can then be used in conjunction ith a fast ion beam as further discussed below.

The second class of techniques is based on the electric field quenching of the metastable $2s_{\frac{1}{2}}$ state. An externally applied static electric field mixes the $2s_{\frac{1}{2}}$ state with $2p_{\frac{1}{2}}$ and $2p_{3/2}$, which then decay by rapid electric dipole photon emission to the ground state. The lifetimes of the perturbed $2s_{\frac{1}{2}}$ state is related to the Lamb shift S by[17]

$$\gamma = \gamma_{2s} + \gamma_{2p}|\vec{E}|^2 \left[\frac{|V|^2}{S^2 + \Gamma^2/4} + \frac{|W|^2}{T^2 + \Gamma^2/4}\right]$$

$$\times\ 1.63721 \left(\frac{\text{MHz}}{\text{V/cm}}\right)^2 + O(|\vec{E}|^4) \tag{13}$$

where

$$T = E(2s_{\frac{1}{2}}) - E(2p_{3/2})$$

$$V = \langle 2s_{\frac{1}{2}}|z|2p_{\frac{1}{2}}\rangle/a_0 = \sqrt{3}/Z[1 - \frac{5}{12}\alpha^2 Z^2]$$

$$W = \langle 2s_{\frac{1}{2}}|z|2p_{3/2}\rangle/a_0 = 6/Z[1 - \frac{1}{6}\alpha^2 Z^2]$$

$|\vec{E}|$ is the electric field strength in V/cm and S, T and Γ are expressed in MHz. The numerical factor in (13) is $(2R_\infty/E_0)^2$, where $E_0 = 5.14225\times10^9$ V/cm is the atomic unit of field strength. The lifetime γ is measured by passing a beam of metastable $2s_{\frac{1}{2}}$ ions through an electric field region and measuring the exponential decrease in Ly-α radiation intensity as a function of position along the beam. The accuracy is limited to about $\pm$ 0.5% by the inherent difficulties of performing high precision lifetime measurements. However, this is still sufficient for a significant test of theory if Z is sufficiently large ($Z \gtrsim 16$).

A variation of the above quench rate technique is the quenching anisotropy method[18]. Instead of measuring the intensity of Ly-α radiation as a function of position along the ion beam, the intensities $I_\parallel$ and $I_\perp$ are measured perpendicular to the applied quenching field (see Fig.3). The anisotropy $R = (I_\parallel - I_\perp)/(I_\parallel + I_\perp)$ is then (approximately) proportional to the Lamb shift, and independent of field strength in the limit of weak fields. The theory is discussed in detail by van Wijngaarden and Drake[18] and Drake et al.[19]. Although many small corrections must be taken into account, the basic physics can be understood as follows. Taking the z-axis as the dc field direction, the adiabatic field peturbed initial state is

$$\psi_i = (1/N^{\frac{1}{2}})[\psi(2s_{\frac{1}{2}}) + \alpha\psi(2p_{\frac{1}{2}}) + \sqrt{2}\beta\psi(2p_{3/2})] \quad (14)$$

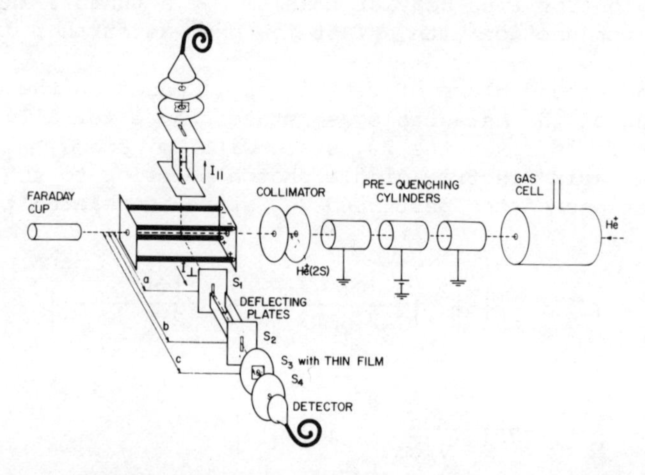

Fig.3. Schematic diagram of the apparatus for the anisotroy method of measuring the Lamb shift by electrostatic quenching (From Drake et al.[19])

where α and $\sqrt{2}\beta$ are mixing coefficients determined by diagonalizing the complete Hamiltonian, including the external field in the $2s_{\frac{1}{2}}$, $2p_{\frac{1}{2}}$, $2p_{3/2}$ basis set. In the electric dipole approximation, the emitted radiation intensity with polarization vector $\hat{e}$ is

$$I(\hat{e}) \propto |\langle\psi_i|\hat{e}\cdot\vec{r}|\psi(|s_{\frac{1}{2}})\rangle|^2 \quad . \tag{15}$$

The total radiation intensity emitted in a direction θ to the dc field is obtained by summing (15) over two perpendicular $\hat{e}$ vectors, both perpendicular to the direction of propagation. The result is an isotropic contribution from the $|2p_{\frac{1}{2}} - 1s_{\frac{1}{2}}|^2$ term and anisotropic contributions from the $|2p_{3/2} - 1s_{\frac{1}{2}}|^2$ term and the $2p_{\frac{1}{2}} - 1s_{\frac{1}{2}} - 2p_{3/2}$ cross term. The sum of all three is

$$I(\theta) \propto 1 + \mathrm{Re}(\rho)(1 - 3\cos^2\theta) + |\rho|^2 (5 - 3\cos^2\theta) \tag{16}$$

with $\rho = \beta/\alpha$. In the limit of weak fields,

$$\rho = \frac{S + i\Gamma/2}{T + i\Gamma/2} + O(|E|^2) \tag{17}$$

where $\Gamma = 99.71Z^4$ MHz is the 2p level width and T is as defined in (13). In this approximation, the anisotropy is therefore

$$R = -\frac{3\mathrm{Re}(\rho) - 3/2|\rho|^2}{2 - \mathrm{Re}(\rho) + 7/2|\rho|^2} \quad . \tag{18}$$

The above result emphasizes that $R \simeq -(3/2)\mathrm{Re}(\rho)$ is determined primarily by the ratio of the Lamb shift to the fine structure splitting. Additional small corrections for perturbation mixing with higher np states, final state perturbations, relativistic effects and magnetic quadrupole transitions are discussed by van Wijngaarden and Drake[18] and Hillery and Mohr[20].

The experimental results obtained with the above techniques are summarized in Table 1 and compared with theory. The theoretical values are essentially the same as those tabulated by Mohr[12] and Erickson[11] except that the revised nuclear radii of 0.862±0.012 fm for hydrogen[21] and 1.674±0.012 fm for helium[22] have been used in place of 0.81 fm and 1.644 fm respectively. The same data are plotted in Fig. 4 in terms of deviations from $\bar{S}$ defined to be

$$\bar{S} = [S(\mathrm{Erickson}) + S(\mathrm{Mohr})]/2 \quad . \tag{19}$$

The deviations are divided by Z^6 so that the theoretical uncertainties and experimental error bars are about the same relative size for all the hydrogenic ions from H to Ar^{17+}. Since the Lamb shift itself increases in proportion to Z^4, it is evident that both the

Table 1. Comparison of Theoretical and Experimental Lamb Shifts in Hydrogenic Ions

Ion	Reference	Technique[a]	Value	Theory[b] Mohr[12]	Theory[b] Erickson[11]
H	Lundeen and Pipkin[23]	SOF	1057.845(9) MHz	1057.883(13)	1057.929(13)
	Newton et al.[24]	rf	1057.862(20)	1057.840(13)	1057.886(13)
	Robiscoe et al.[25]	rf	1057.90(6)		
	Triebwasser et al.[26]	rf	1057.77(6)		
D	Cosens[27]	rf	1059.24(6)	1059.241(27)	1059.287(27)
	Triebwasser et al.[26]	rf	1059.00(6)	1059.137(27)	1059.183(27)
	van Wijngaarden and Drake[28]	A	1059.36(16)		
$^{4}He^{+}$	Narasimham and Strombotne[29]	rf	14046.2(1.2)	14042.36(55)	14045.12(55)
	Lipworth and Novick[30]	rf	14040.2(1.8)	14040.18(55)	14042.94(55)
	Drake et al.[19]	A	14040.2(2.9)		
$^{6}Li^{2+}$	Dietrich et al.[31]	rf	62790.(70)	62737.5(6.6)	62767.4(6.6)
	Leventhal[32]	rf	62765.(21)	62715.5(6.6)	62745.4(6.6)
$^{12}C^{5+}$	Kugel et al.[33]	QR	780.1(8.0) GHz	781.99(21)	783.67(21)
				781.49(21)	783.17(21)
$^{16}O^{7+}$	Curnutte et al.[34]	A	2192.(15)	2196.21(92)	2204.98(92)
	Lawrence et al.[35]	QR	2215.6(7.5)	2194.19(92)	2202.96(92)
	Leventhal et al.[36]	QR	2202.7(11.0)		
$^{19}F^{8+}$	Kugel et al.[37]	LR	3339(35)	3343.1(1.6)	3360.3(1.6)
	Murnick et al.[38]	QR	3405(75)	3339.5(1.6)	3356.7(1.6)
$^{35}Cl^{16+}$	Wood et al.[39]	LR	31190(220)	31347(20)	31965(20)
				31286(20)	31904(25)
$^{40}Ar^{17+}$	Gould and Marrus[40]	QR	38100(600)	38250(25)	39100(25)
				38171(25)	39021(25)

a SOF, separated oscillatory fields; rf resonance; A, anisotropy; QR, quench rate; LR laser resonance

b The second entry of each pair includes the Borie correction term S_B

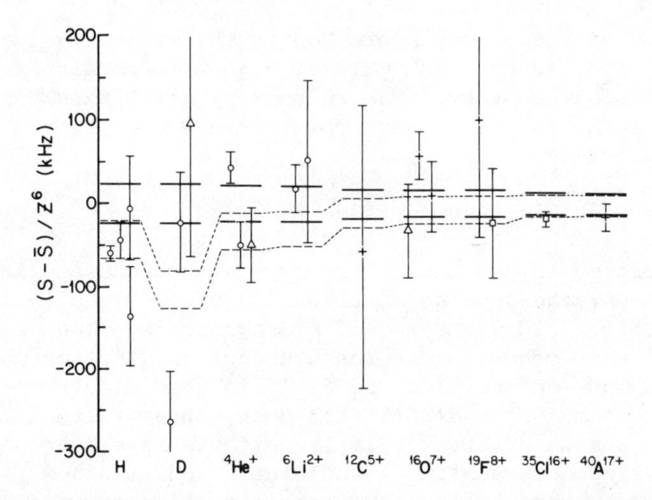

Fig.4. Comparison between scaled theoretical and experimental one-electron Lamb shifts, expressed as deviations from the theoretical mean. The upper horizontal line for each ion is Erickson's[11] theory and the lower horizontal line is Mohr's[12] theory and $\bar{S}$ is the average of the two. The experimental data are labelled by the method of measurement according to: (O) microwave resonance; (△) anisotropy measurement; (+) quench-rate measurement; (□) laser resonance. The dashed lines are the Borie[14] corrections to the theoretical values. (From Drake[6])

theoretical and experimental percentage uncertainties are increasing roughly in proportion to Z^2. The influence of the Borie correction is shown by the dashed line in Fig.4.

Three recent measurements shown in Table 1 and Fig.4 provide particularly significant tests of theory. The first is the high precision (±9 ppm) microwave resonance measurement of Lundeen and Pipkin[23], using the separated oscillatory field method. Their result lies well below both theoretical values when the Borie correction is excluded, but agrees with Mohr's theory when the Borie correction is included. The influence of the Borie correction is particularly large in the case of deuterium, but here the experimental data are insufficiently precise to provide a definitive test. A measurement in deuterium similar in precision to Lundeen and Pipkin's result in hydrogen would provide a valuable check on the validity of the Borie correction.

The other two recent measurements have been at the high-Z end of the isoelectronic sequence. The Ar^{17+} result, obtained by the quench rate technique[40] and the Cl^{16+} result, obtained by laser resonance[39], both tend to support Mohr's theoretical values. The Borie correction becomes too small at high Z to be significant at present levels of accuracy. The anisotropy measurements for He^+ [19] and O^{7+} [34] can be substantially improved in accuracy, and further work is in progress.

II. B Quenching Asymmetries

In addition to the Lamb shift asymmetry discussed in the preceeding section, the angular distribution of Ly-α radiation produced by the electric field quenching of hydrogenic $2s_{\frac{1}{2}}$ ions displays a rich diversity of other interference phenomena. For example, interference between the electric field induced electric dipole (E1) and spontaneous magnetic dipole (M1) decay modes to the ground state produces observable effects. Similar effects have been observed in other systems such as cesium[41], thallium[42] and methane[43]. The work on hydrogenic systems summarized in this section provides high precision tests of the theory of radiative transitions and damping phenomena.

Consider first the quenching of spin-polarized $2s_{\frac{1}{2}}$ ions in the absence of hyperfine structure or magnetic field effects. Then the $2s_{\frac{1}{2}}$ state can decay by allowed M1 transitions, or by electric-field induced E1 or M2 transitions. An emitted photon is characterized by a propagation vector $\vec{k}$ and a polarization vector $\hat{e}$, and the initial atomic state by a spin-polarization vector $\vec{P}$. We choose a coordinate system such that $\vec{k}$ lies along the z-axis, and write $\hat{e}$ for the general case of elliptical polarization in the form

$$\hat{e} = \hat{\imath}\cos\beta + i\hat{\jmath}\sin\beta \tag{20}$$

where $\hat{\imath}$ and $\hat{\jmath}$ are unit vectors in the x- and y-directions. We also introduce an auxillary vector E' defined by

$$\vec{E}' = E_x\hat{\imath} - E_y\hat{\jmath} \tag{21}$$

where E_x and E_y are the x- and y-components of the unit vector $\hat{E}$ in the electric field direction. The intensity emitter per unit solid angle is then[18,20]

$$I(\hat{e},\hat{k},\vec{P}) = \frac{\alpha k}{4\pi}[I_0 + \vec{P}\cdot\vec{J}_0 + \vec{P}\cdot\vec{J}_1\sin2\beta + \vec{E}'\cdot\vec{J}_2\cos2\beta] \tag{22}$$

$$\text{where } I_0 = \tfrac{1}{2}|V_+|^2[1 - (\hat{k}\cdot\hat{E})^2] + \tfrac{1}{2}|V_-|^2[1 + (\hat{k}\cdot\hat{E})^2] + |M|^2 - 2\mathrm{Im}(M^*V_-)(\hat{k}\cdot\hat{E})$$

$$J_0 = (\hat{k}\times\hat{E})\{\mathrm{Re}[M^*(V_+ + V_-)] - \mathrm{Im}(V_-^*V_+)(\hat{k}\cdot\hat{E})\} \tag{23}$$

$$\vec{J}_1 = |V_-|^2(\hat{k}\cdot\hat{E})\hat{E} + \mathrm{Re}(V_-^*V_+)\hat{E}\times(\hat{k}\times\hat{E}) - |M|^2\hat{k}$$
$$- \mathrm{Im}[M^*(V_+ + V_-)]\hat{E} - \mathrm{Im}[M^*(V_+ - V_-)](\hat{k}\cdot\hat{E})\hat{k} \quad (24)$$

$$\vec{J}_2 = \tfrac{1}{2}(|V_+|^2 - |V_-|^2)\hat{E} + \mathrm{Im}(V_-^*V_+)\hat{E}\times\vec{P} + \mathrm{Re}[M^*(V_+ + V_-)]\vec{P}\times\hat{k} \quad (25)$$

and the V and M coefficients are

$$V_+ = V_{\frac{1}{2}} + 2V_{3/2} \quad (26)$$

$$V_- = V_{\frac{1}{2}} - V_{3/2} + M_{3/2} \quad (27)$$

$$M = M_{\frac{1}{2}} + 2i(\hat{k}\cdot\hat{E})M_{3/2} \quad (28)$$

where, in the nonrelativistic limit,

$$V_j = \frac{|\vec{E}|}{3}\frac{\langle 1s|z|2p\rangle\langle 2p|z|2s\rangle}{E(2s_{\frac{1}{2}}) - E(2p_j) + i\Gamma/2}, \quad j = \frac{1}{2}, \frac{3}{2} \quad (29)$$

$M_{\frac{1}{2}}$ is the relativistic M1 matrix element for direct transitions to the ground state, and $M_{3/2}$ is a small M2 correction to the field-induced E1 transitions to the ground state via the $2p_{3/2}$ intermediate state. The numerical values are

$$\langle 1s|z|2p\rangle = 2^8/(3^5\sqrt{2}Z) \quad (30)$$

$$\langle 2p|z|2s\rangle = -3/Z \quad (31)$$

$$M_{\frac{1}{2}} = -8\alpha^3 Z^2/(81\sqrt{2}) \quad (32)$$

$$M_{3/2} = -9(Z\alpha)^2 V_{3/2}/32 \quad (33)$$

in atomic units. Relativistic and finite field corrections are discussed by van Wijngaarden and Drake[18]. If the detectors are insensitive to photon polarization, then the sin 2β and cos 2β terms in (22) average to zero, leaving only the I_0 and $\vec{P}\cdot\vec{J}_0$ terms.

The I_0 term, which is all that survives for an unpolarized beam, gives rise to the Lamb shift asymmetry discussed in the previous section. The two parts of the polarization dependent term $\vec{P}\cdot\vec{J}_0$ are

$$T_M = \vec{P}\cdot\hat{k}\times\hat{E}\ \mathrm{Re}[M^*(V_+ + V_-)] \quad (34)$$

$$T_\Gamma = -\vec{P}\cdot\hat{k}\times\hat{E}(\hat{k}\cdot\hat{E})\,\mathrm{Im}(V_-^*V_+) \quad . \quad (35)$$

These have both recently been measured in He^{+}.[18,44] T_M can be found by choosing $\vec{P}$, $\hat{k}$ and $\hat{E}$ all perpendicular, and observing the asymmetry in the radiation intensity when the direction of observations is reversed ($\hat{k} \rightarrow -\hat{k}$). Defining

$$A_M = \frac{I(\hat{k}) - I(-\hat{k})}{I(\hat{k}) + I(-\hat{k})} \tag{36}$$

then
$$A_M = \frac{\mathrm{Re}[M^*(V_+ + V_-)]}{I_0(\hat{k})} \quad . \tag{37}$$

thus, A_M provides a direct measure of the relativistic magnetic dipole matrix element M. At $|\vec{E}| = 38.14$ V/cm, the measured asymmetry is[18] $A_M = (0.323 \pm 0.085) \times 10^{-3}$. This corresponds to

$$M_{\frac{1}{2}} = -\ (0.262 \pm 0.069)\,\alpha^2 e\hbar/mc \tag{38}$$

in agreement with the theoretical value $-\ 0.2794\alpha^2 e\hbar/mc$ obtained from (32). This is the only measurement of $M_{\frac{1}{2}}$ in a hydrogenic system.

T_Γ can be measured by choosing $\vec{P}$ perpendicular to the $\hat{k}$, $\hat{E}$ plane, and comparing the intensities when $\theta_{\hat{k},\hat{E}} = 45^{\circ}$ and $\theta_{\hat{k},\hat{E}} = 135^{\circ}$. Defining

$$A_\Gamma = \frac{I(45^{\circ}) - I(135^{\circ})}{I(45^{\circ}) + I(135^{\circ})} \tag{39}$$

then
$$A_\Gamma = \frac{3\mathrm{Im}(V^*_{\frac{1}{2}}V_{3/2})}{I_0(45^{\circ})} + O(\alpha^2 Z^2) \quad . \tag{40}$$

Thus, A_Γ provides a direct measure of the level width Γ of the 2p state in (29). The measured asymmetry is[44] $A_\Gamma = 0.00769 \pm 0.00010$. This corresponds to $\Gamma_+ = 1611 \pm 21$ MHz in agreement with the theoretical value 1595 MHz for He^{+}. The measured lifetime $\tau_{2p} = (0.988 \pm 0.013)$ 10^{-10} sec. is more accurate than the beam foil result[45] $\tau_{2p} = (0.98 \pm 0.05)\ 10^{-10}$ sec.

Lévy and Williams[46] have done a closely related experiment in atomic hydrogen. They measure the suppressed microwave electric dipole transition rate between the α_0 and β_0 hyperfine states in a D.C. magnetic field $\vec{B}$ set near the crossing of the β and e states around 550 G (see Fig.5). The transition occurs through the mixing of the states β_0 and e_1 by a small dc electric field $\vec{E}$ perpendicular to $\vec{B}$. If ε is the microwave field, then the transition rate is[46]

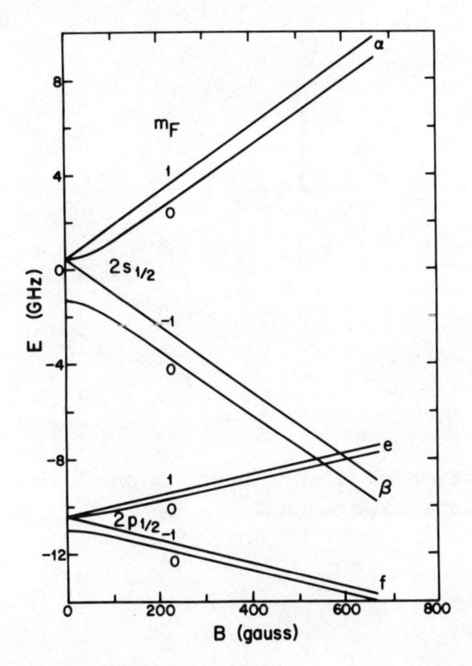

Fig.5. Zeeman diagram of the $2s_{\frac{1}{2}}$ and $2p_{\frac{1}{2}}$ states of hydrogen.

$$R(\alpha_0 \to \beta_0) = a[\vec{\varepsilon}^2 - (\vec{\varepsilon}\cdot\hat{B})^2]\vec{E}^2 + b(\vec{\varepsilon}\cdot\vec{E})^2 + c(\vec{\varepsilon}\cdot\vec{E})(\vec{\varepsilon}\cdot\vec{E}\times\hat{B}) \tag{41}$$

where $a = k|\Delta_1^{-1} + \Delta_2^{-1}|^2 /4$

$b = -k\mathrm{Re}(\Delta_1\Delta_2^*)^{-1}$

$c = k\mathrm{Im}(\Delta_1\Delta_2^*)^{-1}$

and $\Delta_1^{-1} = [\Delta E(\beta_0 - e_1) + \frac{1}{2}i\Gamma]^{-1} + [\Delta E(\alpha_0 - f_{-1}) + \frac{1}{2}i\Gamma]^{-1}$

$\Delta_2^{-1} = [\Delta E(\beta_0 - f_{-1}) + \frac{1}{2}i\Gamma]^{-1} + [\Delta E(\alpha_0 - e_1) + \frac{1}{2}i\Gamma]^{-1}$.

With field directions chosen as shown in Fig.6, the last term of (41), which is formally odd under time reversal, reverses sign if $\vec{B} \to -\vec{B}$, $\psi \to \pi - \psi$ or $\psi \to -\psi$. The resulting maximum asymmetry at $\psi = 45^\circ$ is

$$A_1 = c/(2a + b) \tag{42}$$

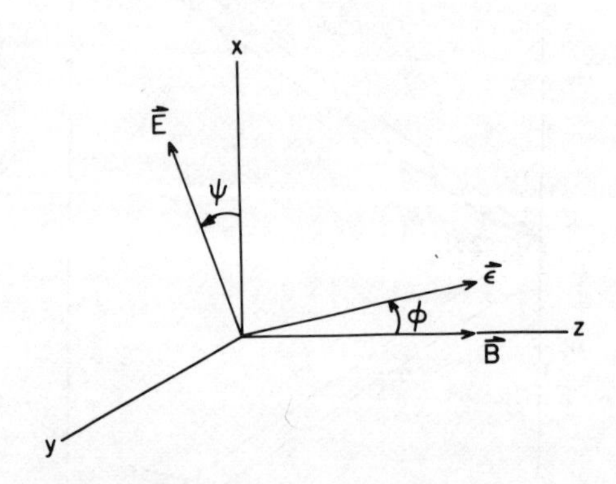

Fig.6. Configuration of electric and magnetic fields in the Lévy and Williams experiments[46,47].

which is about -0.13 at 550 G. Since c is proportional to Γ, the measured asymmetry yields a level width of $\Gamma = 99.3 \pm 1.0$ MHz, a result comparable in accuracy to beam-foil measurements. Since the above experiments are sensitive to the imaginary parts of the perturbation coefficients, they provide an accurate test of the Bethe-Lamb phenomenological damping theory used to interpret the results.

The same experimental set-up has been used by Lévy and Williams[47] to look for evidence of parity nonconserving (PNC) atomic interactions such as the electron-proton weak interaction. The presence of such an interaction would mix the close-lying states β_0 and e_0 and introduce into (41) an additional interference term of the form $(\vec{\varepsilon}\cdot\vec{E})(\vec{\varepsilon}\cdot\vec{B})E1^*_{PC}E1_{PNC}$ + c.c, where $E1_{PC}$ is the parity conserving part of the electric dipole transition moment and $E1_{PNC}$ the parity nonconserving part. The above term reverses sign under reversal of either $\vec{E}$ or $\vec{B}$, giving rise to a PNC asymmetry in $R(\alpha_0 \to \beta_0)$.[47] The asymmetry is

$$A_{PNC} \simeq \frac{2|E1_{PNC}|}{E1_{PC}} \cos\phi \tag{43}$$

where ϕ is the relative phase of $E1_{PC}$ and $E1_{PNC}$. The choice $\psi = 0$ in Fig.6 suppresses $E1_{PC}$ and thereby enhances the PNC asymmetry signal. The measurement yields $A < 2.5\ 10^{-3}$, corresponding to a zero field $2s_{\frac{1}{2}} - 2p_{\frac{1}{2}}$ PNC mixing of less than $1.5\ 10^{-8}$. Although this limit is 4 orders of magnitude lower than the best previous measurement, it is still several orders of magnitude larger than the expected mixing due to the weak interaction.

III. TWO-ELECTRON SYSTEMS

High-precision calculations for two-electron atoms and ions are complicated by the necessity of simultaneously taking into account relativistic, QED and electron correlation effects. There is no unique way of specifying an exact relativistic two-electron Hamiltonian analagous to the Dirac Hamiltonian without at the same time including QED effects to all orders. The rigorous starting point is generally accepted to be the fully covariant Bethe-Salpeter equation derived from the Feynman form of QED by Salpeter and Bethe[48], and from quantum field theory by Gell-Mann and Low[49]. Exact solutions to this equation have not been obtained. All practical calculations are based (explicitly or implicitly) on expansions in powers of α, αZ and Z^{-1} [50]. Unlike the Lamb shift in one-electron systems, there are no QED effects which manifest themselves directly in lowest order. It is first necessary to subtract the nonrelativistic two-electron energies and relativistic corrections before specifically QED effects are revealed.

Interest in higher order QED corrections has recently been stimulated by measurements of transition energies of the type $1s2s\ ^3S_1 - 1s2p\ ^3P_0$ and $^3P_2$ in high-Z two-electron ions[51-57]. Since the Lamb shift increases approximately in proportion to Z^4, while the nonrelativistic energy increases only as Z, the ratio is given approximately by

$$\Delta E_L/\Delta E_{NR} \simeq 1.4\ 10^{-6}\ Z^3 \tag{44}$$

Thus, for Fe^{24+}, the Lamb shift contributes about 2% of the total energy difference. We first discuss below the nonrelativistic and relativistic contributions to the energies, and then turn to a discussion of the two-electron Lamb shift.

For high values of Z, it becomes advantageous to expand in powers of Z^{-1} as well as αZ. Early calculations of this type were done by Dalgarno and Stewart[58], Layzer and Bahcall[59], Collins[60] and Doyle[61]. Extensive numerical results have been obtained by Labzovskii[62], Klinchitskaya et al.[63], Ermalaw and Jones[64], Ivanov et al.[65], and Vainshtein and Safranova[66]. A useful review of this work and the experimental data for helium-like ions in the range Z = 11-18 has been given by Martin[67]. The discussion below is based on a formulation by Drake[68].

The nonrelativistic and relativistic contributions to the

energy can be expanded in the form

$$\begin{aligned} E = {} & Z^2E_0^{\,0} + \alpha^2Z^4E_0^{\,2} + \alpha^4Z^6E_0^{\,4} + \ldots \\ & + Z\,E_1^{\,0} + \alpha^2Z^3E_1^{\,2} + \alpha^4Z^5E_1^{\,4} + \ldots \\ & + Z^0E_2^{\,0} + \alpha^2Z^2E_2^{\,2} + \ldots \\ & + Z^{-1}E_3^{\,0} + \alpha^2Z\,E_3^{\,1} + \ldots \end{aligned} \tag{45}$$

excluding Lamb shift and nuclear motion type corrections. The first column sum is the exact nonrelativistic energy E^0. The second column sum is $\alpha^2\langle B_p\rangle$, where B_p is the two-component Pauli form of the Breit interaction. The first row sum is the sum of the exact single-particle Dirac energies for the two electrons, and the second row sum is the electron-electron interaction correction $\langle e^2/r_{12} + B\rangle$, where B is the full 16-component Breit interaction and the matrix element is evaluated with hydrogenic products of Dirac spinors as wave functions. There are in addition off-diagonal mixing terms between states which are degenerate in zero order and have the same total angular momentum J and parity P. (for example, the states $1s2p\ ^3P_1$ and $1s2p\ ^1P_1$). Both the rows and columns of (45) are summed to infinity, and off-diagonal mixing effects are included, by diagonalizing the matrix.

$$\underline{H} = (\underline{H}^0 + \underline{B}_p)_{LS} + \underline{R}(\underline{H}_D + \underline{V}_{12} + \underline{B})_{jj}\underline{R}^{-1} - \underline{\Delta} \tag{46}$$

in the basis set of states which are degenerate in zero order for given J and P. Here $\underline{H}_D$ is the sum of single-particle Dirac Hamiltonians, $V_{12} = e^2/r_{12}$ and $\underline{\Delta}$ is a double counting correction for the four terms in the upper left hand corner of (45) which are otherwise counted twice. The subscripts LS and jj indicate the natural coupling schemes for the calculation of matrix elements and $\underline{R}$ is the $jj \rightarrow LS$ recoupling transformation. For the LS terms, high precision matrix elements obtained from correlated variational wave funtions[69] or 1/Z expansion techniques[68,70,71] are available. The jj terms involve simple analytic integrations over Coulomb-Dirac wave functions. Thus (46) provides an efficient method of adding relativistic effects to high precision nonrelativistic calculations. The eigenvalues tend to the correct limit in both the low Z and high Z extremes. Extensive numerical calculations based on (46) have now been done for all the low-lying states of two-electron ions up to and including 1s3d[55,68,71].

Although a unique separation of relativistic and QED effects is not well defined, energies as calculated above can be taken as representing the relativistic but non-QED contributions to the total energy. After adding anomalous magnetic moment and mass polarization corrections and subtracting from the experimental energy, what remains is the Lamb shift of relative order $\alpha^3 Z^4 \ln(\alpha Z)$, together with terms not included in (46), the leading one being $\alpha^4 Z^4 E_2$. One can therefore expect about 1% accuracy in the Lamb shift if the E_2^4 term is not taken into account.

The leading terms in the Lamb shift for two-electron systems have been discussed by several authors[72-74]. The results can be expressed in the form

$$E_{L,2} = E'_{L,2}(nLS) + E''_{L,2}(nLS) \tag{47}$$

where $E'_{L,2}$ is the proper Lamb shift and vacuum polarization correction given by

$$E'_{L,2} = \frac{1}{6}(8\alpha^5 Zmc^2)\langle\delta^3(\vec{r}_1) + \delta^3(\vec{r}_2)\rangle \times [\ln(Z\alpha)^{-2} + \frac{19}{30} + \ln\frac{Z^2 R_\infty}{\varepsilon_{nLS}} + 3\alpha\pi Z\left(\frac{427}{384} - \tfrac{1}{2}\ln 2\right)] \tag{48}$$

and $E''_{L,2}$ is an electron-electron interaction correction given by

$$E''_{L,2} = (\alpha^5 mc^2)[\langle\delta^3(r_{12})\left(\frac{28}{3}\ln\alpha + \frac{178}{15} - \frac{40}{3}\vec{s}_1\cdot\vec{s}_2\right)\rangle - \frac{28}{3}Q] \tag{49}$$

where Q is the principal part of $\langle r_{12}^{-3}\rangle$. Some authors also include the anomalous magnetic moment term $E'''_{L,2}(nLSJ)$ in (45), whereas Accad et al.[69] include it in the matrix elements of B_p. If $\langle\delta^3(\vec{r}_1)\rangle$ in (48) is replaced by its hydrogenic value $Z^3/(\pi n^3 a_0^3)$, then (48) reduces to the corresponding expression for the one-electron Lamb shift. The $\langle\delta^3(\vec{r}_{12})\rangle$ terms vanish for triplet states.

The principal uncertainty in evaluating $E_{L,2}$ for the low-lying states of helium-like ions is the value of the Bethe logarithm $\ln \varepsilon_{nLS}$ defined in analogy with (8). Calculations of reasonable accuracy have been done for the ground state of helium by Schwartz[75] and for the helium-like ions up to Z = 10 by Aashamar and Austvik[74]. The latter calculation is further discussed below. For the excited states, explicit calculations have been done only for He[76] and Li^+[73]. Emolaev's[73] calculated Lamb shift contribution to the $2\,^3S_1 - 2\,^3P_1$ transition frequency of 1.316±0.0016 cm^{-1} agrees with the value 1.2543±0.0016 cm^{-1} derived from the measuremnts of Holt et al.[77].

For the higher Z ions DeSerio et al.[55] suggest defining the two-electron Bethe logarithm to be

$$\lim_{Z\to\infty} \ln\left(\frac{\varepsilon(1sn\ell)}{R_\infty}\right) = \ln\left(\frac{\varepsilon(1s)}{R_\infty}\right) + n^{-3}\ln\left(\frac{\varepsilon(nl)}{R_\infty}\right) \tag{50}$$

for a $1sn\ell$ configuaration so that the hydrogenic Lamb shift is recovered in the limit of large Z. This leads to

$$\varepsilon(1s2s) = 28.095Z^2R_\infty \tag{51}$$

$$\varepsilon(1s2p) = 19.695Z^2R_\infty \ . \tag{52}$$

For Z = 2, this gives $\ln[\varepsilon(1s2s)/R_\infty] = 4.72$, in poor agreement with the value 4.38±0.02 calculated by Suh and Zaide[76]. However, at high Z, DeSerio et al. find reasonably good agreement with experiment as shown by the dashed curves in Fig.7.

A better approximation for high Z can be obtained by using Z^{-1} expansion techniques. If the two-electron Bethe logarithm is written in the form

$$\ln \varepsilon_{nLS} = A/B \tag{53}$$

where A and B are the numerator and denominator of (8) for the two-electron case, then the energies and wave functions can be expanded in powers of Z^{-1} to obtain

$$A = Z^4[A_0 + 2B_0\ln Z + A_1Z^{-1} + \ldots] \tag{54}$$

$$B = Z^4[B_0 + B_1Z^{-1} + \ldots] \tag{55}$$

The expansion of B can easily be obtained from the identity (in atomic units)

$$B = 2\pi Z\langle\delta(\vec{r}_1) + \delta(\vec{r}_2)\rangle \tag{56}$$

Thus $B_0 = 4$ and $B_1 = -2.6736$ for the ground state. Since A_0/B_0 is the Bethe logarithm $\ln\varepsilon(1s) = 2.984129 - \ln 2$ for hydrogen, this determines $A_0 = 9.1639$. The remaining coefficient A_1 can be written down in terms of sums over dipole transitions involving first order perturbed wave functions and energies. Since the dipole operator is a one-electron operator, only singly excited states contribute. The sums have been evaluated by Goldman and Drake[78] in a one-electron finite basis set representation, with the preliminary result

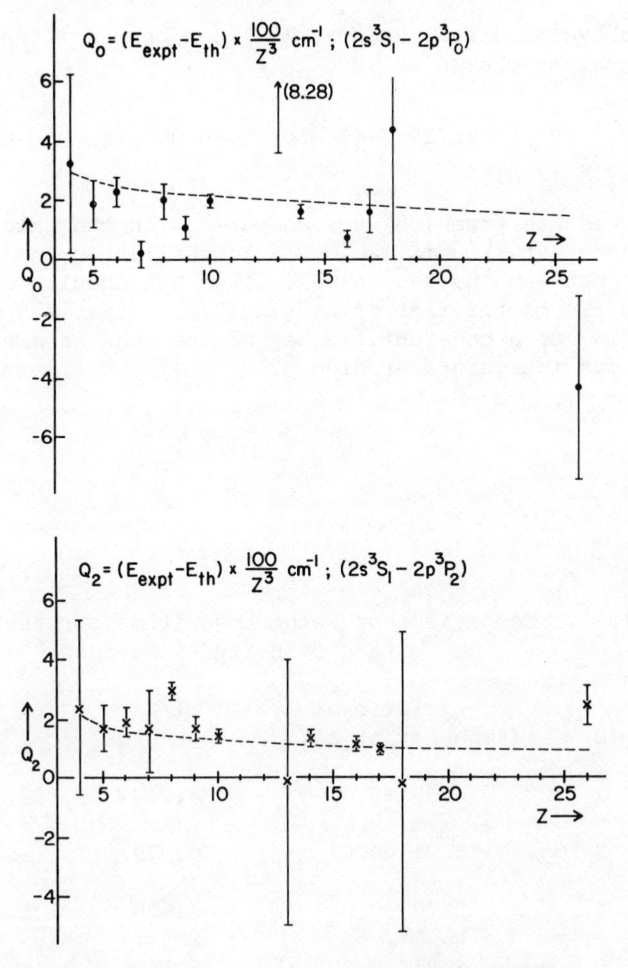

Fig.7. Comparison of theory and experiment for the QED corrections to the $2s\,^3S_1 - 2p\,^3P_2$ and $2s\,^3S_1 - 2p\,^3P_0$ transitions of helium-like ions. E_{th} includes only one-electron QED terms, and the dashed curve represents an estimate of the two-electron corrections. (From DeSerio et al.[55])

$A_1 = -7.30 \pm 0.05$. Finally, using (54) and (55), the expansion of (53) is

$$\ln(\varepsilon_{100}/R_\infty) = \frac{A_0}{B_0} + 2\ln Z + \ln 2 + \frac{(A_1 - A_0B_1)}{B_0} Z^{-1} + O(Z^{-2}) \quad (57)$$

With the above numerical values, (57) can be rewritten in terms of a screened nuclear charge as

$$\ln(\varepsilon_{100}/R_\infty) \quad \ln[19.77(Z - 0.150 \pm 0.007)^2] \tag{58}$$

The results from (58) are compared with the calculations of Aashamar and Austvik[74] in Table 2. In the last column, Δ is the difference between the two values. Thus $Z^2\Delta$ should tend to a constant related to the coefficient of Z^{-2} in (57). Since $Z^2\Delta$ does not appear to be tending to a constant, it may be that the Aashamar and Austvik values become inaccurate at high Z.

Table 2. Comparison of Bethe Logarithms for the Ground State of Helium-like ions.

Z	Variational (Aashamar et al.[74])	1/Z Expansion[a]	$Z^2\Delta$
2	4.37 ±0.01	4.214	0.62
3	5.21 ±0.01	5.079	1.18
4	5.777±0.003	5.680	1.55
5	6.214±0.003	6.142	1.80
6	6.565±0.002	6.517	1.73
7	6.864±0.002	6.833	1.54
8	7.115±0.002	7.105	0.63
9	7.334±0.002	7.345	-0.89
10	7.525±0.002	7.559	-3.41

[a] $\ln(\varepsilon_{100}/R_\infty) = \ln[19.77(Z-0.15)^2] + O(Z^{-2})$

It is also of interest to compare theory and experiment for the fine structure splittings of the 1s2p $^3P_0$, $^3P_1$ and $^3P_2$ states in helium-like ions. These do not depend on the Lamb shift, but are sensitive to one- and two-electron relativistic corrections of $O(\alpha^6 mc^2)$. The leading terms $\alpha^4 Z^6 E_0^4$ and $\alpha^4 Z^5 E_1^4$ are included in (45), but the higher oder terms $\alpha^4 Z^4 E_4^2$ ... can only be obtained from a systematic reduction of the Bethe-Salpeter equation as done by Douglas and Kroll[79]. Their results can be expressed symbolically in the form

$$E_J = E^0 + \alpha^2 \langle B_p \rangle_J + \alpha^4 \langle B_p (E^0 - H^0)^{-1} B_p \rangle_J + \alpha^6 \langle H_6 \rangle_J \qquad (59)$$

where E^0 is the nonrelativistic energy, B_p is the usual Breit interaction, the third term represents the Breit interaction taken to second order, and the fourth term containing H_6 is the sum of higher order spin-dependent QED corrections to B_p obtained by Douglas and Kroll. A full evaluation of (59) has been carried through only for He I. The results of these extremely lengthy calculations are reviewed by Lewis and Serafino[80]. When anomalous magnetic moment and relativistic recoil corrections are added, they obtain 1.35 ppm and 37 ppm agreement with experiment[81,82] for ν_{01} and ν_{12} respectively. Alternatively, if the theoretical value for ν_{01} is taken as correct, then the derived value for α^{-1} is 137.03608(13).

The above calculation in effect sums to infinity the entries in the third column of (45), as well as the first two columns. For higher values of Z, the contributions beyond $\alpha^4 Z^5 E_1^4$ in the third column become decreasingly important, and (45) may yield useful results as it stands. High precision experimental measurements of the fine structure splittings are now available for comparison in Li^+ [77,83] and F^{7+} [84]. Other less accurate values can be obtained by taking differences of the 1s2s $^3S_1$ - 1s2p ^3P_J, J = 0,1,2 transition frequencies tabulated by DeSerio et al.[55] and Stamp et al.[57]. Table 3 shows the comparison between theory and experiment for Li^+. Here, the values of Schiff et al.[69], which include anomalous magnetic moment but not higher order corrections, are already in reasonable agreement with experiment. The addition of

$$\Delta E_{rel} = \alpha^4 Z^6 E_0^4 + \alpha^4 Z^5 E_1^4 \qquad (60)$$

makes the agreement much worse. The source of the discrepancy appears to come from higher order terms in Z^{-1}. For example, ΔE_{rel} for the $0 \rightarrow 2$ transition is approximately

$$\Delta E_{rel}(0 \rightarrow 2) = -1.215\times10^{-5} Z^6 (1 - 7.419Z^{-1}) \text{ cm}^{-1} \qquad (61)$$

Table 3. Comparison of Theory and Experiment for the Fine-Structure Splitting of the 1s2p $^3P_{0,1,2}$ State[a] in Li^+

	$\Delta\nu_{02}$	$\Delta\nu_{01}$	$\Delta\nu_{21}$	Reference
Theory	3.1046	5.1944	2.0898	Schiff et al.[69]
ΔE_{rel}[b]	0.0131	0.0131	-0.0027	
Total	3.1177	5.2048	2.0871	
Experiment	3.1051(12)	5.1948(12)	2.0897(12)	Holt et al.[77]
	3.1028(2)	5.1934(8)	2.0906(8)	Bayer et al.[83]

[a] In cm^{-1}.

[b] Relativistic corrections of order α^4Z^6 and α^4Z^5.

The large contributions from the leading terms will presumably be reduced by higher order terms in the series. The accuracy for low Z could be improved by rewriting (61) in the form of a screening approximation.

More satisfactory results are obtained for F^{7+}. Here, the ΔE_{rel} terms in (60) increase the 2 → 1 transition frequency calculated by Schiff et al.[69] from 955.26 to 957.51 cm^{-1}, in much better agreement with the high precision measurement, 957.88±0.03 cm^{-1}.[84] However, the difference is still more than 10 times the experimental error. The difference could be accounted for if the term $\alpha^4Z^4E_2^4$ contributed an additional $0.2\alpha^4Z^4$ Ry. The less accurate experimental results for other ions are compared with theory in Table 4. The agreement is satisfactory except for N^{5+}, F^{7+} and S^{14+}. For all of these, the differences arise primarily from the location of the J = 0 level relative to the other two.

Table 4. Summary of Theoretical and Experimental Data for the Fine-Structure Intervals of the 1s2p $^3P_{0,1,2}$ from Data Tabulated in refs. 55 and 57.

Ion	Transition	Theory[a]	Experiment	Difference
Be^{2+}	2-1	14.89	14.8(3)	0.09(30)
B^{3+}	2-0	36.35	36.3(8)	0.05(80)
	1-0	-16.29	-16.0(8)	0.29(80)
C^{4+}	2-0	123.04	122.6(1.4)	0.4(1.4)
	1-0	-12.72	-12.9(1.4)	-0.2(1.4)
N^{5+}	2-0	299.18	305.6(1.4)	-6.4(1.4)
	1-0	8.26	15.1(1.4)	-6.8(1.4)
	2-1	290.92	290.5(0.6)	0.4(0.6)
O^{6+}	2-0	609.48	610.1(1.5)	-0.6(1.5)
	1-0	58.07	58.4(1.5)	-0.3(1.5)
F^{7+}	2-0	1107.56	1114.4(3.0)	-6.8(3.0)
	1-0	150.05	157.8(3.0)	-7.8(3.0)
	2-1	957.51	957.88(3)[b]	-0.37(3)
Ne^{8+}	2-0	1856.05	1853.6(2.4)	2.4(2.4)
	1-0	298.94	299.4(2.4)	-0.5(2.4)
Al^{11+}	2-0	6366.00	6180.(150)	186.(150)
Si^{12+}	2-0	8924.38	8931.(5)	-6.6(5.0)
S^{14+}	2-0	16261.5	16295.(11)	-33.(11)
Cl^{15+}	2-0	21286.2	21280.(40)	6.(40)
Ar^{16+}	2-0	27395.5	27150.(400)	245.(400)

[a] In cm^{-1}

[b] High-precision measurement by Myers et al.[81]

ACKNOWLEDGEMENT

Research support by the National Research Council of Canada and the hospitality of the Ford Motor Company are gratefully acknowledged.

REFERENCES

1. W. E. Lamb Jr. and R. C. Retherford, Phys. Rev., 72:241(1947).
2. H. A. Bethe, Phys. Rev., 72:339(1947).
3. B. E. Lautrup, A. Peterman and E. de Raphael, Phys. Rep., 36:193 (1972).
4. H. W. Kugel and D. E. Murnick, Rep. Prog. Phys., 40:297(1977).
5. S. J. Brodsky and P. J. Mohr, Vol. 5, p. 3, in "Structure and Collisions of Ions and Atoms", I. A. Sellin, ed., Springer-Verlag, Berlin and New York 1978.
6. G. W. F. Drake, Adv. At. Mol. Phys., 18:399(1982).
7. G. W. Erickson and D. R. Yennie, Ann. Phys. (N. Y.), 35:271 and 447(1965).
8. B. N. Taylor, W. H. Parker and D. N. Ladenberg, Rev. Mod. Phys., 41:375(1969).
9. G. Klarsfeld and A. Maquet, Phys. Lett. B, 43:201(1973).
10. A. Poquerusse, Phys. Lett. A, 82:232(1981).
11. G. W. Erickson, Phys. Rev. Lett., 27:780(1971) and J. Chem. Phys., Ref. Data,6:831(1977).
12. P. J. Mohr, pp.89-95, in "Beam Foil Spectroscopy", I. A. Sellin and D. J. Pegg, eds., Plenum, New York, 1976.
13. J. Sapirstein, Phys. Rev. Lett., 47:1723(1981).
14. E. Borie, Phys. Rev. Lett., 47:568(1981).
15. G. P. Lepage, D. R. Yennie and G. W. Erickson, Phys. Rev. Lett., 47:1640(1981).
16. N. F. Ramsey, p. 124, in "Molecular Beams", Oxford University Press, London and New York, 1956.
17. W. E. Lamb, Jr. and R. C. Retherford, Phys. Rev., 79:549(1950).
18. A. van Wijngaarden and G. W. F. Drake, Phys. Rev. A, 25:400(1982).
19. G. W. F. Drake, S. P. Goldman and A. van Wijngaarden, Phys. Rev. A, 20:1299(1979).
20. M. Hillery and P. J. Mohr, Phys. Rev. A, 21:24(1980).
21. G. G. Simon, R. Borkowski, Ch. Schmitt and V. W. Walther, Z. Naturforsch, 35:1(1980).
22. E. Borie and G. A. Rinker, Phys. Rev. A, 18:324(1978).
23. S. R. Lundeen and F. M. Pipkin, Phys. Rev. Lett., 46:232(1981).
24. G. Newton, D. A. Andrews and P. J. Unsworth, Phil. Trans. Roy. Soc. London, 290:373(1979).
25. R. T. Robiscoe and T. W. Shyn, Phys. Rev. Lett., 24:559(1970).
26. S. Triebwasser, E. S. Dayhoff and W. E. Lamb, Jr., Phys. Rev., 89:98(1953).
27. B. L. Cosens, Phys. Rev., 173:49(1968).
28. A. van Wijngaarden and G. W. F. Drake, Phys. Rev. A, 17:1366 (1978).
29. M. Narasimham and R. L. Strombotne, Phys. Rev. A, 4:14(1971).
30. E. Lipworth and R. Novick, Phys. Rev., 108:1434(1957).
31. D. Dietrich, P. Lebow, R. deZafra and H. Metcalf, Bull. Am. Phys. Soc, 21:625(1976).

32. M. Leventhal, Phys. Rev. A, 11:427(1975).
33. H. W. Kugel, M. Leventhal and D. E. Murnick, Phys. Rev. A, 6: 1306(1972).
34. B. Curnutte, C. L. Cocke and R. D. Dubois in "Proc. Int. Conf. Fast Ion Beam Spectrosc.", 6th. Laval (1981). (To be published in Nucl. Instr. Methods).
35. G. P. Lawrence, C. Y. Fan and S. Bushkin, Phys. Rev., 28:1612 (1972).
36. M. Leventhal, D. E. Murnick and H. W. Kugel, Phys. Rev. Lett., 28:1609(1972).
37. H. W. Kugel, M. Leventhal, D. E. Murnick, C. K. N. Patel and O. R. Wood II, Phys. Rev. Lett., 35:647(1975).
38. D. E. Murnick, M. Leventhal and H. W. Kugel in "Proc. Int. Conf. Beam-Foil Spectrosc.", 3rd. Boulder, Colorado (1972).
39. O. R. Wood II, C. K. N. Patel, D. E. Murnick, E. T. Nelson, M. Leventhal, H. W. Kugel and Y. Niv, Phys. Rev. Lett., 48: 398(1982).
40. H. Gould and R. Marrus, Phys. Rev. Lett., 41:1457(1978).
41. M. A. Bouchiat and C. Bouchiat, J. Phys. (Paris), 36:493(1975); M. A. Bouchiat and L. Pottier, J. Phys. Lett. (Paris),37:79 (1976).
42. S. Chu, E. D. Commins and R. Conti, Phys. Lett., 60A:96(1977).
43. W. M. Itano, Phys. Rev. A, 22:1558(1980).
44. A. van Wijngaarden, R. Helbing, J. Patel and G. W. F. Drake, Phys. Rev. A, 25:862(1982).
45. L. Lundin, H. Oona, W. S. Bickel and I. Martinson, Phys. Scr., 2:213(1970).
46. L. P. Lévy and W. L. Williams, Phys. Rev. Lett., 48:1011(1982).
47. L. P. Lévy and W. L. Williams, Phys. Rev. Lett., 48:607(1982).
48. E. E. Salpeter and H. A. Bethe, Phys. Rev., 84:1232(1951).
49. M. Gell-Mann and F. Low, Phys. Rev., 84:350(1951).
50. J. Sucher, Phys. Rev., 109:1010(1958)and pp. 1-26 in "Proc. Workshop Found. Relativist. Theory Atom. Struct.", ANL 80 - 126(1981).
51. W. A. Davis and R. Marrus,Phys. Rev. A, 15:1963(1977).
52. H. G. Berry, R. DeSerio and A. E. Livingston, Phys. Rev. Lett., 41:1652(1978) and Phys. Rev. A, 22:998(1980).
53. R. O'Brian, J. D. Silver, N. A. Jelley, S. Bashkin, E. Träbert and P. H. Heckmann, J. Phys. B, 12:L41(1979).
54. I. A. Armour, E. G. Myers, J. D. Silver and E. Träbert, Phys. Lett. A, 75:45(1979).
55. R. DeSerio, H. G. Berry, R. L. Brooks, H. Hardis, A. E. Livingston and S. Hinterlong, Phys. Rev. A., 24:1872(1981).
56. J. P. Buchet, M. C. Buchet-Poulizac, A. Denis, J. Desesquelles, M. Druetta, J. P. Grandin and X. Husson, Phys. Rev. A, 23: 3354(1981).
57. M. F. Stamp, I. A. Armour, N. J. Peacock and J. D. Silver, J. Phys. B, 14:3551(1981).

58. A. Dalgarno and A. L. Stewart, Proc. Phys. Soc., London, 75:441 (1960).
59. D. Layzer and J. Bahcall, Ann. Phys. (N. Y.), 17:177(1962).
60. P. D. Collins, Astrophys. J., 140:1206(1964).
61. H. T. Doyle, Adv. Atom. Molec. Phys., 5:377(1969).
62. L. N. Labzovskii, Zh. Eksp. Teor. Fiz., 59:168(1970) [Sov. Phys. JETP, 32:94(1971)].
63. G. L. Klimchitskaya and L. N. Labzovskii, Zh. Eksp. Teor. Fiz., 60:2019(1971) [Sov. Phys. JETP 33:1088(1971)].
64. A. M. Ermolaev and M. J. Jones, J. Phys. B, 7:199(1974).
65. L. N. Ivanov, E. P. Ivanova and U. I. Safranova, J. Quant. Spectrosc. Radiat. Transfer., 15:553(1975).
66. L. A. Vainshtein and U. I. Safranova, Atom. Data Nucl. Data Tables, 21:49(1978).
67. W. C. Martin, Phys. Scr., 24:725(1981).
68. G. W. F. Drake, Phys. Rev. A, 19:1387(1979).
69. Y. Accad, C. L. Pekeris and B. Schiff, Phys. Rev. A, 4:516(1971) and B. Schiff, Y. Accad and C. L. Pekeris, Phys. Rev. A, 8: 2272(1973).
70. K. Aashamar, Nucl. Instr. Meth., 90:263(1970).
71. G. W. F. Drake, Nucl. Instr. Methods, in press.
72. H. Araki, Prog. Theo. Phys. Japan, 17:619(1957).
73. A. M. Ermolaev, Phys. Rev. A, 8:1651(1973) and Phys. Rev. Lett., 34:380(1975).
74. K. Aashamar and A. Austvik, Phys. Norv., 8:229(1976).
75. C. Schwartz, Phys. Rev. A, 123:1700(1961).
76. K. S. Suh and M. M. Zaidi, Proc. Roy. Soc. London, Ser. A, 296: 94(1965).
77. R. A. Holt, S. D. Rosner, T. D. Gaily and A. G. Adam, Phys. Rev. A, 22:1563(1980).
78. S. P. Goldman and G. W. F. Drake, to be published.
79. M. Douglas and N. M. Kroll, Ann. Phys. (N.Y.), 82:89(1974).
80. M. L. Lewis and P. H. Serafino, Phys. Rev. A, 18:867(1978).
81. A. Kpanou, V. W. Hughes, C. E. Johnson, S. A. Lewis and F. M. J. Pichanick, Phys. Rev. Lett., 26:1613(1971).
82. S. A. Lewis, F. M. J. Pichanick and V. W. Hughes, Phys. Rev. A, 2:86(1970).
83. R. Bayer, J. Kowalski, R. Neumann, S. Noehte, H. Suhr, K. Winkler and G. zu Putlitz, Z. Phys. A, 292:329(1979).
84. E. G. Myers, P. Kuske, H. J. Andrä, I. A. Armour, N. A. Jelley, H. A. Klein, J. D. Silver and E. Träbert, Phys. Rev. Lett., 47:87(1981).

NEW RESULTS ON MUONIUM AND MUONIC HELIUM

Michael Gladisch

Physikalisches Institut der Universität Heidelberg
D-6900 Heidelberg
Federal Republic of Germany

1 INTRODUCTION

The muon (negative or positive) is a member of the lepton family. Like the electron it interacts only weakly or electromagnetically and acts as a point-like particle in all experiments carried out so far. Therefore the muon is a high precision probe for electromagnetic and weak interactions. No finite size effects or hadronic interactions cloud the theoretical description. The muon behaves like a heavy electron.

The muon-electron interaction is an ideal test of the point-like behavior of the two particles. Any deviation between the theoretical calculation and the measurement would be a sign for new types of interactions. A problem which is of high interest in the light of the gauge theories is the difference between muon and electron. Is there only the difference in mass or are there other differences? What are the conservation laws, how well are they fullfilled, why do they exist?

To answer these questions and to test quantum electrodynamics and weak interaction theories, three different types of experiments[1] are carried out: 1)high energy collisions ($e^+e^-, \mu^+e^-, \mu^-e^-, \ldots$), 2)decay experiments (neutrinoless muon decay,...), 3)bound state experiments (muonium, muonic helium, muonic atoms,...).

This paper describes experiments on muonium and neutral muonic helium. New results in this field were made possible by new types of muon beams and high-intensity accelerators. The muons produced at the high-intensity medium energy (500MeV-800MeV) proton accelerators originate from pion decay,

$$\pi^+(\pi^-) \rightarrow \mu^+(\mu^-) + \nu_\mu(\bar{\nu}_\mu),$$

$$\tau_\pi = 2.2 \cdot 10^{-8} \text{ sec.}$$

The muons are 100% longitudinally polarized with respect to their momentum. The pions are produced by proton-nucleon interactions. The muons decay into an electron and two neutrinos,

$$\mu^+(\mu^-) \rightarrow e^+(e^-) + \nu_e(\bar{\nu}_e) + \bar{\nu}_\mu(\nu_\mu),$$

$$\tau_\mu = 2.2 \cdot 10^{-6} \text{ sec.}$$

The decay is asymmetric with respect to the muon spin. Both the pion decay and the muon decay are induced by the weak interaction.

Three main accelerators were specially designed to provide high pion/muon fluxes, the Swiss Institute for Nuclear Research (SIN) in Switzerland, the Clinton P. Anderson Meson Physics Facility (LAMPF) in the USA and the Tri University Meson Facility (TRIUMF) in Canada. The proton current is in the range of 100μA (TRIUMF) up to 700μA (LAMPF) and the number of muons available is up to a few $10^8 \mu$'s/sec.

The muon lifetime is long compared to the atomic time scale so that experiments in bound atomic systems can be carried out just as in "normal" atoms.

2 MUONIUM

In the following section new results on muonium spectroscopy are presented. Muonium (μ^+e^-) is a hydrogen-like atom consisting of two leptons. It provides an ideal system to determine muon properties and measure muon-electron interactions. μ^+e^- is one of the theoretically best known atomic systems. Another purely leptonic system is positronium (e^+e^-). Muonium is easier to calculate because of two main reasons: a)there are no annihilation diagrams, b)the system permits expansions in powers of m_e/m_μ.

The very precise measurement of the ground state hyperfine structure (hfs) is described. A new successful technique for producing muonium in vacuum has been developed and possible future experiments using this technique are presented in the second part.

2.1 MUONIUM GROUND STATE HYPERFINE STRUCTURE

In this chapter a new high precision measurement[2] of the hfs of muonium in the ground state and of the muon magnetic moment is described. The coupling of the magnetic moments of the muon and the electron splits the ground state into triplet and singlet states. The Zeeman energy levels of the ground state (see Fig.1) are described by the Hamiltonian

$$\mathcal{H} = h\Delta\nu \vec{I}_\mu \cdot \vec{J} - \mu_B^\mu g_\mu' \vec{I}_\mu \cdot \vec{H} + \mu_B^e g_j \vec{J} \cdot \vec{H},$$

in which $\Delta\nu$ is the hfs, $\vec{I}_\mu$ is the muon spin operator, $\vec{J}$ is the electron total angular momentum operator, $g_\mu'(g_j)$ is the muon(electron) gyromagnetic ratio in muonium, $\vec{H}$ is the external static magnetic field and $\mu_B^\mu(\mu_B^e)$ is the muon(electron) Bohr magneton. The quantities g_μ' and g_j are related to the free muon and electron g values g_μ and g_e by[3]

$$g_\mu' = g_\mu\left(1 - \frac{\alpha^2}{3} + \frac{\alpha^2}{2}\frac{m_e}{m_\mu}\right)$$

and

$$g_j = g_e\left(1 - \frac{\alpha^2}{3} + \frac{\alpha^2}{2}\frac{m_e}{m_\mu} + \frac{\alpha^3}{4\pi}\right).$$

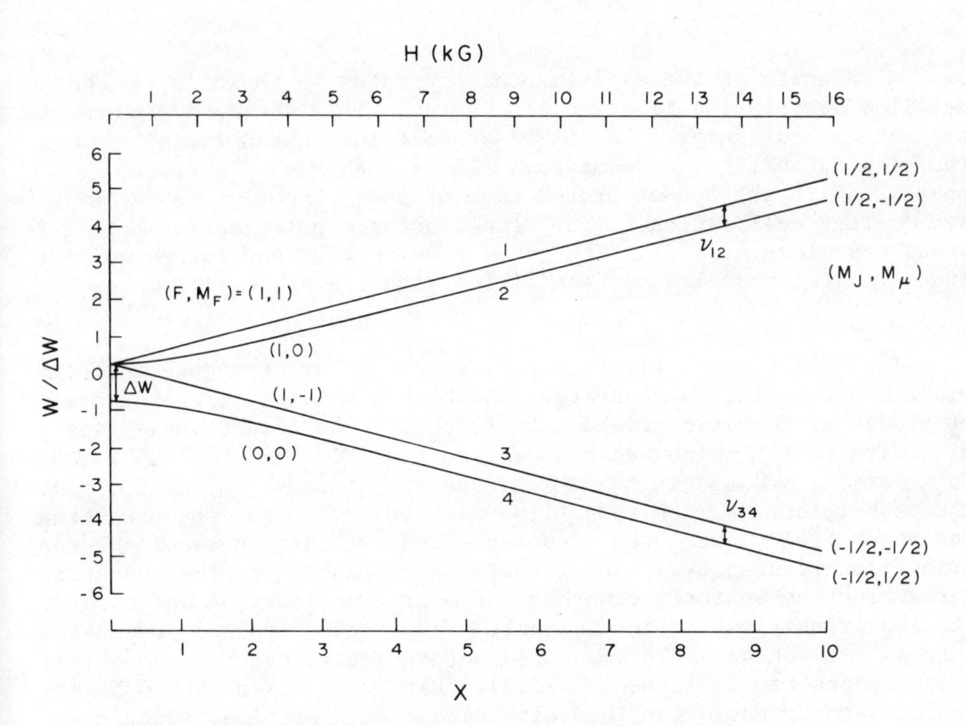

Fig.1. Energy-level diagram for ground-state muonium as a function of magnetic field; a ≃ 4463 MHz.

Measurement of the two Zeeman transitions ν_{12} and ν_{34} (Fig. 1) by the microwave magnetic resonance method yields the values for the hfs, $\Delta\nu$, and the muon magnetic moment μ_μ in units of the proton magnetic moment μ_p,

$$\Delta\nu = \nu_{12} + \nu_{34},$$

$$\mu_\mu/\mu_p = \left[\frac{\Delta\nu^2 - 2\beta\Delta r_e' - \Delta^2}{2\beta(2\beta r_e' + \Delta)}\right]\left[1 - \frac{\alpha^2}{3} + \frac{\alpha^2 m_e}{2\, m_\mu}\right]^{-1},$$

where

$$r_e' = \frac{g_j \cdot \mu_B^e}{\mu_p},$$

$$h\beta = \mu_p \cdot H,$$

$$\Delta = \nu_{34} - \nu_{12}.$$

A diagram of the experimental apparatus is shown in Fig.2. A detailed description is given elsewhere[2]. Only the main components are described here. The LAMPF stopped muon channel was tuned to the high intensity low momentum (25 - 28MeV/c) "surface" muon beam[4]. With the 800MeV proton beam of 300μA the flux was $2 \cdot 10^6$ μ^+ sec^{-1} after collimation. The longitudinal polarization of the muons was close to 100%. The muons came to rest and formed muonium in the target vessel which was filled with 1/2 atm or 1 atm of Kr gas.

The positrons originating from muon decay were detected with plastic scintillation counters. The high precision solenoid magnet provided a magnetic field of 13.6kG. The target vessel was a microwave cavity which was resonant in the TM_{110} mode at 1.918 GHz (ν_{12} at 13.6kG) and in the TM_{210} mode at 2.545GHz (ν_{34} at 13.6kG). The transitions (a muon spin flip) were observed by sweeping the magnetic field over the resonance and switching between the two transition frequencies. The relative difference in the counting rates in the positron counters for microwave power on and off for the two transitions is the signal. In Fig.3 typical resonance curves are shown. The measured points are fitted to theoretical lines shapes (solid lines in Fig.3). Data were taken at 0.5 atm and 1 atm Kr gas. Combined with data from a previous experiment[5] at 1.7atm and 5.2atm and extrapolating to zero gas density the results are:

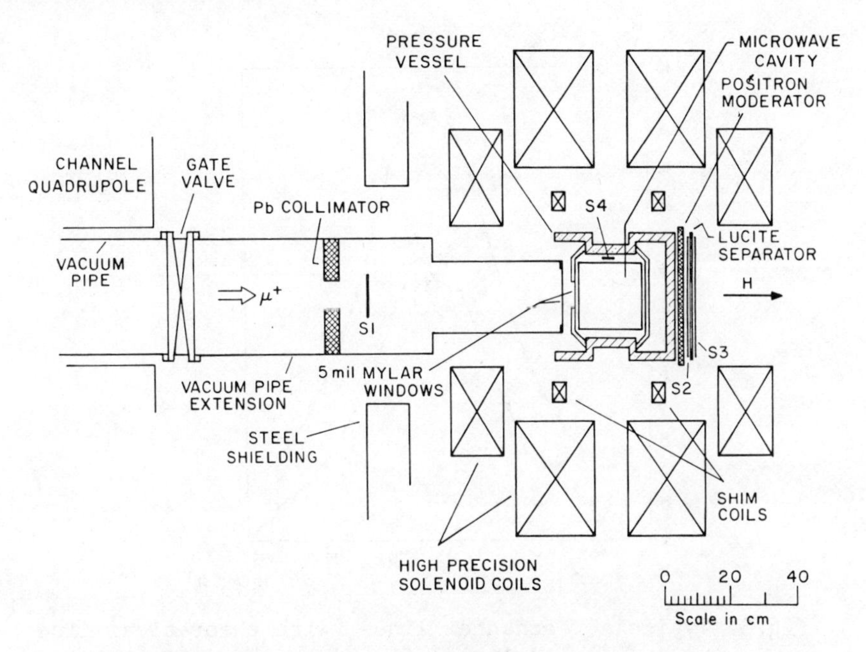

Fig.2. Schematic diagram of the experimental setup used for the precision measurement of muonium $\Delta\nu$ and μ_μ/μ_p. S1-4 are plastic scintillators.

$$\Delta\nu = 4\ 463\ 302.88(16)\ \text{kHz}\ (0.036\text{ppm}),$$

$$\mu_\mu/\mu_p = 3.183\ 3461(11)\ (0.36\text{ppm}).$$

The one standard deviation errors include systematic and random errors. A variety of systematic effects increases the total error by a factor of two compared to the statistical alone.

The value of μ_μ/μ_p agrees with the one obtained[6] from the entirely different technique of measuring the muon spin rotation in liquid bromide which gave

$$\mu_\mu/\mu_p = 3.183\ 344\ 1(17)\ (0.53\text{ppm}).$$

Combining the two moment measurements gives the world avarage

$$\mu_\mu/\mu_p = 3.183\ 345(9)\ (0.30\text{ppm}).$$

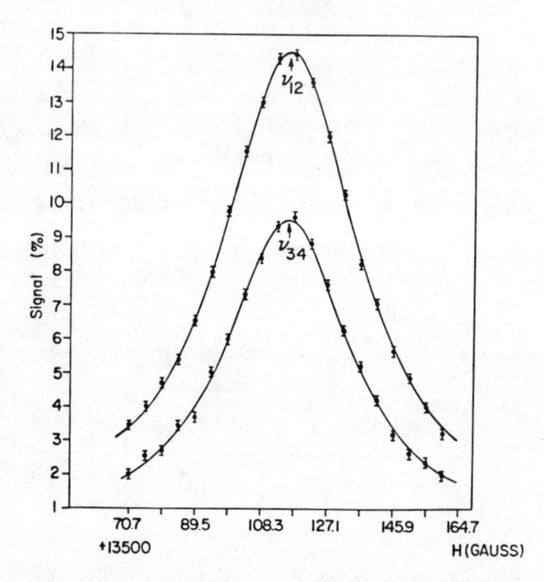

Fig.3. Typical resonance lines with theoretical line shapes (solid lines) fitted to the data points. Krypton pressure 1 atm, data taking time for each pair of resonance lines was less than 2 h.

This value and the experimental values of $g_{\mu}+$ (ref.7) and of μ_p/μ_B^e (ref.8) determines the mass of the muon as follows:

$$m_{\mu}/m_e = (g_{\mu}/2)(\mu_p/\mu_{\mu})\mu_B^e/\mu_p = 206.768\ 259(62)\ (0.30\text{ppm}).$$

The value of $\Delta\nu$ agrees with an earlier measurement[5] but is more precise.

The theoretical value[9] for $\Delta\nu$ is

$$\Delta\nu = 4\ 463\ 303.7(1.7)(3.0)\ \text{kHz}.$$

The 1.7kHz uncertainty comes from combining a 1.3kHz uncertainty from μ_{μ}/μ_p with a 1.0kHz uncertainty in α. The 3.0kHz theoretical uncertainty is due to uncalculated terms in the nonrecoil radiative corrections.

Comparison of the experimental and theoretical values of $\Delta\nu$ gives

$$\Delta\nu_{exp} - \Delta\nu_{theor} = -0.8 \pm 3.4\text{kHz},$$

where the dominant error comes from the theoretical value. The agreement is excellent and is a strong test for the validity of bound state QED.

An alternate approach is to use the measurement of $\Delta\nu$ and μ_μ/μ_p and to accept the theory and determine α. The result is

$$\alpha^{-1} = 137.035\,974(50),$$

which is in good agreement with the values obtained from the ac-Josephson effect[10], the quantum Hall effect[11] and from the electron moment anomaly[12].

Effects of the weak interaction[13] are expected to shift $\Delta\nu$ by +0.07kHz which is half of the experimental error but the theory is far away from isolating this shift from other effects.

Improvement in the precision of the measurement of $\Delta\nu$ and μ_μ/μ_p is only possible with line narrowing techniques. The development of high intensity pulsed muon beams in the near future (the proton storage ring at LAMPF) make it feasable to use the "old muonium" technique for line narrowing[14]. Improvement in the theoretical value requires calculation of the nonrecoil radiative correction.

2.2 MUONIUM IN VACUUM

The observation of muonium in vacuum[15] makes it possible to do experiments on muonium which require a collision-free environment.

Attempts to find thermal muonium in vacuum failed. An early experiment at the Space Radiation Effects Laboratory (SREL) seemed to show[16] that μ^+ stopped in thin gold foils emerges as muonium at thermal energies. But more sensitive experiments at LAMPF[17] and SIN[18] observed no muonium formation.

However with a technique similar to beam neutralization in proton beam-foil spectroscopy[19] energetic muonium in vacuum was observed[15].

The apparatus used is shown in Fig.4. The stopped muon channel at LAMPF was tuned for the surface μ^+ mode with a momentum of 27 MeV/c (4MeV energy). The μ^+ are degraded in energy by the windows and polyethylene placed in the air between the windows. The μ^+ impinge on a thin metal foil inside the vacuum chamber. Because of the energy spread in the μ^+ beam and straggling of the energy loss in the degrader the μ^+ beam emerging from the foil target has an energy spread of 1-2MeV. The expected charge state distribution can be scaled from calculations for protons[20]. The

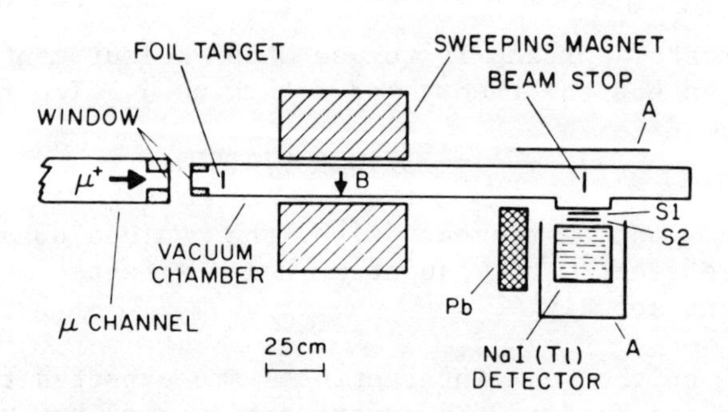

Fig.4. Apparatus for the observation of muonium in vacuum.

sharp decrease of the neutral muonium fraction as a function of velocity suggests that mainly muonium with energies less than ~20keV is formed.

The 5 kG magnet sweeps all charged particles away. Only neutral particles can reach the beam stop. The beam stop is viewed by a large NaI(Tl) detector. Observation of positrons with the characteristic energy spectrum (Michel spectrum) indicates the presence of a neutral containing a muon, muonium.

The measurement showed a small dependence of the amount of muonium formed on the atomic number of the foil target. The measured rates at the NaI were between 0.016(4) sec^{-1} for gold and 0.034(4) sec^{-1} for beryllium with $3 \cdot 10^6$ sec^{-1} incoming μ^+. Assuming isotropic angular distribution for the emerging muonium atoms and taking the efficiency of the positron detector into account the projected rates for muonium production are in the range of 10^3-$3 \cdot 10^3$ μ^+e^- sec^{-1} or about 10^{-3} of the incident beam.

Calculations show that the fraction of muonium formed as a function of the energy of the incident beam increases as the available rate of muons decreases. The absolute number of muonium atoms formed is nearly constant. Hence to go to very low momentum (7-10MeV/c) increases the relative fraction of muonium atoms formed and decreases background. The stopped muon channel at LAMPF can be tuned down to less than 3MeV/c but at less than 5MeV/c the intensity of the μ^+ beam decreases drastically due to decay of muons in the beamline (length ~30 m).

The fact that μ^+e^- can now be studied in vacuum is very important for a number of fundamental experiments. One is the measurement of the Lamb shift in the first excited state of muonium (Fig.5). The beam-foil method produces not only the n=1 state but excited states as well[21] with a probability roughly as $1/n^3$. We expect that ~15% of the muonium formed is in the 2S state.

The major motivation for measuring the Lamb shift is that it is a pure QED test. In hydrogen the uncertainty in the proton size limits the accuracy in the theory[22] to 10ppm, which is the same as for the experimental value[23]. This uncertainty does not exist in muonium. Because of the difference in the masses of the proton and the muon different recoil corrections must be taken into account[24].

An experiment has been proposed[25] at LAMPF to observe the 2S state and to measure the Lamb shift. To get to 10 - 1ppm errors requires line narrowing techniques like the ones used in the hydrogen Lamb shift measurement.

Another important experiment is the sensitive search[26] for the

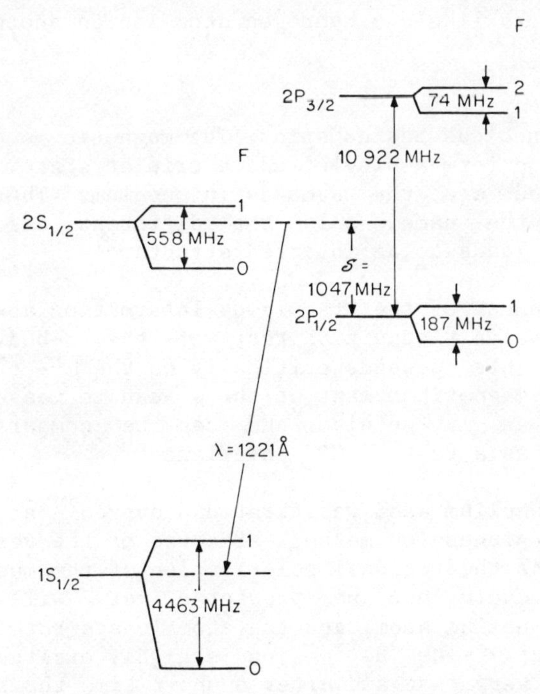

Fig.5. Ground and first exited states of muonium

conversion of muonium to antimuonium ($\mu^+e^- \rightarrow \mu^-e^+$). This conversion would violate muon number conservation[27] but is predicted by some of the gauge theories[28].

Other possible experiments include inducing optical transitions in muonium, measurement of the ionization of muonium in gases and the search for the negative ion $\mu^+e^-e^-$.

3. MUONIC HELIUM

In muonic helium, three particles, the helium nucleus, the muon and the electron are bound in one system, all of them of different masses and not affected by the Pauli exclusion principle. This system provides a sensitive testing ground for the 3-body Schrödinger equation.

The muon is bound to the doubly charged helium nucleus. It is in a hydrogen like 1S state, but the Bohr radius,

$$a \simeq \frac{a_0 m_e}{2m_\mu} \simeq a_0/400 \simeq 1.3 \cdot 10^{-11} \text{cm},$$

is much smaller than in hydrogen and halfway between atomic and nuclear dimensions. The muon shields one charge. Hence the electron sees a large pseudo-nucleus with unit charge. The electron will therefore be bound in a hydrogenic 1S state. The whole system looks like one hydrogen atom inside another.

3.1 $^4He^{++}\mu^-e^-$

The ^4He nucleus has no spin. The magnetic moments of the μ^- and e^- couple and form a singlet and a triplet state. The magnetic moments involved are the same as in muonium. Therefore the hfs will be roughly the same, only the different sign of the μ^- magnetic moment leads to an inverted structure.

The measurement of the hfs yields information about the $e^- - \mu^-$ interaction and a very sensitive test of the 3-body Schrödinger equation. The hfs depends critically on the $\mu^- - e^-$ correlation. In addition the magnetic moment of the μ^- can be measured with the Zeeman effect as in muonium and can be compared with the μ^+ magnetic moment as a test of CPT invariance.

The muonic helium atom was first discovered[29] at SREL by means of the Larmor precession method. Because of the very complicated formation process the residual polarisation of the muons in neutral muonic helium could not be predicted very well. The μ's are captured by the helium atom, and the atom looses both electrons due to Auger effect. The $^4He^{++}\mu^-$ ion is highly excited ($n \simeq 14$). The deexcitation is very fast and after a short time the ion is in its ground state. The binding energy for electrons is like in hydrogen. Therefore it cannot capture an electron from the neighboring helium atoms. An electron donor is needed. In all experiments so far Xe is used in concentrations of 1-2%, which is sufficient to form ≃100% neutral muonic helium.

The residual polarisation is suprisingly low (5%), which makes it much harder to measure the hfs compared to muonium (100% polarisation). In addition the μ^- fluxes at the accelerators are much lower than the μ^+ fluxes and one must run at higher momentum (30-50MeV/c) to get polarized negative muons. Helium has a very low density, to form a large fraction of $^4He^{++}\mu^-e^-$ out of the incoming muons, one must use high helium densities, that means high pressures.

The hfs was first measured[30] at zero magnetic field at SIN. A

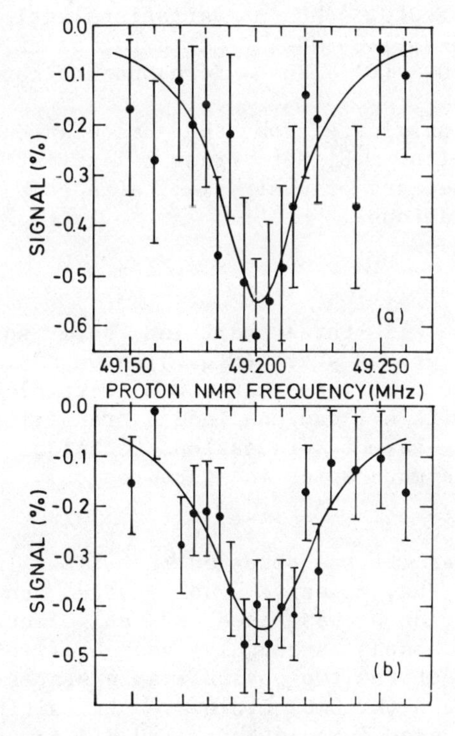

Fig.6. Typical resonance curves for the ν_{12} transition at (a) 15atm and (b) 5 atm. The data for these curves were obtained in (a) 24h and (b) 100h.

more precise measurement[31] at high magnetic fields was done at LAMPF. A report on these experiments is given elsewhere[32], therefore only new results are presented here. At magnetic fields of 11.5 and 13.6 kG, ν_{12} and ν_{34} transitions were observed (Fig.6). Data were taken at two different pressures (5 and 15 atm). Extrapolating to zero density gives (using $\mu_{\mu^+} = \mu_{\mu^-}$)

$$\Delta\nu = 4\ 465\ 004(29)\ \text{kHz}\ (6.5\text{ppm}).$$

Treating both $\Delta\nu$ and μ_{μ^-} as free parameters gives for the magnetic moment

$$\mu_{\mu^-}/\mu_p = 3.183\ 28(15)\ (47\text{ppm}),$$

in agreement with the more accurate value for μ_{μ^+}/μ_p as required by CPT invariance. The quoted errors are mainly statistical.

The theoretical calculations have been carried out by several authors, their latest results are:

4465.0(0.3) MHz[33]	variational calculation,
4464.3(1.8) MHz[34]	perturbation calculation,
4460 (?) MHz[35]	Born-Oppenheimer calculation.

The uncertainties originate from the calculation of the 3-body Schrödinger equation. Higher order relativistic and QED effects cannot be tested because of their small size compared to the error bars in the calculations.

3.2 $^3He^{++}\mu^-e^-$

In $^3He^{++}\mu^-e^-$ all three paricles have spin and interact magnetically. The ground state (Fig.7) shows a large muonic hfs and a smaller electronic hfs in the muonic triplet state. Only in the ground state with G=1 can one induce transitions which flip the muon spin. The Larmor precession technique to observe the polarization of the muon spins is of course only sensitive to the G=1 state.

This atom was first observed[36] at SIN. In a target vessel filled with 20 atm ^{3}He, negative muons were stopped and formed $^3He^{++}\mu^-$. Again Xenon was used as an electron donor. The polarization of the muons is roughly one half of $^4He^{++}\mu^-e^-$, as expected because of the two possible spin states G=0 and G=1. At "zero" magnetic field (H<20mG) microwave transitions between the F=1/2 and F=3/2 states could be observed (Fig.8).The preliminary result for the hfs is:

$$\Delta\nu = 4\ 166.3(2)\ \text{MHz}.$$

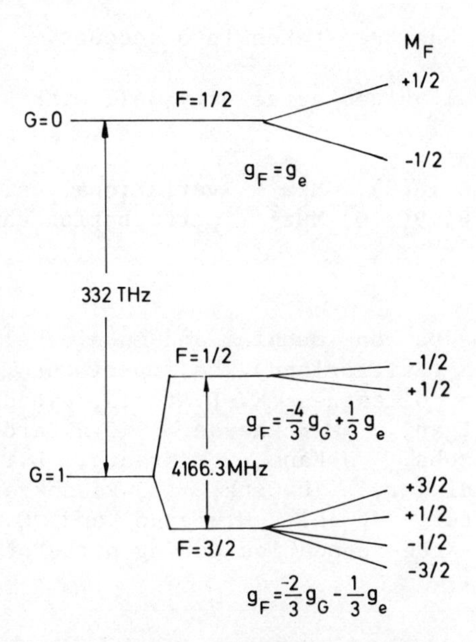

Fig.7. Ground-state hfs of $^3He^{++}\mu^-e^-$

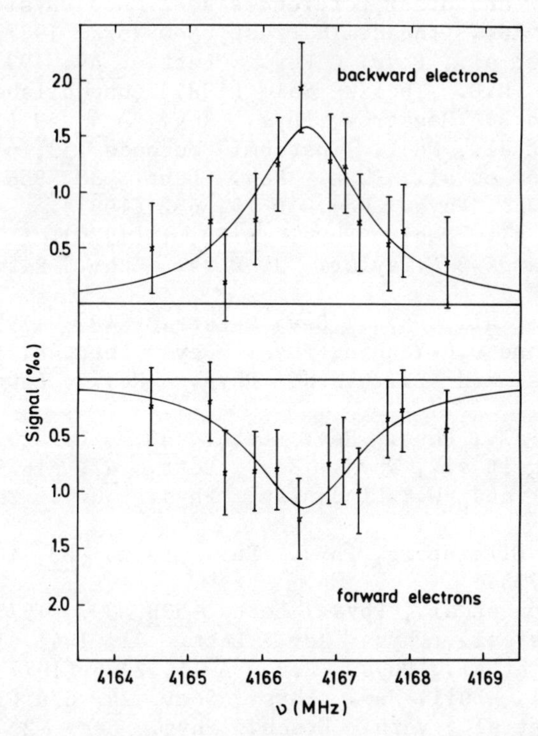

Fig.8. Hfs resonance signal obtained from μ-decay electrons in and opposite to the direction of the polarization of the muon

The pressure shift has been taken into account.

The theoretical values agree very well with the experimental result.They are:

4 166.8(3) MHz[32] variational calculation,
4 164.9(3.0) MHz[3/] perturbation calculation.

ACKNOWLEDGEMENTS

The measurements on muonium and muonic helium where done at LAMPF (USA) and SIN (Switzerland).The experiments where carried out with the following colleagues: K.-P.Arnold, A.Badertscher, W.Beer, P.R.Bolton, P.O.Egan, M.Eckhause, C.J.Gardner, M.Greene, V.W.Hughes, W.Jacobs, J.Kane, M.Krauth, D.C.Lu, F.G.Mariam, U.Moser, H.J.Mundinger, H.Orth, J.Rosenkranz, W.Schäfer, P.A.Souder, J.Vetter, W.Wahl, M.Wigand and G.zu Putlitz. The support of the Max-Kade foundation during parts of my stay in the US is acknowledged.

REFERENCES

1) For a review and other references see: Muon Physics, V.W.Hughes and C.S.Wu, Eds. (Academic Press, New York, 1977) Vol.1-3.
2) F.G.Mariam et al., Phys. Rev. Lett. 49, 993 (1982) and F.G.Mariam, Ph.D. Thesis, Yale (1981) (unpublished).
3) H.Grotch and R.A.Hegstrom, Phys. Rev. A 4, 59 (1971).
4) H.W.Reist et al., Nucl. Instrum. Methods 153, 61 (1978).
5) D.E.Casperson et al., Phys. Rev. Lett. 38, 956, 1504 (1977).
6) E.Klempt et al.,Phys. Rev. D 25, 652 (1982).
7) J.Bailey et al., Nucl. Phys. B150, 1 (1979).
8) E.R.Cohen and B.N.Taylor, J. Phys. Chem. Ref. Data 2, 663 (1973)
9) G.T.Bodwin et al., Phys. Rev. Lett. 48, 1799 (1982) and E.A.Terray and D.R.Yennie, Phys. Rev. Lett.48, 1803 (1982).
10) E.R.Williams and P.T.Olsen, Phys. Rev. Lett. 42, 1575 (1979).
11) D.C.Tsui et al., Phys. Rev. Lett. 48, 3 (1982).
12) P.Schwinberg et al., Phys. Rev. Lett. 47, 1679 (1981) and T.Kinoshita and W.B.Lindquist, Phys. Rev. Lett. 47, 1573 (1981).
13) M.A.Bég and G.Feinberg, Phys. Rev. Lett. 33, 606 (1974) and 35, 130 (1975).
14) D.E.Casperson et al., Phys. Lett. 59B, 397 (1975).
15) P.R.Bolton et al., Phys. Rev. Lett. 47, 1441 (1981).
16) B.A.Barnett et al., Phys. Rev. A 15, 2246 (1977).
17) W.Beer et al., Bull. Am. Phys. Soc. 24, 675 (1979).
18) K.P.Arnold et al., Verh. Dtsch. Phys. Ges. 2, 552 (1979).
19) H.G.Berry, Rep. Prog. Phys. 40, 155 (1977).
20) J.A.Phillips, Phys. Rev. 97, 404 (1955).

21) G.Gabrielse, Phys. Rev. A 23, 775 (1981).
22) P.Mohr, Phys. Rev. Lett 34, 1050 (1975) and G.W.Erickson, Phys. Rev. Lett. 27, 780 (1971).
23) S.R.Lundeen and F.M.Pipkin, Phys. Rev. Lett. 46, 232 (1981).
24) D.A.Owen, Phys. Lett. 44B, 199 (1973).
25) P.O.Egan et al., LAMPF Proposal 724, Los Alamos National Laboratory Report LA-7444-SR, Rev. 3, 397 (1982).
26) J.J.Amato et al., Phys. Rev. Lett. 21, 1709 (1968).
27) B.Pontecorvo, Zh. Exsp. Teor. Fiz. 33, 549 (1957) and G.Feinberg and S.Weinberg, Phys. Rev. Lett. 6, 381 (1961).
28) A.Halprin, Phys. Rev. Lett. 48, 1313 (1982).
29) P.A.Souder et al., Phys. Rev. A 22, 33 (1980).
30) H.Orth et al., Phys. Rev. Lett. 45, 1483 (1980).
31) C.J.Gardner et al., Phys. Rev. Lett. 48, 1168 (1982).
32) P.O.Egan, Atomic Physics 7, D.Kleppner and F.M.Pipkin, Eds. (Plenum Publishing Corporation, New York) 373 (1981).
33) K.-N.Huang and V.W.Hughes, Phys. Rev. A 26, 2330 (1982).
34) S.D.Lakdawala and P.J.Mohr, Abstracts of the Eighth International Conference on Atomic Physics, A59 (1982).
35) R.J.Drachman, Phys. Rev. A 22, 1755 (1980).
36) K.P.Arnold et al., Abstracts of the Eighth International Conference on Atomic Physics, A58 (1982).
37) S.D.Lakdawala and P.J.Mohr, Phys. Rev. A 24, 2224 (1981).

STRUCTURE AND DYNAMICS OF ATOMS PROBED

BY INNER-SHELL IONIZATION

Werner Mehlhorn

Fakultät für Physik
Universität Freiburg
D-7800 Freiburg, FRG

INTRODUCTION

The aim of this talk is to review the results obtained in the studies of inner-shell ionization and decay by means of photoelectron (PES) and Auger electron spectrometry (AES). Since the scope of such a talk would be much too broad I will consider AES only in connection with photon or electron impact ionization. AES associated with ion-atom collisions, important for studies of multi-ionized atoms, will not be discussed. The angular anisotropy of photo- and Auger electrons will also not be treated in spite of the fact that for inner-shells the ß-parameter may depend sensitively on the structure and the dynamics of the atom, in particular for heavier atoms (e.g. Xe(4d)[1a,b]). Therefore my talk will be more selective than exhaustive. By means of selected examples of inner-shell photo-and Auger electron spectra I will demonstrate the sort of information one can obtain for atoms. The main emphasis will be laid on effects caused by the electron correlation. Reviews on parts of this subject have been given earlier by Siegbahn et al.[2], Krause[3a,b], Mehlhorn[4a,b], Wendin[5] and most recently by Siegbahn and Karlsson[6].

Electron correlation changes the structure and the dynamics of an independent-particle model atom considerably and can thus be studied in various ways, e.g. 1) experimental values of energies, intensities (cross sections), branching ratios and line widths of diagram transitions deviate from the theoretical ones obtained for the best independent particle model, the HF model, 2) satellite lines, which would be forbidden in the independent-particle model, occur. The intensities of these satellite lines relative to the corresponding diagram lines are a direct measure of the strength of the electron correlation.

By means of PES and AES the following inner-shell ionization and decay processes are studied

$$\gamma + A \rightarrow A^{+} + e^{-}_{Ph}\ (E_1) \qquad \text{PES} \qquad (1a)$$

$$\hookrightarrow A^{++} + e^{-}_{A} \qquad \text{AES} \qquad (2a)$$

$$\gamma + A \rightarrow A^{+*} + e^{-}_{Ph} \qquad \text{PES} \qquad (1b)$$

$$\hookrightarrow A^{++} + e^{-}_{A} \qquad \text{AES} \qquad (2b')$$

$$\hookrightarrow A^{++*} + e^{-}_{A} \qquad \text{AES} \qquad (2b'')$$

$$\gamma + A \rightarrow A^{++} + 2e^{-}_{Ph} \qquad \text{(PES)} \qquad (1c)$$

$$\hookrightarrow A^{+++} + e^{-}_{A} \qquad \text{AES} \qquad (2c)$$

Here the processes (1a) and (2a) are the normal or diagram transitions. In process (1a) exactly one electron participates and in process (2a) exactly two electrons are involved according to the one-electron operator in the photoprocess and the two-electron Coulomb operator $1/r_{12}$ in the Auger transition, respectively. Processes (1b) and (1c), where a second electron is either excited or ionized in the photoprocess, are only possible through electron correlation and manifest themselves by satellite lines in PES (1b) or AES (2b', 2b'', 2c). In the process (1c) two electrons are emitted simultaneously which leads to a continuous energy distribution of electrons and can thus hardly be studied by means of PES (but see e.g. Krause et al.[7]). Rather, inner-shell ionization accompanied by outer-shell ionization (process (1c)) has been studied most conveniently through Auger satellites of process (2c) (e.g. Krause et al.[8,9], Schmidt[10], Chattarji et al.[11]).

Electron correlation, conventionally treated by configuration interaction (CI), can enter through the initial and/or the final state of the transition. CI may be divided into initial state configuration interaction (ISCI) and final state configuration interaction (FSCI). The latter is usually subdivided into final ionic state CI (FISCI) and final continuum state CI (FCSCI). Since the final state of the photoionization is the initial state of the Auger transition it follows that FSCI studied by means of PES is identical to ISCI studied by means of AES.

Electron correlation in the final Auger state can be investigated through transition (2a) or, more directly, through transitions like

$$A^{+} \rightarrow A^{+*} + e_A^{-} \qquad \text{AES} \qquad (2a')$$

$$\rightarrow A^{+++} + 2e_A^{-} \qquad \text{(AES), Ion charge spectr.} \qquad (2a'')\ ,$$

following the single inner-shell ionization of process (1a). Processes (2a') and (2a''), where actually three electrons are involved in the Auger transition, can occur only via electron correlation. Process (2a') leads to satellite lines in AES, process (2a'') leads to a continuous energy distribution of the two ejected electrons[12] and has been studied mostly by means of ion charge spectrometry (see e.g. Åberg[13] and references therein).

The usual mode of inner-shell ionization for AES is by electron impact where neither the scattered nor the inner-shell ejected electron is detected. The total information stems therefore from processes (2a, 2a', 2a''), (2b', 2b'') and (2c).

So far we have assumed that the inner-shell ionization and the following Auger transition can be treated independently as a two-step process. This is true only if the following conditions are fulfilled. Firstly, the photoelectron is emitted with large enough energy E_1 and will therefore not interact with the following Auger electron, i.e. one can neglect what is called post-collision interaction. Secondly, the interaction of the core hole state with the Auger continuum is weak, in which case the Auger decay can be treated by the Golden Rule giving a certain level width to the core hole state. However, if the interaction is very strong, then both the core hole state and the continuum are strongly modified. In this case the Auger decay and the inner-shell ionization have to be treated in a unified manner[5,13b,c,14a-c].

EXPERIMENTAL

In Fig. 1 the general scheme of an experimental apparatus for PES and AES is shown (for a detailed and most recent review see Siegbahn and Karlsson[6]). Primary beams of either characteristic X-radiation (without or with monochromatization), or monochromatized synchrotron radiation or electrons are used. The atoms being studied are either in gaseous or in vapour form (in case of metals). The ejected electrons are energy-analyzed by means of a high-resolution electrostatic electron spectrometer, where presently an energy resolution of $2 \cdot 10^{-4}$ can be obtained in special cases. The electron detection is by means of a channeltron or, better still, by means of a multi-channel detection system[6].

The progress made in AES during the last two decades can be seen for example in the $L_{2,3}MM$ Auger spectrum of Ar. In Fig. 2a

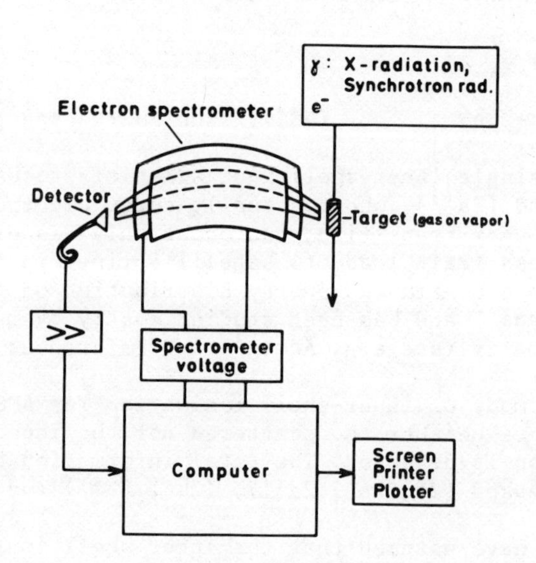

Fig. 1. Principal setup of apparatus for PES and AES

the first Auger spectrum of Ar, measured in 1960[15] for a gaseous target and a resolution of $2.4 \cdot 10^{-2}$,[16] is shown. Fig. 2b and 2c show the same spectrum, measured in 1973 with resolution $5 \cdot 10^{-4}$, and the high energy group, measured in 1981[17a,b] with resolution $3 \cdot 10^{-4}$, respectively.

CORRELATION EFFECTS STUDIED BY MEANS OF PES

Shift and distortion of main photoelectron line

The binding energy $E_B(n\ell j)$ of an inner-shell electron $n\ell j$ can be measured most directly through the energy E_1 of the photoelectron in process (1a): $E_B = h\nu - E_1$. Fig. 3 shows as example the photoelectron spectrum of free Hg atoms ionized by monochromatized Al Kα radiation[20]. Theoretically, the best values for binding energies are given by relativistic (Dirac-Slater[18] or Dirac-Fock[19]) ΔSCF calculations that include complete relaxation, Breit-energy contribution and vacuum polarization. Good agreement is found (generally within 1 eV) with experimental binding energies of free atoms[6,21-23], for which solid-state[24] and chemical shifts are absent (see Table 1). Also, for inner-shell binding energies the correlation effects tend to be small since the correlation in the ground as well as in the inner-hole state are very nearly the same.

On the other hand it has been found that for certain inner-

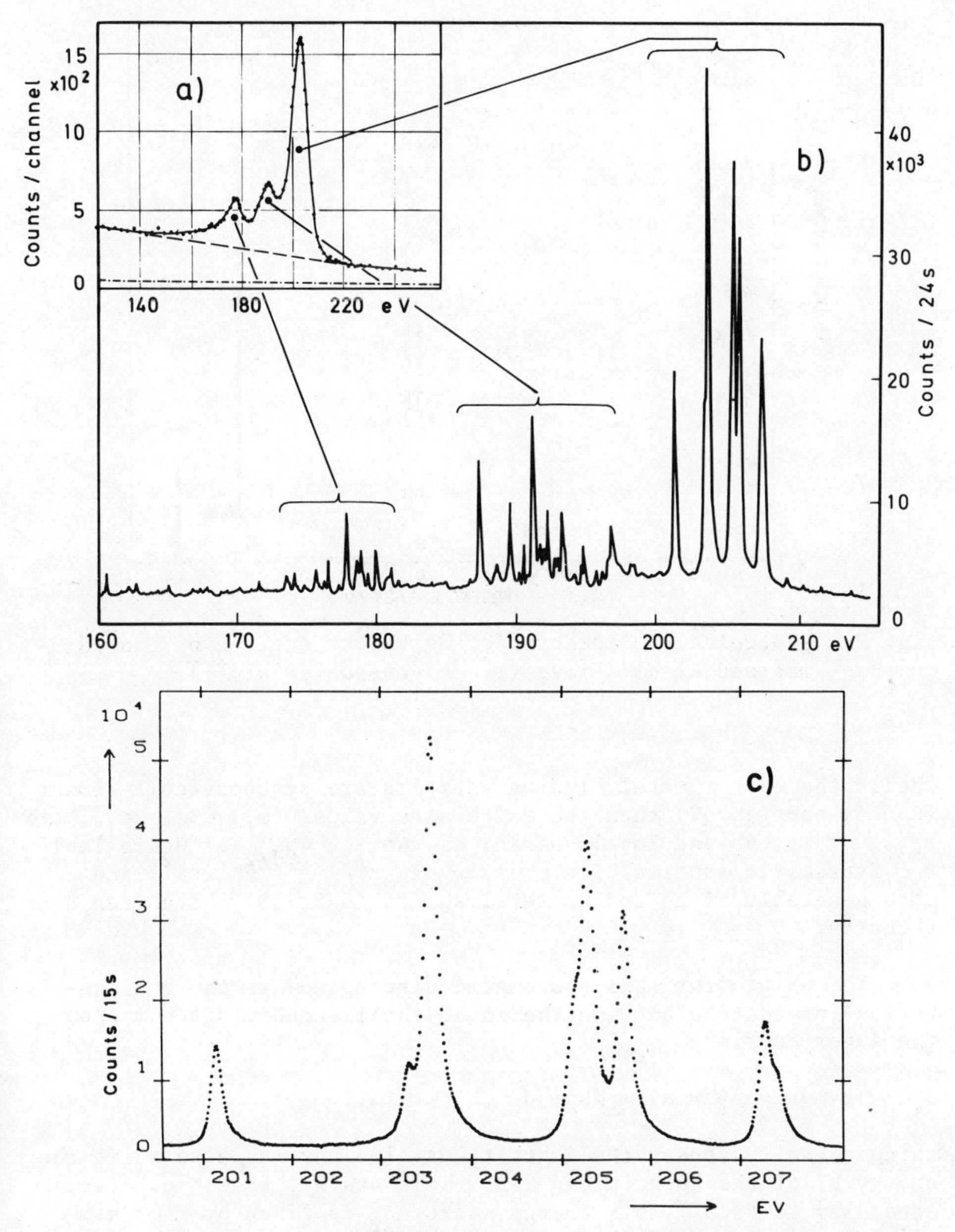

Fig. 2. Progress in AES demonstrated at the $L_{2,3}MM$ Auger spectrum of Ar. a) First spectrum of gaseous Ar target excited by X-ray bremsstrahlung and taken at 2.4% resolution of the spectrometer (Mehlhorn 1960[15]). b) Spectrum excited by 3 keV electrons and taken at $5\cdot10^{-4}$ resolution (Werme et al. 1973[16]). c) High-energy part of the spectrum excited by 2 keV electrons and taken at $3\cdot10^{-4}$ resolution (Huster 1981[17b]).

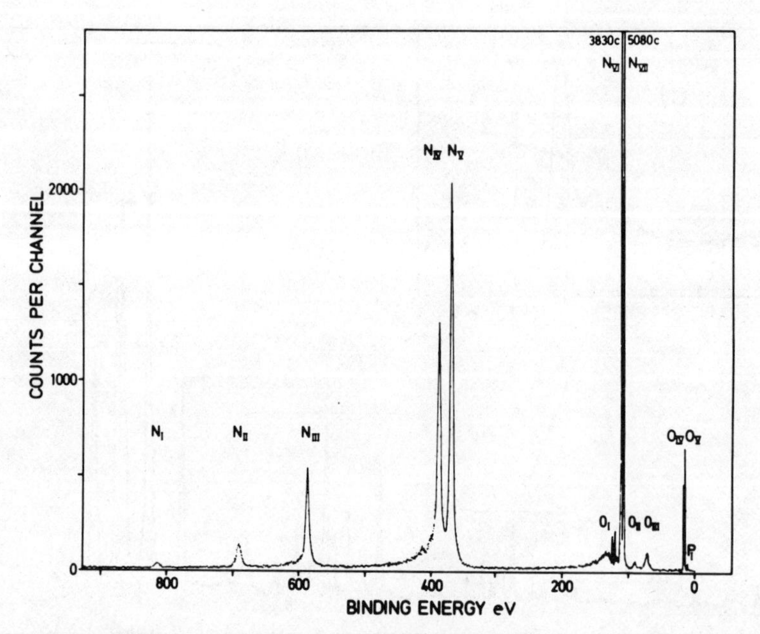

Fig. 3. Photoelectron spectrum of Hg vapour excited by monochromatized Al Kα X-rays (from Svensson et al.[2o]).

shells the experimental binding energies are systematically lower (up to about 10 eV) than the calculated values (e.g. 3s, $3p_{1/2}$ $3p_{3/2}$ of Kr, Rb and Xe; 4s of Xe; 4s, $4p_{1/2}$, $4p_{3/2}$ of Hg in Table 1). Systematic studies of the 2s level shifts[27] and more generally of ns and np level shifts (n=3,4,5)[5] have been done recently.

The origin for this systematic discrepancy is the configuration interaction between the inner-shell vacancy state Φ and the Auger continuum ψ_ε[13b,c]

$$\underline{\psi} = a\Phi + \int b_\varepsilon \psi_\varepsilon d\varepsilon ,$$

which actually causes the Auger transition but may also shift the energy E_Φ of the unperturbed inner-hole state Φ (see Fig. 4). According to Fano[28] this energy shift ΔE_r is given by (for simplicity we assume only one Auger continuum)

$$\Delta E_r = E_r - E_\Phi = P \int_0^\infty \frac{|V_{\Phi\varepsilon}|^2 d\varepsilon}{E_\Phi - (E_f + \varepsilon)} \qquad (1)$$

where $V_{\Phi\varepsilon} = \langle\psi_\varepsilon|H|\Phi\rangle$ is the matrix element between the discrete state Φ and the continuum state $\psi_\varepsilon = \psi_\varepsilon(A^{++}(E_f) + e^-(\varepsilon))$.

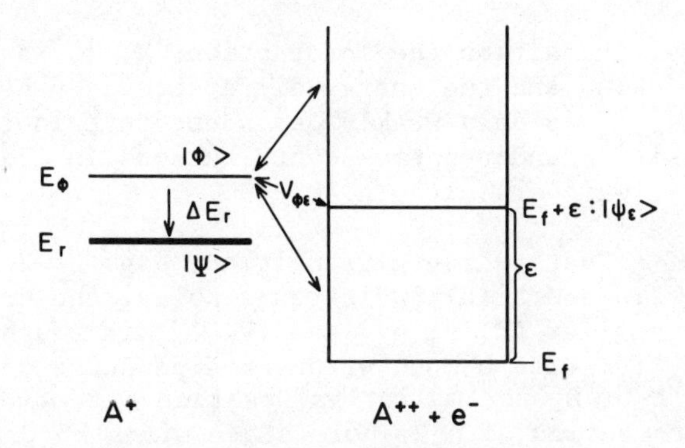

Fig. 4. Shift ΔE_r of inner-hole state Φ due to the interaction $V_{\Phi\varepsilon}$ of Φ with the Auger continuum ψ_ε.

Table 1. Examples of experimental binding energies of inner-shell electrons in eV and comparison with theoretical values obtained by Dirac-Slater (ΔSCF) calculations.

Element and shell	Experiment	Ref.	Theory[18]	Δ(Exp-Th)
Ne 1s	870.2(1)	25	869.16	1.0
Mg 1s	1311.3(3)	21	1310.6	-0.9
2s	96.5(3)	21	97.59	-1.1
Ar 2s	326.3(1)	2	326.66	-0.4
$2p_{1/2}$	250.6(1)	2	249.95	0.6
$2p_{3/2}$	248.5(1)	2	247.80	0.7
Kr 3s	292.8(1)	26	295.85	-3.1
$3p_{1/2}$	222.2(1)	26	224.82	-2.6
$3p_{3/2}$	214.4(1)	26	216.81	-2.4
Rb $3p_{1/2}$	254.3(1)	23	257.10	-2.8
$3p_{3/2}$	245.4(1)	23	247.86	-2.5
$3d_{3/2}$	117.4	21	117.3	0.1
$3d_{5/2}$	116.0	21	115.8	0.2
Xe 3s	1148.7(5)	26	1152.4	-3.7
$3p_{1/2}$	1002.1(3)	26	1005.3	-3.2
$3p_{3/2}$	940.6(2)	26	943.1	-2.5
$3d_{3/2}$	689.35(8)	26	688.8	0.6
$3d_{5/2}$	676.70(8)	26	676.2	0.5
4s	213.32(3)	26	222.67	-9.4
Hg 4s	809.4	20	817.6	-8.2
$4p_{1/2}$	686.4	20	693.2	-6.8
$4p_{3/2}$	583.8	20	588.4	-4.6

In an Auger transition the interaction $|V_{\Phi\varepsilon}|^2_A$ is in the order of 10^{-4} to 10^{-3} a.u. and the energy difference $E_A = E_\Phi - E_f$ is large. Since $|V_{\Phi\varepsilon}|^2_A$ is only weakly dependent on ε in the neighborhood of E_A, positive and negative contributions in (1) cancel almost completely, which results in $\Delta E_r \cong 0$.

In case of a Coster-Kronig transition, e.g. $\underline{2s}$-$\underline{2p}$M,$\underline{3p}$-$\underline{3d}$N (where an underlined orbital indicates a hole), the transition energy is much smaller ($\lesssim$ 1 a.u.) and $|V_{\Phi\varepsilon}|^2_{CK}$ is roughly 10 times larger than $|V_{\Phi\varepsilon}|^2_A$ with a much stronger dependence on ε. This results in a shift of E_r to smaller values (see Fig. 4). The shift ΔE_r for the interaction of a 2s-hole state with the Coster-Kronig continuum has been recently calculated by Chen et al.[27] which removed almost all of the existing discrepancy between the calculated and measured 2s-binding energies. Fig. 5 demonstrates this clearly. Here the theoretical $\underline{2s}$-$\underline{2p}_{3/2}$ energy splittings are compared with the experimental results for Z < 50. With the 2s binding energies, corrected for the energy shift ΔE_r, these energy splittings are now within 1 eV.

In case of a super Coster-Kronig (sCK) transition, where at least three levels from the same main shell are involved, the above picture may change dramatically as has been found for the

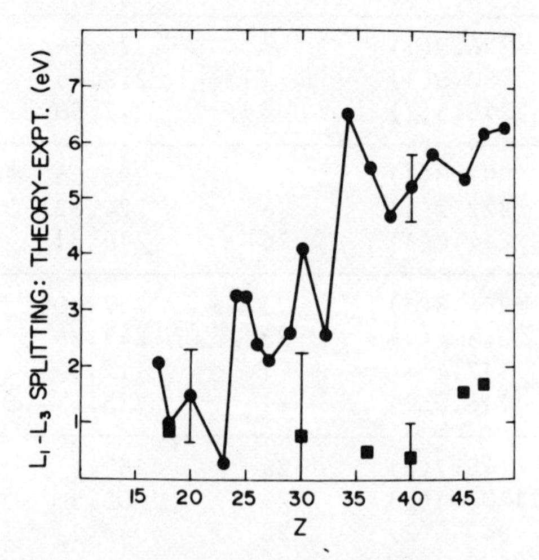

Fig. 5. Energy splitting between atomic $\underline{2s}$ and $\underline{2p}_{3/2}$ vacancy states as a function of atomic number Z. ●: Differences between experimental results and Dirac-Slater calculations[18], ■: the same energies but corrected with the shift ΔE_r of the 2s level (from Chen et al.[27]).

4p levels in the range of atomic number Z=48-70[29,30,31a,b,5]. In the case of super CK transitions 4p-4d$^2\varepsilon$f,nf the interaction $|V_{\Phi\varepsilon}|^2_{sCK}$ is especially large (the term giant CK transitions has been used[5]) and the 4p-hole state energy is very close to the 4d^2 state energies. This results in a large shift ΔE_r of about 10 eV. In addition, the discrete 4p states are completely diluted into the continuum 4d$^2\varepsilon$f and into the adjacent Rydberg states 4d^{2}nf. In Fig. 6 this situation is shown for the 4p states of Xe[31a,b,5].

The level diagram on the left side of Fig. 6 represents the 4p hole states together with the super Coster-Kronig continuum 4d$^2\varepsilon$f and Rydberg states 4d^{2}nf before interaction. The central level diagrams illustrate how the 4p hole levels shift and broaden when the interaction is turned on. The influence of the weaker coupled Coster-Kronig continuum 4d5pεd,nd on the spectral distribution (right side of Fig. 6) of photoelectrons is also shown. In particular the 4d^{2}4f level picks up a large fraction of the intensity of the $4p_{3/2}$ hole and gives rise to the large peak in the experimental spectrum. In Fig. 7 the experimental spectrum of 4s and 4p photoelectrons of Xe is compared with the theoretical spectrum calculated by Wendin and Ohno[31a]. As can be seen, the main features of the experimental spectrum, i.e. the shift of the 4s level, the complete smearing of the 4p levels and the sharp peak at 145.5 eV due to the localized state 4d^{2}4f, are well reproduced by the theory. In this case, the idea

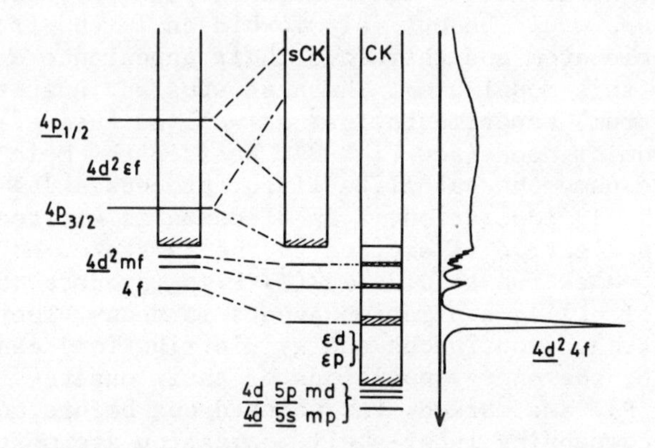

Fig. 6. Energy level diagram for the interacting $4p_{1/2}$ and $4p_{3/2}$ levels of Xenon. Left side: Level energies including relaxation but before interaction with the final states. Centre: When the interaction is turned on, the levels $4p_{1/2}$ and $4p_{3/2}$ broaden and shift. Right side: Predicted photoelectron spectrum, schematically only (from Wendin and Ohno[31b]).

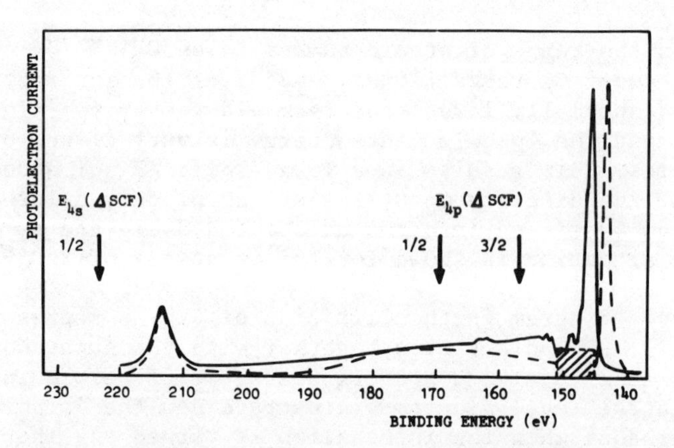

Fig. 7. The photoelectron spectrum for atomic Xe in the 4s-4p region. ——— = Experiment[26,29], ----- = Theory[31a,b] (from Wendin and Ohno[31b]).

of localized $\underline{4p}_j$ levels and of corresponding $\underline{4p}_j$ binding energies is therefore meaningless. Similar situations have also been found for the $\underline{4p}$ levels of atoms with Z=48 to Z=58[5].

Satellites in PES

Another source of information on electron correlation are the two-electron transitions in the photoabsorption process. Two-electron transitions would be entirely forbidden in the frozen independent particle atom and therefore their appearance signals departures from this model atom. The best studied inner-shell photoelectron spectrum, experimentally[29] as well as theoretically[32], is the 1s spectrum of neon (see Fig. 8). Besides the main 1s photopeak there are numerous satellite lines (processes 1b) where in addition to the 1s ionization, a 2p electron is excited to np states or a 2s electron is excited to the 3s state. Moreover ionization of a 2p electron to $1s2s^22p^5(^{1,3}P)\varepsilon p$ (process 1c) occurs in about 15% of all 1s ionization events in neon[8]. The corresponding photoelectrons (continuous energy distribution) cannot be seen in Fig. 8, the energy positions of their onsets $(1s2s^22p^5(^3P,^1P))$ are marked. (As pointed out before, outer-shell ionization accompanying inner-shell ionization are most conveniently studied by means of AES (see e.g.[8])). Here we are mainly interested in the intensities of the $1s2s^22p^5np(^2S)$ satellite lines relative to the diagram $1s2s^22p^6(^2S)$ photopeak.

According to the shake theory and the sudden approximation the probability of excitation of the j th final state $\Psi_j(N-1)$ is given by[33,34]

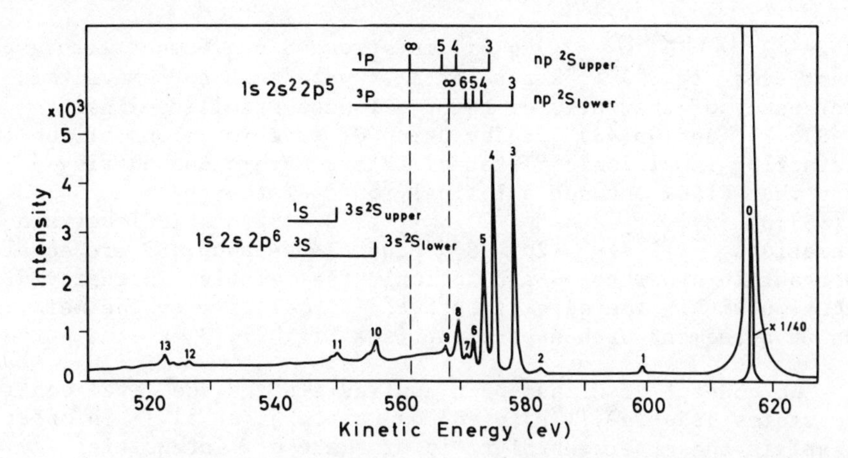

Fig. 8. Ne 1s photoelectron spectrum excited by monochromatized AlK_α X-rays (from Gelius[29]). The various final shake up states np and ns have been indicated by the tick marks. The two series np $^2S_{upper}$ and $^2S_{lower}$ converge to the limits $1s2s^22p^5(^1P)$ and (^3P), the onsets of the continuous photoelectron spectra $1s2s^22p^5(^{1,3}P)\varepsilon p\ ^2S$.

$$P_j \sim |\langle \Psi_j(N-1;{}^2S)|\Psi_i(N,\underline{1s};{}^2S)\rangle|^2 \qquad (2)$$

Here $\Psi_i(N,\underline{1s};{}^2S)$ is given by the ground state wave function with one of the 1s-orbitals removed and $\Psi_j(N-1;{}^2S)$ is a relaxed final state wave function. In eq. (2) both, the initial and final, wave functions are in general CI wave functions. With eq. (2) one would obtain the correct results if the ground state wave function $\Psi_i(N)$ could be reasonably approximated by the antisymmetrized product wave function $\varphi(1s)\cdot\Psi(N,\underline{1s};{}^2S)$. In the conventional shake theory the initial and final state wave functions of eq. (2) are described by Slater determinants. Then eq. (2) has to be replaced by

$$P_j \sim |\langle \overline{1s}\ \overline{2s}^2\ \overline{2p}^5(^{1,3}P)\ \overline{np}(^2S)|\ 1s2s^22p^6(^2S)\rangle|^2 \qquad (3)$$

which reduces to expressions in terms of one-electron overlap matrix elements[34] $\langle \overline{n_2\ell}|n_1\ell\rangle$, the bars indicate relaxed orbitals. From these overlap matrix elements the selection rules $\Delta\ell = \Delta m_\ell = \Delta m_s = 0$ follow for shake transitions.

Application of eq. (3) to the 1s-shake spectrum of neon would yield an intensity ratio of $I(1s2s^22p^5(^3P)np(^2S)_{lower})$:

$I(1s2s^22p^5(^1P)\,np(^2S)_{upper}) = 3:1$ in strong disagreement with experiment (see Fig. 9). Krause et al.[7] were able to remove this discrepancy of the ratio of the two lowest satellite lines $3p(^2S)_{lower}$ and $3p(^2S)_{upper}$ by using CI wave functions of the two interacting final ionic states. Later, Martin and Shirley[32] have shown that FISCI between all final ionic states $2p^6(^2S), np(^2S)_{lower}$ and $np(^2S)_{upper}$ (n=3,4,5) and ISCI between configurations $1s^22s^22p^6(^1S)$ and $1s^22s^22p^5np(^1S)$ are equally important in order to obtain not only the relative intensity distribution within the satellites but also relative to the main 1s-line in agreement with experiment (see Fig. 9).

The concept of CI has been applied also to the final continuum states $1s2s^22p^5(^{1,3}P)\varepsilon p(^2S)$ by Chattarji et al.[11] in order to explain the experimental ratio of shake off intensities $I(1s2s^22p^5(^3P))/I(1s2s^22p^5(^1P)) = 1.36$, which again is far off the expected value of 3 in the independent particle model. The ratio of intensities $^3P/^1P$ of the two-hole configuration 1s2p was derived from the $KL_{2,3}$-$LLL_{2,3}$ satellite spectrum[10].

Recently, FISCI has been used to calculate also the satellite structures in the argon 2s and 2p photoelectron spectra[35,36] and better agreement has been obtained with the experiment than by use of the independent particle model[37].

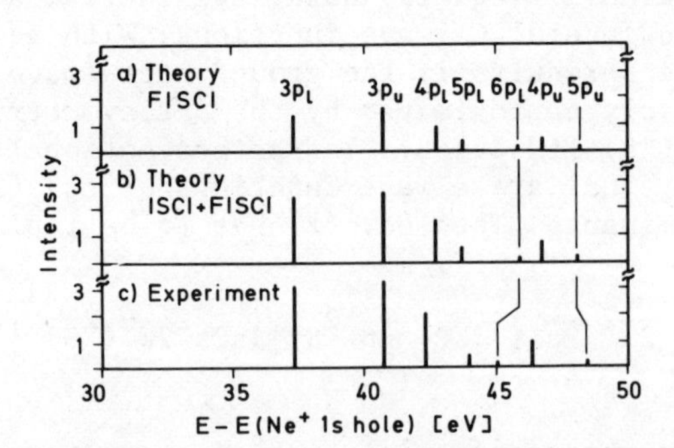

Fig. 9. Bar diagram of $2p \rightarrow np_{lower}$, np_{upper} intensities of the Ne 1s photoelectron spectrum. The intensities are in percent relative to the 1s main line. a) Theory: Only HF ininitial state, FISCI between $1s2s^22p^6$ and $1s2s^22p^5np$ (n=3,4,5,6). b) Theory: With ISCI between $1s^22s^22p^6$ and singly and doubly excited states in ns and np (n=3,4,5,6), FISCI as in a). c) Experiment (from Martin and Shirley[32]).

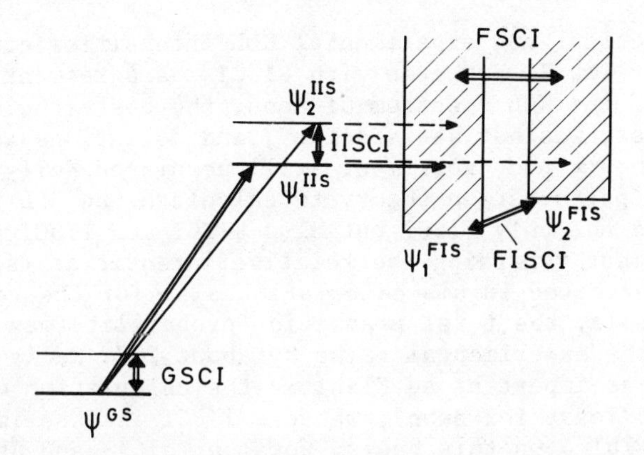

Fig. 10. Various channels of CI in the Auger transition: GSCI, IISCI, FISCI and FSCI. Diagram and CI satellite Auger transitions are given by the solid and broken lines, respectively.

CORRELATION EFFECTS STUDIED BY MEANS OF AES

According to Fig. 10 the various channels of electron correlation which can be studied by means of AES are conveniently divided into ground state CI (GSCI), initial ionic state CI (IISCI), final ionic state CI (FISCI) and final continuum CI (FCSCI). The CI which may enter into the Auger transition matrix element (e.g. of process 2a) and thus will alter the relative intensities and widths of diagram Auger lines are IISCI, FISCI and FCSCI. On the other hand, GSCI and IISCI lead to additional initial Auger states whereas FISCI and FCSCI are responsible for additional final Auger states, all giving rise to satellite transitions (broken lines in Fig. 10). It has been the custom to distinguish between these electron correlation satellite lines and the Auger satellite lines caused by the initial multiple ionization (e.g. shake off)[38,4a,b]. Although we have already seen that CI is also important in the outer shell ionization accompanying inner shell ionization (shake off process), at least when finer details of branching ratios are of interest[11], we will not consider the Auger satellites following initial multiple ionization in the present talk (see e.g. ref.[4a,b]).

Electron correlation in diagram Auger spectra[4a,b]

Asaad[39] was first in 1965 to show that CI between the final ionic states $\underline{2s}^2\ ^1S_o$ and $\underline{2p}^2\ ^1S_o$ (FISCI) improves the agreement between theoretical and experimental KLL intensities considerably, especially for low Z. But even with FISCI the agreement for the intensities of the KLL spectrum of neon, the best studied Auger spectrum so far, was not satisfactory, and deviations as large as 20% between theory and experiment still persisted. Kelly[40], using the many-body perturbation theory to calculate the KLL rates of neon, included not only FISCI but also FCSCI and IISCI and got perfect agreement regarding the relative intensities (see Table 2). Although he corrected in his calculation also for the relaxation of final orbitals, the total transition probability was still smaller than the experimental value by about 20%. Kelly showed that FCSCI is as important as FISCI in the calculation of transition rates (at least for neon), whereas IISCI is less important but non-negligible. On this basis, Howat et al.[41] and Åberg and Howat[13c] developed a new approach of the Auger theory by using the Fano-Prats formalism of the interaction of a discrete state with several continua. This approach includes explicitly FCSCI between the various continuum channels. They also modified the Auger theory for non-orthogonal initial and final state wave functions and could therefore built in the orbital relaxation effect. For the continuum orbitals they used either a potential generated by restricted HF initial $\underline{1s}\,^2S$ state orbitals (i in Table 2) or by restricted transition-operator orbitals based on $\underline{1s}$-$\underline{2s2p}$ configuration (ii in Table 2). The agreement with the experimental values for neon is much better than from that calculations

Table 2. Comparison of theoretical relative and total absolute KLL transition rates of neon with experiment. The relative rates are normalized to the $KL_{2,3}L_{2,3}(^1D)$ rate, the total rates are in units of 10^{-4} a.u.

Transition	Kelly[40] HF	Kelly[40] MBPT	Howat et al.[41] (i)	Howat et al.[41] (ii)	Exp.[9]
$KL_1L_1\ ^1S$	16.7	9.9	9.9	8.2	9.95(10)
$KL_1L_{2,3}\ ^1P$	35.8	27.8	32.1	27.3	28.2(4)
^3P	8.8	9.9	16.7	12.7	10.2(3)
$KL_{2,3}L_{2,3}\ ^1S$	8.0	15.6	15.7	15.1	15.6(4)
^1D	100.0	100.0	100.0	100.0	100.0
KLL (total)	99.1	80.5	105.7	89.8	99 ± 8[26]

including only FISCI (although not as good as Kelly's values), which again demonstrates that IISCI is not negligible. On the other hand, the total KLL rate calculated by Howat et al.[41] is in better agreement with the experimental value.

Howat[42] included FISCI and FCSCI also for the K Auger rates of Mg and obtained results in excellent agreement with experimental values for free Mg atoms[43], also here FCSCI was as important as FISCI.

The foregoing examples have shown that full inclusion of electron correlation (at least FISCI and FCSCI) brings theory in good accord with experiment for Auger transitions, where the outermost or the next inner shell is involved in the final state. For KLL transitions and higher Z one would expect that FCSCI becomes smaller because of the smaller interaction between the final ionic core and the Auger electron, the latter having increasingly larger energy. Thus the experimental relative intensities of KLL lines of argon are in reasonable agreement with theoretical values where only FISCI is included[44]. On the other hand, recent experimental values by Asplund et al.[45] deviate from the earlier experimental values and are thus in less good agreement with theory. For $Z > 18$ relativistic calculations of the KLL rates in the intermediate coupling scheme with FISCI are in reasonable agreement with experimental values[46,47].

For other than K Auger spectra, where both the initial and final state are inner-shell states, it has been found that electron correlation plays only a minor role if relative intensities are considered, e.g. $M_{4,5}N_{4,5}N_{4,5}$ of Xe[48,49], Cd[50], Ag[51] and Sb_4[52].

Correlation satellites due to final ionic state CI

Because electron correlation is largest in the outermost shell we expect large satellite intensities if the final Auger state involves electrons from this shell. If, in addition, the energies of the final Auger diagram state and of an CI excited state are close, then we expect very intense FISCI satellites. It has been found, that this is indeed the case when d electrons are involved in the final excited states. For example, the final diagram states $sp^5(^{1,3}P)$ interact with the excited states $s^2p^3\{^d_{s'}\}(^{1,3}P)$ according to

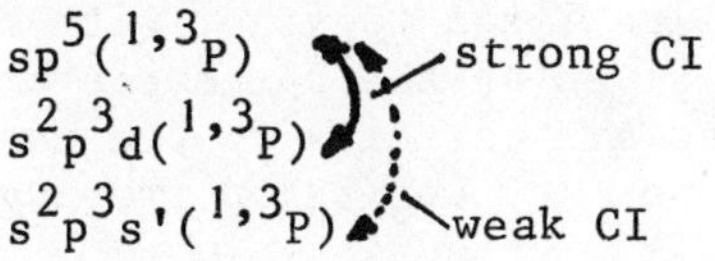

$$\begin{array}{l} sp^5(^{1,3}P) \\ s^2p^3d(^{1,3}P) \\ s^2p^3s'(^{1,3}P) \end{array} \quad \begin{array}{l} \text{strong CI (between } sp^5 \text{ and } s^2p^3d\text{)} \\ \text{weak CI (between } s^2p^3d \text{ and } s^2p^3s'\text{)} \end{array} \qquad (4)$$

This results in an energy shift of the diagram line and an intensity transfer from the diagram line to the satellite lines.

Strong FISCI satellites of the kind mentioned above have been found in the $3d$-$4s4p^5(^{1,3}P)5s^m$ spectra of Kr (m=0)[53-55], Rb (m=1)[56] and Sr (m=2)[21] and in the $2p$-$3s3p^5(^{1,3}P)4s^m$ spectra of Ar (m=0)[16,57a], K (m=1)[58] and Ca (m=2)[58]. For example, the full $L_{2,3}$ Auger spectrum of Ar is displayed in Fig. 11 (see also Fig. 2). The diagram lines ($L_{2,3}$-MM) are the black colored lines of the spectrum. Most of the other lines between 185 to 200 eV are due to Auger transitions $L_{2,3}M$-MMM. The initial double ionization $L_{2,3}M$ is either due to the shake off of an M electron in the $L_{2,3}$ ionization or given by the final state of a preceding Coster-Kronig transition L_1-$L_{2,3}M$. These $L_{2,3}M$-MMM transitions have been

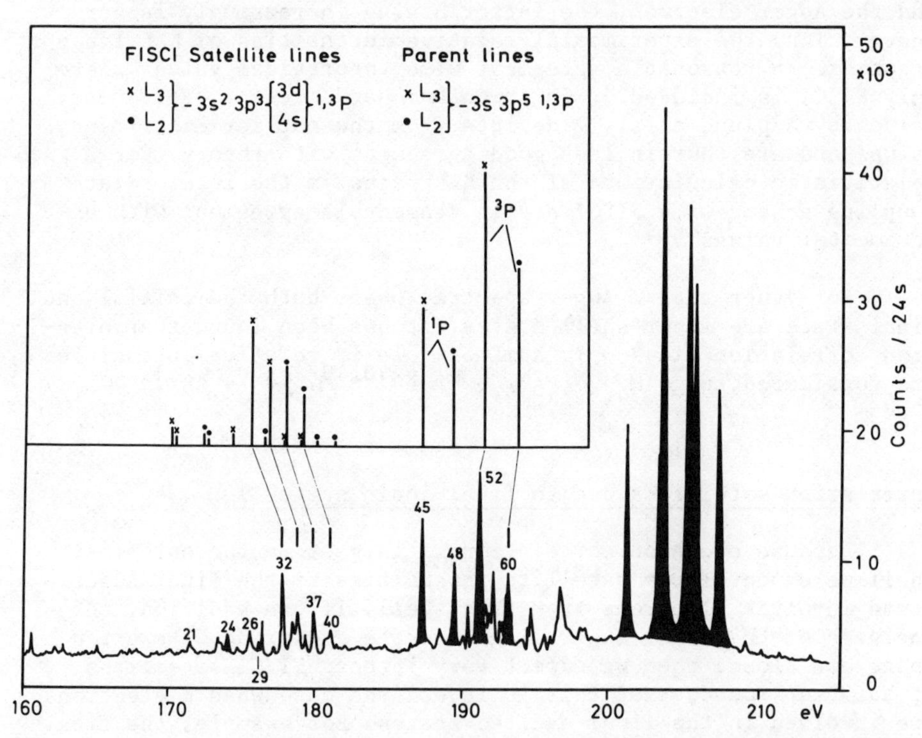

Fig. 11. $L_{2,3}$-MM spectrum of Ar excited by 3 keV electrons (from Werme et al.[16]). The diagram lines are the black full lines. The FISCI satellite lines together with their parent diagram lines are given by the bar spectrum. Their intensities and energies are relative to the experimental $L_{2,3}$-$3s3p^5$ ^1P parent lines and are taken from Dyall and Larkins[57b].

discussed in detail by McGuire[57a]. Here we are interested in the lines No. 21 to 41 between 172 to 182 eV. In a multi-CI calculation Dyall and Larkins[57b] obtained the positions and intensities of FISCI satellite lines due to the interaction of configurations (4). The corresponding satellite and diagram line spectrum is also shown in Fig. 11 as bar spectrum and agrees well with the experimental spectrum. A similar calculation was done by McGuire[57a], who in particular reassigned the lines No. 24/29 to the diagram transitions $L_{2,3}$-$\underline{3s}^2$ instead of lines No. 32/37[59] improving the agreement with theoretical intensities.

In Fig. 12 we compare the $L_{2,3}$-MM Auger spectra of Ar, K and Ca, where the energy axis for K and Ca is multiplied by constant factors in order to match the diagram Auger groups of the different elements[58]. It is interesting to note, that the total relative intensity of FISCI satellites $\underline{2p}$-$3s^2 3p^3 \{{nd \atop ns}\}$ remains approximately unchanged when going from Ar to Ca which indicates that this kind

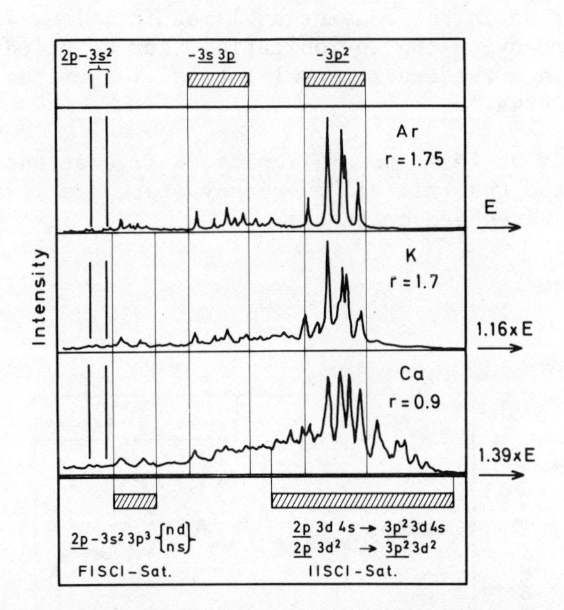

Fig. 12. Comparison of experimental $L_{2,3}$-MM spectra of Ar, K and Ca excited by 2 keV electrons (from Breuckmann[58]). The energy scales of the K and Ca spectra have been changed in order to have the Auger groups $\underline{3s}^2$, $\underline{3s3p}$ and $\underline{3p}^2$ falling on top of each other. The energy regions of FISCI satellites in Ar, K and Ca and of strong IISCI satellites in Ca are indicated by the shaded areas. The ratio r of diagram line intensities $I(L_3\text{-MM})/I(L_2\text{-MM})$ is also given.

of electron correlation is still strong even when the final diagram state is not in the outermost but in the penultimate shell.

Correlation satellites due to initial ionic state CI

In Fig. 12 the intensity at the high energy side of the $\underline{2p}$-$\underline{3p}^2$ Auger group changes dramatically when going from K to Ca. At the same time the intensity ratio of corresponding fine-structure components $I(\underline{2p}_{3/2}-M_iM_k)/I(\underline{2p}_{1/2}-M_iM_k)$ varies from $r = 1.75$ for Ar and $r = 1.7$ for K to $r = 0.9$ for Ca. These two facts have been interpreted so far as due to a very strong CI in the initial ionic state $\underline{2p}_{3/2}$ of Ca[58]. In Fig. 13 the experimental $\underline{2p}$-$\underline{4s}^2$ spectrum of Ca is also shown[58]. In this case the final state is a closed shell ion $3s^2 3p^6$ ^1S and we expect exactly one doublet according to the $\underline{2p}_{1/2,3/2}$ splitting. At least 5 lines with comparable intensity have been found, where the lines a/c and b/d have the fine structure splitting of 3.6 eV of the $\underline{2p}_{1/2,3/2}$ vacancy states. Again, the additional satellite lines can only be explained through a very strong CI in the initial $\underline{2p}$ vacancy state[58]. Shake up in the conventional meaning, i.e. $4s \rightarrow 5s$ excitation accompanying the 2p-ionization, can be ruled out as a main source since the experimental intensities of the satellites are much too large.

Due to CI at least the following configurations in the ground state and the initial $\underline{2p}$ vacancy state are mixed (passive closed shells have been omitted):

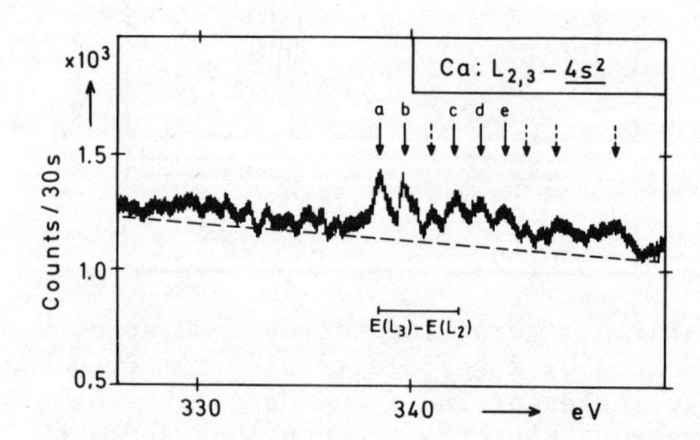

Fig. 13. $L_{2,3}$-$\underline{4s}^2$ Auger spectrum of Ca excited by 2 keV electrons (from Breuckmann[58]). The energy splitting of the pairs of lines a/c and b/d correspond to the fine structure splitting of the initial L_3 and L_2 levels.

GS: $2p^6 4s^2(^1S)$, $2p^6 3d^2(^1S)$ (5a)

IIS: $2p^5 4s^2(^2P)$, $2p^5 3d^2(^1S)(^2P)$, $2p^5 4s3d(^2P)$. (5b)

Then the initial excited state $\Psi_2^{IIS}(2p^5 3d^2(^1S)^2P)$ can be reached from the ground state via two paths (see Fig. 14), either via the admixture of $3d^2(^1S)$ in the ground state or via the admixture of $2p^5 4s^2$ in the excited state. (This is exactly the same mechanism of excitation as discussed in connection with the 1s photoelectron spectrum of neon[32]). A MCHF calculation of the two configurations $4s^2$ and $3d^2(^1S)$ mixed into the ground state (see (5a)) or into the 3p or 2p inner-vacancy state functions

$$\Psi = a \cdot \Phi((....)4s^2) + b \cdot \Phi((....)3d^2(^1S)) \qquad (6)$$

yield as admixtures b^2 the values 0.02, 0.10 and 0.43, where the configuration (....) is the ground state core, the 3p and the 2p vacancy core, respectively. This shows clearly that for a 2p vacancy IISCI dominates by far over the GSCI. The shaded area below the high-energy part of the Ca spectrum in Fig. 12 indicates the expected energy range of Auger transitions to final states $3p^4 3d^2$ and $3p^4 3d4s$ due to HF calculations.

The very large admixture of about 50% indicates quasi-degeneracy of $2p4s^2$ ^2P and $2p3d^2(^1S)^2P$ for the 2p vacancy. This can be seen also from Fig. 15 where the relevant HF-energies of ^1S and ^2P states of the outer shell configurations $4s^2$, $3d^2$ and 3d4s are given for the neutral atom and the 3p and 2p vacancy atom. The important point in Fig. 15 is that with increasing effective Z seen

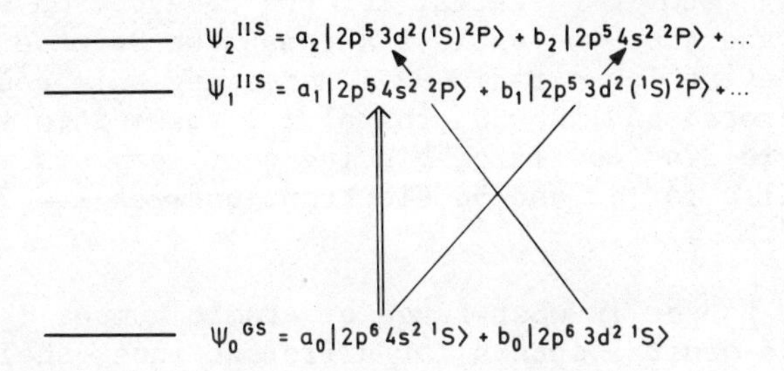

Fig. 14. Model to illustrate the various paths to reach the excited inner-shell vacancy state Ψ_2^{IIS} either by GSCI or by IISCI. The zero order CI transition amplitude (⇑) leads only to state Ψ_1^{IIS} and is prop. to $a_o a_1$. The first order CI transition amplitudes (↑) exciting Ψ_2^{IIS} are proportional to $a_o b_2$ and $b_o a_2$.

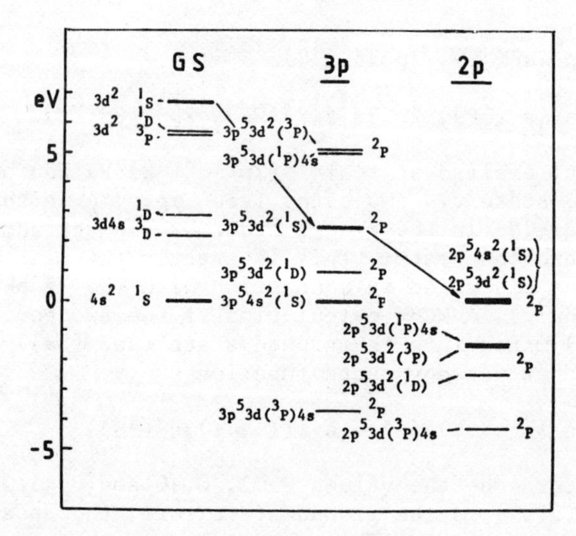

Fig. 15. HF term energies of outer-shell configurations 4s3d and $3d^2$ relative to the $4s^2$ configuration for the neutral and the 3p and 2p vacancy Ca atom. For the 3p and 2p vacancy configurations only the ^2P terms, which can mix with $np4s^2\ ^2P$ are given.

by the 4s and 3d orbitals the energy of the 3d orbital changes relative to the 4s orbital and crosses the 4s energy. This is known as collapse of the 3d-orbital[60,61]. It has its origin in the shape of the effective atomic potential for the 3d electron which has a small barrier and thus two valleys keeping the 3d wave function in the outer valley at large radius[62]. Increasing Z_{eff} causes the barrier to decrease and the 3d orbital collapses into the deep inner valley yielding now large binding energies. A similar collapse occurs also for 4d and 5d electrons between Z = 36-38 and Z = 56-58[63-67].

In order to see in what range of atomic number Z the collapse of the 3d-orbital occurs for different inner-shell vacancy states we calculate the mixing coefficient b^2 of the configuration $3d^2(^1S)$ into wave functions (6). The coefficients b^2 are plotted against atomic number Z in Fig. 16. From this figure it can be seen that the collapse occurs in a rather small range of ΔZ and for a 2p vacancy it occurs just for Z = 20, i.e. the Ca atom. From this we conclude that the unexpected strong excitation of excited 2p-vacancy states in Ca can be considered as an accident and will not occur in the neighboring elements K and Sc. Rather we expect here the strength of IISCI and GSCI to be less than 0.1. If we now assume that the strong IISCI occurs for the $\underline{2p}_{3/2}$ vacancy states and not for the $\underline{2p}_{1/2}$ vacancy states, then this would explain also the small ratio r=0.9 of $\underline{2p}_{3/2}$-MM transitions to $\underline{2p}_{1/2}$-MM transitions. This small ratio r should be found also for the intensities of the 2p photoelectron lines of Ca, but no such spectrum of free Ca atoms has been measured up to date.

Although a quantitative MCDF calculation of the relevant $\underline{2p}_j$-vacancy states has not yet been done, the above assumption is supported by a similar strong IISCI for $\underline{5p}_{3/2}$ vacancy states (but not for $\underline{5p}_{1/2}$) of Ba, where the collapse of the 5d electron occurs[66,67]. In the lower part of Fig. 17 the $\underline{5p}(^2P_{3/2,1/2})$-$\underline{6s}^2(^1S_0)$ Auger spectrum of Ba is shown[21]. Although one expects only one Auger doublet the spectrum is rich in satellite lines. The ratio r of the doublet lines $I(^3P_{2/2})/I(^2P_{1/2})$ is 0.8 (1) and also much smaller than the expected statistical value of 2. The quantitative interpretation of this spectrum was given by Connerade et al.[66,67]. On the basis of a MCDF calculation[67] for the J = 1/2 and J = 3/2 levels of configurations $5p^56s^2$, $5p^56s5d$ and $5p^55d^2$ they calculated the positions of the J = 1/2 and J = 3/2 levels and could assign these levels to the series limits found in the 5p absorption spectrum of Ba [66]. These series limits are identical with the initial states of the $\underline{5p}(^2P_{3/2,1/2})$ Auger spectrum and are given in the upper part of Fig. 17 as tick marks where the J value is indicated according to the calculation. From this we note, that strong CI occurs only for the J = 3/2 levels, their total intensity is distributed into 5 intense components, whereas for the J = 1/2 levels there exist only two small satellites. If one now adds the intensities of the J = 3/2 and the J = 1/2 components,

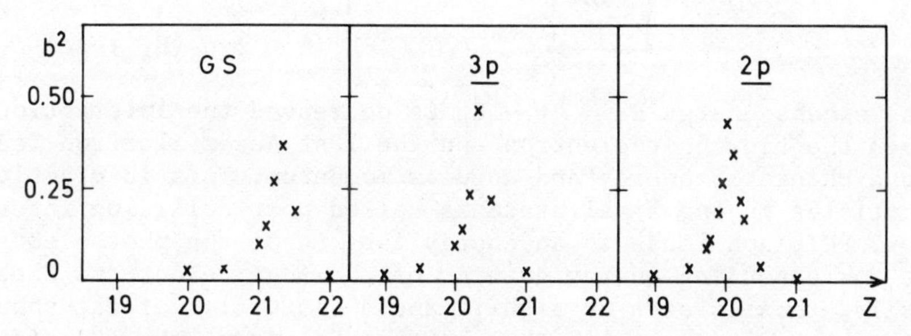

Fig. 16. Admixture b^2 of basic state $3d^2(^1S)$ into the ground state, $\underline{3p}$ and $\underline{2p}$ vacancy states according to eq. (6) (from Breuckmann[58]).

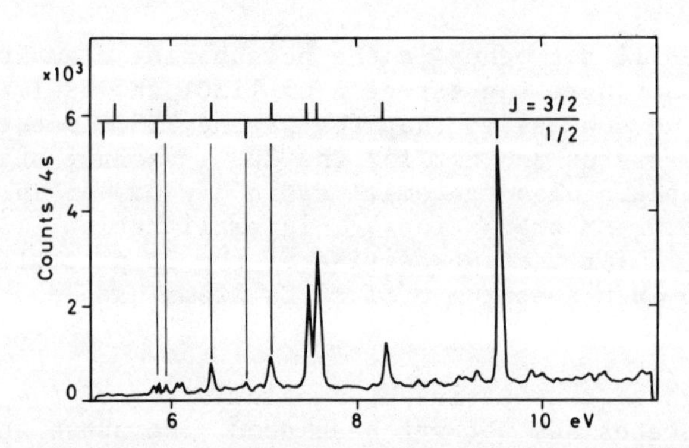

Fig. 17. Experimental 5p-6s^2 Auger spectrum of Ba excited by 2 keV electrons (from Mehlhorn et al.[21]). The experimental BaI 5p-photoabsorption series limits are given as tick marks where the assignment of series limits to J = 1/2 or 3/2 levels based on a MCDF calculation is also indicated (Connerade et al.[66], Rose et al.[67]).

then the ratio r is 1.8(1) and thus close to 2. Furthermore, using the sudden approximation and calculating the overlap probabilities between the wave function Ψ (N 5p) and the CI wave function of the final 5p relaxed states (according to eq. (2)), Rose et al. have obtained a 5p spectral strength distribution in good agreement with the experiment of Fig. 17.

POST-COLLISION INTERACTION IN INNER-SHELL IONIZATION

Consider the following inner-shell photoionization process

$$\gamma(h\nu) + A \rightarrow A^{+}(E_B) + e^{-}(E_1) \longrightarrow e^{-}(E_1 - \Delta\varepsilon), \quad A^{+}(E_B) \longrightarrow A^{++} + e^{-}_{A}(E^{o}_{A}) \longrightarrow A^{++} + e^{-}_{A}(E^{o}_{A} + \Delta\varepsilon) \tag{7}$$

As the excess energy $E_1 = h\nu - E_B$ is decreased the interaction between the slow photoelectron and the fast Auger electron leads to an exchange of energy and angular momentum. This interaction of particles in the final state is called post-collision interaction, PCI, and leads to an energy loss $\Delta\varepsilon$ of the photoelectron and a corresponding energy gain $\Delta\varepsilon$ of the Auger electron. Experimentally, this is seen as a shift and a distortion of the shape of the corresponding lines in the electron spectrum. The PCI effect in inner-shell ionization processes has found great experimental (for photon impact see Schmidt et al.[68] and references therein, for electron impact see Huster and Mehlhorn[69] and references therein) and theoretical[70-78,69] interest in the last few years.

In the case of inner-shell photoionization Niehaus[70] has predicted within a semiclassical framework the PCI shifts $\Delta\varepsilon_m$ and shapes of Auger lines. Both, the shift and the shape depend - apart from the life time τ (or width Γ) of inner-shell vacancy - only on the excess energy E_1. Also quantum mechanical treatments have been formulated[71-78], but to date neither shifts nor line shapes have been calculated quantitatively. As example, in Fig. 18 the experimental Xenon $\underline{4d}_{5/2}$-$\underline{5p}^2$ $^1S_0$ Auger peak following the photoionization of a $4d_{5/2}$ electron by monochromatized synchrotron radiation for different excess energies E_1 are compared with the theoretical line shapes (solid lines in Fig. 18). In Fig. 19 the corresponding experimental PCI shifts $\Delta\varepsilon_m$ and those obtained by Southworth[1b] are compared to the theoretical ones. As can be seen from Fig. 18 and 19, the theoretical shapes and shifts are in good agreement with the experimental results.

In the case of inner-shell ionization by electron impact we have two slowly moving electrons e_1 and e_2 (sharing the excess energy E_1) which interact with the Auger electron. No general theoretical treatment of the PCI effect for inner-shell ionization by

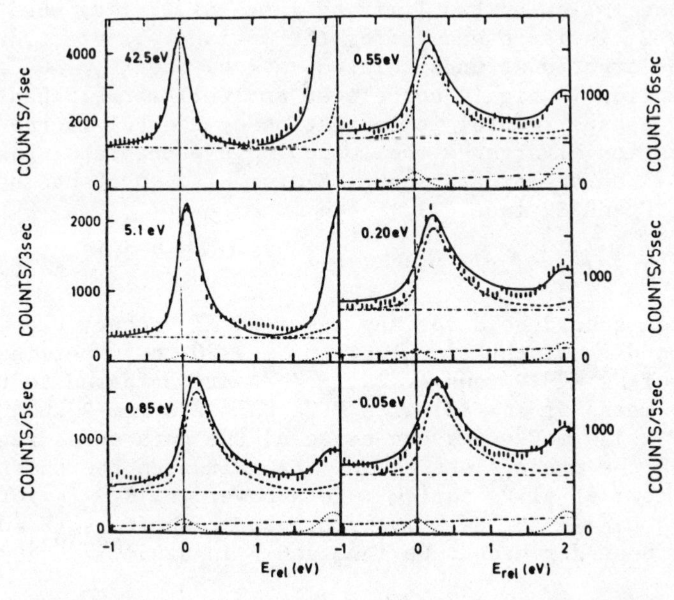

Fig. 18. Experimental PCI effect (shift and shape) of the Xe $\underline{4d}_{5/2}$-$\underline{5p}^2(^1S_0)$ Auger line following the photoionization of a $4d_{5/2}$ electron (with E_B = 67.55 eV) for various excess energies E_1. The solid line is the calculated PCI intensity distribution (from Schmidt et al.[68]).

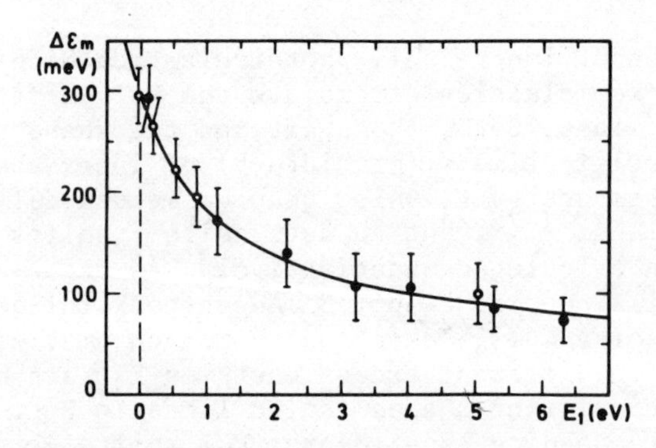

Fig. 19. PCI shifts $\Delta\varepsilon_m$ of the Xe $\underline{4d}_{5/2}$-$\underline{5p}^2(^1S_o)$ Auger line after photoionization as function of the excess energy E_1. The theoretical curve has been obtained according to Niehaus[70] with Γ=110 meV. The experimental data are from Schmidt et al.[68] (O) and from Southworth[1b] (●).

electron impact has yet emerged, i.e. an adequate formulation of the interaction of an electron with two slowly moving and correlated electrons is still missing. Only very recently the PCI shifts and shapes of Auger lines were calculated with a classical model for the following two limiting cases of E_1[69]: a) When the excess energy E_1 is large enough for the motion of electrons e_1 and e_2 to be treated as uncorrelated, and b) when $E_1 \rightarrow 0$, for which the motion is highly correlated and well established[79]. Recently, in an experimental investigation of the PCI in the L_3 shell ionization of argon[69] the following dependence of the PCI shift $\overline{\Delta\varepsilon_m}(E_1)$ on the excess energy E_1 for $E_1 > 10$ eV has been found ($\overline{\Delta\varepsilon_m}$, Γ and E_1 in a.u.)

$$\overline{\Delta\varepsilon}_m(E_1) = (1.2 \pm 0.1) \cdot \Gamma \cdot E_1^{(-0.45 \pm 0.04)} \qquad (8)$$

This relation should hold for any inner-shell vacancy provided the corresponding width Γ is used. In Fig. 20 the experimental PCI shifts $\overline{\Delta\varepsilon_m}$ of Ne K[80,81] and Ar L_2,L_3[69,81] are compared to those obtained by means of eq. (8) with Γ(Ne K) = 270 meV[26] and Γ(Ar L_2,L_3) = 130 meV[82]. As can be seen, the data agree reasonably well with the predictions of eq. (8), although the shifts measured by Wakiya et al.[81] follow more a power law $E_1^{-0.72}$ for Ar L_2,L_3 and $E_1^{-0.65}$ for Ne K ionization. The shift $\overline{\Delta\varepsilon_m}(E_1=0)$ for $E_1 = 0$ has been derived to be ($\overline{\Delta\varepsilon_m}$ and Γ in a.u.)[69,83,84]

$$\overline{\Delta\varepsilon}_m(E_1 = 0) = (8 \cdot \sqrt{3} \cdot \Gamma/15)^{2/3} \quad , \qquad (9)$$

which yields 1.19 eV and 0.84 eV for Ne K and Ar L_2,L_3 ionization. These values are also plotted in Fig. 20, they are larger

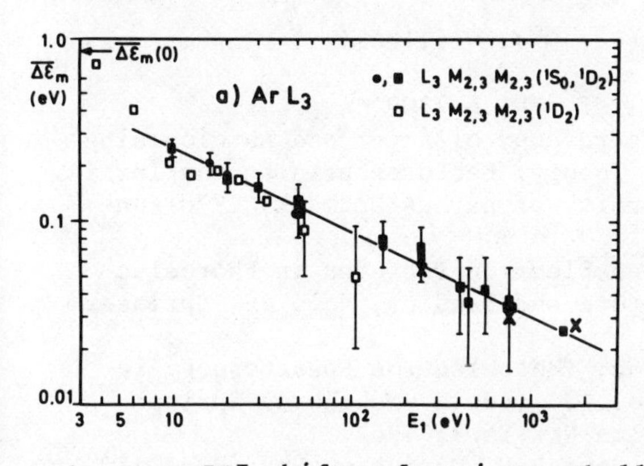

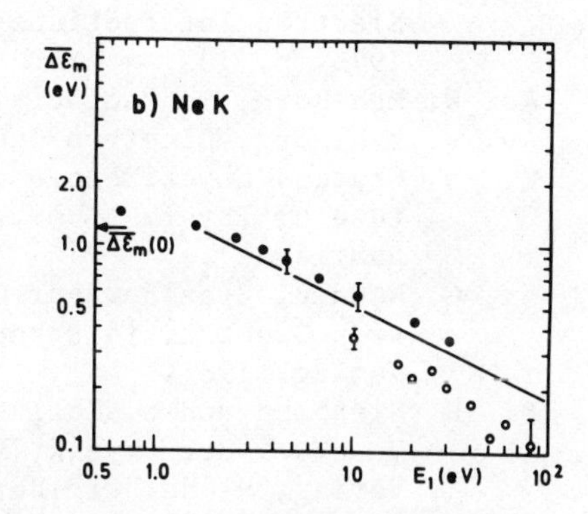

Fig. 20. PCI shifts after inner-shell ionization by electron impact as function of excess energy E_1.
a) Shift of L_3-$M_{2,3}M_{2,3}(^1S, ^1D)$ lines of argon. Experiment: ● , ■ = Ref.69, ■ = Ref. 81. The solid line is due to eq. (8) with Γ = 130 meV, the crosses (x) are theoretical values of the PCI shifts[69], the arrow indicates the predicted shift for $E_1 = 0$.
b) Shift of K-$L_{2,3}L_{2,3}(^1D)$ line of neon. Experiment: ● = Ref.80 , O = Ref. 81 . The solid line is due to eq. (8) with Γ = 270 meV. The predicted shift[69] for $E_1 = 0$ is indicated by the arrow.

by a factor $(32/3)^{1/3} = 2.20$ compared to the threshold value $\Delta\varepsilon_m(E_1 = 0)$ for photoionization[70]. Also the PCI distorted Auger line shapes can be rather well reproduced by theoretical PCI distributions[69]. Recently, PCI has also been treated as a natural link between inner-shell ionization and inner-shell excitation processes[85].

REFERENCES

1a. S.H. Southworth, P.H. Kobrin, C.M. Truesdale, D. Lindle, S. Owaki, and D.A. Shirley, Phys. Rev. A 24, 2257 (1981).
1b. S.H. Southworth, Ph.D. Thesis, Lawrence Berkeley Laboratory, University of California, 1982.
2. K. Siegbahn, C. Nordling. G. Johansson, J. Hedman, P.F. Heden, K. Hamrin, U. Gelius, T. Bergmark, L.O. Werme, R. Manne, and Y. Baer, ESCA Applied to Free Molecules, North-Holland, 1969.
3a. M.O. Krause, in Atomic Inner-Shell Processes (ed. B. Crasemann), Academic Press, 1975 Vol. II, p. 34.
3b. M.O. Krause, in Photoionization and other Probes of Many-

Electron Interactions (ed. F.J. Wuilleumier) Plenum Press 1976, p. 133.
4a. W. Mehlhorn, in loc. cit. ref. 3b, p. 309.
4b. W. Mehlhorn, Electron Spectroscopy of Auger and Autoionizing States: Experiment and Theory, Lectures held at the Institute of Physics, University of Aarhus Denmark, 1978 (unpublished).
5. G. Wendin, Breakdown of One-Electron Pictures in Photoelectron Spectra, in Structure and Bonding, Vol. 45, Springer Verlag, 1981.
6. H. Siegbahn and L. Karlsson, Photoelectron Spectroscopy in Handbuch der Physik, Vol. 31 (ed. W. Mehlhorn) Springer-Verlag, Heidelberg-Berlin-New York, 1982.
7. M.O. Krause, T.A. Carlson, and R.D. Dismukes, Phys.Rev. 170, 37 (1968).
8. M.O. Krause, J.Physique (Paris) 32, C4-67 (1971)
9. M.O. Krause, T.A. Carlson, and W.E. Moddeman, J.Physique (Paris) 32, C4-139 (1971).
10. V. Schmidt, Proceedings of Intern.Conf. Inner-Shell Ionization Phenomena and Future Applications, Atlanta, 1972. Report No. CONF-720404, USAEC Technical Information Center, Oak Ridge, Tenn. 1973, p. 548.
11. D. Chattarji, W. Mehlhorn and V. Schmidt, J.Electr.Spectrosc. 13, 97 (1978).
12. T.A. Carlson and M.O. Krause, Phys.Rev.Lett. 17, 1079 (1966).
13a. T. Åberg in Atomic Inner-Shell Processes (ed. B. Crasemann), Academic Press, 1975, Vol. I, p. 353
13b. T. Åberg, in loc. cit. ref. 3b, p. 273.
13c. T. Åberg and G. Howat, Theory of the Auger Effect, in Handbuch der Physik, Vol. 31 (ed. W. Mehlhorn), Springer-Verlag, 1982.
14a. Y. Yafet, Phys.Rev. B 21, 5023 (1980)
14b. F. Combet Farnoux and M. Ben Amar, Phys.Rev. A 21, 1975 (1980).
14c. F. Combet Farnoux, Phys.Rev. A 25, 287 (1982).
15. W. Mehlhorn, Z.Physik 160, 247 (1960).
16. L.O. Werme, T. Bergmark and K. Siegbahn, Phys.Scripta 8, 149 (1973).
17a. K. Siegbahn, private communication, 1981.
17b. R. Huster, Univ. Freiburg, private communication, 1981.
18. N. Huang, M. Aoyagi, M.H. Chen, B. Crasemann, and H. Mark, At.Data Nucl.Data Tables 18, 243 (1976).
19. J.P. Desclaux, Comput.Phys.Comm. 9, 31 (1975), and private communication, 1977.
20. S. Svenson, N. Mårtensson, E. Basilier, P.A. Malmquist, U. Gelius and K. Siegbahn, J.Electr.Spectrosc. 9, 51 (1976).
21. W. Mehlhorn, B. Breuckmann, D. Hausamann, Phys.Scripta 16, 177 (1977)
22. M. Breinig, M.H. Chen, G.E. Ice, F. Parente, B. Crasemann,

and G.S. Brown, Phys.Rev. A 22, 520 (1980).
23. R.J. Key, M.S. Banna, and C.S. Ewig, J.Electr.Spectrosc. Relat.Phenom. 24, 173 (1981).
24. B. Johansson and N. Mårtensson, Phys.Rev. B 21, 4427 (1980).
25. H. Ågren, J. Nordgren, L. Selander, C. Nordling and K. Siegbahn, J.Electr.Spectrosc. 14, 27 (1978).
26. S. Svensson, N. Mårtensson, E. Basilier, P.A. Malmquist, U. Gelius and K. Siegbahn, Phys.Scripta 14, 141 (1976)
27. M.H. Chen, B. Crasemann and H. Mark, Phys.Rev. A 24, 1158 (1981) and references therein.
28. U. Fano, Phys.Rev. 124, 1866 (1961).
29. U. Gelius, J.Electr.Spectrosc. 5, 985 (1974).
30. S.P. Kowalczyk, L. Ley, R.L. Martin, F.R. McFeely, and D.A. Shirley, Faraday Disc. 60 (1975).
31a. G. Wendin and M. Ohno, Phys.Scripta 14, 148 (1976).
31b. G. Wendin and M. Ohno, Proc.2nd Int.Conf. on Inner Shell Ionization Phenom., Invited Papers (ed. W. Mehlhorn and R. Brenn), Freiburg 1976, p. 166.
32. R.L. Martin and D.A. Shirley, Phys.Rev. A 13, 1475 (1976).
33. T. Åberg in loc. cit. ref. 3b, p. 49.
34. T. Åberg, Ann.Acad.Sci.Fenn. A VI, 308, 1 (1969).
35. K.G. Dyall, F.P. Larkins, K.D. Bomben, and T.D. Thomas, J. Phys. B 14, 2551 (1981).
36. K.G. Dyall and F.P. Larkins, J.Phys. B 15, 1021 (1982).
37. D.J. Bristow, J.S. Tse, and G.M. Bancroft, Phys.Rev. A 25, 1 (1982).
38. W. Mehlhorn, Physica Fennica 9, Suppl. S1, 223 (1974).
39. W.N. Asaad, Nucl.Phys. 66, 494 (1965).
40. H.P. Kelly, Phys.Rev. A 11, 556 (1975).
41. G. Howat, T. Åberg, and O. Goscinski, J. Phys. B 11, 1575 (1978).
42. G. Howat, J.Phys. B 11, 1589 (1978).
43. B. Breuckmann, J.Phys. B 12, L 609 (1979).
44. M.O. Krause, in loc. cit. ref. 31b, p. 184.
45. L. Asplund, P. Kelfve, B. Blomster, H. Siegbahn, and K. Siegbahn, Phys.Scripta 16, 268 (1977).
46. W.N. Asaad and D. Petrini, Proc.Roy.Soc. London A 350, 381 (1976).
47. M.H. Chen, B. Crasemann, and H. Mark, Phys.Rev. A 21, 442 (1980).
48. S. Hagmann, G. Hermann, and W. Mehlhorn, Z.Physik 266, 189 (1974).
49. S. Aksela, H. Aksela, and T.D. Thomas, Phys.Rev. A 19, 721 (1979).
50. H. Aksela and S. Aksela, J.Phys. B 7, 1262 (1974).
51. J. Väyrynen, S. Aksela, M. Kellokumpu, and H. Aksela, Phys. Rev. A 22, 1610 (1980).
52. H. Aksela, J. Väyrynen, and S. Aksela, J.Electr.Spectrosc. 16, 339 (1979).

53. W. Mehlhorn, W. Schmitz, and D. Stalherm, Z.Physik 252, 399 (1972).
54. L.O. Werme, T. Bergmark, and K. Siegbahn, Physica Scripta 6, 141 (1972).
55. E.J. McGuire, Phys.Rev. A 11, 17 (1975).
56. W. Menzel and W. Mehlhorn, in Inner-Shell and X-Ray Physics of Atoms and Solids (ed. D.J. Fabian, H. Kleinpoppen, L. M. Watson) Plenum Press 1981, p. 319.
57a. E.J. McGuire, Phys.Rev. A 11, 1880 (1975).
57b. K.G. Dyall and F.P. Larkins, in loc. cit. ref. 56, p. 265.
58. B. Breuckmann, Ph.D. Thesis, Universität Freiburg, 1978, unpublished.
59. W. Mehlhorn and D. Stalherm, Z.Physik 217, 294 (1968).
60. J.P. Connerade, Proc.Roy.Soc. London A 347, 575 (1976).
61. M.W.D. Mansfield, Proc.Roy.Soc.London A 348, 143 (1976).
62. A.R. P. Rau and U. Fano, Phys.Rev. 167, 7 (1968).
63. J.P. Connerade and M.W.D. Mansfield, Proc.Roy.Soc. London A 348, 239 (1976).
64. J.P. Connerade, M.A. Baig, W.R.S. Garton, and G.H. Newsom, Proc.Roy.Soc. London A 371, 295 (1980).
65. J.P. Connerade, J.Phys. B 11, L 381 (1978).
66. J.P. Connerade, M.W.D. Mansfield, G.H. Newsom, D.A. Tracy, M.A. Baig and K. Thimm, Phil.Trans.Roy.Soc. London 290, 327 (1979).
67. S.J. Rose, I.P. Grant and J.P. Connerade, Phil.Trans.Roy. Soc. London 296, 527 (1980).
68. V. Schmidt, S. Krummacher, F. Wuilleumier, and P. Dhez, Phys.Rev. A 24, 1803 (1981).
69. R. Huster and W. Mehlhorn, Z.Physik 307, 67 (1982).
70. A. Niehaus, J.Phys. B 10, 1845 (1977).
71. M.Ya. Amusia, M.Yu. Kuchiev, S.A. Sheinerman, and S.I. Sheftel, J.Phys. B 10, L 535 (1977).
72. M.Ya. Amusia, M.Yu. Kuchiev and S.A. Sheinerman, in Coherence and Correlation in Atomic Collisions (ed. H. Kleinpoppen and J.F. Williams) Plenum Press, 1980, p. 297.
73. V.N. Ostrowskii, Zh.Eksp.Theor.Fiz. 72, 2079 (1977) (Sov. Phys. - JETP 45, 1092 (1977)).
74. G. Wendin in Photoionization of Atoms and Molecules, Proceedings of the Daresbury Meeting (ed. B.D. Buckley), Report No. DL/SCI/R11 (1978), p. 1.
75. M. Ohno and G. Wendin, J.Phys. B 12, 1305 (1979).
76. G. Wendin, in Proceedings of the VIth Int.Conf. on VUV Radiation Physics, Charlottesville, 1980, extended abstracts II-87.
77. T. Åberg, Phys.Scripta 21, 495 (1980).
78. T. Åberg, in loc. cit. ref. 56, p. 251.
79. G.H. Wannier, Phys.Rev. 90, 817 (1953).
80. W. Hink, L. Kees, H.P. Schmitt, and A. Wolf, in loc. cit. ref. 56, p. 237.

81. K. Wakiya, H. Suzuki, T. Takayanagi, M. Muto, S. Ito, V. Iketaki, and S. Ohtani, Abstracts of XII. ICPEAC, Gatlinburg 1981, p. 247.
82. M.O. Krause and J.H. Oliver, J.Phys.Chem.Ref. Data 8, 329 (1979).
83. J. Mizuno, T. Ishihara and T. Watanabe, Abstracts of XII. ICPEAC, Gatlinburg 1981, p. 253.
84. T. Watanabe, private communication, 1982.
85. V. Schmidt, Invited Paper at Intern. Conference on X-Ray and Atomic Inner-Shell Physics, Eugene, University of Oregon, 1982.

SOME CURRENT PROBLEMS IN ELECTRON SPECTROSCOPY

Kai Siegbahn

Institute of Physics
University of Uppsala, Box 530
S-751 21 Uppsala, Sweden

INTRODUCTION

When atomic systems are excited to sufficiently high energies, i.e. above the ionization limit, a competitive mode to photon deexcitation is electron emission. This latter mode is quite dominant in most cases and electron spectroscopy has therefore become of increasing importance during the last two decades. It is typically a vacuum spectroscopy, and for certain studies of surfaces it even requires ultra high vacuum (UHV). With essentially the same spectroscopic techniques one can cover a very large range of excitation energies, from the eV region towards the keV region. It is possible to apply this spectroscopy to gases, liquids and solids (in particular to surfaces) under variable conditions. Fig.1 summarizes the situation[1].

The emission of electrons from atoms is accompanied by electronic relaxation due to the creation of a hole state. Fig.2 illustrates an atom bound to other atoms or to a surface. One distinguishes between core electrons and valence electrons, the former being localized to a certain atom and the latter being more or less delocalized. The core electrons can only be emitted by means of sufficiently hard radiation whereas the valence electrons are conveniently emitted by means of VUV radiation. In interpreting the electron spectra the first step is to consider the electron structure as 'frozen' under the photoelectron emission process. In this approximation the measured electron binding energies can be identified with the Hartree-Fock energy eigenvalues of the orbitals. One then disregards the fact that the remaining electronic structure, at electron emission, is relaxing to a core hole state. This relaxation energy is by no means negligible, and so accurate calculation of the relevant binding energy must be made by taking the difference between the hole state and the ground state energies. Semi-empirical models can also be used. Inclusion of relativistic effects in this treatment is essential for inner-core ionization and heavier

ELECTRON SPECTROSCOPY
FOR
ATOMS, MOLECULES AND CONDENSED MATTER

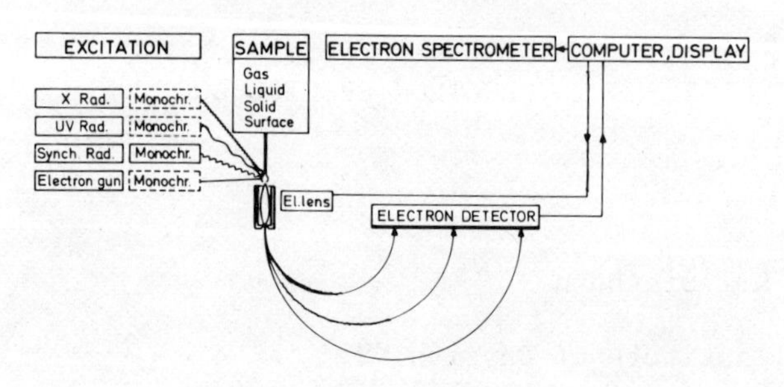

Fig. 1. Scope of electron spectrometry.

elements. More recently, methods have been devised to describe photc electron emission by means of a transition operator which properly accounts for the relaxation process. Various conceptual models compl ment the computational procedures on an ab initio level.

For chemical shifts in free molecules, it is usually sufficient accurate to consider only the ground state properties. This is becau relaxation energies for series of similar electronic systems vary or marginally. This can be described by division of the relaxation ene into two contributions, one connected to the atomic contraction at ionization, the other to the 'flow' of charge from the rest of the molecule. The atomic part, which is nearly constant for a specific ment, is the dominating contribution to the relaxation energy. The flow part generally varies marginally for free molecules of similar structure, leading to constant relaxation energies. There are cases however, where the flow part can significantly change from one situ ation to another. For example, when a molecule is adsorbed on a met surface, the flow of conduction electrons from the metal substrate will contribute to the relaxation of the core hole. This can increa the relaxation energy by several electron volts. Other cases are pu metals and alloys, where the conduction electrons are responsible f the screening of the hole.

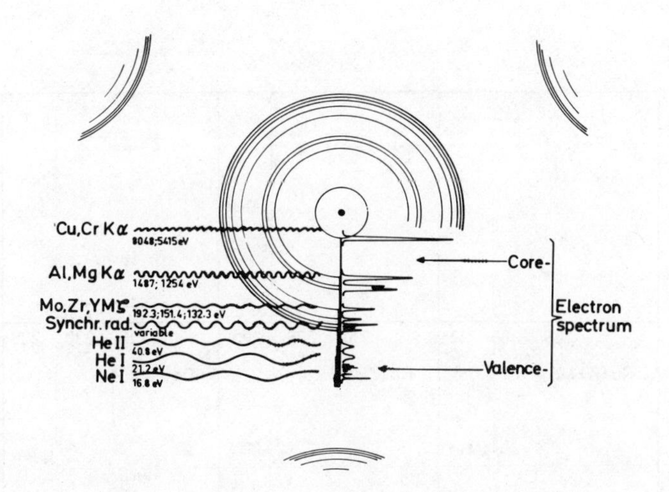

Fig. 2. Excitation of core and valence electron spectra.

SOLID STATE AND SURFACE STUDIES

The experimental accuracy in the determination of photoelectron or Auger electron peak positions have gradually improved to such an extent that the binding energies which can be extracted from electron spectroscopic data in principle can be given with errors in the 0.1 eV range and even much less for gases. It is then of importance to carefully consider problems related to the choice of. for example, reference levels in order to connect such spectroscopic data with other data. Fig.3 is an illustration of a series of binding energy measurements[2] of the 3d core electrons for Z between 40 and 52. Experimentally, the core level binding energies for a metal are measured relative to the Fermi level. If one wishes to transfer this binding energy to that referred to the vacuum level of the free atoms one has to add a metallic work function ϕ to the experimental metallic core-level binding energy. There are some uncertainties in doing this which have been lively discussed in literature during recent years. One can avoid this difficulty, however, by means of a thermochemical approach and

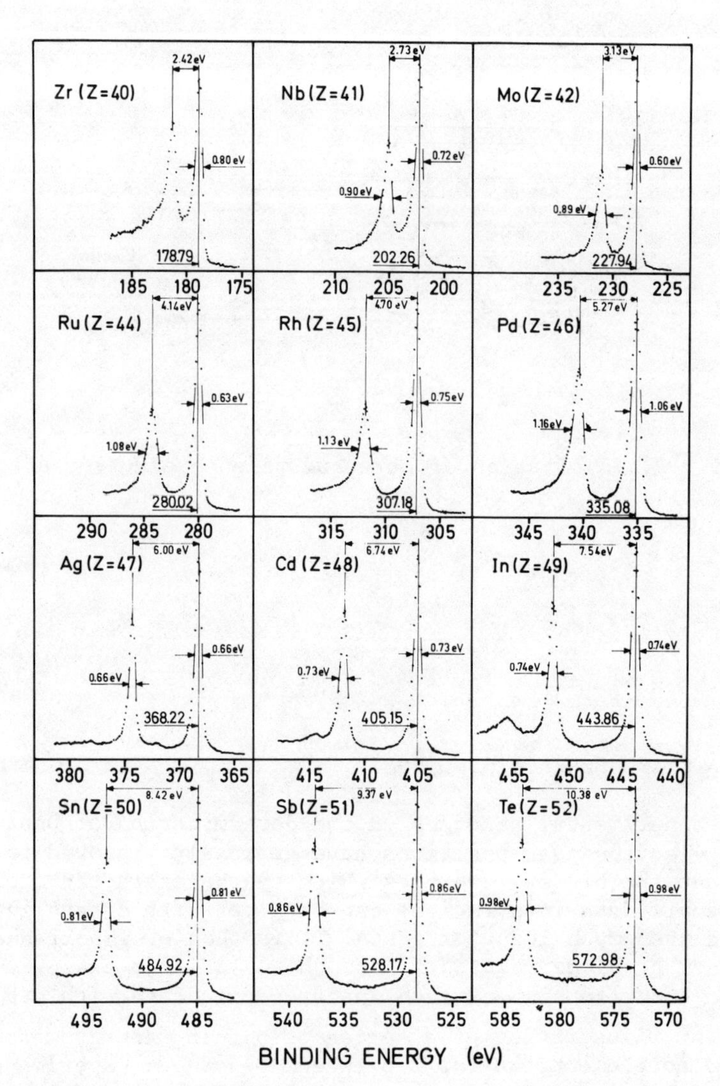

Fig. 3. 3d core electron spectra taken with HP ESCA 5950A.

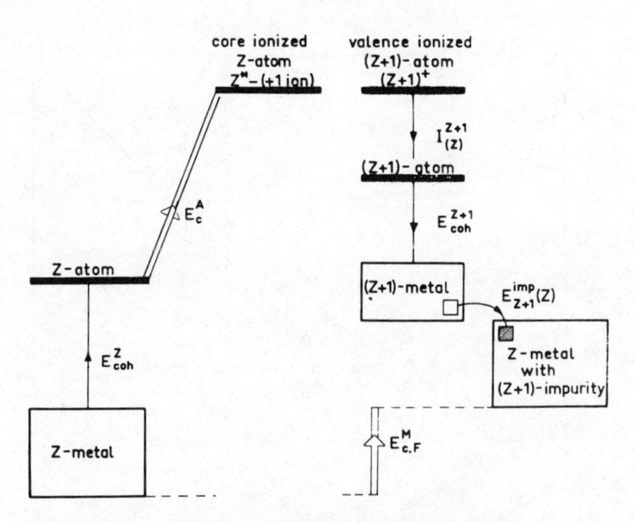

Fig. 4. Born-Haber cycle for metal.

by making a few very reasonable assumptions regarding the relaxation in metals. We will follow the treatment by Johansson and Mårtensson[3-7] which for many purposes has turned out to be particularly suitable. This theory can be generalized so that surface effects on core level binding energies can also be incorporated as well as implantation of for instance rare gases in noble-metal hosts. Furthermore, core level shifts in alloys and intermetallic compounds can be treated in the same fashion. One assumes a complete screening of the positive hole left behind in the metal lattice at the photoelectron emission due to the good mobility of the conduction electrons. One makes use of the so called 'equivalent core' model which means that the core ionized atom is treated as an impurity atom with nuclear charge Z+1. Under these assumptions a Born-Haber cycle is performed according to Fig.4. The result is:

$$E^M_{C,F} = E^Z_{coh} + E^A_C - I^{Z+1}_{(Z)} - E^{Z+1}_{coh} + E^{imp}_{Z+1}(Z)$$

Here M denotes the metal, A the atom, C the core, F the Fermi level, Z the atomic number and I is an appropriate valence ionization energy (in most cases the first ionization energy). E_{coh} is the cohesive energy and $E^{imp}_{Z+1}(Z)$ is the solution energy of impurity atom (Z+1) in (Z) metal.

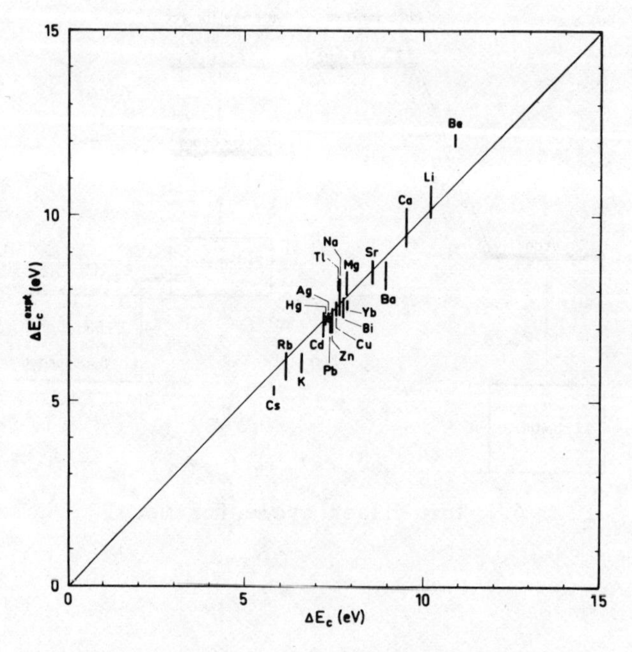

Fig. 5. Test of the thermochemical model.

The binding energy shift between the metal core level referred t the Fermi level and the corresponding free atom referred to the vacuu level is obtained as

$$\Delta E_C = E_C^A - E_{C,F}^M$$

We therefore can write

$$\Delta E_C = I_{(Z)}^{Z+1} + E_{coh}^{Z+1} - E_{coh}^{Z} - E_{Z+1}^{imp}(Z)$$

Thus in this treatment the dominating contributions to the shifts are the difference in cohesive energy between the (Z+1) and Z metal and the ionization energy of the (Z+1) atom. Fig.5 is a comparison betwee the experimentally measured and theoretically deduced core level bind ing energy shifts according to the above formula. Details of the results and explanations for small deviations from the straight line are given in the quoted paper.

An extension of the above Born-Haber cycle treatment for the core energy shift between the metal (referred to the Fermi level) and the free atom (referred to the vacuum level) can easily be made to describe the core level binding energy shift of an atom A in its metallic state when dissolved in a metallic matrix B. The result is

$$\Delta E_{calc} = E(A;\underline{B}) + E(A+1;A) - E(A+1;\underline{B})$$

$E(A;\underline{B})$ is the energy of solution of a metallic atom A in a matrix B. A+1 is the (Z+1) element relative to A. Fig.6 shows the result of a study of some forty dilute alloys $A\underline{B}$ where the concentration of the dilute component A is 10% or less. The experimental shifts are analysed in terms of alloy heat of formation data according to a semi-empirical scheme due to Miedena. Again, the agreement with the above formula is good. For details, see ref. 8.

The same sort of treatment as for ordinary metallic bulk matter can also be applied to surface core-level shifts. The surface atoms experience a different potential compared to the layers below because of the lower coordination number. This results in somewhat different core level binding energies. One can extend the previous Born-Haber cycle model to account for the surface-bulk core level shift. Empirically, the surface cohesive energy is approximately 80% of the bulk value. The impurity term can then be written as

$$E_{Z+1}^{imp,surf} \approx 0.8\ E_{Z+1}^{imp}(Z)$$

One then gets:

$$DE_{C}^{S,B} = E_{C}^{surf} - E_{C}^{bulk} = 0.2\ [E_{coh}^{Z+1} - E_{coh}^{Z} - E_{Z+1}^{imp}(Z)]$$

This equation obviously relates the surface chemical shift and the heat of surface segregation of a (Z+1) substitutional impurity in the Z metal.

As a recent example[9], where the surface and bulk core lines can be completely resolved we show in Fig.7 the As3d electron spectrum from UAs(100) excited by synchrotron radiation at $h\nu$=80 eV. The two spin doublets, each with a spin-orbit splitting of 0.66 eV, are shifted from each other by 0.3 eV. By exposing the surface to 0.5 L O_2 one can observe that only the surface doublet is affected. In most cases the bulk-surface shift is smaller and requires both high resolving power and a careful line profile investigation to enable one to extract the above surface information. A comprehensive study of these problems with recommended procedures have been published recently by Citrin, Wertheim and Baer[10,11].

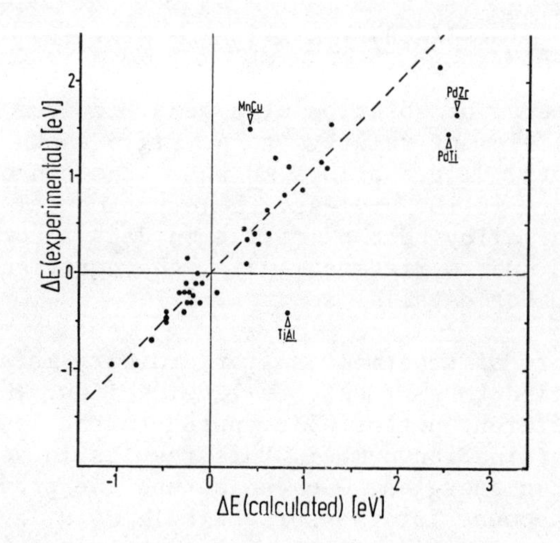

Fig. 6. Test of model for 40 alloys.

Some recent studies of surface shifts can be mentioned in which the above model has come into use. In the case of $EuPd_2Si_2$ one can s how the magnitude of the surface shift depends on the chemical surrounding for a certain element[12]. One can furthermore show that this surface effect has the consequence that the valence is changed in th surface compared to the bulk. In the system $EuPd_x$ one can follow how the valence 2+ in the bulk changes to 3+ whereas the surface remains divalent[13]. In this system the Eu4f electron emission can be well separated from the Pd4d valence band, since it is possible to choose a photon energy where at the same time the Eu4f intensity is high and the Pd4d emission is low due to a Cooper minimum.

A special class of impurities is provided by rare gases implant in noble metals, a case which has been studied by Citrin and Hamann[1 and by Watson, Herbst and Wilkins[15]. The above Born-Haber cycle can

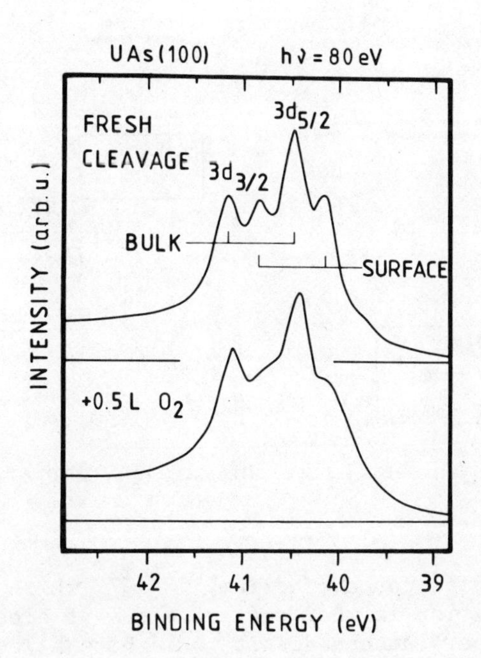

Fig. 7. Surface and bulk As3d core lines for UAs(100).

also be applied here[3] (see Fig.8). The result is

$$\Delta E_C^{implant} = E_C^A - E_{C,F}^{implant} = I_{(Z)}^{Z+1} + E_{coh}^{Z+1} - E^{implant}(Y) - E_{Z+1}^{imp}(Y)$$

$E^{implant}(Y)$ is the energy required to implant a rare gas atom into the metal host of atomic number Y.

In some other works thermochemical quantities have been determined by means of the chemical shifts for alloys and metallic compounds, CuPd and USbTe. In the former case[16,17] both single hole and Auger, i.e. double hole, ionizations have been treated. In the systems[18,19] USbTe and UAsSe the surface shifts can be resolved and thus a separation between the surface and bulk contribution can be made. The analysis could be performed although the shifts were only a few tenths of an eV.

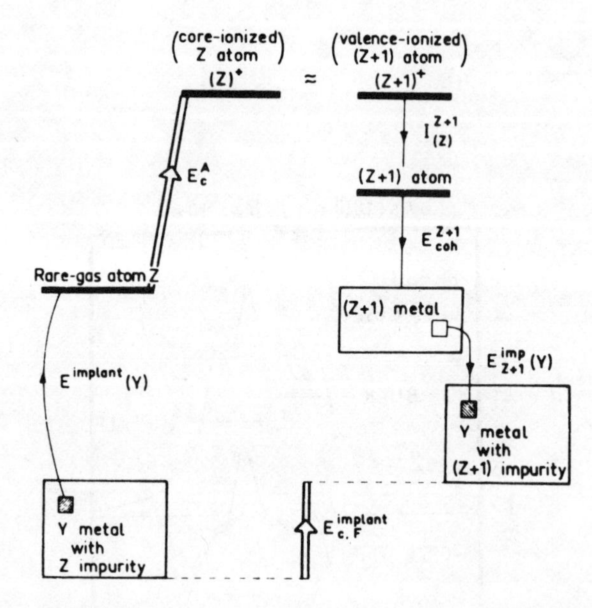

Fig. 8. Born-Haber cycle for rare gas implanted in metal.

Several studies concern cerium[20,21,22], which is the first 4f-element. The question is what character the 4f electron has. If Ce behaves as most other lanthanides it would have a localized 4f electron and three valence electrons in 5f and 6s, which determine the chemistry for Ce. The 4f electron is localized because it is inside the valence electrons. We know from chemistry that Ce also can take part in tetravalence compounds (e.g. cerium dioxide). The system gains energy by exciting the 4f electron to the valence shell. For the atom this costs energy which, however, is compensated for by extra chemical energy in the tetravalent state. At room temperature the Ce metal is trivalent but on cooling or at increased pressure it exhibits a phase transition. This has been interpreted as a transition towards a tetravalent state. This α-phase should then have a mixed valence. Johansson had suggested before that the γ-α transition is a Mott transition where the 4f electron becomes band-like (itinerant) in the α-phase. An electron spectroscopic study was recently performed on the γ-α transition (CeTh and not on Ce because of certain practical reasons). If mixed valence had occurred in the α-phase the 4f line position would be at the Fermi level, which is a result of the total screening. If the 4f electron is localized (which is assumed in the mixed-valence picture) and if a trivalent Ce atom is ionized a 4f electron is ejected which is screened by the valence electrons, i.e. the atom is still neutral after ionization. After being a trivalent atom it is now a tetravalent final state atom and the 4f-ionization energy which is measured, corresponds to the energy difference between the two valence states. For a mixed valence state the tri and tetra valent configurations should be degenerate and the 4f peak situated at the Fermi level. However, the main 4f electron emission occurs at 2 eV from the Fermi level also in α-Ce, which excludes the traditional mixed-valence description of this system. It was furthermore observed that in the spectrum from γ-cerium (as in several cerium compounds) two 4f peaks occur, both with 4f character, a new and puzzling result.

A surface-bulk phenomenon of technical importance is the diffusion of an evaporated metal surface layer into the bulk of the backing metal. Fig.9 shows how such a process can be followed[6] in an electron spectrum as a function of time. The time evolution of the Au4f lines is observed as the evaporated gold diffuses into the various substrates, Zn, Cd and In.

Often in surface science not only the top layer but also the composition and chemical bonds between atoms close to and below the top surface layer are of importance. When some element has a concentration profile within the electron escape depth, i.e. when there are surface layers of different composition with a thickness less than the escape depth, the peak amplitude of that element shows exit angle dependence. By varying the exit angle, surface concentration profiles can be investigated by means of ESCA[23,24]. A recent study[25], of medical interest, dealt with blood-compatible surfaces. The systems studied were colloidal heparin, or dextran sulphate stabilized with hexadecyl ammonium chloride, deposited onto steel substrates, and chemically related substances. Using the angular dependence techniques it was then found that the intensity ratio for the S2p peaks from disulfide and sulphate exhibit exit angular dependences for albumin covered

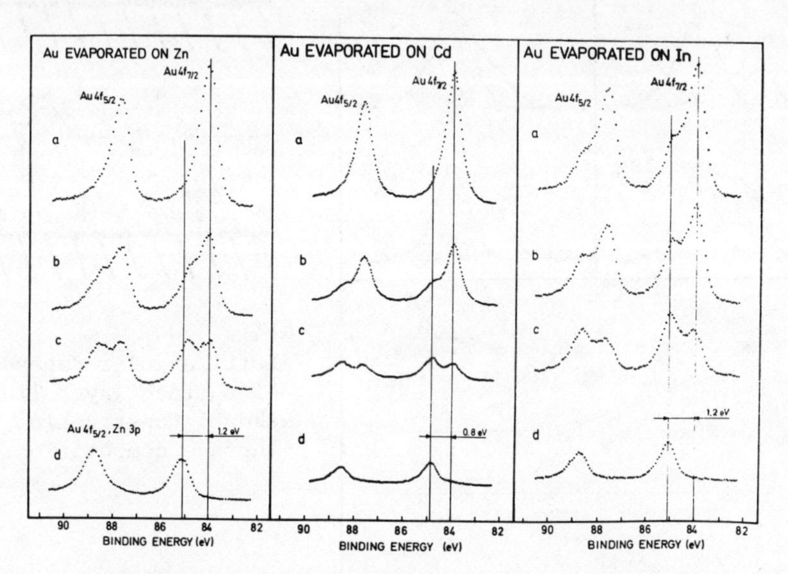

Fig. 9. Diffusion of evaporated Au into Zn, Cd and In as a function of time.

heparin-glutar and dextran sulphate-glutar surfaces, which indicates that the disulfide groups are positioned closer to the external surface than the sulphate groups. In a series of experiments[26] on a similar problem cationic polyethyleneimine was adsorbed on sulphated polyethylen surfaces at different pH, varying from 4.0 to 9.0. From the angular dependence of the amine/protonated amine peak ratio (see Fig.10) it was possible to conclude that there was an accumulation of charged amino groups towards the sulphate surface at high pH. The angular dependence of the intensity ratio N/N^+ (neutral amine/protonated amine) furthermore shows that adsorption at pH 4.0 gives a higher relative amount of charged groups and that this amount is independent of the exit angle. Adsorption at pH 9.0 gives a relatively larger amount of neutral amine groups and a N/N^+ ratio which is dependent on the exit angle. A straightforward interpretation of these results is that the configuration of PEI when adsorbed at pH 4.0 is essentially flat on the surface while adsorption at pH 9.0 gives 'layered' configuration with the charged groups (N^+) closer to the sulphated surface and the neutral groups further out (see Fig.10). The fact that the number of charged amino groups remains constant leads to the conclusion that the adsorption could be considered as an ion exchange reaction. The results also imply that polymer surfaces with different densities of sulphate groups would adsorb different amounts

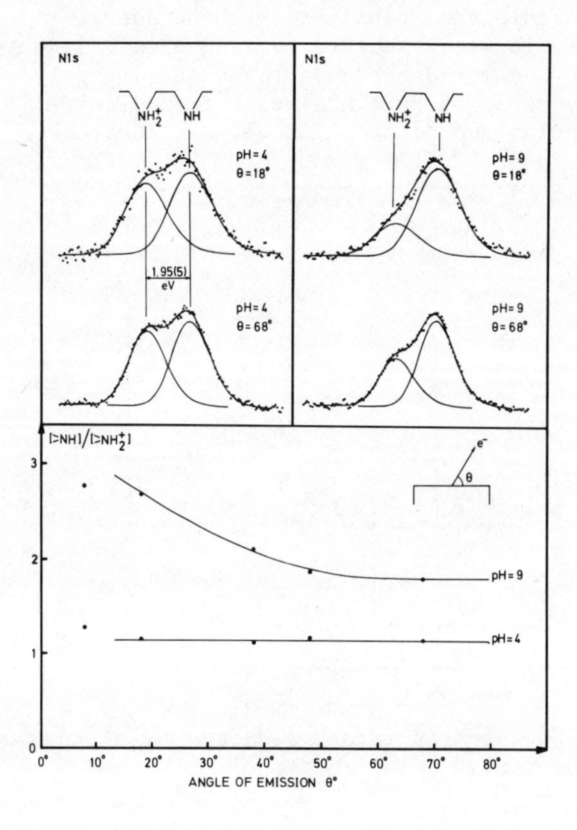

pH=4

pH=9

Fig. 10.
Left: Angular dependence of surface layer lines.
Above: Construction of surface composition.

of PEI. Thus, it would be possible to make polymer surfaces with different densities of amine groups not only by adsorption at different pH but also by varying the sulphate group density.

LIQUID PHASE STUDIES

Let us now leave the solid state with its bulk and surface properties for the liquid state. From our laboratory the first results in this field were reported at the electron spectroscopy conference at Asilomar in 1971[27] and the development has since then proceeded[28-36] to the extent that one can now safely conclude that the study of liquids by ESCA will provide new interesting information on this state of aggregation. The techniques started with liquid 'beams' in vacuums and signals from core levels were then recorded from both the liquid and the gas phase as 'doublets'. Also valence levels were studied and different alternative schemes for achieving continuously renewable liquid sources were considered and tested. One of these is shown in Fig.11 consisting of a wetted wire which passes in front of the spectrometer slit from the liquid. Differential pumping on the gas phase is a prerequisite in the study of liquids. Fig.12 shows a wetted disc arrangement where a monochromatic AlKα beam is focused onto the disc in front of the spectrometer slit[45]. The vapour pressure of the liquid is reduced and controlled by a variable temperature liquid nitrogen bath which cools the liquid. In the cases when the vapour pressure cannot be reduced or otherwise when the extra gas peaks complicate the observed

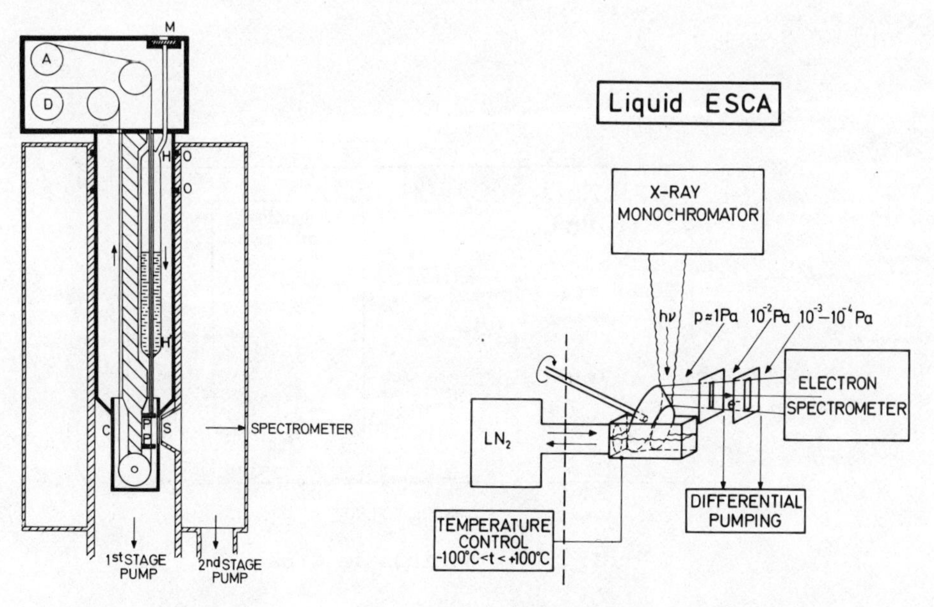

Fig. 11. Wetted wire for ESCA.

Fig. 12. Wetted disc with monochr. AlKα.

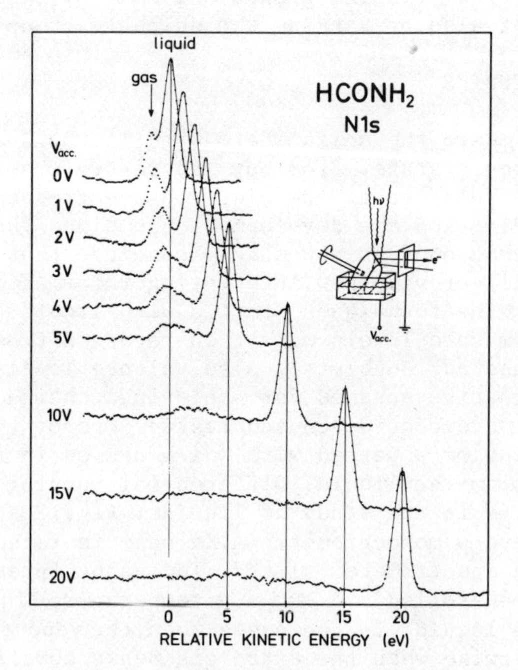

Fig. 13. Elimination of gas line.

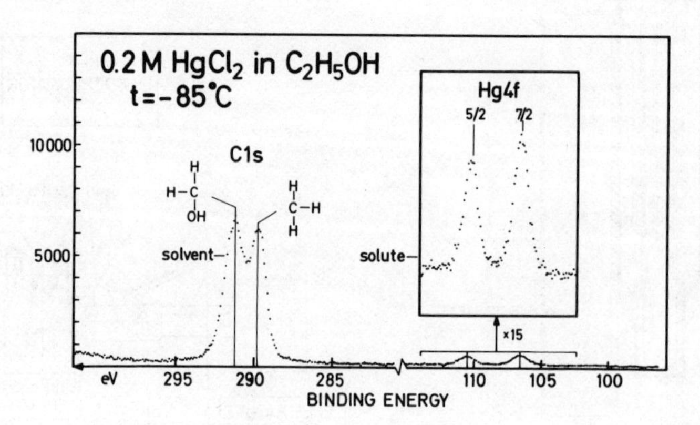

Fig. 14. Hg ions in ethanol.

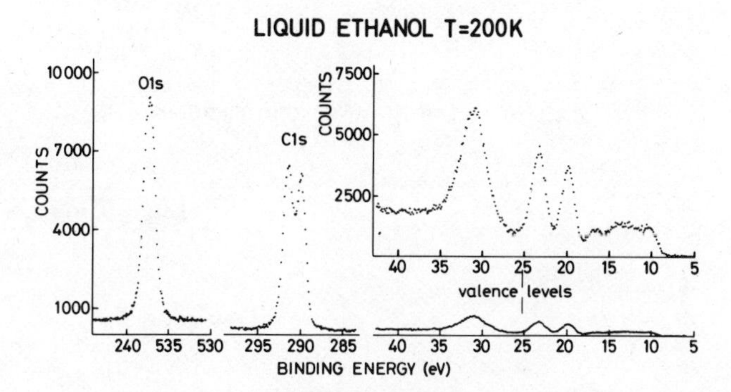

Fig. 15. Liquid ethanol, core and valence spectrum.

structures it was found that the gas lines can be eliminated by a very simple arrangement[46], see Fig.13. A potential of ~10 eV is applied between the disc and the slit. The gas line which is excited over this volume will then be broadened to the extent that it essentially disappears, whereas the liquid line profile is unaffected and only shifted corresponding to the applied voltage. A clean liquid spectrum can then be recorded.

Figs.14, 15 and 16 show three examples of liquid spectra[45]. In these cases the solvent has been ethanol at a temperature of -85°C. One observes the chemical splitting of the C1s line in CH_2CH_3OH due to the different carbons. In Fig.15 also the valence spectrum of ethanol is recorded. Fig.16 shows that I_3^- ions are formed in the solution when NaI and I_2 are dissolved in ethanol. Each of the I spin 3d components is split in two chemically shifted lines in the intensity ratio of 2:1, provided that the shake-up line is attributed to the externally situated two iodine atoms in the (linear) I_3^- ion.

As for metals and alloys it is now convenient to be able to correlate the positions of the measured liquid lines to a reference level, preferrably the vacuum level, to arrive at a consistent picture which enables direct comparison to be made between the gas phase and liquid phase binding energies[47]. In order to study this question closer a Born-Haber cycle can be performed. The liquid sample plus the metal trundle (in Fig.12 or the wire in Fig.11) which is used to create the liquid film together form an electrochemical half cell. One can then express the electromotive force of this half cell in

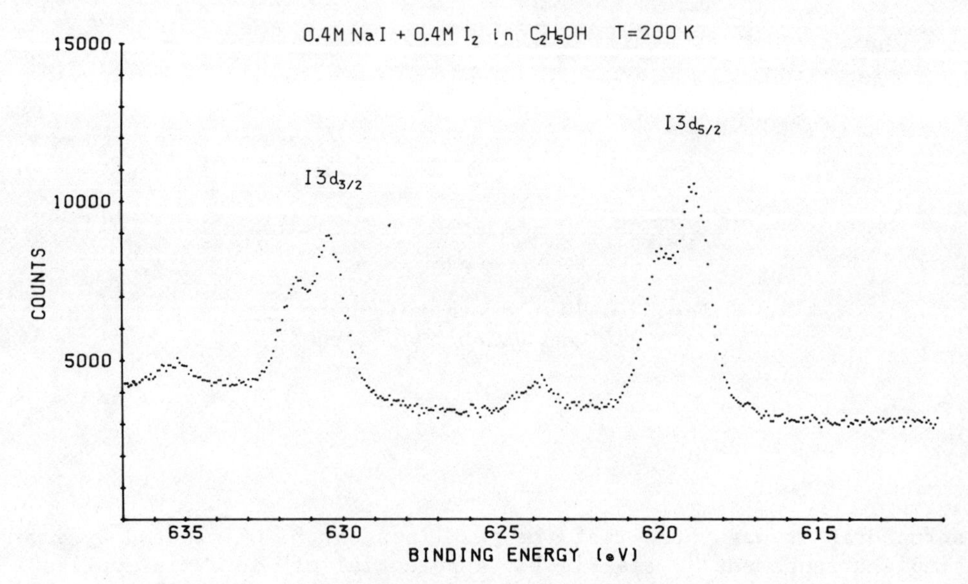

Fig. 16. I_3^- ions formed in ethanol.

terms of thermochemical entities (heat of evaporation and solvation) related to the creation of an ion and its transfer from the metal electrode into the bulk of the solution. The emf of this half cell can be experimentally determined by connecting it to a reference electrode. The solution surface and the reference electrode thus form a condensor pair whose potential (Volta potential) can be determined by means of vibrating the reference electrode (vibrating condensor method). The situation at hand is illustrated in Fig.17 which is equivalent to the liquid ESCA arrangement except that in the latter case the reference electrode is replaced by the spectrometer entrance slit. From Fig.17 one can deduce the relation between the liquid and gaseous phase ESCA spectra. For the liquid lines (say the core lines of a metal ion or the C1s line of an organic solvent), one has:

$$E_{kin}(\text{liq}) = h\nu - E_B^V(\text{liq}) + \phi_{Volta}$$

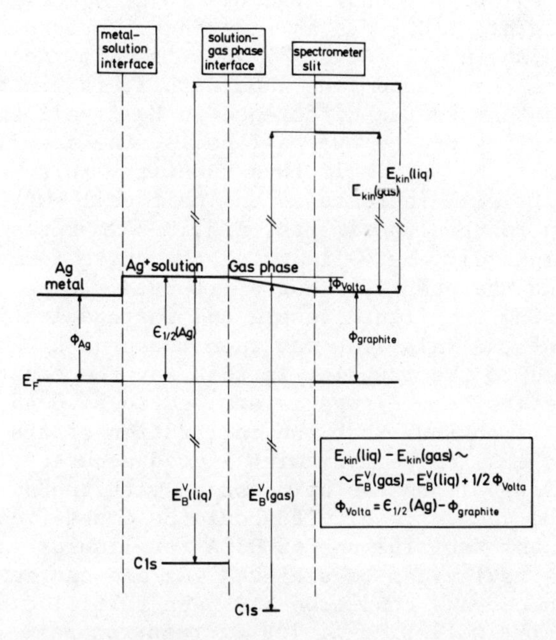

Fig. 17. Relation between liquid and gas ESCA.

The gaseous phase photoelectrons are created within the Volta potential field and they will thus not experience its full effect but only about half on the average. With this in mind we get for the gaseous phase:

$$E_{kin}(gas) = h\nu - E_B^V(gas) + k \cdot \phi_{Volta}$$

where k is approximately equal to $\frac{1}{2}$. This quantity can, however, be directly measured. Combined, the two equations give:

$$E_{kin}(liq) - E_{kin}(gas) = E_B^V(gas) - E_B^V(liq) + \frac{1}{2}\phi_{Volta}$$

and

$$\phi_{Volta} = \varepsilon_{1/2}(Ag) - \phi_{graphite}$$

if the trundle is made of silver and the spectrometer and the slit are painted with aquadag. These are the expressions required to refer the measured liquid lines, using a gas phase line as a reference line to the vacuum level. Fig.18 shows the C1s in ethylene glycol using a simultaneous inlet of CO_2 gas as reference. Three different half cells were studied: (Zn,Zn^{2+}), (Cu,Cu^{2+}) and (Ag,Ag^{+}). According to the above equations the liquid line should shift in kinetic energy from one case to another by the difference in half cell emf, i.e. by the cell emf consisting of the two half cells. The gas lines on the other hand should shift by only half this quantity. This means that the gas-liquid shift will increase as the half cell emf increases. This is nicely seen in the experiments. Fig.19 summarizes the experiments and verify completely the validity of the above formulae. The straight line connecting the points is not a fit but is experimentally determined by removing the liquid sample and instead letting in Ne gas. The position of the Ne1s line was then measured as a function of potential applied to the trundle. In this way the constant k (in the formula assumed to be $\sim \frac{1}{2}$) was determined to be 0.45 in the arrangement used. It is obvious that the calibration of the liquid ESCA arrangement is easy to perform with a good accuracy and that the core and valence liquid lines can be measured with the same accuracy and referred to the vacuum level. It should be emphasized that the basic condition for any fruitful use of ESCA for liquids is that high quality electron lines really can be achieved. As one can see from the above spectra this necessary condition has been fulfilled here. Under such circumstances the following fields of research seem feasible to enter:

1. Ions in solution
   a. Electronic structure
   b. Solvation and reaction energies
   c. Complex formation
   d. Selective solvation
   e. Ion pairs
2. Intermolecular forces
   a. Valence level effects
   b. Core level effects
3. Liquid-vacuum interface
   Adsorption
4. Molten substances

Part of this program is now under way and the results look promising. Fig.20 illustrates an application of core electron spectroscopy to the investigation of metal complexes in solution. The figure shows how measured core electron binding energy shifts and Auger electron energy shifts between the gaseous and the liquid phase can be related to solvation energies via a Born-Haber cycle[48]. Ions in an electrolyte mostly have closed shell rare-gas electronic configurations

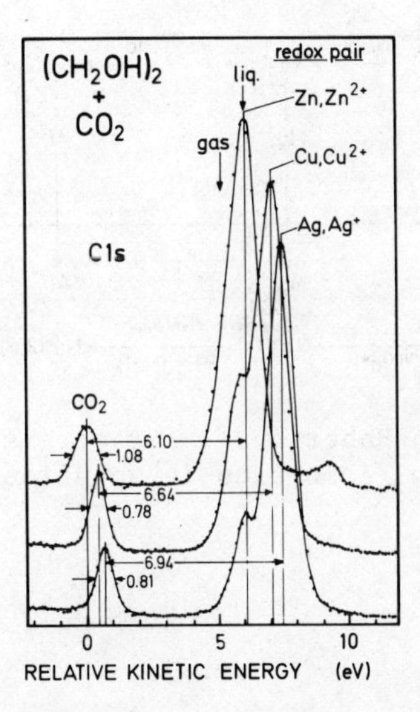

Fig. 18. C1s core lines in ethylene glycol with CO_2 as reference gas. Three different half cells.

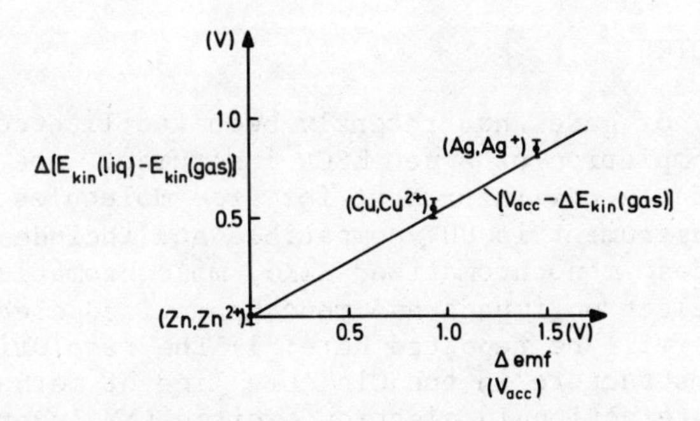

Fig. 19. Verification of formula according to Fig. 17.

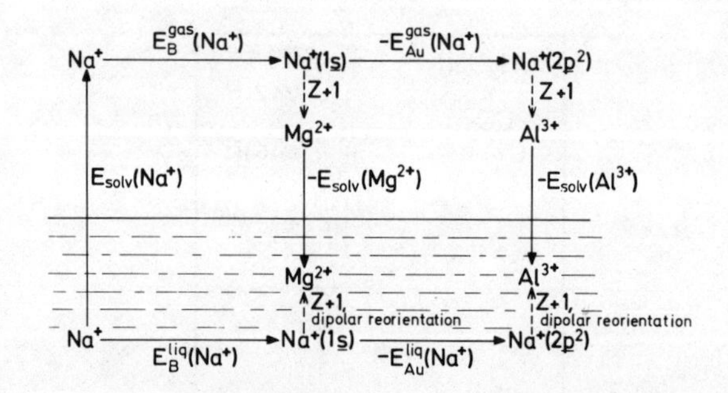

Fig. 20. Born-Haber cycle for core spectroscopy of metal complexes in solution.

Core electron spectroscopy is therefore particularly advantageous sin[ce] it allows comparison to be made between isovalence-electronic species such as Na^+, Mg^{2+}, Al^{3+} etc., for which accurate thermodynamic data have been measured. In the particular case illustrated here, 1s-ionization of Na^+, accurate figures can be obtained for the reorientation energies associated with the change from a Na^+ to a Mg^{2+} or a Al^{3+}-io[n]. This, in turn, leads to information concerning the relative magnitude of electronic and ionic contributions to the solvation energies. Here the Auger energy shifts play an important role in that the electronic reorganization energy in going from Na^+ to Mg^{2+} can be obtained by co[m]paring with the corresponding binding energy shift.

GAS PHASE STUDIES

The study of gases has recently been facilitated in our laboratory by the completion of a new ESCA instrument, see Fig.21. It shows a side view of the new instrument for free molecules and condensed matter. The instrument is UHV-compatible and includes four different excitation modes: monochromatized AlKα, monochromatized and polarized ultraviolet, electron impact and monochromatized electron impact. Two recent studies will be reported here: 1) The resolution of the vibrational fine structure in the C1s core line of methane[48] and 2) Post collisional interaction in electron excited LMM Auger electron emissi[on] from argon[49,50].

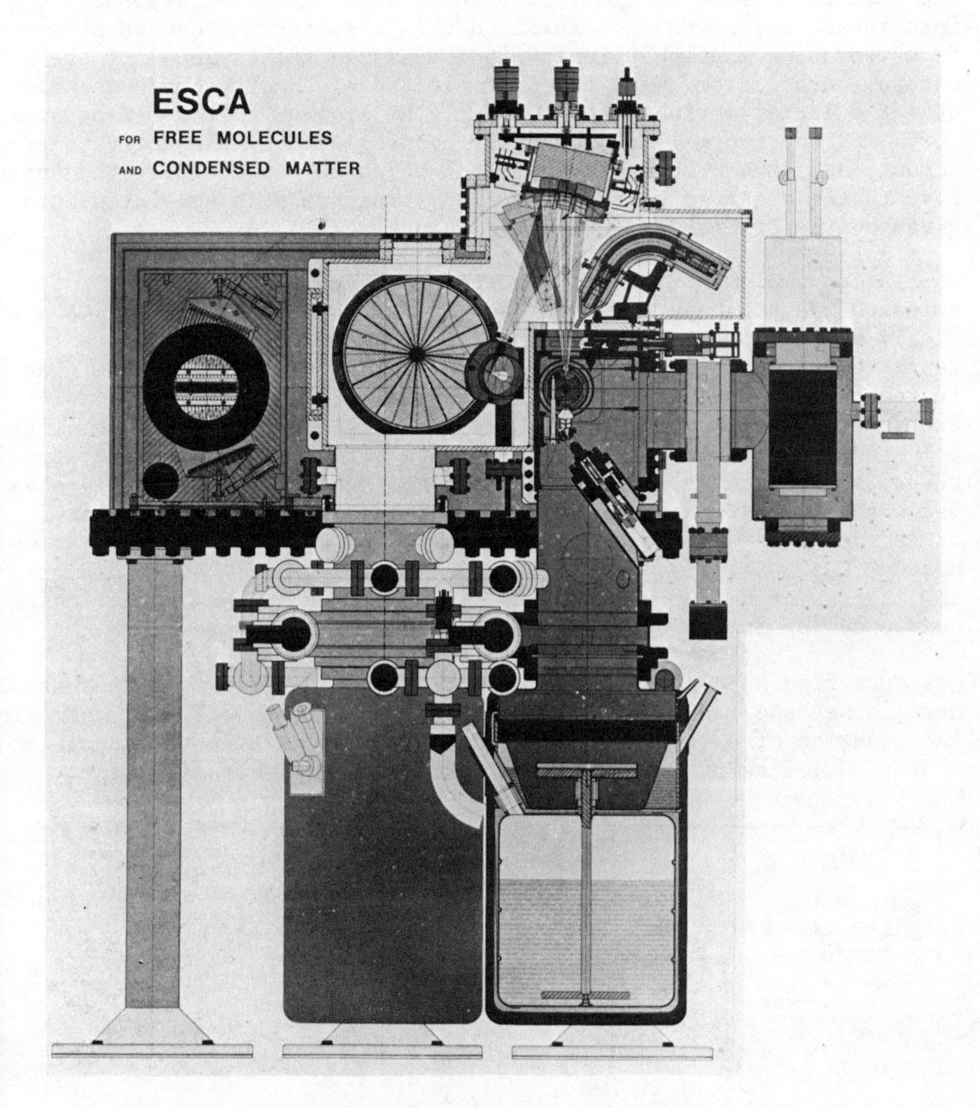

Fig. 21. The new ESCA instrument for gases and solids.

Some years ago[49-52] we found that at increased resolving power and by using monochromatized AlKα radiation, approaching the limit set by the rocking curve of the quartz crystals used, the C1s line of methane appeared to have an internal structure, see Fig.22. We separated this into three components and explained this structure as caused by symmetric vibration when photoionization occurs in the 1s level of the central carbon atom. When the photoelectron leaves the methane molecule the latter shrinks about 0.05 Å. The minimum of the new potential curve for the ion will consequently be displaced by the corresponding amount, and Franck-Condon transitions which take place will then give rise to the observed vibrational fine structure with the intensities given by the Franck-Condon factors (see Fig.23).

As a complement to electron spectroscopy molecular ultra soft X-ray emission spectroscopy was developed in our laboratory in 1971. A specially designed instrument for this purpose was constructed[53] and reported at the Asilomar conference[54]. This high resolution 3m grazing incidence spectrometer with a differentially pumped electron excitation arrangement was built in order to observe - if possible - not only individual molecular orbitals[55] but also vibrational fine structures (see ref.54, p.40-45). Such a fine structure was actually observed with the instrument first in N_2[56] and then later on in several other cases[57-62] with additional information on molecules such as CO, NO, O_2, CO_2, CS_2, N_2O, H_2O, NH_3 and SF_6. Later on, a larger dispersion 10m instrument has been constructed[63] and is in the process of being developed for future use.

Our finding discussed above on the core electron line fine structure in methane is related to the molecular ultra soft X-ray work since the creation of a core hole obviously will affect also the resulting vibrational fine structure of the following X-ray emission line.

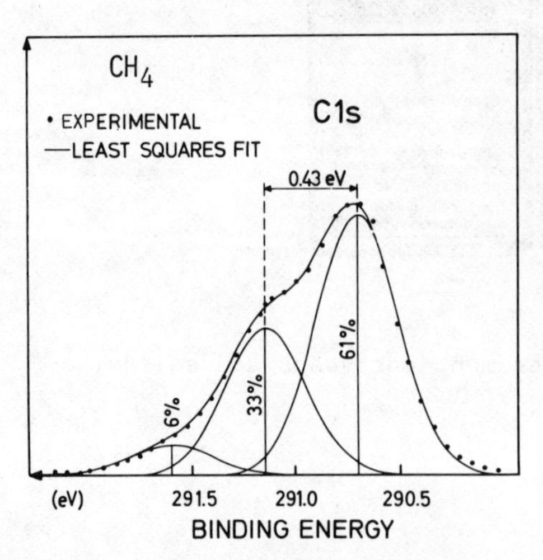

Fig. 22. First observed vibrational core line structure.

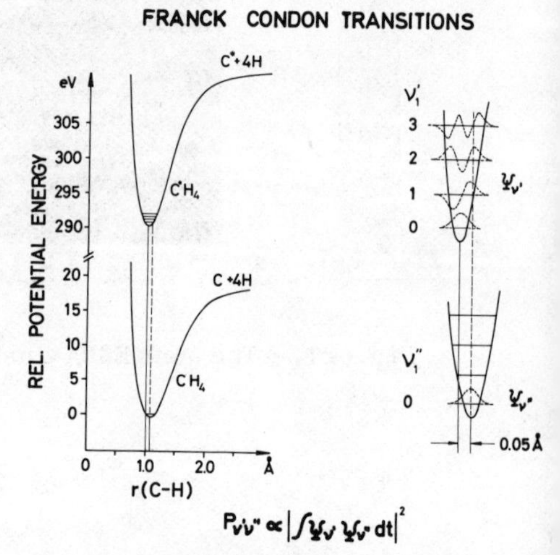

Fig. 23. Explanation of CH_4 core line structure.

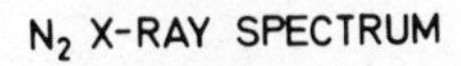

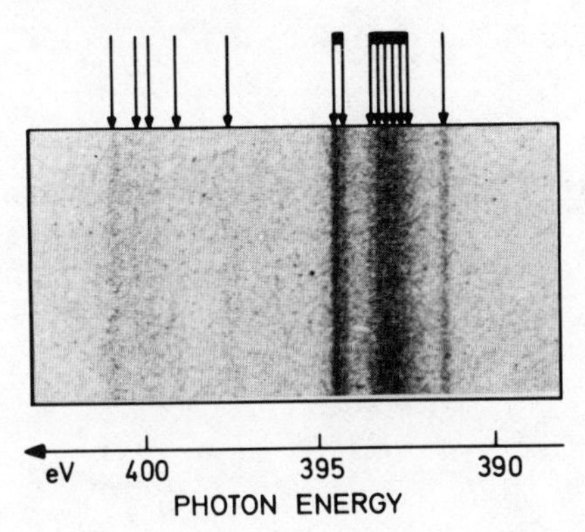

Fig. 24. First observed vibrational structure of an X-ray emission line (N_2).

Our first observation[56] of vibrational fine structure in N_2 is shown in Fig.24. Fig.25 is a similar, more recent[64] study of CO_2. Since vibrations are thus found in both core and valence electron spectra, X-ray emission lines from such molecules must be expected to get contributions from both the initial and final states as a superposition (see Fig.26, valid for CO). One or the other may be small and the fine structure in the soft X-ray lines may then in favourable cases be observed. The resolution is set by the width of the core level, taking part in the X-ray transition. This is around 100 meV for C1s.

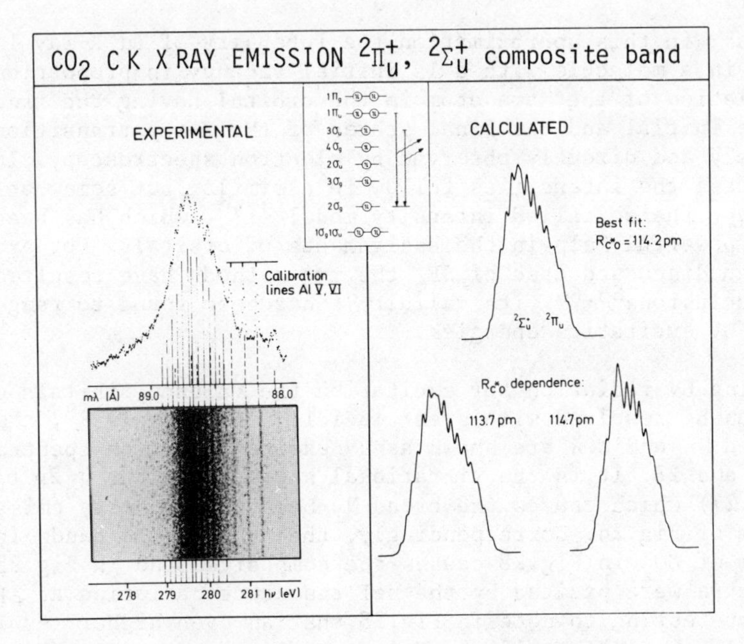

Fig. 25. Vibrational structure of C K X-ray of CO_2.

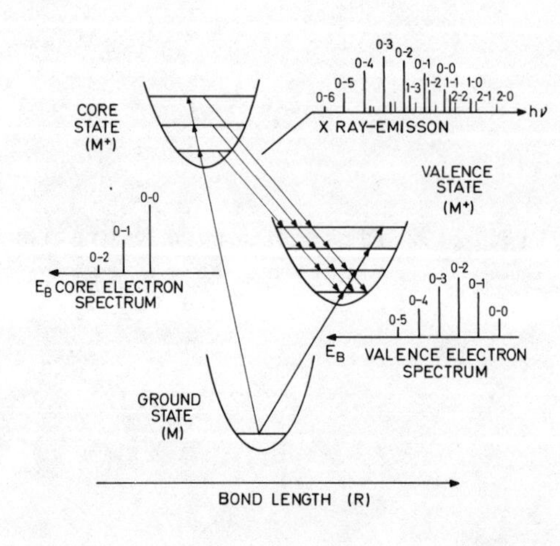

Fig. 26. Vibrational overlap of initial and final states in soft X-ray emission from CO.

Because of the strong dipole selection rules for X-ray transitions one can easily show that in the LCAO picture the intensities of the ultra soft X-ray lines for a molecule where the 1s and valence orbitals are involved can be written:

$$I \propto |\langle\chi(1s)|\ \underline{r}\ |\chi(2p)\rangle|^2 \propto C_{2p}^2$$

where χ is the atomic wave function and C_j is the expansion coefficient of the final state hole ϕ_f in atomic valence orbitals, i.e.

$$\phi_f = \Sigma_j\ \chi_j\ C_j$$

Thus, in this approximation the intensity of an X-ray line from an atom in a molecule with a 1s initial vacancy is proportional to the 2p population of the same atom in the orbital having the final vacancy Both the initial and the final states of the X-ray transitions can be separately and directly observed by electron spectroscopy. In the latter case the intensities follow in a similar but somewhat less rigid way, the so called intensity model[49,50], which has been shown to be a powerful help in the assignments of orbitals. For example, in the often discussed case of SF_6 the two methods have resulted in the same conclusions[65,66]. Its validity[67] has been found to range even down to UV excitation energies.

Using UV radiation for excitation the valence orbitals of molecules can be resolved with great lucidity. As examples[68], the two mentioned N_2 and CO_2 are shown as UV excited electron spectra in Figs.27 and 28. It is the vibrational sequence of the π_u 2p band in N_2 (Fig.27) which causes the broad N_2 band in the X-ray emission spectrum of Fig.24. Correspondingly, the $^2\Pi_u$ and $^2\Sigma_u^+$ bands in the spectrum of CO_2 in Fig.28 causes the composite band in Fig.25. These two spectra were excited by the HeI resonance radiation at 21.22 eV. It is interesting to note in Fig.28 that an even higher resolution is attainable when NeI at 16.85 eV is used for excitation. The splitting

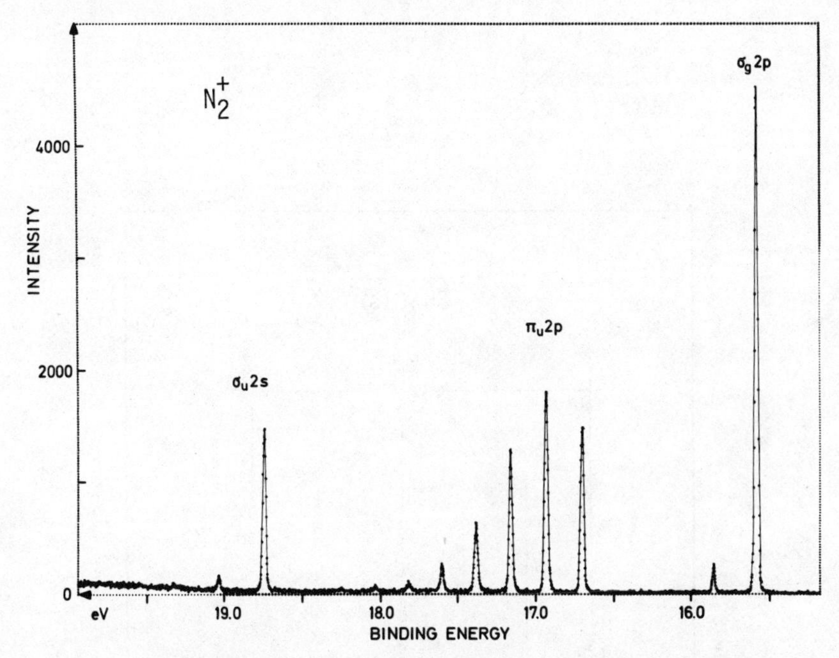

Fig 27. HeI excited valence electron spectrum of N_2. Compare X-ray emission spectrum of CO_2 in Fig. 24.

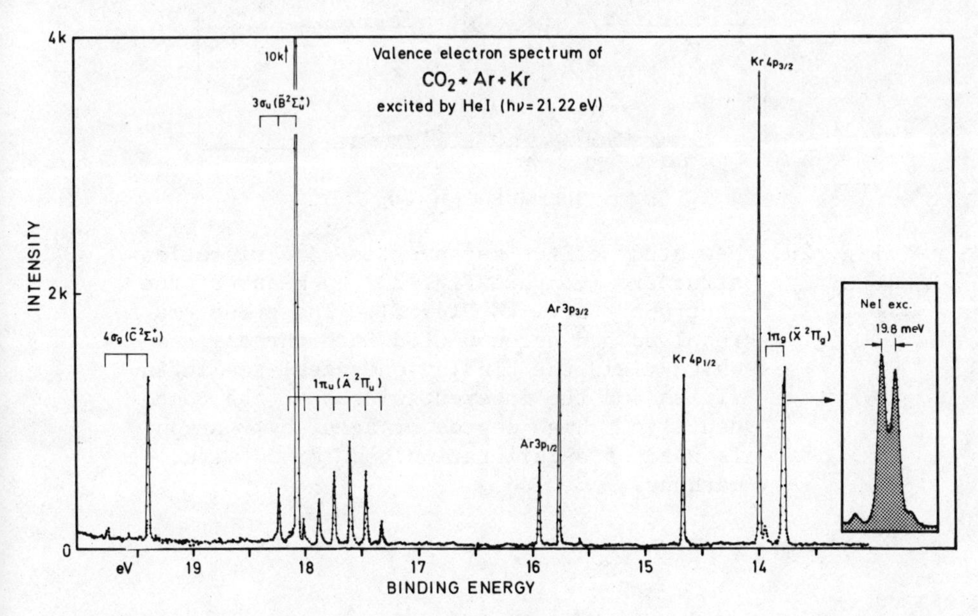

Fig. 28. HeI (and NeI) excited valence electron spectrum of CO_2. Compare X-ray emission spectrum of CO_2 in Fig. 25.

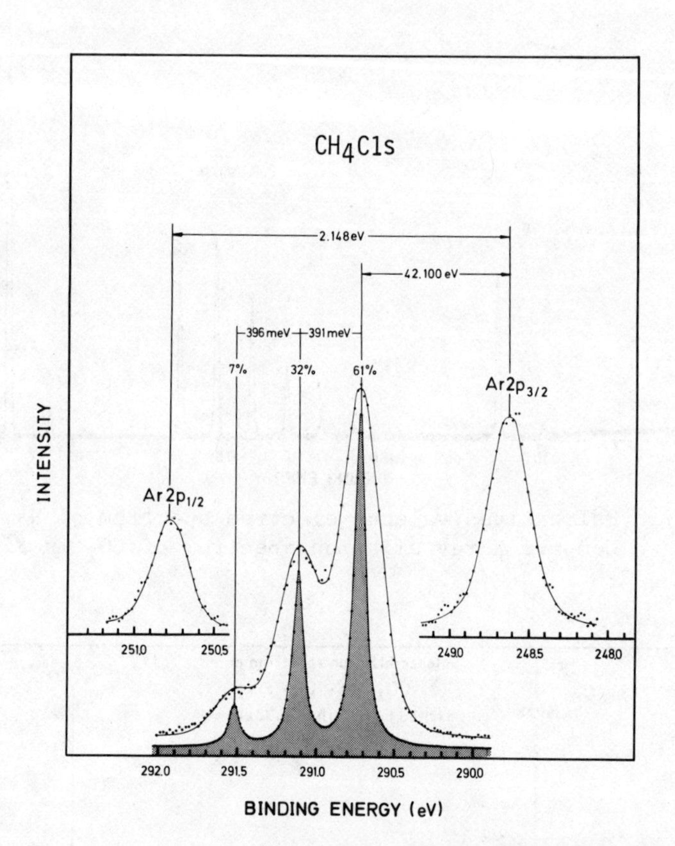

Fig. 29. New study of the methane C1s core vibrational structure (compare Fig. 22) by means of the instrument acc. to Fig. 21. The structure is resolved and deconvoluted into three lines which yield the binding energies, the intensities and the inherent widths of the components to a high degree of accuracy. Argon is used as a calibration gas, mixed with methane.

of $1\pi_g(\tilde{X}\ ^2\Pi_g)$ into two separate lines is caused by spin-orbit interaction. Doppler broadening and rotational degrees of freedom are main causes for the remaining narrow widths. The possible elimination of these factors is further discussed in the last section.

Let us now return to the vibrational splitting in core electron spectra as first observed in the methane 1s core line (Fig.22). This spectrum has now been studied again in our new instrument and the result[69] is shown in Fig.29. For calibration purposes argon was simultaneously introduced together with the methane. Our instrument automatically records in a repetitive manner the three lines and in this arrangement any line position relative to selected standard lines can be determined with some confidence. As one can see the C1s fine structure is now very nicely resolved and by means of a curve fitting procedure and deconvolution program the three components can be obtained with their intensities and inherent widths. The widths are just around 100 meV, a figure which will soon be given by fairly small limits of error. Soon the line positions will also be given with argon as a reference (these lines are precisely known from optical and other data) with a much increased accuracy in the meV range. At binding energies around 300 eV this approaches an accuracy of $\sim 1:10^5$.

Auger electron or autoionization electron emission can be excited by photons or electrons. At the emission of Auger electrons we have for the latter case in principle three electrons available for interaction with each other: the primary inelastically scattered electron, the expelled atomic electron and the subsequently emitted Auger electron. If the energy of the incoming electron is near the threshold of the event and the lifetime of the hole state is short, there will be two slow electrons around the ionized atom when the (fast) Auger electron is emitted. In this case the post collision interaction (PCI) will be strong, whereas if the lifetime is long and the incoming electron has much excess energy the PCI will be small. In the Auger decay the effect of PCI generally manifests itself as an energy shift as well as a modified and broadened peak profile. PCI shows up in different connections and has been the subject of several studies during recent years. Since our instrument is particularly suited for problems where high resolving power is an advantage we undertook a study [70,71] of the PCI at the LMM Auger electron emission excited by electrons of different energies, from 2750 eV to 6.4 eV above threshold for L_3 emission. Fig.30 shows such a series of Auger electron spectra. One observes the gradual changes of the line forms and the shifts towards higher kinetic energies when the incoming electron comes closer to threshold energy. This is a consequence of the fact that the doubly ionized argon ion left behind by the fast Auger electron still has the other two, slow electrons nearby, which can then screen the ion. Furthermore, an extra structure at the high-energy side of the $L_2M_{2,3}M_{2,3}(^1D_2)$ peak and an edge at 209 eV are observed. These are interpreted to be caused by electron shake-down from the electron continuum to ionic Rydberg states. In order to account for the precise line shapes which could be recorded at high resolution and good statistics, we have developed a new theory for PCI (for photo-excited Auger spectra). This can be applied to electron excited Auger spectra if one makes the reasonable assumption that the energy is distributed such that one of the two remaining electrons is slow and the other correspondingly faster. For electron excitation the photoexcited theory should then be applicable to somewhat higher energy events. Our new theory, like the preceeding one due to Niehaus[72], expresses the energy shift of the peak position, $\Delta\varepsilon$, as a function of the excess energy E_1 and the lifetime $\hbar\Gamma^{-1}$ according to

$$\Delta\varepsilon = c \cdot \Gamma \cdot E_1^{\alpha}$$

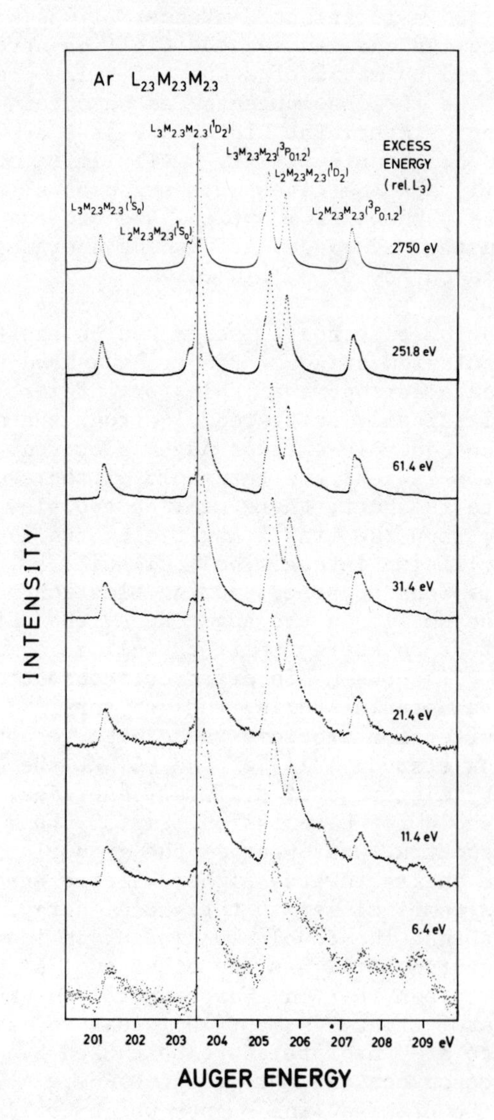

Fig. 30. Part of ArLMM Auger electron spectrum with ~140 observed lines related to the exciting electron excess energy, showing post collision interaction (PCI).

where c and α are constants. From our measurements we obtain the best fit, see Fig.31, with the constants c=4.9 and α=-0.37. For Γ we used the value 120 meV. From this point of view our treatment and measurements do not differ much from the latest studies by Huster and Mehlhorn[73] who obtain the values c=5.3 and α=-0.45. Our new theory is, however, consistent with our more detailed investigations of the shape of the PCI line over the entire energy region, which Niehaus' theory is

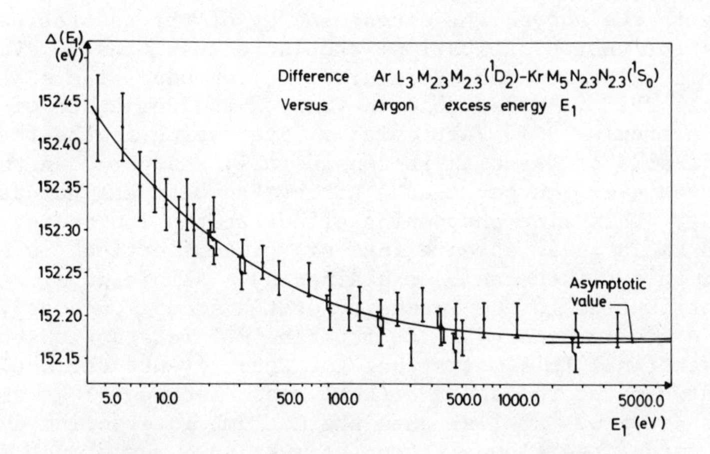

Fig. 31. Shift of line due to PCI.

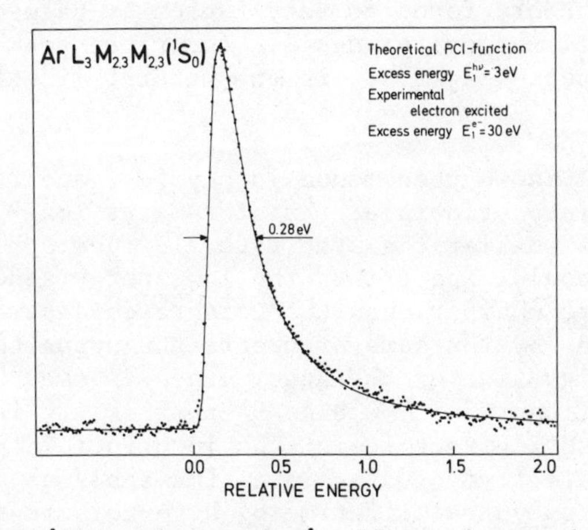

Fig. 32. 'Line sharpening' effect observed in PCI.

not. In the latter theory the line shape at high excess energies approaches a delta function instead of a Lorentz function with the proper natural linewidth. We have used WKB wave functions to describe the slow receding electron, before and after the Auger decay. The transition amplitude can then be expressed as a one particle overlap integral. The new theory describes the line shape continuously from low to high excess energies. In particular this theory can account well for the steep left-side of the peaks at low excess energies and the proper Lorentzian shape in the high energy limit. As an interesting example Fig.32 shows the experimental $ArL_3M_{2,3}M_{2,3}(^1S_0)$ peak obtained at 30 eV excess energy compared with a line calculation according to our theory According to the above, the excess energy of the theoretical line (which is for photon excitation) should be less than 30 eV. We have chosen this value to be 3 eV in order to reproduce the experimental line shape. This is reasonable in view of the dominance of the low energy electron at PCI. Particularly interesting is the fact that the steep left-side of the peak is considerably sharper than the Voigt function and even sharper than a Lorentzian with the natural half-width (Γ_{Ar}). This line sharpening effect at the low-energy side of the peaks can be seen as an effect of 'time-selection' in PCI. Qualitatively, this effect can be explained by a simple model. Auger electrons contributing to the intensity at the steep low-energy side have acquired a smaller energy shift from the PCI interaction than most of the electrons. This means that for these events the Coulomb interaction between the ion and the (two) slow electron(s) at the time of the Auger decay was smaller than the Coulomb interaction on an average In other words, the slow electron(s) must have moved much further away from the ion when the Auger decay took place. But the excess energy is the same for all Auger transitions. Hence these Auger transitions must have occurred at a much later time, i.e. the corresponding primary ions must have been extra long-lived. This is also seen from the semiclassical theory by Niehaus in which the PCI energy is directly related to the decay time. The decay time at the steep low energy side is by this theory found to vary typically between about $10\tau_0$ (at 1% of the peak intensity) and $2\tau_0$ (near the peak), being about $4\tau_0$ at half the peak height. τ_0 is the natural lifetime of the primary hole state.

It is a well-known phenomenon in physics, and fully in line with the uncertainty principle, that for extra long-lived states the linewidth will be smaller than the natural linewidth for that state. Therefore, one should expect that the low-energy side of the Auger lines actually are sharper than the Lorentzian distribution with the natural linewidth. By the same arguments the opposite must be true for the high energy side of the Auger lines. These PCI effects become strikingly significant for low excess energies. It is interesting to note that by the PCI effect it is thus in principle possible to make a kind of time-selective studies in a time scale as short as 10^{-15} to 10^{-16} s which is a typical lifetime of core-hole states.

DEVELOPMENTS IN PROGRESS

There is a need for further improvements of the resolution in electron spectroscopy. In the low energy region <50 eV the most convenient light sources are the rare gas discharge lamps. Ultimately, the resolution is set by the widths of the UV lines themselves. There are, however, further causes for additional line broadenings coming from Doppler broadening due to thermal motions of the gas molecules, residual electric fields over the gas sample volume and the finite spectrometer resolution. Fig.33 summarizes the situation concerning the contribution from the Doppler broadening due to thermal motion in the sample gas. The natural way to get around this is to use molecular beams and, particularly supersonic jet beams. Not only can the contribution due to the translational thermal motion be practically eliminated in this way but also the rotational degrees of freedom of the molecules can be strongly depressed. For future work in the 'ultra' high resolution field it is of primary importance to get these two factors under control. Attempts in this direction are at present under way in several laboratories. Improvements have thus been achieved but even when these precautions have been taken the resolution actually attained has often been modest. This shows that there are factors which have still not been properly taken into account. One of these is no doubt the lamp conditions. Most of these are of the capillary dis-

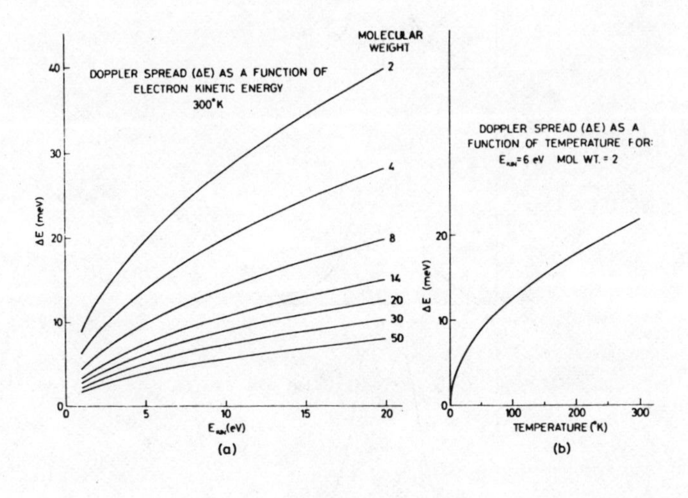

$$\Delta E = 0.723 \left(\frac{T \cdot E}{M}\right)^{1/2} ,$$

where M is the molecular weight, T is the absolute temperature, and E the kinetic energy in eV of the emitted photoelectron.

Fig. 33. Doppler broadening of UV excited electron lines.

charge type with a fairly long discharge length - a factor associated with the working conditions of such lamps. Previous studies[74] have show that the rare gas resonance lines get strongly selfabsorbed along the discharge column, increasing the width at half maximum height, see Fig.34. The accelerated ions in the discharge may furthermore transfer part of their energies to neutral atoms, increasing the Doppler broadening. High intensity microwave discharges at low pressure and small gas absorption distances may alternatively be the natural light source to develop further for high resolution work around 5 meV or, hopefully below that. At present the HeI resonance radiation only rarely produce electron spectra with a resolution below 10 meV. NeI is better suited in this respect as seen in the spectrum of CO_2 according to Fig.28. Fig.35 shows the electron spectrum of argon excited with the Ne resonance line doublet at 16.8 eV. Even without molecular beam techniques this recording[75] shows that a linewidth as small as 5.7 meV can be obtained. The complication - apart from its fairly low energy - is its

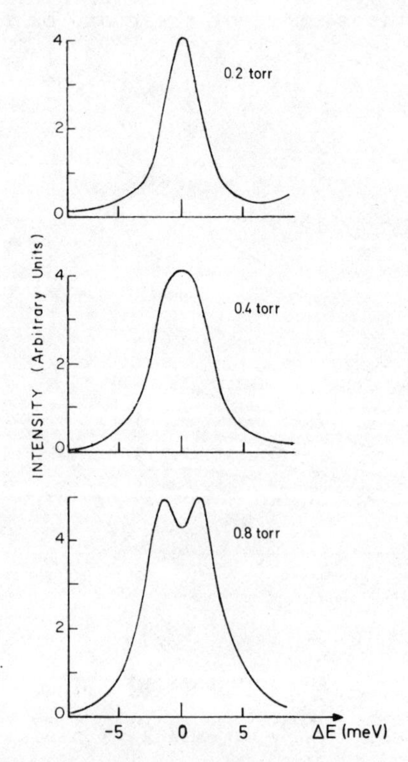

Fig. 34. Selfabsorption in the He lamp.

doublet structure which greatly complicates the interpretation of more complex spectra. In order to overcome this limitation a toroidal grating monochromator (see Fig.36a and b) with a good light collecting ability and a sufficiently high resolving power can be designed[76] to eliminate one of the two Ne lines (see Fig.37).

The problem of designing such a toroidal grating monochromator for use together with a discharge lamp is quite different and much easier from that encountered in the monochromatization of the continuous

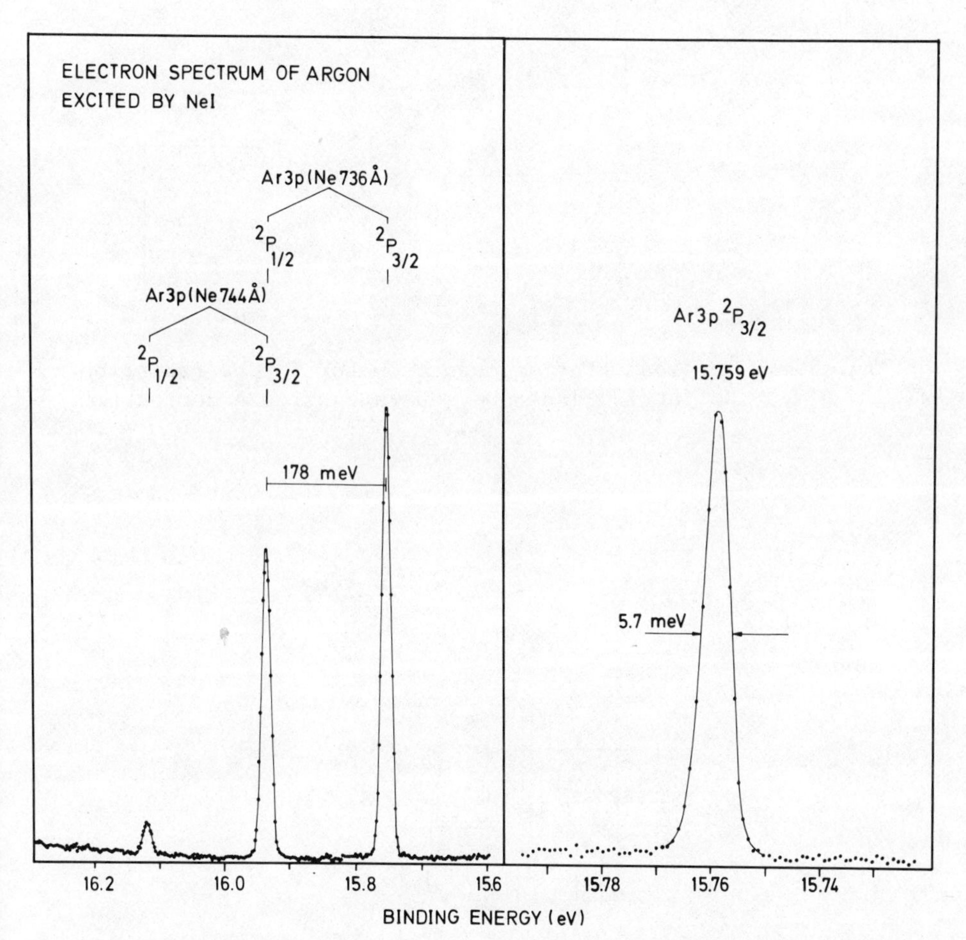

Fig. 35. 3p doublet electron lines from Ar gas studied by means of NeI radiation. Light source and spectrometer adjusted for max resolving power of 5.7 meV.

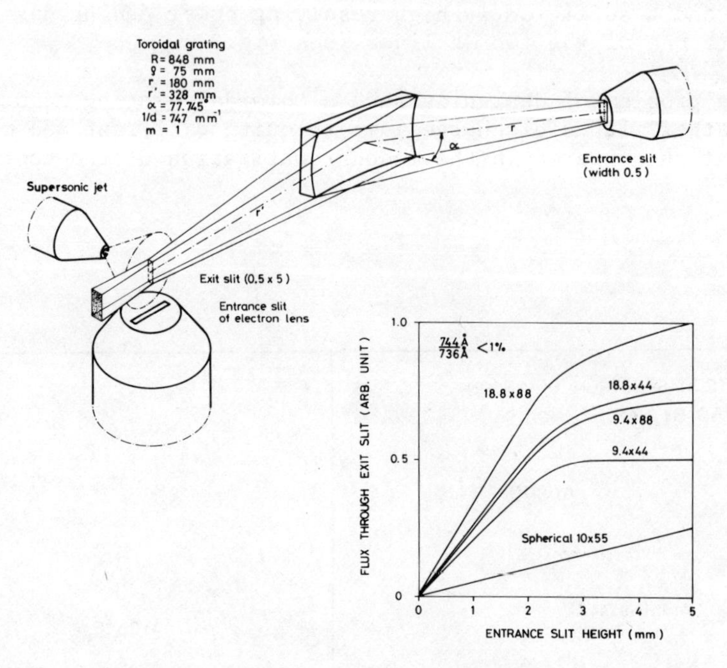

Fig. 36a. Toroidal grating monochromator for UV radiation. NeI at 736 Å can be separated from Å component.

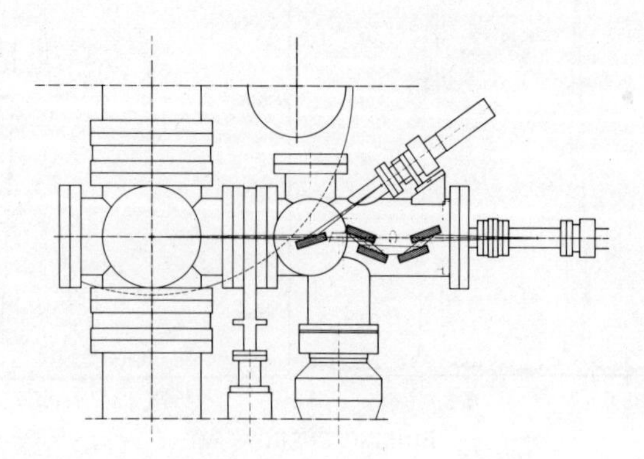

Fig. 36b. UV monochromator and polarizer for instrument acc. to Fig. 42.

synchrotron radiation. The four main contributions to the optical aberrations are the so called focusing term, the coma, the astigmatism and the astigmatic coma. In the case of a primarily continuous radiation these impose rather severe restrictions on the maximum collecting power at ultra high resolving power. If a UV lamp is used as a light source, however, the resolving power is limited by the inherent widths of the extremely sharp UV lines and the purposes of the toroidal grating are to separate neighbouring lines from each other and to focus one or the other onto the sample in front of the spectrometer slit. The background is also reduced in this way. The beam optics and the calculation of the toroidal grating can be conveniently performed by ray tracing computer techniques. Fig.37 is the result of such a 15 sec. run comprising some 20 000 rays, showing that the actual input parameters just permit a complete separation to be made between the two NeI lines at 736 Å and 744 Å. The resolution actually obtainable is then much better than this, only set by the inherent linewidth.

For high intensity work a spherical light collecting mirror[76] can alternatively be used to reflect and focus a very large solid angle of the UV light to a position in front of the spectrometer slit where the gas beam is passing, see Fig.38.

For core electron spectroscopy the way to higher resolution is ultimately limited by the rocking curve of the X-ray or synchrotron radiation monochromator. For synchrotron radiation, which is well collimated, gratings at grazing incidence is the natural choice at excitation energies of several hundred eV. For still higher energies and in particular for the diverging X-radiation from a swiftly rotating anode at high power levels the best choice is spherically bent focusing crystal monochromators at nearly normal incidence, which have a very high collecting efficiency. The rotating anode provided with a high power, well focusing electron gun is capable of producing an extremely high X-ray flux and since this is distributed over a large crystal area at the diffraction there is no risk for radiation damage of the crystal surfaces. The design of reliable rotating anodes for continuous running at speeds of 20 000 rpm is technically a demanding task. Our present rotors have been working satisfactorily. Based on the long experience from these a new, technically much improved version has recently been constructed in our laboratory by Gelius, capable of higher speeds, better cooling and, hopefully, longer lifetime of bearings etc. Like our present ones, the new ones are designed according to the turbo pumping principle with no friction sealings, see Fig.39.

We have at present a maximum of 25 spherically bent quartz crystals (ϕ=30 mm) mounted for simultaneous focusing of the AlKα radiation onto the sample in front of the spectrometer slit. The result of the monochromatization is shown in Fig.40. It is fortuitous that at an almost identical Bragg angle (~78°) as first order focusing of AlK$\alpha_{1,2}$ occurs, second order focusing of AgLα_1, third order ScK$\beta_{1,3}$ and TiKα, and finally fourth order MnKα and CrK$\beta_{1,3}$ occur. Using the X-ray diffraction dynamical theory the rocking curves for quartz (010)

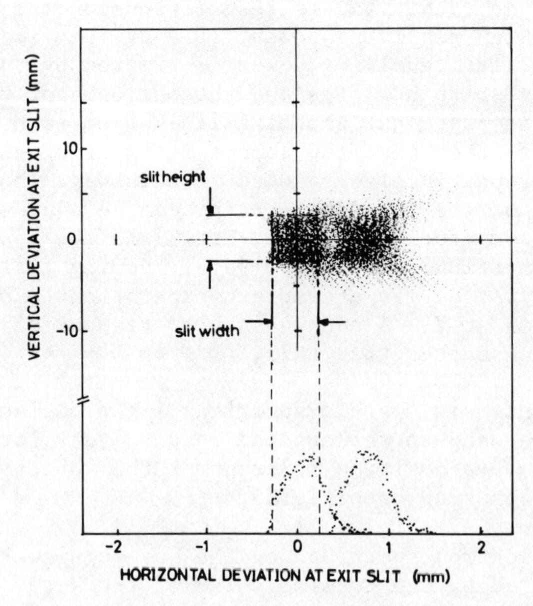

Fig. 37. Ray tracing to achieve separation of UV lines.

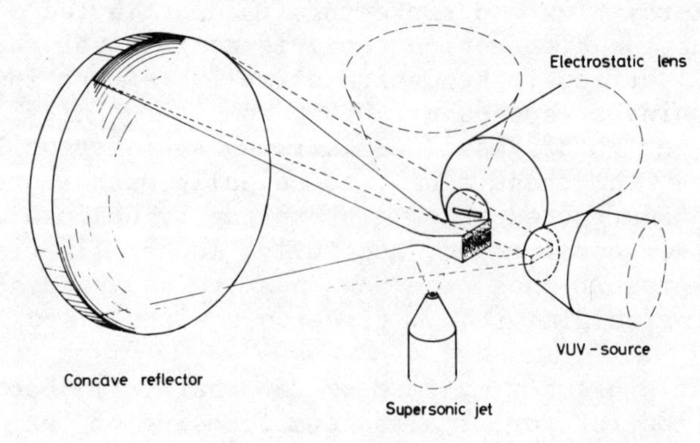

Fig. 38. High intensity UV arrangement.

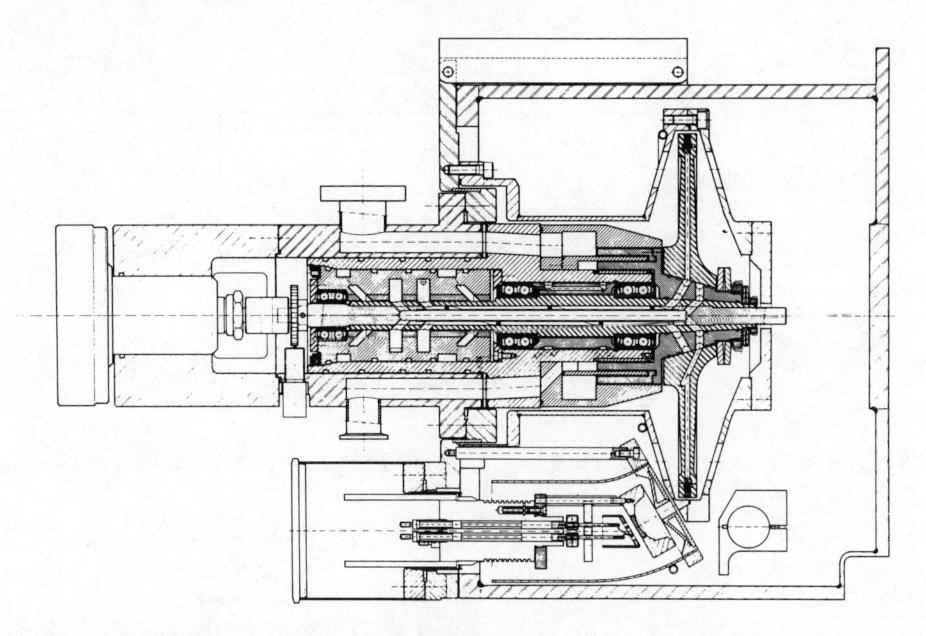

Fig. 39. New ESCA high intensity X-ray generator by means of rotating water-cooled anode.

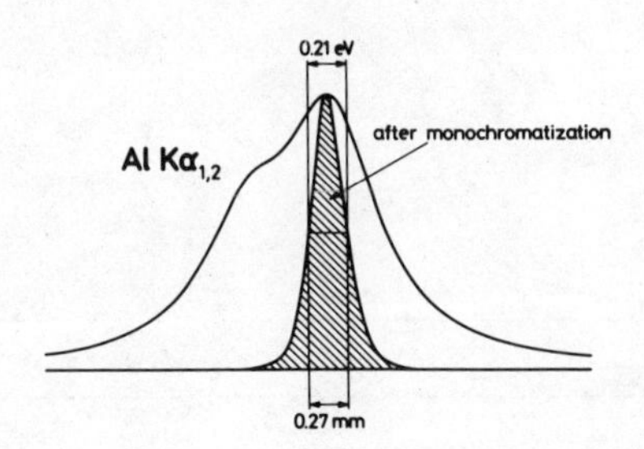

Fig. 40. Monochromatization of AlKα from spherically bent quartz crystal (010).

have been calculated[77] for these radiations. According to Fig.41a they have widths of 135 meV for AlKα at 1486.65 eV, 77 meV for AgLα_1 at 2984.41 eV and about 25 meV for the other radiations at 4500 eV and 5900 eV respectively. For certain purposes these are evidently alternatives to the hitherto used AlKα radiation, both because of the longer mean free paths of the produced photoelectrons and because of the theoretically achievable higher resolution. This can only be attained at a great expense in intensity, in particular because of the lower cross sections and because of the increasing demands for higher spectrometer resolution. The crystal reflectivity is still acceptable, in particular for the AgLα radiation which for certain purposes is an interest-

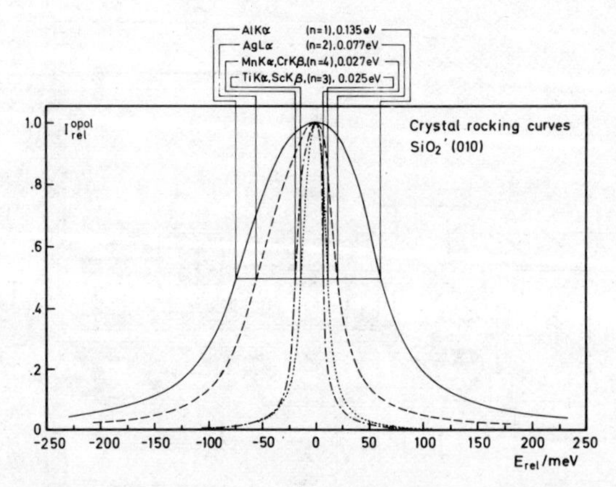

Fig. 41a. Calculated rocking curves for several X-rays reflected from SiO_2 (010) near 78°.

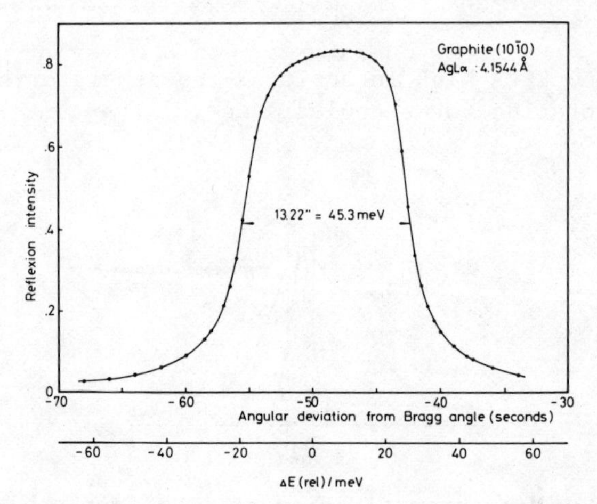

Fig. 41b. Calculated rocking curve for a perfect single crystal of graphite (0.10) and AgLα.

ing alternative to the generally much more useful AlKα radiation. The possibility of using two successive reflections at spherically bent quartz crystals to achieve higher resolution has not yet been investigated. Since the reflectivity is quite high, around 50% for AlKα, it is an interesting future possibility. A new alternative which has appeared during the course of these rocking curve calculations is graphite (010) for first order AgLα. The theoretical rocking curve is then only 45 meV with a very good reflectivity. The rectangular curve form is intersting from the point of view of possibly introducing a double crystal arrangement for (Wiggler) synchrotron radiation (see Fig.41b). The use of diffraction from a crystal in order to achieve the high resolution set by the theoretical rocking curve requires a perfect single crystal structure. In many cases, graphite is one example, available crystals exhibit mosaic structure. Unless improved fabrication in the future can remove this type of structure the attainable resolution is much reduced. Quartz is known to have excellent properties in this respect. The calculations performed in ref.77 indicate further interesting possibilities using quartz in the future.

In order to introduce several of the improvements discussed above and other concepts a new instrument is now being developed. A cross section is shown in Fig.42. Supersonic beams and cryopumping

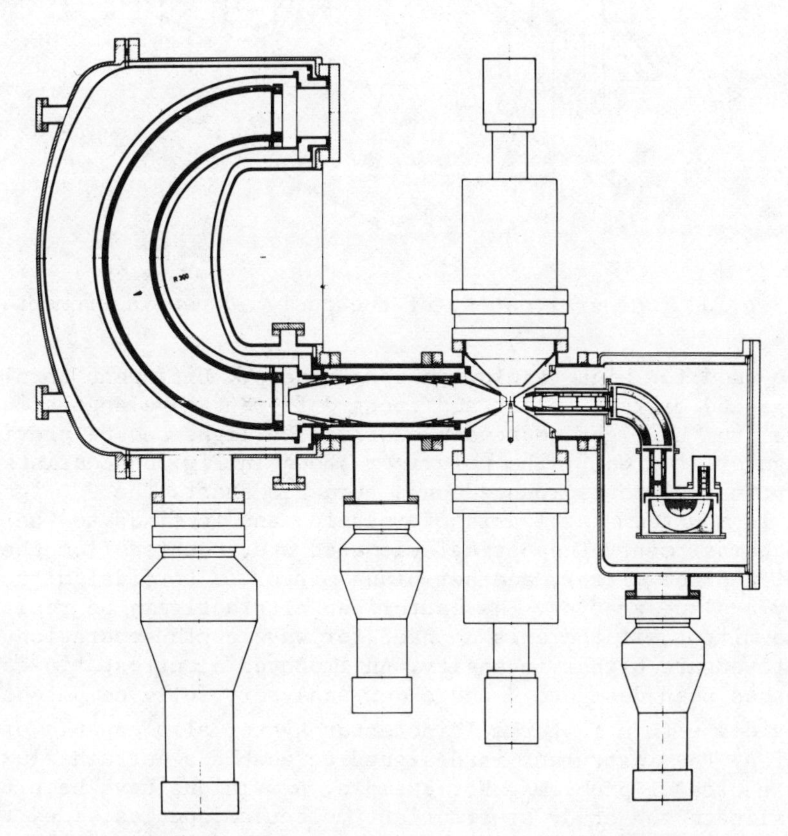

Fig. 42. Design of a new spectrometer for multipurpose use. This section shows one of several alternative modes: electron scattering in forward direction from a molecular jet beam.

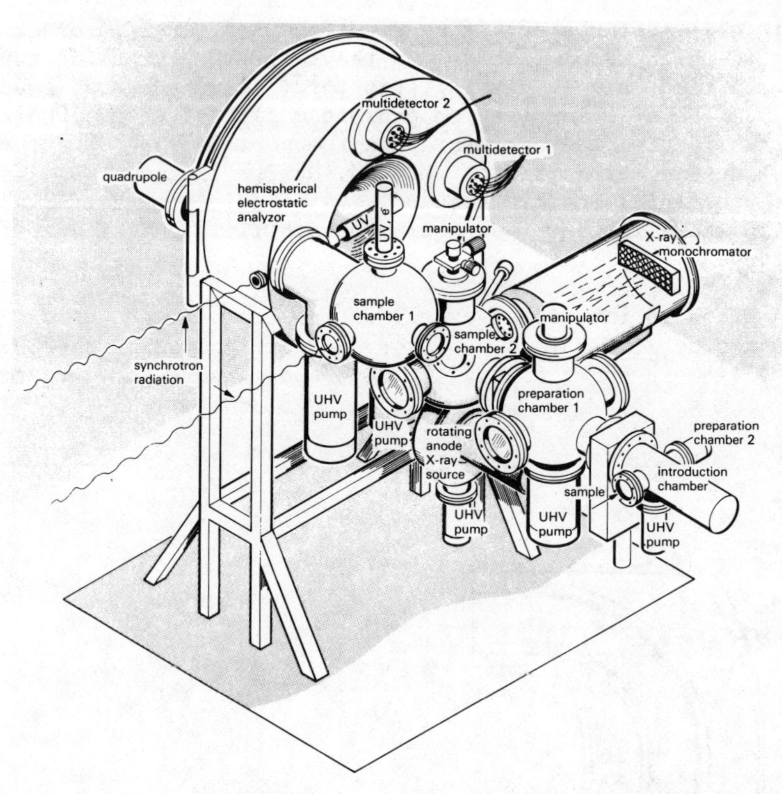

Fig. 43. General outlay of the designed new instrument.

will be used for high resolution spectroscopy. Different toroidal gratings and mirrors select and focus different wavelengths in the spectra from UV light sources. Polarized UV light can be provided for angular studies[78]. The polarizer shown in Fig.36b consists of four optical elements mounted on a turnable shaft. The UV light source is mounted at the axis of rotation and is fixed to the light source compartment. The optical elements are, counted from the light source, a plane mirror, another plane mirror, a toroidal mirror and finally a plane grating. The latter can alternatively be replaced by a plane mirror, if there is no need for wavelength separation in order to ensure higher intensity. An improved electrostatic lens system has been designed[79] and a new analyzer, fully computerized[80], is provided with a faster multidetector system, also capable of pulse operation. The instrument is designed to enable a certain flexibilit in the choice of problems. For example, provisions have been taken to facilitate the study of transient molecular species, i.e. the source compartment can easily be dismounted and cleaned. The instrument can be baked out and even constantly run at an elevated tempera ture. Experiments on excited or fragmented molecules would be feasib to perform. An electron monochromator (see Fig.42) can be attached t the sample house either along or perpendicular to the emission direction of the electrons. Experiments with monochromatized X-rays or UV light or electron beams with simultaneous laser irradiation are some of the goals for the new instrument. Fig.43 shows its general structure. Because of its symmetric construction two independent spectrometer ports are built into the instrument so that different experiments can be prepared simultaneously. External radiation sources, such as synchrotron radiation, laser beams or other laboratory light sources can be used alternatively to or in combination with the radiation sources built into the instrument.

REFERENCES

1. H. Siegbahn and L. Karlsson, Photoelectron Spectroscopy, Handbook of Physics 31, editor: W. Mehlhorn, Springer 1982.
2. R. Nyholm and N. Mårtensson, J. Phys. C, 13, L279 (1980).
3. B. Johansson and N. Mårtensson, Phys. Rev. B, 21, 4427 (1980).
4. J.S. Jen and T.D. Thomas, Phys. Rev. B, 13, 5284 (1976).
5. J.Q. Broughton and D.C. Perry, J. Electron Spectrosc. 16, 45 (1979).
6. N. Mårtensson, Atomic, Molecular and Solid State Effects in Photoelectron Spectroscopy, Thesis, Uppsala 1980.
7. R. Nyholm, Electronic Structure and Photoionization Studied by Electron Spectroscopy, Thesis, Uppsala 1980.
8. P. Steiner, S. Hüfner, N. Mårtensson and B. Johansson, Solid State Commun. 37, 73 (1981).
9. N. Mårtensson, B. Reihl and O. Vogt, Phys. Rev, B, 25, 824 (1982).
10. P.H. Citrin and G.K. Wertheim, Photoemission from Surface Atoms Core Levels, Surface Densities of States, and Metal Atoms Clusters. A Unified Picture, Manuscript 1982.
11. P.H. Citrin, G.K. Wertheim and Y. Baer, Surface Atom X-ray Photoemission from Clean Metals. Cu, Ag and Au. Manuscript 1982.
12. N. Mårtensson, B. Reihl, W.-D. Schneider, V. Murgai, L.C. Gupta and R.D. Parks, Phys. Rev, B, 25, 1446 (1982).
13. V. Murgai, L.C. Gupta, R.D. Parks, N. Mårtensson and B. Reihl, Valence Instabilities, editor: P. Wachter, North Holland, 1981.
14. P.H. Citrin and D.R. Hamann, Phys. Rev. B, 10, 4948 (1974).
15. R.E. Watson, J.F. Herbst and J.W. Wilkins, Phys. Rev. B, 14, 18 (1976).
16. N. Mårtensson, R. Nyholm and B. Johansson, Phys. Rev. Lett. 45, 754 (1980).
17. N. Mårtensson, R. Nyholm, H. Calén, J. Hedman and B. Johansson, Phys. Rev. B, 24, 1725 (1981).
18. B. Reihl, N. Mårtensson, P. Heimann, D.E. Eastman and O. Vogt, Phys. Rev. Lett. 46, 1480 (1981).
19. N. Mårtensson, B. Reihl and O. Vogt, Phys. Rev. B, 25, 824 (1982).
20. A. Franciosi, J.H. Weaver, N. Mårtensson and M. Croft, Phys. Rev. B, 24, 3651 (1981).
21. N. Mårtensson, B. Reihl and R.D. Parks, Solid State Commun. 41, 573 (1982).
22. R.D. Parks, N. Mårtensson and B. Reihl, Valence Instabilities, editor: P. Wachter, North Holland, 1982.

23. C.S. Fadley, J. Electron Spectrosc. 5, 725 (1974).
24. M.F. Ebel, J. Electron Spectrosc. 14, 287 (1978).
25. B. Lindberg, R. Maripuu, K. Siegbahn, R. Larsson, C.-G. Gölander and J.C. Eriksson, UUIP-1066, 1982.
26. N. Larsson, P. Stenius, J.C. Eriksson, R. Maripuu and B. Lindberg, J. Colloid Interface Sci. 75 (1982) in print.
27. K. Siegbahn, Perspectives and Problems in Electron Spectroscopy, Proc. of the Asilomar Conf. 1971, editor: D Shirley North Holland (1972).
28. H. Siegbahn and K. Siegbahn, J. Electron Spectrosc. 2, 319 (1973).
29. H. Siegbahn, L. Asplund, P. Kelfve, K. Hamrin, L. Karlsson and K. Siegbahn, J. Electron Spectrosc. 5, 1059 (1974).
30. K. Siegbahn, Electron Spectroscopy - An Outlook (secs. 3 and 4, figs. 5-10) in the Proceedings of the Namur Conference, editors: R. Caudano and J. Verbist, Elsevier 1974.
31. H. Siegbahn, L. Asplund, P. Kelfve and K. Siegbahn, J. Electron Spectrosc. 7, 511 (1975).
32. H. Fellner-Feldegg, H. Siegbahn, L. Asplund, P. Kelfve and K. Siegbahn, J. Electron Spectroscopy 7, 421 (1975).
33. B. Lindberg, L. Asplund, H. Fellner-Feldegg, P. Kelfve, H. Siegbahn and K. Siegbahn, Chem. Phys. Lett. 39, 8 (1976).
34. H. Siegbahn, ESCA Studies of Electronic Structure and Photoionization in Gases, Liquids and Solids, Thesis, Uppsala 1974.
35. L. Asplund, Electron Spectroscopy of Liquids and Gases, Thesis, Uppsala 1977.
36. P. Kelfve, Electronic Structure of Free Atoms, Molecules and Liquids Studied by Electron Spectroscopy, Thesis, Uppsala 1978.
37. R.E. Ballard, S.L. Barker, J.J. Gunnell, W.P. Hagen, S.J. Pearce, R.H. West and A.R. Saunders, J. Electron Spectrosc. 14, 331 (1978)
38. R.E. Ballard, J.J. Gunnell and W.P. Hagan, J. Electron Spectrosc. 16, 435 (1979).
39. H. Aulich, L. Nemec, L. Chia and P. Delahay, J. Electron Spectrosc 8, 37 (1976).
40. L. Nemec, H.J. Gaehrs, L. Chia and P. Delahay, J. Chem. Phys. 66, 4450 (1977).
41. I. Watanabe, J.B. Flanagan and P. Delahay, J. Chem. Phys. 73, 2057 (1980).
42. K. Burger, F. Tshismarov and H. Ebel, J. Electron Spectrosc. 10, 461 (1977).
43. K. Burger, J. Electron Spectrosc. 14, 405 (1978).
44. S.C. Avanzino and W.L. Jolly, J. Am. Chem. Soc. 100, 2228 (1978).
45. H. Siegbahn, S. Svensson and M. Lundholm, J. Electron Spectrosc. 24, 205 (1981).
46. H. Siegbahn and M. Lundholm, J. Electron Spectrosc., in course of publication, 1982.
47. H. Siegbahn, M. Lundholm, M. Arbman and S. Holmberg, UUIP-1070, Uppsala 1982.
48. H. Siegbahn, M. Lundholm, M. Arbman and S. Holmberg, UUIP-1071, Uppsala 1982.
49. U. Gelius and K. Siegbahn, Far. Disc. Chem. Soc. 54, 257 (1972).
50. U. Gelius, Molecular Spectroscopy by means of ESCA; Experimental and Theoretical Studies, Thesis, Uppsala 1973.

51. U. Gelius, S. Svensson, H. Siegbahn, E. Basilier, Å. Faxälv and K. Siegbahn, Chem. Phys. Lett. 28, 1 (1974).
52. U. Gelius, E. Basilier, S. Svensson, T. Bergmark and K. Siegbahn, J. Electron Spectrosc. 2, 405 (1974).
53. K. Siegbahn, L.O. Werme, B. Grennberg, S. Lindeberg and C. Nordling, UUIP-749, Uppsala 1971.
54. K. Siegbahn, Perspectives and Problems in Electron Spectroscopy, Proc. of the Asilomar Conf. 1971, editor: D. Shirley, North Holland, 1972.
55. K. Siegbahn, L. Werme, B. Grennberg, J. Nordgren and C. Nordling, Phys. Lett. 41A, 111 (1972).
K. Siegbahn, Electron Spectroscopy for Chemical Analysis, Proc. of Conf. on Atomic Phys. Vol.3, p. 507-509, editors: S.J. Smith and G.K. Walters, Plenum, 1972.
56. L.O. Werme, B. Grennberg, J. Nordgren, C. Nordling and K. Siegbahn, Phys. Rev. Lett. 30, 523 (1973); Nature 242, 453 (1973).
57. L.O. Werme, Electron and X-ray Spectroscopic Studies of Free Molecules, Thesis, Uppsala 1973.
58. K. Siegbahn, Electron Spectroscopy - An Outlook, Proc. Namur Conf. 1974, p. 70-83, editors: R. Caudano and J. Verbist, Elsevier 1974.
59. J. Nordgren, X-ray Emission Spectra of Free Molecules, Thesis, Uppsala 1977.
60. H. Ågren, Decay and Relaxation of Core Hole States in Molecules, Thesis, Uppsala 1979.
61. J. Nordgren, H. Ågren, C. Nordling and K. Siegbahn, Ann. Acad. Reg. Scientiarum Upsaliensis 21, 23 (1978).
62. L. Selander, USX Emission Spectroscopy of Atoms and Molecules, Thesis, Uppsala 1982.
63. J. Nordgren, H. Ågren, L. Pettersson, L. Selander, S. Griep, C. Nordling and K. Siegbahn, Physica Scripta 20, 623 (1979).
64. J. Nordgren, L. Pettersson, L. Selander, C. Nordling, K. Siegbahn and H. Ågren, J. Phys. B: At. Mol. Phys. 15, L153 (1982).
65. U. Gelius, J. Electron Spectrosc. 5, 985 (1974).
66. H. Ågren, J. Nordgren, L. Selander, C. Nordling and K. Siegbahn, Physica Scripta 18, 499 (1978).
67. V.G. Yarzhemsky, V.I. Nefedov, M.Y. Amusia, N.A. Cherepkov and L.V. Chernysheva, J. Electron Spectrosc. 19, 123 (1980).
68. I. Reineck, R. Maripuu, H. Veenhuizen, S. Al-Shamma, L. Karlsson and K. Siegbahn, to be published.
69. U. Gelius, K. Helenelund, S. Hedman, L. Asplund, B. Finnström and K. Siegbahn, to be published.
70. S. Hedman, K. Helenelund, L. Asplund, U. Gelius and K. Siegbahn, J. Phys. B: At. Mol. Phys. 1982 and references therein.
71. K. Helenelund, S. Hedman, L. Asplund, U. Gelius and K. Siegbahn, Physica Scripta 1982 (in the course of publication).
72. A. Niehaus, J. Phys. B, 10, 1845 (1977); Proc. 10th Int. Conf. on Electr. and Atomic Collisions, Paris 1977, editor G. Watel, p. 105, North Holland 1978.
73. R. Huster and W. Mehlhorn, Z. Physik A, 307, 67 (1982).
74. J.A.R. Samson, Rev. Scient. Instr. 40, 1174 (1969).

75. R. Maripuu, I. Reineck, C. Nohre, P. Lodin, L. Karlsson and K. Siegbahn, to be published.
76. L. Mattsson and K. Siegbahn, UUIP-1068, Uppsala 1982.
77. B. Finnström, UUIP-1069, Uppsala 1982.
78. L. Mattsson, High Resolution Valence Electron Spectroscopy. Development of a VUV-Polarizer for Angular Distributions, Thesis, Uppsala 1980.
79. B. Wannberg, UUIP-1072, Uppsala 1982.
80. L. Asplund, UUIP-1073, Uppsala 1982.

COLLECTIVE EFFECTS IN ISOLATED ATOMS

(MANY-BODY ASPECTS OF PHOTOIONIZATION PROCESS)

M.Ya. Amusia

Ioffe Physical-Technical
Institute of the Academy of
Sciences of the USSR, Leningrad

"Еще, быть может, каждый атом
Вселенная, где сто планет"...

(Still, perhaps each atom
is a universe, where a hundred
planets . . .)
V. Briusov, World of the Electron,
1922 (in Russian)

1. INTRODUCTORY REMARKS

About half a century has passed after the publication of Hartree's[1] and Fock's[2] papers on the self-consistent field method and Bloch's and Jensen's[3,4] - on the collective sound-like vibrations of electrons in atoms. For more than fifty years our ideas have developed on single-particle and collective aspects of electron motion in atoms.

The last decade is marked by especially deep and careful investigation of electronic structure of complex atoms and by prominent bringing near of those two approaches.

Although sound-like oscillations were not observed, it appeared that in a number of cases atoms behave as a complex many-electron system. The ordered, collective (or correlated) motion of electrons clearly manifests itself in the processes of ionization and excitations and in a number of atomic characteristics[5-8].

2. COLLECTIVE EFFECTS IN OUTER SHELLS

The investigation of outer and intermediate many-electron shells demonstrate that photoionization is of collective nature because in the atomic reaction to the external electromagnetic field at least all electrons of the ionized subshell take part. The calculation of complex atom photoionization is performed using one of the three methods: RPAE[6,7], many-body perturbation theory MBPT[5] and R-matrix[8]. The MBPT achievements will be discussed by prof. H.P. Kelly at this conference. In our report we concentrate on RPAE and mainly on its generalization. In RPAE[6] the ionization amplitude is presented as a sum of two terms, describing the direct knock-out and the induced one which is connected with a variation of the self-consistent field, caused by polarization of atomic shells under the action of the external field:

$$(i|D(\omega)|\varepsilon) = (i|d|\varepsilon) + (i|D_{in}(\omega)|\varepsilon),$$

$$(i|D_{in}(\omega)|\varepsilon) = \sum_{\substack{\varepsilon>F \\ j\leqslant F}} \left[\frac{(j|D(\omega)|\varepsilon')\ (i\varepsilon'|u|\varepsilon j)}{\omega - \varepsilon' + \varepsilon_j + i\delta} - \right.$$

$$\left. - \frac{(\varepsilon'|D(\omega)|j)\cdot(ij|u|\varepsilon\varepsilon')}{\omega + \varepsilon' - \varepsilon_j - i} \right] \tag{1}$$

where ω is the photon energy, d - is the dipole moment operator in one-electron approximation, $(i\varepsilon'|u|\varepsilon j) = (i\varepsilon'|V|\varepsilon j) - (i\varepsilon'|V|j\varepsilon)$, V being the Coulomb potential and ε, ε', ε_j - Hartree-Fock one--electron states and energies. The summation in (1) is performed over all vacant (including continuous spectrum integration) ($\varepsilon>F$) and occupied ($\leqslant F$) states. The photoionization cross section is expressed via the amplitude D: $\sigma = 4\pi^2\alpha\omega \sum_{i\leqslant F} |(i|D|\varepsilon)|^2$ *), α being the fine structure constant, $\alpha \approx 1/137$.

Mainly the electrons of the same subshell correlate but there are cases in which two or even three subshells interact strongly. As an example serves the xenon, cesium and barium 5s subshell which is completely collectivized under the action of the outer $5p^6$ and inner $4d^{10}$ many electron subshells.

The collective effects strongly manifest themselves also in angular distribution of photoelectrons, which for unpolarized light is determined by the well known expression:

$$\frac{d\sigma}{d\Omega} = \frac{\sigma_{nl}}{4\pi}\left[1 - \frac{1}{2}\beta_{nl}(\omega)\ P_2(\cos\vartheta)\right] \tag{9}$$

* The system of units $\hbar = m = e = 1$ is used in this paper.

where β is the anisotropy parameter and σ_{nl} is the absolute photoionization cross section of the nl-subshell. The energy dependence of β is sensitive to electron correlations. This is illustrated by the example of krypton, the β_{4p} parameter of which with and without the action of $3d^{10}$ shell is quite different[9]. The experimental data[10] definitely supports the many electron picture, as it is evident from Fig. 1.

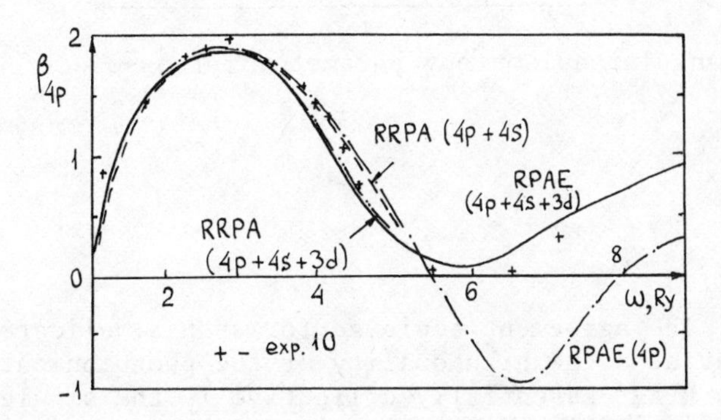

Fig. 1. Angular anisotropy parameter for 4p-electrons in Kr.

The relativistic effects are essential in heavy atoms. Their account is possible using the relativistic version of the Random Phase Approximation - RRPA[11]. The recent achievements in RRPA are discussed in Prof. W. Johnson's talk at this conference.

Therefore we mention only two examples. The nonrelativistic β-parameter for s-electrons is equal to 2, while the inclusion of both relativistic and correlational corrections leads in the case of 5s-electrons in Xe to a very complicated curve[12,11], given in Fig. 2 together with recent experimental data[13]. The action of both $5p^6$ and $4d^{10}$ excitations upon 5s-electrons is very essential. The inclusion of many-electron correlations leads to a very complex behaviour of $D(\omega)$ as a function of ω, $D(\omega)$ even changing its sign several times. The detailed information on $D(\omega)$ may be obtained in a so called complete experiment: For photoionization it includes measurement of partial subshell cross section and angular distribution, and of electron spin direction, i.e. its polarization. The polarization follows all variations of $D(\omega)$. Recently, extensive calculations of polarization parameters for noble gases were performed[14] and the RPAE predictions are now confirmed expe-

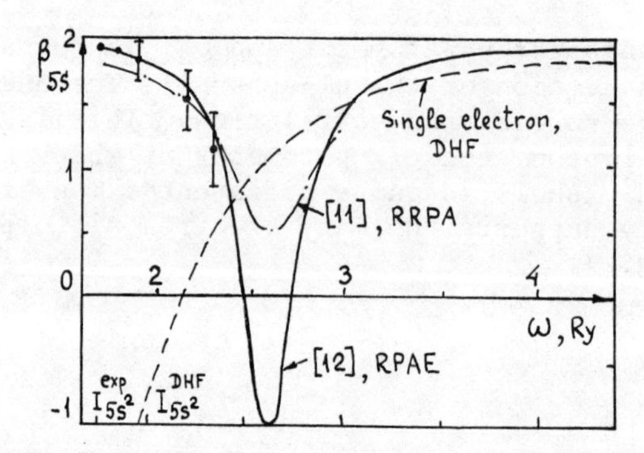

Fig. 2. Angular anisotropy parameter for 5s-electrons in Xe.

rimentally[15]. The agreement achieved for such a delicate characteristic demonstrates the high quality of the photoionization amplitude given by RPAE. Essentially collective is the single and doubly charge ion formation in photoionization of outer shells near inner shell thresholds. Here the Xe atom serves as a very good example with its easily deformable $4d^{10}$ shell. The main contribution to the cross section comes not from the direct outer shell ionization. More essential is the dynamical polarization of the $4d^{10}$ shell by the incoming photon. Therefore the self-consistent field acting upon outer electrons becomes time-dependent, and easily ionizes them. As a result the single charge ion yield has a powerful maximum of collective nature in the 4d-threshold region. The double electron ionization cross section also increases, even below the 4d-threshold, which may be explained by dynamical polarization of the $4d^{10}$ subshell[16].

Because correlations in open shells may be even stronger, generalizations of RPAE were proposed recently [17,18], which however are too complicated and different. But it proved to be easy to generalize RPAE for atoms with a semifilled shell[19] using the so called later spin polarized method. According to the Hund rule all electron spins in this shell have the same direction, let us call it "up". Then each subshell splits into two levels with the same ("up") and opposite ("down") spin directions. It looks like the atom has two types of electrons, "up" and "down", the exchange between them being forbidden. In the absence of spin-spin and neglecting the spin-orbit interaction the "up" and "down" states do not mix. Therefore the respective "up" and "down" ionization potentials of the same subshell become different and inclusion of RPAE correlations are required to generalize equation (1) for closed shell system with two types of particle. The photoionization cross section of each "up" ("down") level must be considered separately, the amplitudes D (u,d) being determined by the matrix equation:

$$\widehat{D_u\ D_d} = \widehat{d_u\ d_d} + \widehat{D_u\ D_d}\begin{pmatrix}\chi_{uu} & 0\\ 0 & \chi_{dd}\end{pmatrix}\begin{pmatrix}U_{uu} & U_{ud}\\ U_{du} & U_{dd}\end{pmatrix} \tag{3}$$

where indexes u(d) represent particle-hole state (e.g. nl; $\varepsilon, \ell + 1$) in Hartree-Fock approximation, in which however "up" and "down" splitting is included; χ_{uu} and χ_{dd} represent energy denominators and summation over states >F and $\leqslant$F in (1), all of them having indexes "u" or "d" ("up" and "down"). In (3) u is the Coulomb interaction matrix and $u_{ud(du)}$ differs from $u_{uu(dd)}$ because in u_{ud} the exchange is neglected (see (1)). The cross section is proportional to $(|D_u|^2 + |D_d|^2)$. It is worth to mention that there are twenty atoms in the periodic table which may be studied using "up" and "down" separation. As an example let us consider σ for Mn in the region of the main maximum which is near the $3p_u^3$ threshold. The ground state configuration of Mn is $3d_u^5\ 4s_u\ 4s_d$ above an argon-like core. The "up-down" splitting is as large as 1 Ry for the 3p-subshell: $I_{3p_d} - I_{3p_u} \approx 1$ Ry. Even 4s-electron splitting is observable, being 1.3 eV. The cross section σ is dominated by the discrete transition $3p_d \to 3d_d$, which decays very fast into continua $3d_u \to \varepsilon p_u, \varepsilon f_u$. The results of calculations[19] and measurements[20] given in Fig. 3 are at least in satisfactory agreement. The same type of calculations were performed for Technetium*) where even a broader autoionization

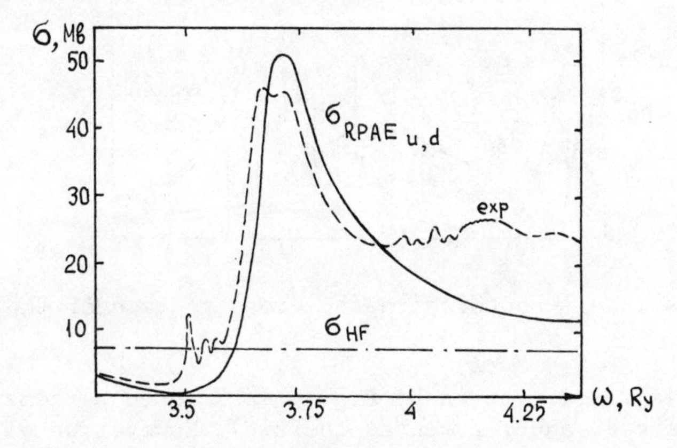

Fig. 3. Photoionization cross section of the Mn atom.

*) Obtained together with V.K. Dolmatov and V.K. Ivanov.

maximum is obtained. Just as for closed shell atoms, the angular anisotropy parameter is calculated in which the autoionization maximum leads to a complex energy dependence.

The "up/down" RPAE version was applied also to Eu[21]. In the sequence Xe, Cs, Ba, La and then to Eu a powerful maximum dominates the experimental cross section which is explained as $4d^{10}$-subshell ionization in Xe, Cs, Ba and to our opinion - in La. The nature of this maximum in Eu is quite different - it is a result of very fast autoionization of $4d_u \rightarrow 4f_u$ level into outer shell continua. The difference between these two mechanisms is easy to observe using the electron spectroscopy methods because in E_u the photoelectron energy must be much higher.

3. DEVIATION FROM RPAE PREDICTION IN OUTER SHELLS

Usually the photoionization cross section of 5s-electrons in xenon serves as an example of the very significant role of correlations. As is known, 5s-electrons are strongly effected by their neighbours $5p^6$ and $4d^{10}$ subshells and are completely collectivised[22], their photoionization cross section being even qualitatively different from the single-particle picture - Fig. 4.

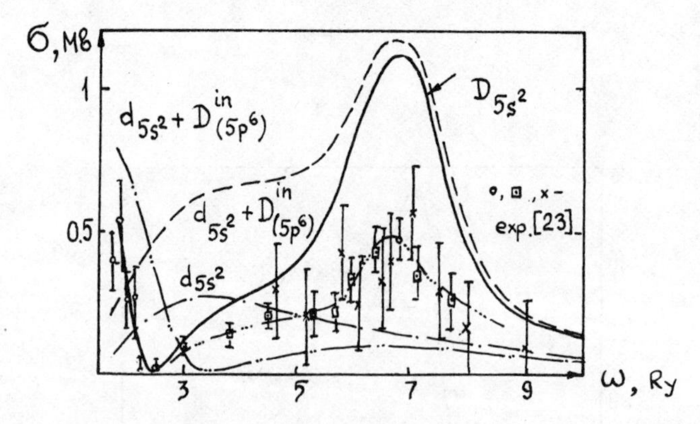

Fig. 4. 5s-subshell photoionization cross section of the Xe atom.

The general impression given by Fig. 4 is in good agreement with experiment (see 23 and references therein). However we will focus our attention on the disagreement prominent in the region of the $4d^{10}$ ionization threshold, where the ratio $\sigma_{5s}^{RPAE}/\sigma_{5s}^{exp}$ is as large as 2.3. There is also prominent difference for $3s^2$ electrons of argon far from the threshold: $\sigma_{3s}^{exp} \approx 0.6\sigma_{3s}^{RPAE}$. It seems that with increase of ω this difference between experiment and theory is conserved. We shall demonstrate that the origin of this deviation is the neglect of the so called spectroscopic factor F which characterises the mixing of pure vacancy state with more complex excitations. So, for the $3s^{-1}$ - vacancy in Ar the interaction with $3p^{-1}$ 3d state is most significant.

Let us take into account F in photoionization cross section. In the many-body theory F is less than 1 only out of RPAE frame due to the energy dependence of the self-energy part Σ. The simplest diagrams which include Σ in photoionization amplitude are given in Fig. 5 and use standard notations: the dotted line denotes the photon, the line with the arrow to the right is an electron, the same directed to the left is a vacancy, the wavy line being the Coulomb interaction.

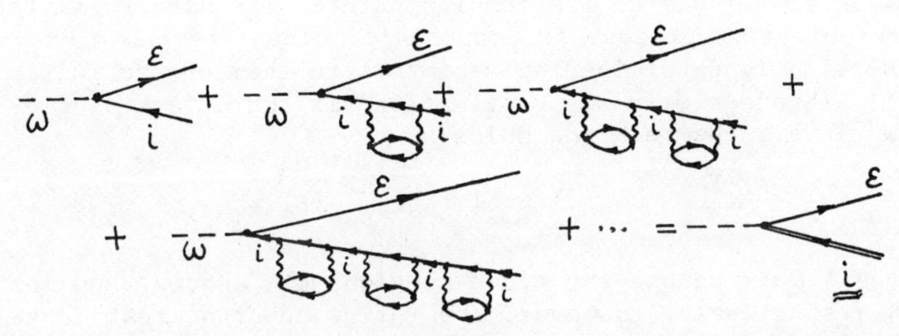

Fig. 5. Self-energy corrections to the photoionization amplitude.

The lowest order approximation in interelectron interaction for Σ (Fig. 5a) proves to be usually sufficient. The same sequence of diagrams permits to eliminate the difference between Hartree--Fock and experimental values of ionization potentials, because it includes the ions rearrangement due to vacancy creation as well as the mixing of a pure vacancy with more complex excitations, e.g. $\underline{i}$ and $\underline{j}\ \underline{k}$ n *) in Fig. 5. As a result the new $\underset{\approx}{i}$ state contains the pure vacancy $\underline{i}$ with the weight F_i, F_i being less than 1. The diagrams of Fig. 5 lead to σ(ω), given by the following relation

$$\sigma(\omega) = \sigma_i(\omega)\ (1 - \partial\Sigma/\partial\varepsilon\,|_{\varepsilon=\varepsilon_i})^{-1} \equiv F_i \sigma_i(\omega) \qquad (4)$$

where $\varepsilon_i = -I_i$, I_i being the i-level ionization potential and σ_i - the one-electron approximation cross section.

*) The absence of an electron in the state i is denoted as $\underline{i}$.

The ns-electron photoionization cross section is affected by the adjoining manyelectron shells. If their own spectroscopic factors are close to one the F_s2 may be taken into account very easily by the following relation: $\sigma_{ns}2(\omega)=\sigma_{ns}2^{RPAE}(\omega)F_{ns}$. With ω increasing all correlations except those accounted for by Σ of the hole become inessential. Thus for $\omega >> I_{ns}2$

$$\sigma_{ns}2\ (\omega) = F_{ns}2\ \sigma^{HS}_{ns2} < \sigma^{HS}_{ns2} \tag{6}$$

For example, in Ar the transitions $3s^{-1}\rightarrow 3p^{-1}\rightarrow 3d$ (nd) give the main contribution to $F_{ns}2$ and accounts for about 45% of the energy difference between Hartree-Fock and experimental potential. The mixing of states is large for 3s in Argon and especially for 5s in Xe, the F-values being 0.6 and 0.34, respectively. The remaining part (1-F) of the level's strength is transferred to the states, the mixing with which is strong. They acquire some properties of the initial vacancy becoming its shadow and therefore we call it "shadow level"*. Due to its complex nature, its direct excitation in photoionization process is improbable. Thus the photoionization cross section is determined by admixture to them of the initial s-level. In one-electron approximation, if there is a single "shadow" level, the relation holds:

$$\sigma_{\tilde{ns}}\ (\varepsilon) = \frac{1-F_{ns}}{F_{ns}}\ \sigma_{ns}(\varepsilon) = (1-F_{ns})\sigma^{HF}_{ns}\ (\varepsilon) \tag{7}$$

where $\sigma_{\tilde{ns}}$, σ_{ns} denote the cross sections of "shadow" and initial levels, respectively. However, the influence of nearest manyelectron shells upon ns^2 is strong and because I_{ns} and $I_{\tilde{ns}}$ are different, the relation between $\sigma_{\tilde{ns}}$ (ε) and σ_{ns} (ε) is not so simple as (7). The results of calculations for Ar are given in Fig. 6**).

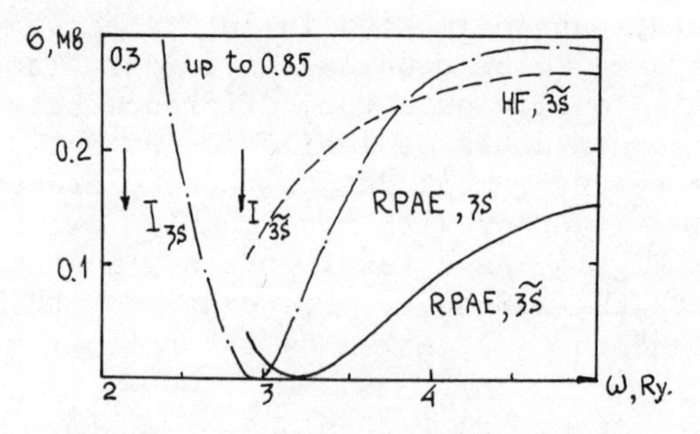

Fig. 6. Photoionization of the $\tilde{3}$s-level in Ar.

*) We discern "satellite" and "shadow" levels. "Satellite," contrary to "shadow," includes only those complex states in which the initial level also takes part, e.g. $3s^{-1}3p^{-1}$ 4p is a "satellite" of $3s^{-1}$.

**) This is considered together with M.Yu. Kuchiev and S.A. Sheinerman.

Experimentally, the 3s Ar level with the energy 2.9 Ry was observed in (e, 2e) reaction[24]. If the initial level mixes with several complex excitations all they commonly acquire the properties of a "shadow". If the mixing with the continuous spectrum is strong, equality (7) links the single- and double-electron photoionization cross section.

The inclusion of F_{5s} leads to agreement between theory and experiment in the main maximum region of σ_{5s} in Xe (Fig. 7), but destroys it near threshold.

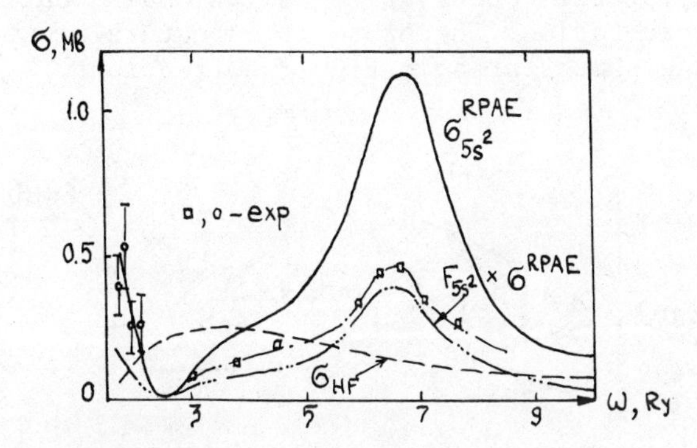

Fig. 7. 5s-electron photoionization cross section of Xe.

The agreement may be restored, only if one takes into account the variation of intershell interaction, which becomes different from pure Coulomb V due to connection with the same states, which lead to F<1. Let us estimate the variation of interaction near threshold where the relation $\sigma_{5s}^{exp} \approx \sigma_{5s}^{RPAE}$ holds. Denoting the ratio of the rigorous value of the induced amplitude to D^{in} in RPAE (see (1) by g, one finds, using the equation $F(d + gD^{in})^2 = (d + D^{in})^2$, where the index 5s is for simplicity omitted:

$$g = \frac{(1-F^{1/2})d + D^{in}}{F^{1/2} \cdot D^{in}} \tag{8}$$

which transforms into $g \approx F^{-1/2}$ if $D^{in}/d \gg 1$. For Xe, because $\sigma_{5s}^{exp} \approx 6\sigma_{5s}^{HF} = \sigma_{5s}^{RPAE}$ and $F_{5s}=0.34$ the equation (8) gives $g_{5s} = 1.49$, the intershell interaction variation being rather large. The prominent deviation of g and F from 1 demonstrates that the RPAE amplitude must be corrected, taking into account the connection with states more complex than the simple RPAE electron-hole excitations.

4. EXCITATIONS "TWO ELECTRONS - TWO HOLES" AND AUTOIONIZING STATES

For ns - vacancies in outer shells of noble gas atoms F_{ns} is prominently less than one, $g \neq 1$ and the excitation energy is shifted from its Hartree-Fock value. The cross section in the vicinity of the ns→(n+1)p level is described in RPAE badly. Let us include "two electron-two hole" excitations of the same type that interact with $\underline{ns}$ strongly, i.e. $\underline{np}$, $\underline{np}$ nd. As an example we shall take the 3s→4p level in Ar^{26}, and include the photoexcitation vertex, given in Fig. 8. It is implied that the insertions of Σ in the s-hole line (see Fig. 5) are also included. All these corrections are of "two electron-two hole type". To take them into account rigorously is impossible because even the motion of two electrons in an ion field is a three-body problem. However to Σ_{3s} the main contribution comes from $\underline{3p}$ $\underline{3p}$ 3d excitation. Therefore it is natural to take into account just the same states in Fig. 8 corrections.

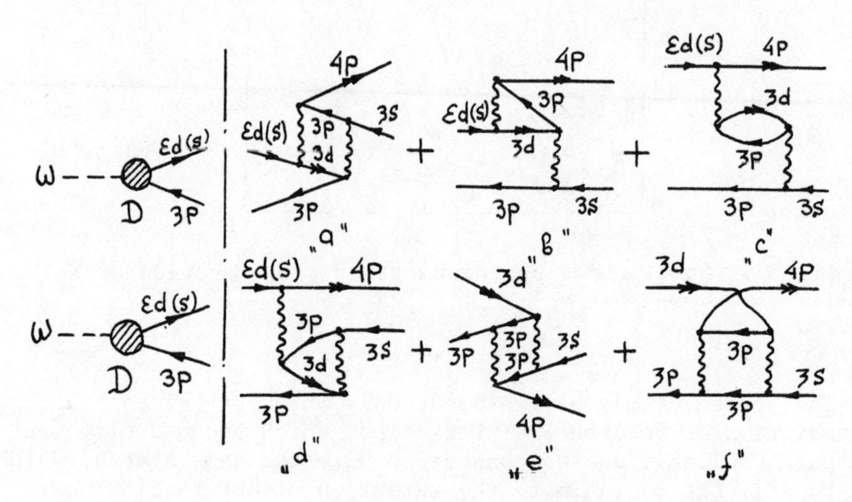

Fig. 8. "Two electron - two hole" excitation corrections to the effective interaction and photovertex for autoionization of the 3s 4p level in Argon.

The autoionization cross section is described by a Fano profile:

$$\sigma(\varepsilon) = \sigma\left[1 - \rho^2 + \rho^2 \frac{(q+\xi)^2}{1+\xi^2}\right], \tag{9}$$

where σ is the cross section far from resonance, $\xi = (\omega - \omega_s)/(\frac{1}{2}\gamma_s)$, ω_s and γ_s being the resonance energy and width respectively, while q and ρ^2 determines its form. They are connected with the amplitudes of 3s→4p level excitations (D_s) and $3p^6$-subshell photoionization (D_t), as well as with the effective interaction Γ_{st} of the discrete 3s→4p level and continuous 3p→εd and 3p→εs spectra:

$$q=-\mathrm{Re}D_s/I_m D_s;\ \rho^2=(I_m D_s)^2/\pi^2(\sum_t|\Gamma_{ts}|^2)(\sum_t|D_t|^2),\qquad (10)$$
$$\gamma=2\pi\sum_t|\Gamma_{ts}|^2$$

If only Σ corrections are included, then q and ρ^2 are the same as in RPAE. But Fig. 8 corrections permit to achieve an agreement with experiment, which is illustrated by the table:

	q	γ,eV	ρ^2
RPAE	2.13	0.024	0.78
RPAE with "2e--2h" corrections	-0.27	0.085	0.89
experiment[26]	-0.22± ±0.05	0.08± ±0.005	0.86± ±0.04

The anisotropy parameter is also strongly altered if Fig. 8 corrections are included[27]. As to "Shadow" autoionization $\widetilde{3}s\rightarrow 4p$ state in Argon it may be presented rudely as double electron dipole excitations $3p^{-2}$ 3d 3p. The interaction with "2 electron-2 hole" excitations is very essential in the continuous spectrum also. Near ns-threshold*), this interaction enhances the influence of adjoining many-electron shells, nearly completely compensating the decrease of the ns^2 cross-section because of the spectroscopic factor F. It means, that the calculation leads in fact to the same value of q, as it follows from (8). With increase of ω the influence of connection with all "2 electron-2 hole" states but those taken into account by Σ rapidly decrease, so that σ_{ns} becomes equal to $F_{ns}\cdot\sigma_{ns}^{RPAE}$. It seems, that almost complete compensation of g and F influence at threshold is a consequence of a relation analogous to Word identity in Quantum electrodynamics**), which connects the photon-electron vertex and the electron self-energy part. Being valid for $\omega\rightarrow 0$, it links in our notations F and q in such a way that the equality between experimental and RPAE cross sections, which leads to (8), must hold.

The correction of intershell interaction affects some other characteristics e.g. β_{ns} (See Fig. 2), where the preservation of the achieved agreement with experiment may require considerable efforts from theorists.

An important virtue of RPAE is the equivalence of length and velocity forms in photoionization cross section calculations, which

*) Derived together with A.S. Kheifetz.

**) I am grateful to M.Yu. Kuchiev for this comment.

may be violated when g and F corrections are included. However, recently a method was developed[28], which permits to maintain length and velocity form equivalence, i.e. the gauge invariance, in any order of perturbation theory.

5. COLLECTIVE EFFECTS IN INNER SHELLS

Collective effects near threshold in inner shell photoionization include along with RPAE corrections also relaxation processes. They are a consequence of the fact that near inner shell threshold the photoelectron leaves the atom slowly and all other electrons have sufficiently time to feel the field of the vacancy created as well as its decay-Auger or radiative. It is very essential that because the field variation takes place after vacancy creation, the photoelectron wave function must be ortogonalized to all other electron states of the atom - otherwise it includes the interaction with the final state of the ion before its formation.

The variation of the self-consistent field acting upon the outgoing electron due to vacancy formation or decay we shall call static and dynamic rearrangement, respectively. The slow photoelectron moves in a modified field which alters the cross section. The process is determined by the time of static rearrangement τ_i which may be estimated as $(I_i + \varepsilon_i)^{-1}$, by the lifetime of vacancy $\underline{i}$ decay $\tau_d \sim \gamma_i^{-1}$ and the ejection time of the photoelectron with energy ε off the atom $\tau_{ej} \sim R/(2\varepsilon)^{1/2}$, where R is the atoms radius, I_i- the i-shell ionization potential and γ_i the ionized level width. The rearrangement is able to complete if τ_{ej} is larger than τ_i and τ_d. If $\tau_{ej} \gg \tau_i$ but for τ_d being sufficiently large, only static rearrangement proceeds which leads to an increase of the ion charge screening and decrease of photoionization cross section near thresholds. This is taken into account by GRPAE[29] which includes together with RPAE also the variation of all single electron states in the field of vacancy $\underline{i}$. Using GRPAE satisfactory agreement with experiment is achieved for intermediate shells, e.g. $4d^{10}$ in Ba and $2p^6$ in Ar. However, near threshold ionization of inner shells is a complicated mixture of static rearrangement, vacancy decay effects and ordering of the photoionization process in time, the latter being achieved to some extent by ortogonalization of the outgoing electron wave function.

For inner shells τ_i is not large and the vacancy decays either by Auger-process or radiatively. For a slow photoelectron τ_{ej} is larger than τ_i and it feels the altered field of the ion. After Auger-decay the ion becomes double-charged and attracts the slow electron much stronger. It is slowed down and may be even captured to a discrete level in the new field, the energy excess $\Delta\varepsilon$ being transferred to the fast Auger-electron. The magnitude $\Delta\varepsilon$ may be easily estimated as

$$\Delta\varepsilon \sim 1/r \sim 1/v\tau_d \sim \gamma_i/\sqrt{\varepsilon} \qquad (12)$$

This phenomenon, called in the literature Post Collision Interaction (PCI)[30], is a strong collective effect, which prominently modifies the photoionization cross section. For not too large γ_i and fast Auger electrons it is sufficient to include the interaction of outgoing electron with the rearranged ion field, which is presented by diagrams in Fig. 9.

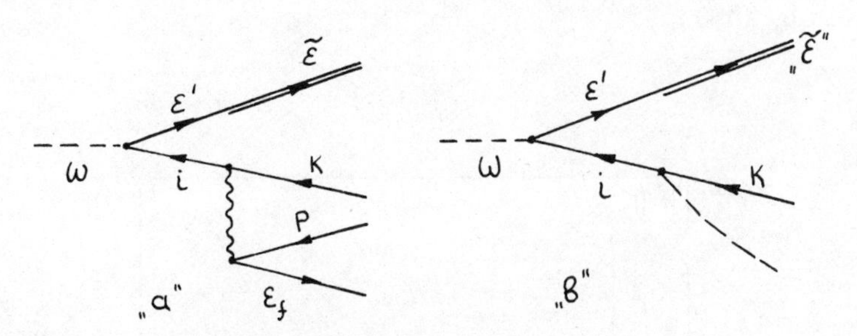

Fig. 9. Diagrammatical representation of the PCI effect.

Here the double line describes the electron motion in the field of k and p vacancies created in Auger-decay of i. The amplitude of such a process is given by the expression[31]:

$$\tilde{D}_{\varepsilon\varepsilon_f} = \sum\!\!\!\!\!\!\int_{\varepsilon'>F} (i|D|\varepsilon') \frac{(kp|u|i\varepsilon_f)(\varepsilon'|\tilde{\varepsilon})}{\omega - \varepsilon' - I_i + i\gamma_i/2} \qquad (13)$$

where summation implies also integration over the continuous spectrum. The first term of the integrand is given by (1), the second is the i-vacancy Auger-decay amplitude, and the last one is an overlap of photoelectron wave functions in the fields of i - $(\varepsilon'|$, and kp - $|\tilde{\varepsilon})$, respectively. The slow electron essentially modifies the fast electron energy distribution, which becomes broader, asymmetric, shifted to the higher energy side by $\approx\Delta\varepsilon$ and oscillative, instead of being a simple Breit-Wigner profile. The asymmetry demonstrates that the formation and decay of vacancy i for small ε is in fact a unified process which can not be divided into two separate steps: at first formation of vacancy i and then its decay. Close to the threshold of i-vacancy creation the spectroscopic factor F_i becomes essential, nontrivially modifying the PCI cross section by altering the potential of vacancy i, in which the electron initially moves.

If $\tau_{ej} \gg \tau_j$ and τ_d is small, the decay leads to a sudden variation of the field just after creation. In fact, if γ_i in (13) is large, the difference between ε and ε' in the denominator is negligible and one has for the cross section:

$$\sigma \sim \int |\tilde{D}_{\tilde{\varepsilon}\varepsilon_f}|^2 \, \delta(\omega - \tilde{\varepsilon} - \varepsilon_f - I_{KP}) d\tilde{\varepsilon} d\varepsilon_f \approx \int d\tilde{\varepsilon} \, |(i|D|\tilde{\varepsilon})|^2 \cdot \gamma_i \left[(\omega - \varepsilon - I_i)^2 + \gamma_i^2/4 \right]^{-1} = \pi \, |(i|D|\tilde{\varepsilon}_o)|^2 \quad (14)$$

where $\tilde{\varepsilon}_o = \omega - I_i$. Thus, σ is determined by a dipole matrix element, in which $|\tilde{\varepsilon}_o)$ stands for the outgoing electron wave function in the statically rearranged double ion field. As an example, Fig. 10 presents the photoionization cross section of 1s-electrons in Ar, calculated using (14).

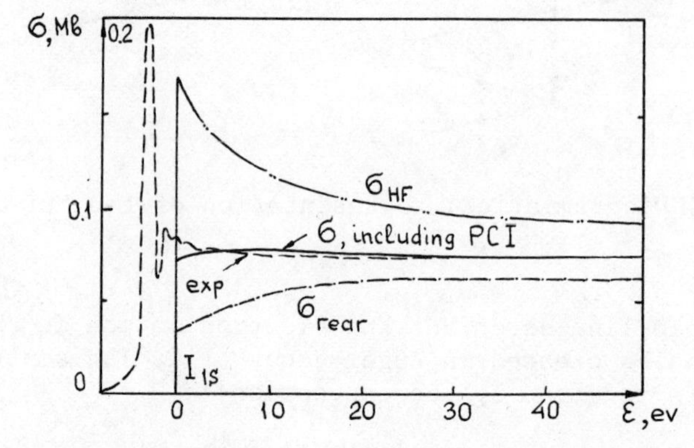

Fig. 10. Photoionization cross section of 1s-electrons in Ar.

It is seen, that the static rearrangement makes the cross section too small, while the 1s-vacancy decay into <u>2s</u> <u>2s</u>, <u>2p</u> <u>2s</u> and <u>2p</u> <u>2p</u> states leads to prominent increase[32] and permits to achieve good agreement with experiment[33]. The PCI effects far from threshold are insignificant. Therefore, the further decay of $2s^{-1}$ and $2p^{-1}$ vacancies does not alter the cross section even very close to the 1s ionisation threshold. The radiative decay of $1s^{-1}$ affects the cross section, but much weaker, than the Auger, because the transition $1s^{-1} \to 2p^{-1}$ does not change the asymptotic behaviour of the vacancy field. Fig. 11 illustrates the influence of radiative decay. The cross section is calculated using (15) $|\tilde{\varepsilon}_o)$ being determined in the $2p^{-1}$ field*).

*) Derived together with V.K. Ivanov and V.A. Kupchenko.

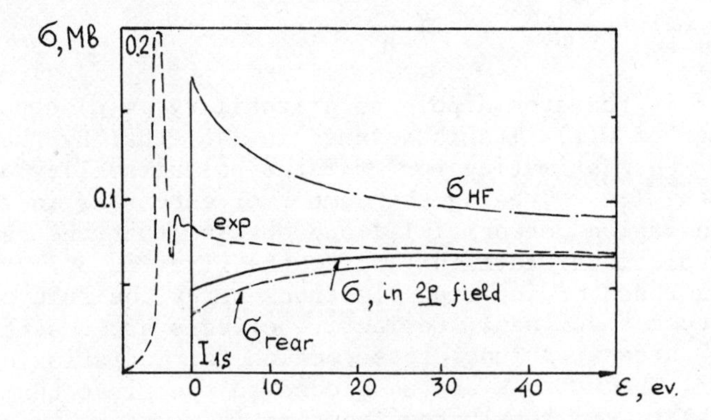

Fig. 11. Effect of 1s-vacancy decay.

Let us note that after the decay $\underline{i} \rightarrow \underline{kp}\varepsilon$ the symmetry of the ion field may change, and therefore the electron wave function acquires on admixture of another angular momentum. It may change the anisotropy parameter, which even for $\underline{1s}$ vacancy will deviate from being equal to 2. However, in the Ar $\underline{1s}$-vacancy case, it is necessary to have in mind that relativistic (spin-orbit) corrections to an ε p-wave function and an admixture of quadrupole transition, which is essential in this frequency region, also leads to $\beta_{1s} \neq 2$.

6. BREMSSTRAHLUNG

The ability of an atom to respond to the external field like a resonator, although without well determined frequencies, is manifested also in the bremsstrahlung[34]. In a rather broad region of ω (of the ionization potentials order) the electron scattered by an atom radiates quite different from its radiation in a static potential. The incoming electron polarizes the atom, mainly dipole, and the virtually excited target radiates. The radiation amplitude is determined by (1), where the state $\underline{i}$ belongs to the continuous spectrum also. The second term proved to be larger than the first for easily polarisable atoms, for frequencies of the order of ionization potentials. With increased energy of the incoming electron, and for small moments, transferred to the atom, the second term becomes larger than the first. Describing the scattered electron with momentum p by a plane wave, we obtain the following expression for the lowest bremsstrahlung amplitude in the static self-consistent potential W(r) of the electron-atom interaction for small transferred momentum q:

$$D_{p,p-q}(\omega) \approx \frac{W(q)q}{\omega} + 4\pi \frac{\omega}{q} \alpha(\omega) \tag{15}$$

where $\alpha(\omega)$ - is the atom dipole polarizability, w(q) denotes the Fourier-image of W(r). Its is assumed in (15) that by the order of magnitude $\omega \sim pq$. Estimating for $\omega \sim I$ the polarizability $\alpha(\omega)$ as $\alpha(\omega) \approx \alpha(0) \approx N_i/I_i^2$, N_i being the number of electrons in the shell with the ionization potential I_i, one obtains that the ratio of the second, correlational term in (15) to the first is $p^2/I_i \gg 1$ and thus with increase of incoming electron energy the role of "atomic" radiation becomes dominant. Collective effects are essential also in the total bremsstrahlung cross section for radiation of frequencies $\omega \sim I_i$ [35]. The ratio k of the second to the first terms contributions in (15) to the total cross section is given by the relations:

$$k \approx \frac{\omega^4 |\alpha(\omega)|^2}{Z^2} \approx \left(\frac{c\omega^2_{max}\sigma^{max}}{4\pi Z}\right)^2 \tag{17}$$

where c is the velocity of light, Z is the nuclear charge, σ^{max} and ω_{max} are the maximum photoionization cross section and its position in energy, respectively. For atoms with strong collective maxima in $\alpha(\omega)$ in the continuous spectrum region, k may be much larger than 1. This is illustrated in Fig. 12 by an example of the La atom, where in photoemission near $\omega \approx 8$ Ry a powerful maximum of collective nature appears.

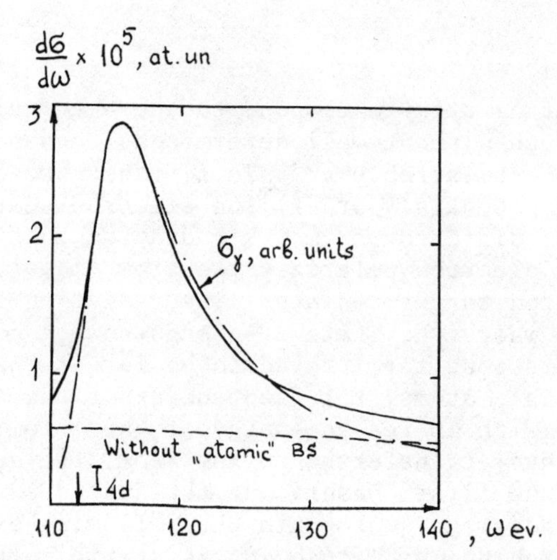

Fig. 12. Bremsstrahlung spectrum of fast (10 keV) electrons on the La atom.

8. CONCLUDING REMARKS

Initially, about half a century ago, the apparently simple idea of collective atomic excitations, analogous to hydrodynamical oscillations of a homogeneous electron gas stimulated the development of extensive experimental and theoretical investigations.

The very separation of electron motion in atoms into collective and one-electron proved to be difficult and to large extent conventional. By correcting and improving the one-electron approximation it becomes possible even in its frame to include much of that what seems to be manyelectron corrections. Instead of Bloch-Jensen type collective oscillations a large number of many-particle effects were observed, which manifest themselves practically in all atomic processes. This has essentially enriched our understanding of the structure of the manyelectron atoms, contributing to the development of their theory.

The author is grateful to M.Yu. Kuchiev and G.N. Ogurtsov for critical reading of the manuscript.

REFERENCES

1. R.D. Hartree, Proc. Camb. Phil. Soc. 24, 89: 111 (1928).
2. V.A. Fock, Zs. f. Phys. 61: 126 (1930).
3. F. Bloch, Z. Phys., 81: 363 (1933).
4. H. Jensen, Z. Phys., 106: 620 (1937).
5. H.P. Kelly, in "Atomic Inner-Shell Processes", B. Craseman, ed., New York, Academic (1972).
6. M.Ya. Amusia and N.A. Cherepkov, Case Studies in Atomic Physics, 5, 2: 47 (1975).
7. G.Wendin, in "Vacuum Ultraviolet Radiation Physics", E. Koch, ed. Pergamon, Vieweg (1975).
8. P.G. Burke, in "Electronic and Atomic Collisions", G. Watel, ed., North-Holland, Amsterdam, New-York, Oxford (1978).
9. M.Ya. Amusia, L.V. Chernysheva and S.A. Sheinerman, Phys. Lett 82A, 4: 171 (1981).
10. D.L. Miller, J.D. Dow, R.G. Houlgate, G.V. Marr and J.B. West, J. Phys. B10: 3205 (1977).
11. W.R. Johnson, C.D. Lin, K.T. Cheng and C.M. Lee, Physica Scripta 21: 409 (1980).
12. N.A. Cherepkov, Phys. Lett. 66A: 204 (1978).
13. M.G. White, S.H. Southworth, E.D. Kobrin, E.D. Poliakoff, R.A. Rosenberg and D.A. Shirley, Phys. Rev. Lett. 43, 22: 1661 (1979).
14. N.A. Cherepkov, J. Phys. B12: 1279 (1979).
15. U. Heinzmann, G. Schönhense and J. Kessler, Phys. Rev. Lett., 42: 1603 (1979).
16. M.J. Van der Wiel and T.N. Chang, J. Phys. B11: L 125 (1978).
17. A.F. Starace and L. Armstrong, Jr., Phys. Rev. A 13: 1850 (1976).

18. N.A. Cherepkov and L.V. Chernysheva, Phys. Lett. 60 A: 103 (1977).
19. M.Ya. Amusia, V.K. Ivanov and L.V. Chernysheva, J. Phys. B 14: L 19 (1981).
20. R. Bruhn, B. Sonntag and M.W. Wolff, Phys. Lett., 69 A: 9 (1978).
21. M.Ya. Amusia, L.V. Chernysheva and S.I. Sheftel, Journal Tech. Fiz. 51: 2411 (1981) (in Russian).
22. M.Ya. Amusia, V.K. Ivanov and L.V. Chernysheva, IV ICAP, abstracts p. 332 (1974).
23. J.B. West, P.R. Woodruff, K. Codling and R.G. Houlgate, J. Phys. B 9: 407 (1976).
24. I.E. McCarthy and E. Weigold, Adv. Phys., 25, 5: 489 (1976).
25. M.Ya. Amusia and A.S. Kheifets, Phys. Lett. 82A: 407 (1981).
26. R.P. Madden, D.L. Ederer and K. Codling, Phys. Rev. 177: 136 (1969).
27. M.Ya. Amusia and A.S. Kheifets, Phys. Lett. A, in press (1982).
28. M.Yu. Kuchiev, VIII ICAP, abstracts of papers, Göteborg (1982).
29. M.Ya. Amusia, V.K. Ivanov, S.I. Sheftel and S.A. Sheinerman, Zhurnal Eksp. Teor. Fiz., 78: 910 (1980) (in Russian).
30. F.H. Read, Radiation Research, 64: 23 (1975).
31. M.Ya. Amusia, M.Yu. Kuchiev and S.A. Sheinerman, in "Coherence and Correlations in Atomic Collisions", H. Kleinpoppen and J.F. Williams, ed., Plenum, New York and London, (1980).
32. M.Ya. Amusia, V.K. Ivanov and V.A. Kupchenko, J. Phys. B 14: L667 (1981).
33. K. Schnopper, Phys. Rev. 131: 2558 (1963).
34. M.Ya. Amusia, Comments At. Mol. Phys., X, 3 - 5: 123 (1982).
35. M.Ya. Amusia, T.M. Zimkina and M.Yu. Kuchiev, Journal of Tech. Fiz., 52, 5: 1045 (1982).

MANY BODY CALCULATIONS OF PHOTOIONIZATION

Hugh P. Kelly

Department of Physics, University of Virginia
Charlottesville, Virginia 22901

INTRODUCTION

There has been great interest in photoionization during the past decade, both experimentally[1] and theoretically.[2] Many interesting properties of photoionization have been measured by laser spectroscopy and particularly by synchrotron radiation.[3,4] During this same period theoretical techniques which can account for electron correlations have been developed and refined. Many calculations have been carried out using many-body perturbation theory (MBPT)[5], R-matrix theory,[6] the random phase approximation with exchange (RPAE),[8] and other related techniques. This article will focus on nonrelativistic calculations since relativistic calculations such as the RRPA will be covered in the article by W. Johnson in this volume.

One of the interesting aspects of photoionization studies is the strong overlap between atomic, molecular, and solid state physics. In addition, there are important applications of photoionization in atmospheric physics, astrophysics, and plasma physics.

The subject of photoionization has been reviewed in a number of articles. Among them are the important review by Fano and Cooper[9] in 1968, the 1975 NATO summer school articles,[10] and the extensive recent reviews of experiments by Samson[1] and of theory by Starace.[2] Other important reviews have been given by Krause,[3] Manson,[11] Wuilleumier,[4] and by Manson and Starace.[12] In addition, there are reviews of RPAE results by Amusia and Cherepkov[7] and by Amusia.[13] Other articles of interest include those by Brandt and Lundqvist[14] who predicted strong correlations and possible collective effects in atomic photoabsorption, by Altick and Glassgold[15] who introduced the random phase approximation in atomic physics and calculated photoionization of alkaline earth atoms. There are also many calculations which have used close-coupling methods to account for electron correlation in continuum states.[16,17] Many-body perturbation calculations were first carried out for Be[18] and Li[19] and later for A and other elements.[5] The time-dependent coupled Hartree-Fock approximation, which is equivalent to the RPAE was introduced in atomic physics by Dalgarno and Victor.[20] Extensive RPAE calculations were carried out for the rare gases by Amusia, Cherepkov, and Chernysheva[21] and by Wendin.[22] The R-Matrix method, which had previously been used in electron-atom scattering, was first applied to atomic photoionization of Ar by Burke and Taylor[6]. The RPAE method was extended to open-shell systems by Armstrong and Starace[23-25] and by Cherepkov and Chernysheva.[26] The transition matrix method of Chang and Fano[27] involves solution of a set of coupled differential equations and has produced results[28] equivalent to the RPAE. A multiconfiguration method has been introduced by Swanson and Armstrong[29] which has produced results for the rare gases equivalent to MBPT or RPAE. It has also been extended to open shells.[30] Multi-channel quantum defect theory (MQDT) has been developed by Ham,[31] Seaton,[32] Fano and coworkers[33-35] to allow a description of photoionization by a few parameters and extrapolation from one energy region to another. This method complements the ab-initio methods which may be used to calculate the parameters used in MQDT. The RRPA has been used in such calculations[36,37] as well as cross section calculations for closed-shell systems.[38] The Stieltjes-Chebycheff method[39] uses discrete states to approximate continuum states and has recently been used[41] to calculate photoionization of $He^{-}1s2s2p\,{}^4P$. In addition to calculations including correlations there have been many calculations of considerable physical interest carried out at the Hartree-Fock or Hartree-Slater level by Manson and co-workers.[11,42,43] In the following sections examples will be given of calculations using most of these methods.

THEORY

Roughly, we expect photoionization cross sections to be in the range $\pi a_o{}^2 = 25 \times 10^{-18}$ cm^2. Although cross sections vary greatly with energy, they are in fact measured in units of 10^{-18} cm^2 (Mb). It is easy to show[9] that the photoionization cross section

$$\sigma(\omega) = 4\pi \frac{\omega}{c} \operatorname{Im}\alpha(\omega), \tag{1}$$

where $\alpha(\omega)$ is the frequency-dependent polarizability, given by

(in atomic units)

$$\alpha(\omega) = -\sum_k |\langle\psi_k|\sum_{i=1}^{N} z_i|\psi_o\rangle|^2 \left\{\frac{1}{E_o-E_k-\omega} + \frac{1}{E_o-E_k+\omega}\right\}, \qquad (2)$$

where $|\psi_o\rangle$ is the exact initial many-particle state and $|\psi_k\rangle$ is an excited many-particle state. Since the second denominator may vanish, we add a small imaginary part in and use the formula

$$\lim_{\eta\to o} (D+i\eta)^{-1} = P\, D^{-1} - i\pi\delta(D). \qquad (3)$$

Replacing $\sum_k$ by $\int dk$, the imaginary part of $\alpha(\omega)$ is proportional to $|\langle\psi_k|\Sigma z_i|\psi_o\rangle|^2$, and Eq. (1) may be shown to follow.

Perturbation theory may be used to calculate the many-particle matrix element $\langle\psi_k|\Sigma z_i|\psi_o\rangle$. We choose a single-particle potential $V_i = V(r_i)$. Then $H = H_o + H'$, where

$$H_o = \sum_{i=1}^{N} (T_i + V_i), \qquad (4)$$

and T_i is the sum of kinetic energy and nuclear interaction for the ith electron.

$$H' = \sum_{i<j} v_{ij} - \sum_i V_i, \qquad (5)$$

and

$$\psi_o = \sum_L \left(\frac{1}{E_o-H_o} H'\right)^n \Phi_o, \qquad (6)$$

where $\sum_L$ indicates a sum over "linked" terms only.[44,45] The expression for ψ_k is similar. The "length" form of the dipole matrix element

$$\langle\psi_k|\sum z_i|\psi_o\rangle = (E_o-E_k)^{-1} \langle\psi_k|\sum_i \frac{d}{dz_i}|\psi_o\rangle, \qquad (7)$$

where ψ_k and ψ_o are eigenstates of H. The matrix element $\langle\psi_k|\Sigma\, d/dz_i|\psi_o\rangle$ is referred to as the "velocity" form. We may use perturbation theory to calculate $\langle\psi_k|\Sigma z_i|\psi_o\rangle$ in which case we calculate a complete set of single particle states ϕ_n which satisfy

$$(T+V)\phi_n = \varepsilon_n \phi_n. \qquad (8)$$

The N orbitals lowest in energy are occupied in Φ_o, and the others are called excited states. It is desirable to choose V as the usual Hartree-Fock potential R_{HF} for ϕ_n in Φ_o. For excited states, the potential is then not unique as is seen by considering[5,46]

$$V = R_{HF} + (1-P)\Omega\ (1-P), \tag{9}$$

where Ω is an arbitrary Hermitian operator and

$$P = \sum_{n=1}^{N} |n\rangle\langle n|. \tag{10}$$

For excited states Ω is chosen so that $R_{HF} + \Omega$ is the appropriate LS-coupled Hartree-Fock potential. As an example, consider argon $3p^6\ {}^1S$. Dipole absorption of a photon leads to final states $3p^5$ (ks or kd)^1P. The usual $3p^5k\,{}^1P$ Hartree-Fock potential results by setting

$$\langle 3p^5k'\,{}^1P|H'|3p^5k\,{}^1P\rangle = 0. \tag{11}$$

The term $-P\Omega(1-P)$ in Eq. (9) ensures orthogonality of excited ks orbitals to the occupied ns orbitals.

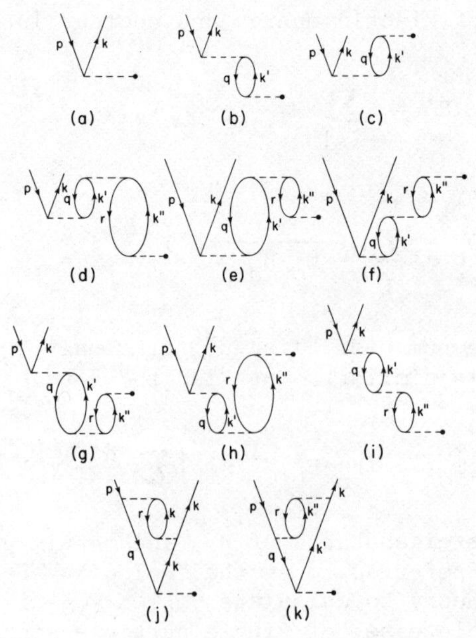

Fig. 1. Diagrams contributing to $\langle\psi_k|\Sigma z_i|\psi_o\rangle$. Dashed line ending in solid dot represents interaction with z. Other dashed lines are Coulomb interactions. Diagrams should be read from bottom to top.

Diagrams for the perturbation expansion of $\langle\psi_k|\Sigma z_i|\psi_0\rangle$ are shown in Fig. 1 for a transition in which orbital p is excited to k. Dashed lines ending in a solid dot represent the dipole interaction z, and other dashed lines represent correlation interactions with H'. The diagram of Fig. 1(a) is $\langle k|z|p\rangle$, and the diagram of Fig. 1(b), which represents final state correlation, is given by

$$\sum_{k'} \frac{\langle kq|v|pk'\rangle\langle k'|z|q\rangle}{\varepsilon_q-\varepsilon_{k'}+\omega+i\eta} . \tag{12}$$

The diagram of Fig. 1(c) represents correlations in the initial state and is given by

$$\sum_{k'} \frac{\langle q|z|k'\rangle\langle kk'|v|pq\rangle}{\varepsilon_p+\varepsilon_q-\varepsilon_k-\varepsilon_{k'}} . \tag{13}$$

When the Hartree-Fock LS-coupled potential is used, it cancels the diagram of Fig. 1(b) when p and q are orbitals of the same subshell, and only the ground state correlation diagram (c) remains. Diagram (b) can represent resonances when q is in an inner shell, k' is a bound excited states, and ω is such that the denominator vanishes. Higher-order diagrams may then be summed geometrically to give a real shift and imaginary shifts $i\Gamma/2$ to the denominator. In the RPAE, an integral equation is solved which sums to all orders in perturbation theory those diagrams which describe the excitation and de-excitation of particle-hole pairs at each interaction. This includes all the diagrams of Fig. 1 except for (j) and (k) which describe relaxation affects when p=q. One difficulty is that any number of particle-hole pairs are excited by RPAE diagrams, whereas physically only N can be excited. There is also a tendency, in some cases to be discussed, for the second-order RPAE diagrams within a given subshell such as (d), (e), (g), and (h) to cancel the relaxation diagrams (j) and (k).

CLOSED SHELL ATOMS

Prior to 1970, calculations for the rare gas atoms carried out in the Hartree-Slater approximation had failed to give qualitatively correct results for the photoionization cross sections.[9] However, in 1971 RPAE calculations were presented by Amusia, Cherepkov, and Chernysheva[21] which gave excellent agreement with experiment for all the rare gases. An example for the $4d^{10}$ subshell of xenon is given in Fig. 2. Subsequent many-body perturbation calculations on argon $3p^6$ found very good agreement in low-order using the Hartree-Fock LS-coupled ^1P continuum states.[47] The length and velocity results were brought into good agreement with experiment by including the ground state correlation diagram of Fig. 1(c). Additional improvement was achieved by including diagrams (d) and (e) of Fig. 1.

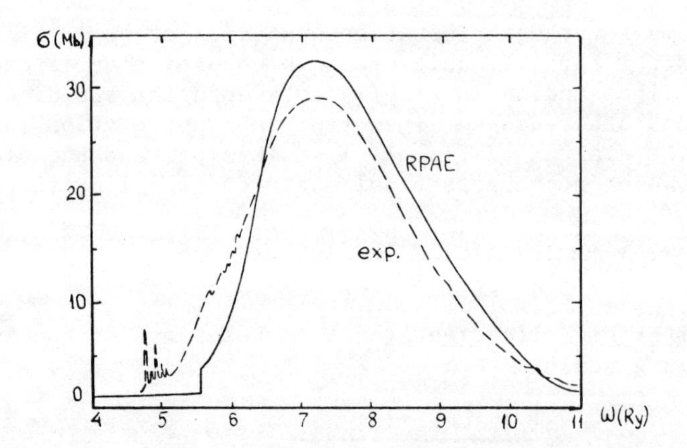

Fig. 2. Photoionization cross section for the $4d^{10}$ subshell of xenon. ______, Amusia et al., ref. 21. - - - -, expt. by Haensel et al., ref. 48.

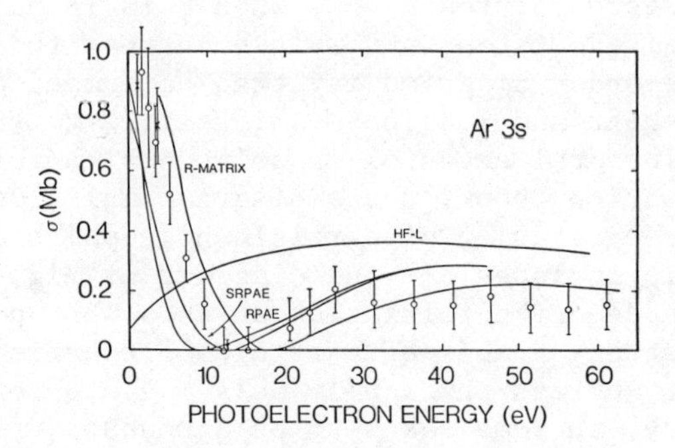

Fig. 3. Photoionization cross section of the $3s^2$ subshell of argon. R-matrix, Burke and Taylor, ref. 6; RPAE, Amusia and Cherepkov, ref. 7; SRPA, approximate RPAE calculation by Lin, ref. 49; HF-L, Kennedy and Manson, ref. 50; expt: o Houlgate et al., ref. 51; • Samson and Gardner, ref. 52.

A particularly striking effect is shown in Fig. 3 which gives the 3s photoionization cross section for argon. It is seen that the Hartree-Fock results are not even qualitatively correct. However, both R-matrix[6] and RPAE results[7] are in reasonable agreement with experiment.[50,51] In the language of perturbation theory, diagram (b) of Fig. 1 with p = 3s and q = 3p dominates the direct diagram (a) with p = 3s. Figure 4 shows a similar situation in the 5s cross section for xenon. Here it is necessary to include coupling between $5s^2$, $5p^6$, and $4d^{10}$ subshells.[7]

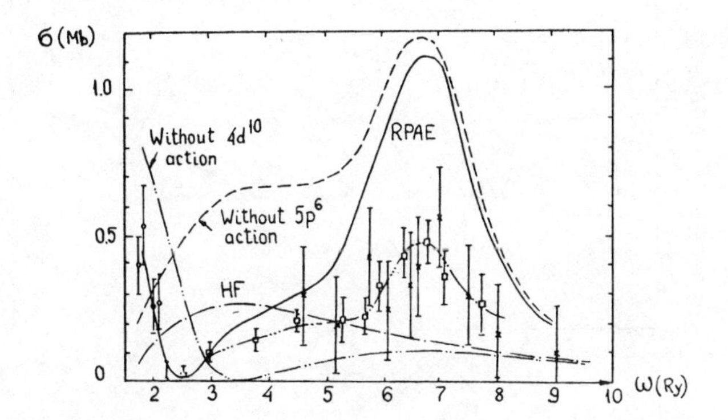

Fig. 4. Photoionization cross section for the $5s^2$ subshell of xenon. Expt: o Samson and Gardner, ref. 52; x West et al, ref. 53; □ Adam et al., ref. 54. ——. . . —— fit to data. All other curves are theory by Amusia and Cherepkov, ref. 7. (figure from ref. 7).

Another property of interest in photoionization is the asymmetry parameter β defined (for linearly polarized radiation) by [7]

$$\frac{d\sigma}{d\Omega} = \frac{\sigma}{4\pi} (1+ \beta P_2(\cos \theta)), \tag{14}$$

where θ is the angle between the polarization direction of the incident light and the direction of the ejected electron.[7] Figure 5 shows a comparison of the β parameter for xenon $5p^6$ calculated[55] with and without coupling with $4d^{10}$ and with experiment.[56,57]

Thus far the effects of relaxation have not been discussed. They were, however, found to be particularly important for calculations of photoionization in the $4d^{10}$ subshell of barium. In Fig. 6 are shown RPAE calculations by Wendin with and without the matrix elements of relaxation.[58] It is seen that the relaxation effect is considerable.

Figure 7 also shows this effect for Ba $4d^{10}$. The curve labelled HFU is the geometric mean of length and velocity results calculated[59] in an unrelaxed field, i.e., neutral barium

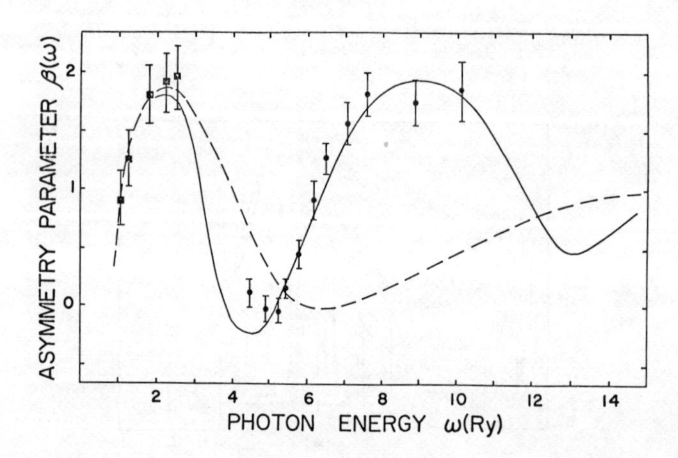

Fig. 5. Angular anisotropy parameter β for the $5p^6$ subshell of xenon. ——— and - - - -, calculations, by Amusia and Ivanov with and without coupling with $4d^{10}$, ref. 55. Expt: ■ Lynch et al, ref. 56; ● Torop et al, ref. 57.

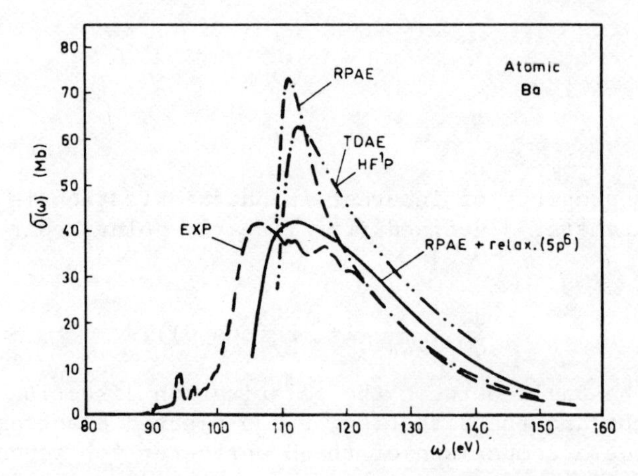

Fig. 6. Ba 4d→kf subshell photoionization cross sections calculated by Wendin, ref. 58. Expt. by Rabe et al., ref. 58a, normalized to contain 10 electrons.

with one 4d electron removed. Taking the geometric mean of length and velocity curves is nearly the same[60] as including ground state correlations, since within a given subshell the ground state correlation diagram of Fig. 1(c) has the effect of bringing length (L) and velocity (V) results together. Since the RPAE results are considerably worse than HFU results, we interpret this as due to the fact that including higher-order RPAE terms in this

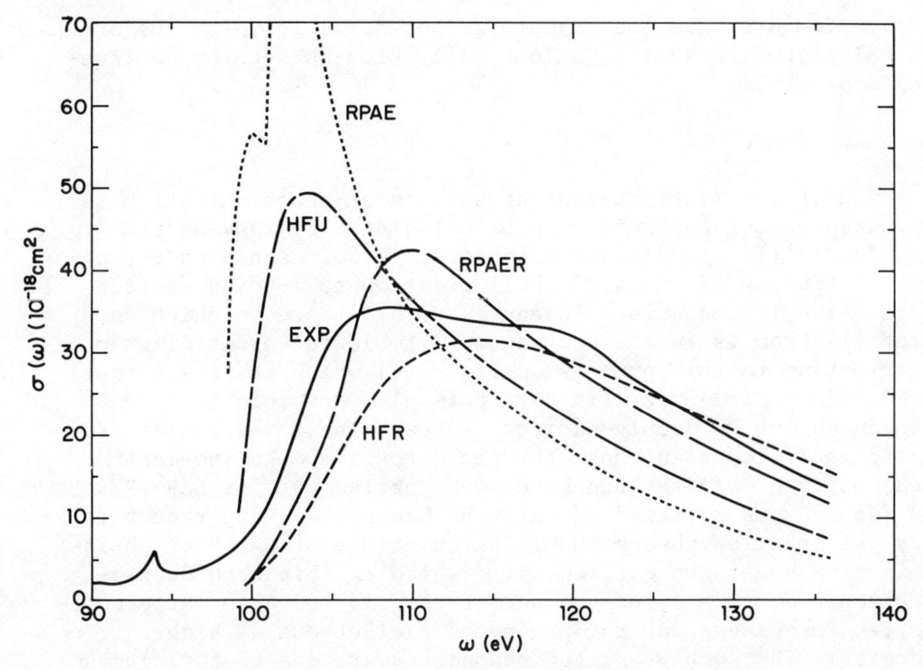

Fig. 7. Photoionization cross section for the $4d^{10}$ subshell of Ba. HFU (Hartree-Fock unrelaxed) and HFR (Hartree-Fock relaxed) curves calculated by Kelly et al., ref. 59. RPAE calculated by Fliflet et al., ref. 64. RPAER calculation by Amusia et al, ref. 63. Expt. by Hecht and Lindau, ref. 65.

case unbalances the perturbations. In particular, there is considerable cancellation between higher-order RPAE diagrams and the relaxation diagrams such as (j) and (k) of Fig. 1. This has been explicitly checked by calculations for barium, and for cadmium.[62] The curve labelled HFR is a Hartree-Fock (geometric mean of L and V) curve with excited states calculated in the field of the relaxed ion so as to approximate relaxation effects.[59] The curve labelled RPAER is an RPAE calculation by Amusia et al[63] in which the excited Hartree-Fock states were calculated in the field of the relaxed ion. The experimental curve was measured by Hecht and Lindau.[65] A recent calculation by Zangwill and Soven[66] using the (LDTDHF) method gives a result in excellent agreement with experiment. The threshold in this case is calculated to be at 93 eV rather than at the experimental value of 99 eV. When the HFR result uses the threshold 93 eV, it gives equally good agreement with experiment. The reasons for the success of the LDTDHF method are unclear since it omits relaxation effects. Instead, it is equivalent to using RPAE but with orbitals calculated in the local density approximation

(LDA). In this case the asymptotic potential in which the orbitals are calculated lacks the Coulomb tail which is seen by Hartree-Fock orbitals.

RESONANCE STRUCTURE

One of the most interesting aspects of photoionization is resonance structure, which may be calculated by many-body methods.[5,47,67] As already discussed, these resonances may arise from excitation of an inner shell electron to a bound excited state which is degenerate in energy with a state in which an outer electron is in the continuum. The lowest-order diagram contributing to this process is Fig. 1(b) which is first-order in H'. In second-order (in H') it is also possible to have resonances due to doubly-excited states. These resonances are particularly important near threshold for the alkaline-earth elements, and calculations have been carried out for Be[68,69] and for Mg[70] as reviewed recently by Greene.[71] Very recent many-body perturbation theory (MBPT) calculations of $Ca3p^64s^2$ photoionization have been carried out[72] which exhibit both double-electron resonance structure near threshold and also large core-excited resonances due to $3p \rightarrow nd,ns$ excitations at higher energies. The doubly-excited resonances are due to the diagrams shown in Fig. 8. In Fig. 8(d) we show a higher-order diagram which is Fig. 8(a) times the segment shown in (d). The horizontal line indicates that the imaginary part of the denominator is taken. In higher-order diagrams, the segment is repeated and leads to a geometric series which accounts for the widths of resonances and interactions among different resonances.[67] In the Hartree-Fock (HF) approximation, the threshold cross-section for $Ca(4s)^2$ photoionization is 15 Mb (length) and 3 Mb (velocity). Including diagrams as shown in Fig. 1 brings length and velocity results into close agreement at 5 Mb . When the double-electron excitation diagrams of Fig. 8 are included, the threshold result is further reduced and much resonance structure appears as shown in Fig. 9 where comparison is made with the experimental results and MQDT calculation of Carter, Hudson, and Breig.[73] Only length results of the MBPT calculation are shown since velocity results are very close. There is experimental discrepancy concerning the normalization since the peak of the 3d5p resonance is reported as 30 Mbn (Ditchburn and Hudson[74]), 23 Mbn (Carter, Hudson, and Breig[73]) and 14 Mbn (Newsom[75]). In addition, McIlrath and Sandeman[76] claim that the results of Carter et al[73] should be increased by a factor 2.2 on the basis of their photo-electric absolute measurements.

In Fig. 10 are shown calculated and experimental results which include the 4pns and 4pnd double electron resonances. The calculations only included 3dnp resonances to n=9. The large

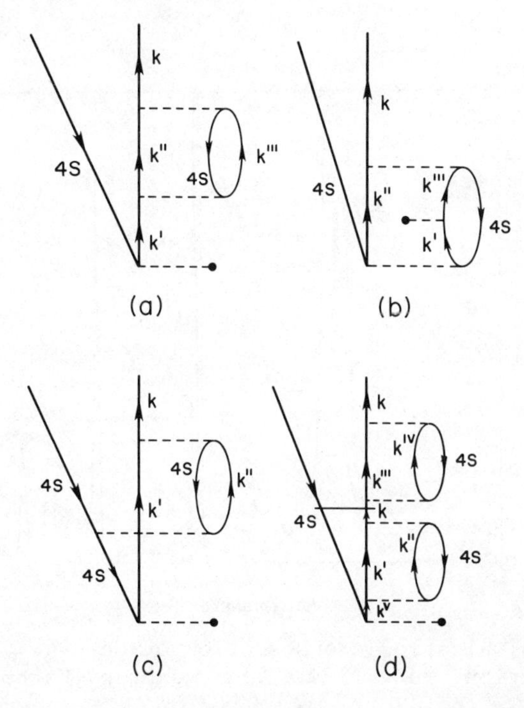

Fig. 8. Resonance diagrams involving doubly excited states of $4s^2$. Solid dot indicates matrix elements of z. Horizontal line in (d) indicates denominator D is treated as $-i\pi\delta(D)$.

calculated 4p4d structure is not seen experimentally. Beyond 7.8 eV, photoionization with excitation of 4s to 3d can occur. This was not included in the calculation and should account for the discrepancy with experiment just beyond 7.8 eV. It may also account for part of the discrepancy near 8.3 eV. Results by Connerade et al.[77] were normalized to Newsom's.[75]

At higher energy, there is resonance structure in the 4s cross section due to 3p→ns,nd excitations. These resonances also exhibit spin-orbit splitting. Calculations[78] of the 4s cross section in the region of these resonances are shown in Fig. 11 and compared with experimental results by Mansfield and Newsom.[79] Spin-orbit effects were only calculated for the six lowest ns and nd resonances. The resonance structure has not been drawn all the way to the $3p_{1/2}$ and $3p_{3/2}$ edges. In addition, it should be noted that the upper experimental scale is not linear so care should be taken in making comparison of theory and experiment. There remains a need to calculate the photoionization with excitation cross sections leaving Ca^+ in the 3d and 4p levels. Also, the doubly-excited states were only

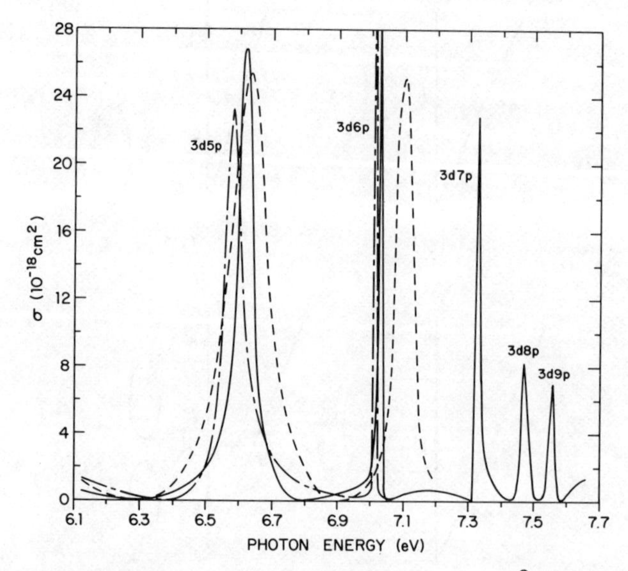

Fig. 9. Photoionization cross section of Ca $4s^2$. ——— Many body length calculation, ref. 72; — · —, expt by Carter et al., ref. 73; - - -, MQDT, ref. 73.

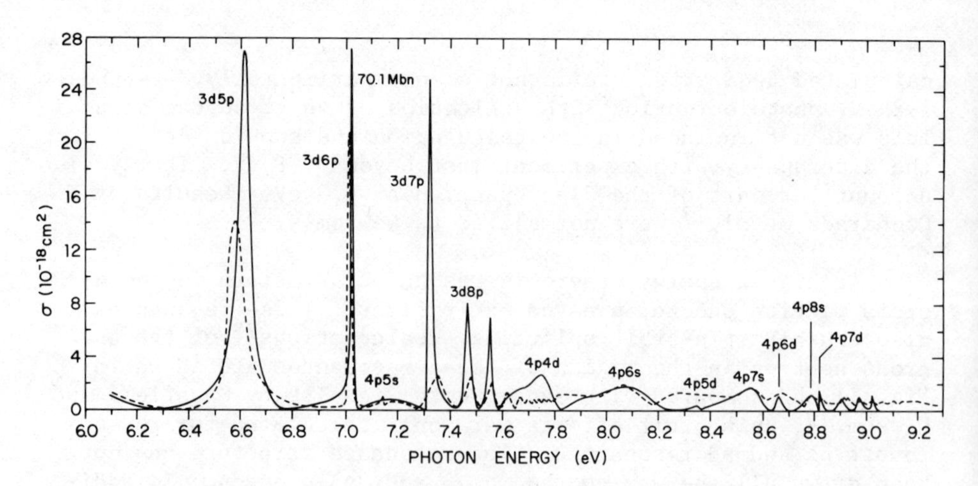

Fig. 10. Calcium $4s^2$. ——— MBPT length calculation, ref. 72; ---, expt by Newson, ref. 75, from 6.11 eV to 7.45 eV and by Connerade et al., ref. 77, from 7.45 eV to 9.2 eV (normalized to Newsom result at 7.45 eV).

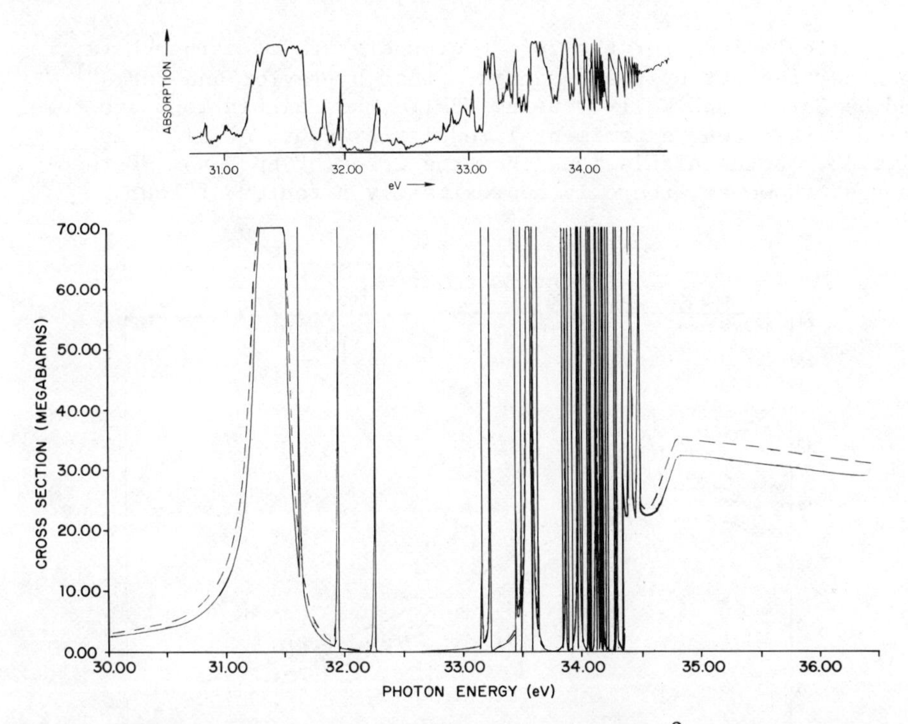

Fig. 11. Photoionization cross section of $Ca4s^2$ showing 3p→ns,nd resonance structure. ——— length calculation and - - - velocity calculation by Altun et al., ref. 78. The upper curve is from expt by Mansfield and Newsom, ref. 79. Note that experimental curve is not linear.

calculated in the Hartree-Fock approximation, and future calculations should account for the direct mixing among these states.

OPEN SHELL ATOMS

The problem of photoionization of open-shell atoms is rather difficult, and, as a result, the number of calculations and particularly of experiments is limited. The R-matrix method[6] has been used to calculate detailed cross sections of open-shell atoms such as Al[80], C[81], N[82], and O[81]. In the R-matrix method, a sphere of radius a centered at the nucleus divides configuration space into two regions. In the inner region the solutions of the Schrödinger equation are expanded in terms of discrete functions satisfying a prescribed value of the logarithmic derivative at $r = a$. In the outer region, a close-coupling continuum solution is obtained and matched onto the solution in the inner region at $r = a$ by means of the R-matrix obtained from the eigenstates and eigenvalues of the Schrodinger equation for $r \leq a$.

Calculations for the $2p^2\ ^3P$ ground state of carbon have been carried out by the R-matrix method by Taylor and Burke[81] and by Carter and Kelly[83] using MBPT. The calculations are compared with recent experimental results by Cantù et al [84] in Fig. 12. Both calculations show the $2s2p^2(^4P)np\ ^3D$ and ^3P resonances. However, there is approximately a factor of four

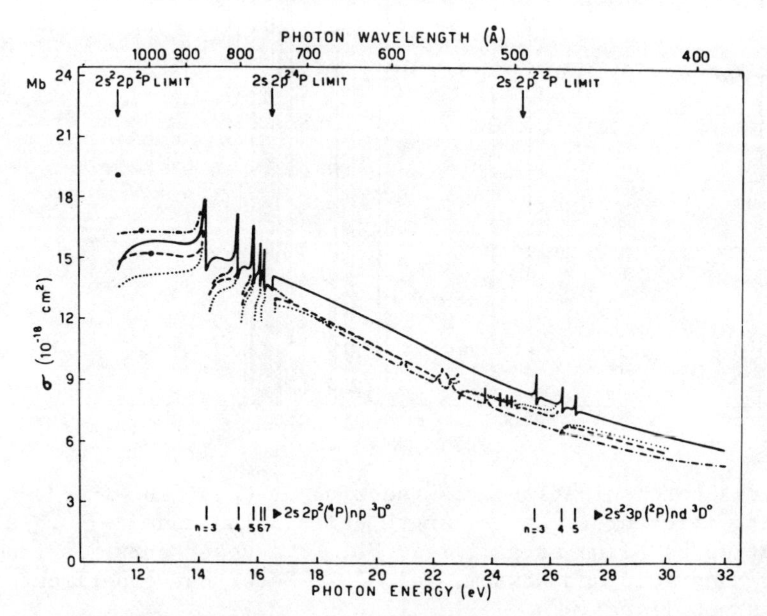

Fig. 12. Photoionization cross section of the neutral carbon atom. —— relative experiment by Cantù et al., ref. 84; — ·— R-matrix calculation by Taylor and Burke, ref. 81. ··· length and - - - velocity MBPT calculation by Carter and Kelly, ref. 83. Dots from previous measurement, described in ref. 84.

discrepancy in the calculated ^3D widths, the R-matrix widths being larger. Since the MBPT calculations are low-order calculations, it is expected that the R-matrix widths are probably more accurate in this case. The discrepancy between widths of the $2s2p^2(^4P)np\ ^3P$ resonances is even greater, but this was due to the inclusion in the MBPT calculation of only one decay channel. Cantù et al [84] find evidence of double electron resonance structure near 26 eV as predicted by Carter and Kelly. The β parameter to describe the angular distribution of photoelectrons ejected from the 2s subshell of carbon has been calculated for the ionic states ^4P, ^2D, and ^2S by Chang and Taylor[85] using matrix elements obtained in R-matrix calculations. Their results, shown in Fig. 13, show considerable structure due to resonances. In this figure, the higher members of resonance series and minor series are not displayed.

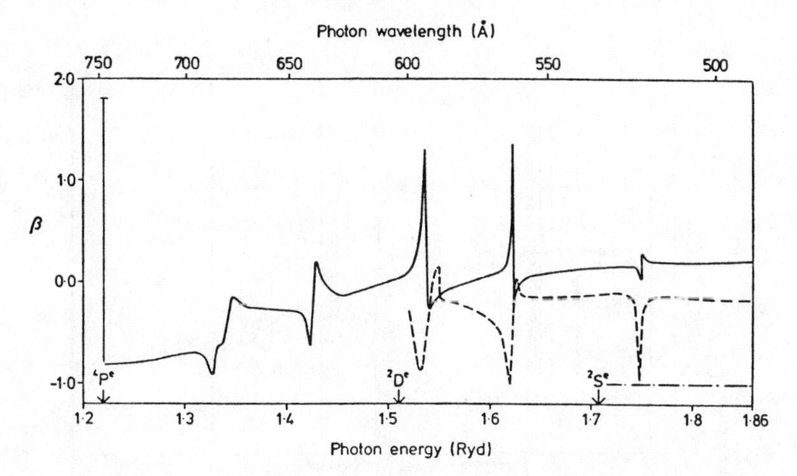

Fig. 13. Angular asymmetry parameter β for the $2s^2$ subshell of carbon. The final state thresholds are $^4P^3$ ——, $^2D^e$ - - - , and $^2S^e$ —.—. Higher members of resonance series are not shown.

R-matrix calculations have also been carried out for the $2p^3\ ^4S$ ground state of atomic nitrogen by Le Dourneuf et al.[82] The final ionic states are $N^+(2s^22p^2\ ^3P;\ 2s2p^3\ ^5S,\ ^3D,\ ^3P,\ ^3S)$ and were calculated in two approximations called CI_o and CI_1. In the CI_o approximation, the ground state of $N^+(^3P)$ is represented by a superposition of $2s^22p^2$ and $2p^4$ and the excited states of $N^+(^5S,\ ^3D,\ ^3P,\ ^3S)$ by a single $2s2p^3$ configuration. In the CI_1 approximation, there is also allowed one virtual excitation from the n = 2 shell to virtual correlation pseudo-orbitals $\bar{3s},\bar{3p},\bar{3d}$ which were chosen to minimise the $N^+(^3P)$ energy. In the CI_1^{exp} calculations, CI_1 results used experimental thresholds. Results of these calculations are shown in Fig. 14 along with close-coupling calculations by Henry[86] and RPAE calculations by Cherepkov et al.[87] The RPAE calculations did not allow for coupling between the $2s^2$ and $2p^3$ subshells and therefore do not show resonance structure. The experimental results by Comes and Elzer[88] are considerably below the R-matrix calculations. However, Le Dourneuf et al. conclude that the normalization of these experimental results is probably in error. The results by Samson and Cairns[89] are one half measured photoionization cross sections for molecular nitrogen. The $N(2s2p^3\ ^5Snp\ ^4P)$ resonance structure is in excellent agreement with the relative experiment of Dehmer et al.[90]

A large number of different theoretical calculations[24,25,30,91-93] have been carried out for the neutral

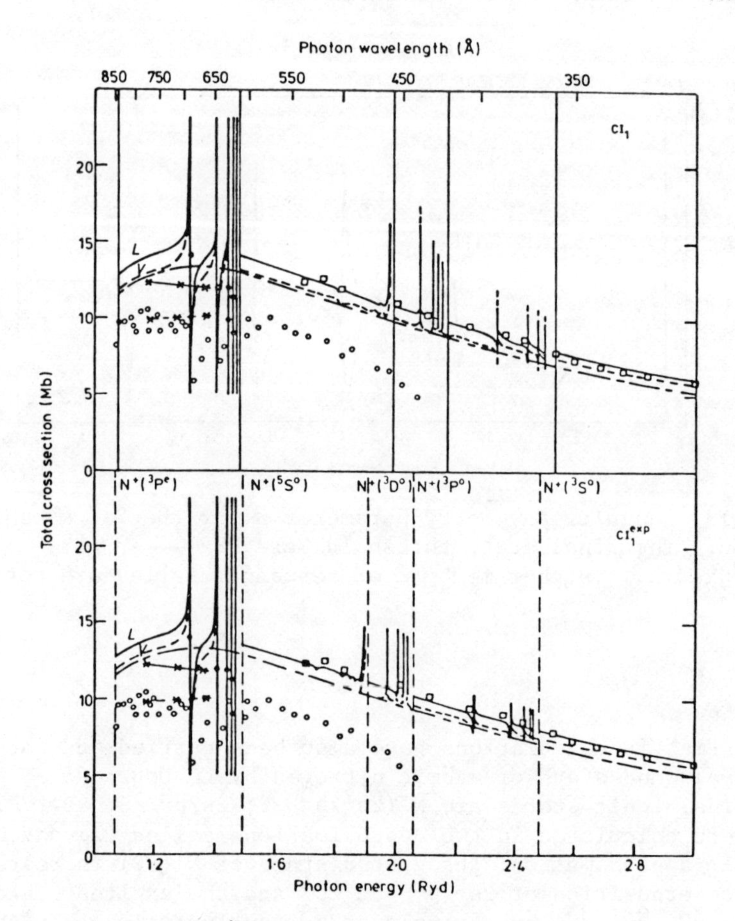

Fig. 14. The total $N(^4S^o)$ photoionization cross section calculated by Le Dourneuf et al., ref. 82, in length (——) and velocity (- - - -) approximations; xxx, close coupling calculations by Henry, ref. 86; —·—, RPAE calculation by Cherepkov et al., ref. 87; O, expt by Comes and Elzer, ref. 88; □, 1/2 of expt for N_2 by Samson and Cairns, ref. 89.

chlorine atom $3p^5\ ^2P$. It has been found[91] that there is very strong coupling between excited states involving different multiplets of the chlorine ion $3p^4\ ^3P$, ^1D, and ^1S. For example, in the ^2D final channel there is strong mixing between $3p^4(^3P)kd(^2D)$, $3p^4(^1D)k'd(^2D)$, and $3p^4(^1S)k''d(^2D)$. Perturbation terms involving these interactions were nearly non-convergent, and a set of coupled integral equations was solved for the MBPT calculation in order to sum these interactions to all orders.[91] Within each subchannel (such as $3p^4(^3P)dk(^2D)$), an LS-coupled potential appropriate for that channel was used. This procedure is equivalent to a close-coupling calculation among the channels. This illustrates some of the difficulty of open-shell systems compared with closed-shell systems. By comparison in argon $3p^6$ there is degeneracy among the 3p electrons and the effective coupling among channels is accounted for by merely using the proper exchange potential, i.e., the correct LS-coupled potential.

The RPAE equations have been generalized by Armstrong,[23] by Cherepkov and Chernysheva[26], and by Starace and Shahabi[25] to be applicable to open-shell atoms. Calculations for Cl have been carried out by Starace and Armstrong[24] omitting the final state couplings among the multiplets of Cl^+ and are shown in Fig. 15 by the dash-dot line. These results are similar to those of Hartree-Fock calculations. The MBPT calculations by Brown et al.[91] are shown by the solid line (length) and the dashed line

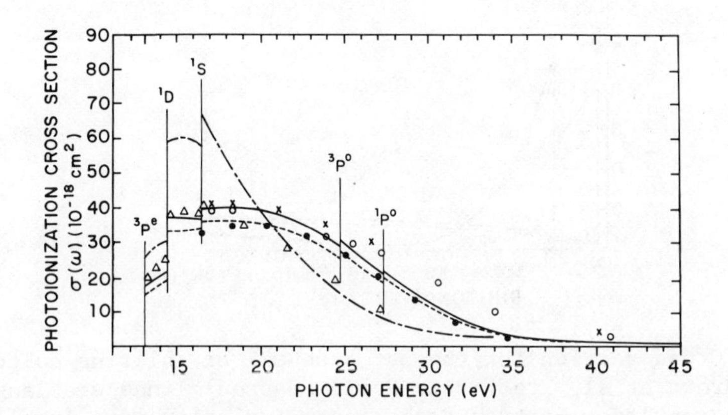

Fig. 15. Photoionization cross section for chlorine. The ^{3}P and ^{1}P edges near 25 eV are due to 3s photoionization.—— (----) is dipole length (velocity) result by Brown et al., ref. 91; —·—·—, Starace and Armstrong, ref. 24; Δ, Cherepkov and Chernysheva, ref. 26; O, Lamoureux and Combet-Farnoux, ref. 92; X, Conneely, ref. 93; ●, Armstrong and Fielder, velocity, ref. 30. In ref. 91 resonances leading to the $^1D^3$, ^{1}S, ^{3}P and ^{1}P edges were calculated.

(velocity). The RPAE calculation by Cherepkov and Chernysheva[26] is shown by the open triangles. This calculation includes the coupling among final state multiplets and is noticeably lower than the Starace and Armstrong curve near threshold. R-matrix calculations by Lamoureux and Combet-Farnoux[92] and close-coupling calculations by Conneely[93] are also shown. Armstrong and Fielder[30] have used the multiconfiguration Hartree-Fock method to correlate the initial state and have devised a method to form a multiconfiguration LS-coupled potential for final states. For open shells, the interchannel interactions are included by use of a K-matrix or close-coupling technique. Their velocity results for chlorine are shown by the solid dots in Fig. 15, with length results slightly lower.

As for argon, the 3s subshell cross section is strongly affected by correlations as shown in Fig. 16. Note that although the Hartree-Fock length and velocity results are close, the more accurate correlated results are qualitatively different. The curves of Fig. 16 were calculated by Brown et al [91] using MBPT.

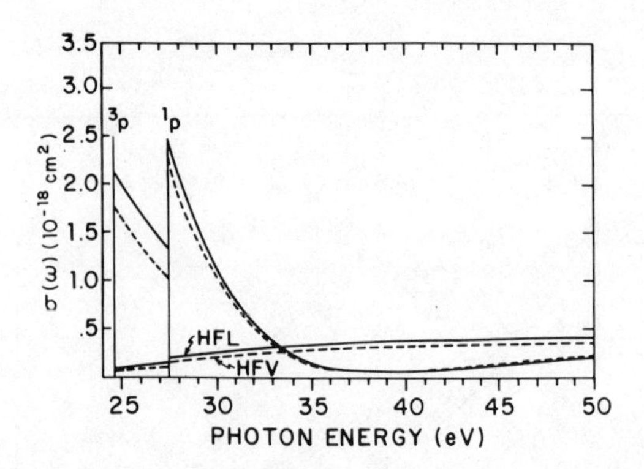

Fig. 16. Cross section for the $3s^2$ subshell of chlorine calculated by Brown et al, ref. 91. Solid (dashed) lines are length (velocity) results. Hartree Fock length (velocity) are labelled HFL (HFV). Resonance structure is given in ref. 91.

A particularly interesting feature of photoabsorption by some open-shell atoms has been demonstrated by experiments on solid and atomic Mn[94,95] and also on atomic Fe, Co, and Ni.[96] For these atoms there is an open 3d shell and very large resonances in the 3d cross section occur due to 3p transitions into the open 3d shell. An example of this structure is shown in Fig. 17. Results are given both for atomic and metallic Mn as measured by Bruhn et al [95] and also for atomic Mn by Connerade et al [94] The dashed line shows a calculation of the photoionization by Davis and Feldkamp[97] using the formalism of Fano.[9] The main resonance near 50 eV is due to the transition $3p^6 3d^5 4s^2\ {}^6S \rightarrow 3p^5 3d^6 4s^2\ {}^6P$. There is, however, noticeable resonance structure due to spin-orbit splitting prior to the main resonance. The

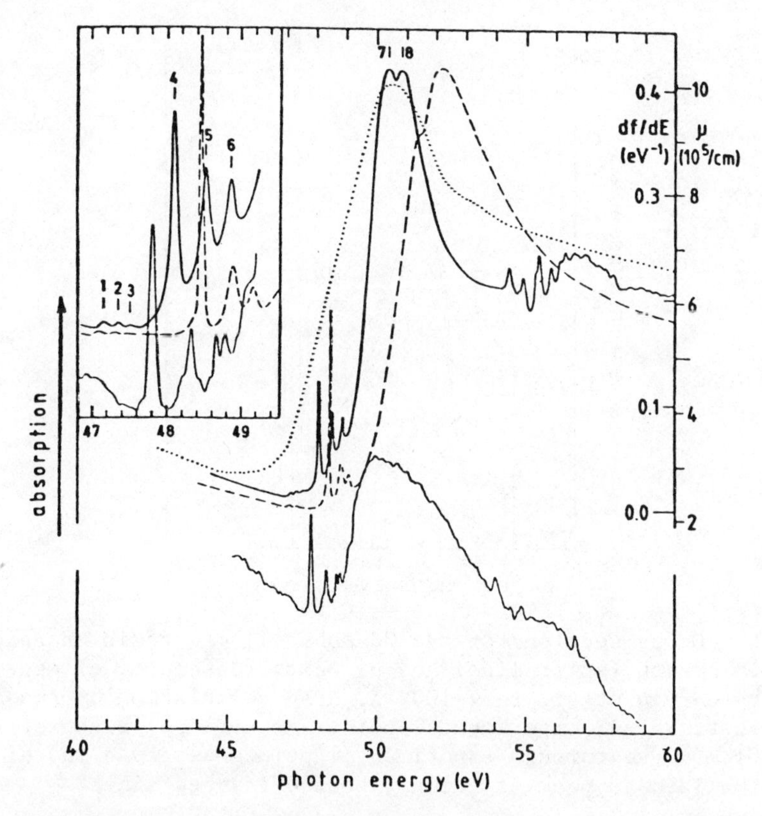

Fig. 17. Absorption of atomic (solid line) and metallic (dotted line) Mn measured by Bruhn et al, ref. 95. The lower solid line is densitometer trace reported by Connerade et al., ref. 94. The oscillator strength df/dE for atomic Mn calculated by Davis and Feldkamp (ref. 97) is given by the dashed line.

calculations by Davis and Feldkamp[97] give a very good description of the spin-orbit resonances near 48 eV. Recently RPAE calculations have been carried out by Amusia, Ivanov and Chernysheva.[98] They use the closed-shell RPAE formalism but take approximate account of multiplet structure by separating electrons into "up" and "down" groups depending on spin projection as was done in earler MBPT calculations[99] for Fe. They did not include spin-orbit splittings. In a recent MBPT calculation, Garvin et al[100] included spin-orbit splitting and also extended the 3d cross section to higher energies so as to include the $3p^5 3d^5 4s^2$ ^7P and ^5P thresholds. Their results are shown in Fig. 18 and show resonance structure due to $3p \rightarrow nd,ns$ transitions leading to the

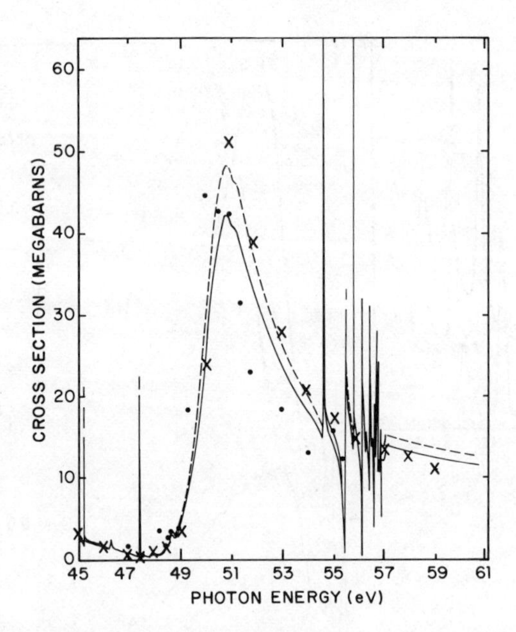

Fig. 18. Cross section for the 3d subshell of atomic Mn calculated in length (solid line) and velocity (dashed line) approximations by Garvin et al, ref. 100 X, RPAE calculation by Amusia et al, ref. 98; •, experimental points measured by Bruhn et al, ref. 101. Curve through experimental points as given in ref. 101 was normalized at peak of velocity curve (dashed line).

$3p^5 3d^5 4s^2\ ^7P$ edge near 57 eV. The results by Amusia et al.[98] are also shown by crosses in Fig. 18. By comparison with the experimental curves in Fig. 17, it is seen that the experimental photoabsorption curve levels off near 53 eV while the theoretical curves continue to drop. In order to study these phenomena in greater detail, Sonntag and co-workers[101] have recently carried out a photoemission experiment which measured both the 3d and 4s partial cross sections with no excitation of the residual ion. Their results for 3d photoemission are also plotted in Fig. 18 with a curve through their points normalized to the peak of the MBPT velocity results. It is seen that the photoemission points agree reasonably with the calculated values. It seems, therefore, that considerable photoionization with excitation is occurring at 52 eV and beyond as shown in Fig. 17. Experimental studies of photoionization with excitation for Mn are now being undertaken by Sonntag and coworkers[102] and also by Kobrin and Shirley[103] and should result in an improved understanding of these processes. Calculations of the $3p \rightarrow 3d$ resonance region in the 3d cross section of Ni have also been carried out by Combet-Farnoux and Ben-Amar[104] using R-matrix theory.

DOUBLE PHOTOIONIZATION

The process of double photoionization in which one photon is absorbed and two electrons ejected is very interesting since it cannot occur in an independent particle model using orthogonal states. As a result, it is difficult to calculate this process, and there has been only a small number of calculations thus far. Early calculations by Brown[105] and by Byron and Joachain[106] used a correlated initial state wave function but neglected correlations in the final state. However, a recent MBPT calculation[107] for He found that at low energies the final state correlations are larger but that there is much cancellation between terms representing correlations in the initial and final states. Calculations have also been carried out for Ne[109-111], Be[112], C[113], and A[111], all using MBPT. The calculations for Ne and A by Carter and Kelly[111] used LS-coupled states which allowed a determination of thresholds corresponding to different multiplets of the doubly charged ion.

In order to calculate double photoionization it is necessary to calculate the dipole length (or velocity) many-particle matrix element $Z(pq \to k'k) = \langle\psi_f|\Sigma z_i|\psi_o\rangle$, where $|\psi_f\rangle$ is a many-particle state with the pair of orbitals pq excited to k'k. Many-body diagrams for $Z(pq \to k'k)$ are given in Fig. 19. Diagrams (a) and (b) are the lowest order diagrams describing ground state correlations and diagrams (c) and (d) are the lowest-order diagrams describing final state correlations. Diagrams (e), (f), and (g) are higher-order diagrams modifying the single-excitation matrix element which were found to be important in calculations of single photoionization. Exchange diagrams are not shown but are understood to be included. Since the outgoing electrons share the energy, we find

$$\sigma^{++}(\omega) = 16\,\frac{\omega}{c}\int_0^{k_{max}} dk\,\frac{|Z(pq \to k'k)|^2}{k'} \quad , \tag{15}$$

using continuum normalization

$$P_k = rR_k \to \cos[kr + \delta_\ell + (q/k)\ln 2kr - \tfrac{1}{2}(\ell-1)\pi] \quad , \tag{16}$$

as $r \to \infty$ with $V(r) \to q/r$. Also

$$k' = [2(\varepsilon_p + \varepsilon_q - \frac{k^2}{2} + \omega)]^{1/2} \quad ,$$

and

$$k_{max} = [2(\varepsilon_p + \varepsilon_q + \omega)]^{1/2} \quad . \tag{17}$$

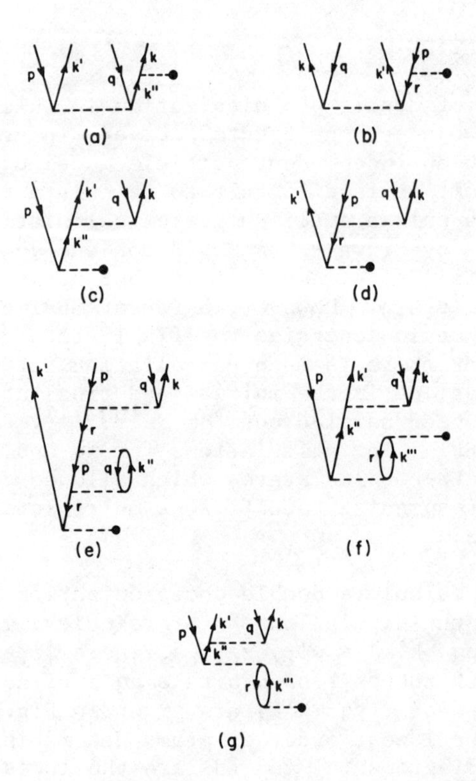

Fig. 19. Diagrams contributing to the double excitation matrix element $Z(pq \rightarrow k'k)$. Solid dots indicate interaction with the dipole operator z. Diagrams (a)-(d) are lowest-order contributions.

For atoms such as neon and argon, an $(np)^2$ pair may be ejected leaving the ion $(np)^4$ ^3P, ^1D, or ^1S. Also, an nsnp pair may also be ejected leaving the ion $ns(np)^5$ ^3P or ^1P. The angular momentum of the outgoing k'k pair couples with the ion to give a total ^1P state and so there is an infinite number of different outgoing waves, contrary to the situation of single photoionization. Calculations[111] for argon giving the contributions of different partial waves are shown in Fig. 20. Although the kpkd contribution is largest it is readily seen that other contributions should be included. It was also found that the 3s3p pair transitions account for more than 20% of the total double cross section. The total cross section σ^{++} for argon is shown in Fig. 21 and compared with experiment. In the range of 100 eV, the double cross section is approximately 25% of the cross section for single photoionization and for Xe it becomes as large as 50% of the single cross section over an appreciable energy range.[114] The calculations used V^{N-1} states and included correlations only among the $3p^6$

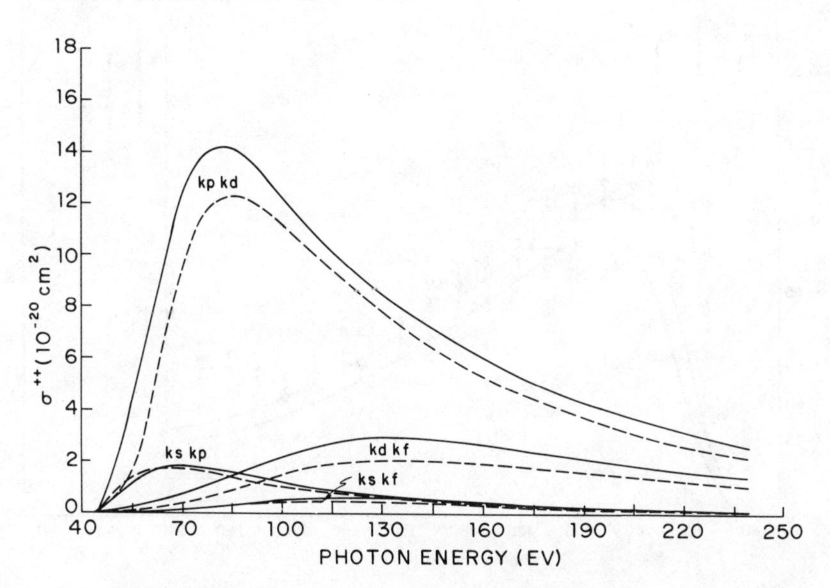

Fig. 20. Double photoionization of argon calculated by Carter and Kelly, ref. 111. Partial-wave cross sections are given for four different $3p^2 \rightarrow k'k$ channels in both dipole length (solid curves) and velocity (broken curves) approximations.

and $3s^2$ subshells. It is noted that the calculated cross section rises too slowly near threshold. This discrepancy was greater for Ne and was attributed to the fact that physically one of the electrons moves in a V^{N-2} type potential.

The rise in the experimental cross section beginning near 220 eV may be due to the approach of the 2p ionization edge. Since correlations with $2p^6$ were omitted in the calculations, the theoretical curve fails to show this rise. It is probably due to interference effects between direct double photoionization and the Auger process described in lowest order by diagram (d) of Fig. 19 with r = 2p. It would be very interesting to study this region further, both theoretically and experimentally.

Recently a careful MBPT study of σ^{++} for helium was carried out using a mixture of V^{N-1} and V^{N-2} single-particle states.[107] The $\ell=1$ states were V^{N-1} states (i.e., calculated in the field of

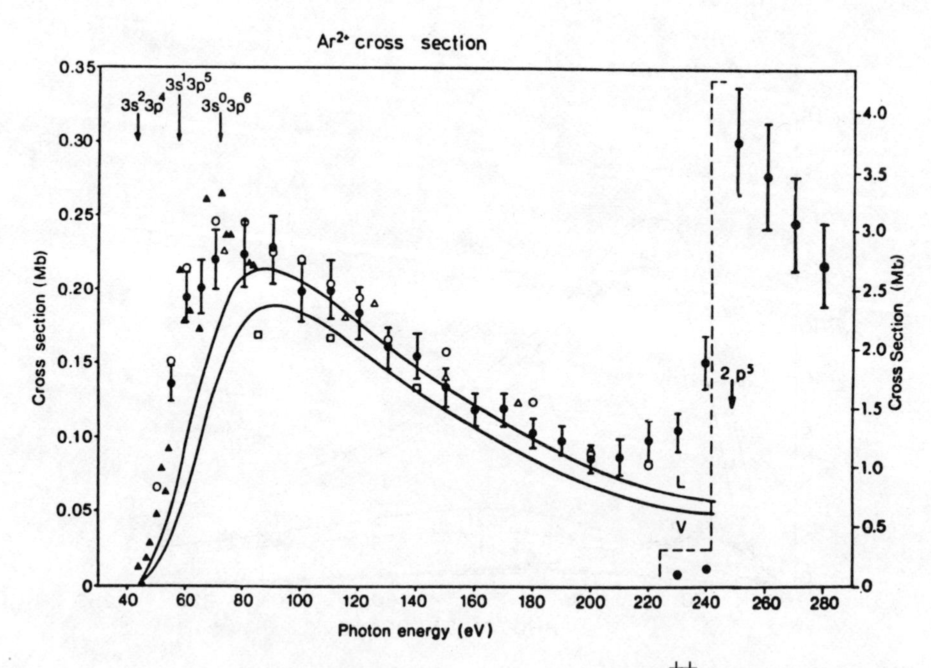

Fig. 21. Double photoionization cross section σ^{++} for argon. Solid curves are dipole length (L) and velocity (V) calculated by Carter and Kelly, ref. 111. Expt: ●, Holland et al., ref. 115; Δ, Schmidt et al., ref. 116; □, Carlson, ref. 117; ■, Lightner et al., ref. 118; ▲, Samson and Haddad, ref. 114; O, Wight and Van der Wiel, ref. 119.

He^+) and excited $\ell = 0$ and $\ell = 2$ states were V^{N-2} (i.e., hydrogenic). It was pointed out that this results in considerable cancellation among higher-order diagrams for final state correlations. The equivalent higher-order diagrams for ground state correlations were calculated to a good approximation by calculating third-order and second-order correlation energy diagrams and using their ratios to estimate effects of higher-order diagrams. Calculations are shown in Fig. 22 and compared with the previous results by Byron and Joachain.[106] The curves labelled L and V contain the higher order corrections to the ground state correlations and bring length and velocity results into close agreement. The velocity curve appears to be less sensitive to higher-order effects. Subsequent experimental results[120] were found to be in excellent agreement with the calculations as is shown in Fig. 23. Beyond 290 eV, the calculations agree reasonably well with the asymptotic result by Amusia et al [121] who predict σ^{++}/σ^{+} as .049.

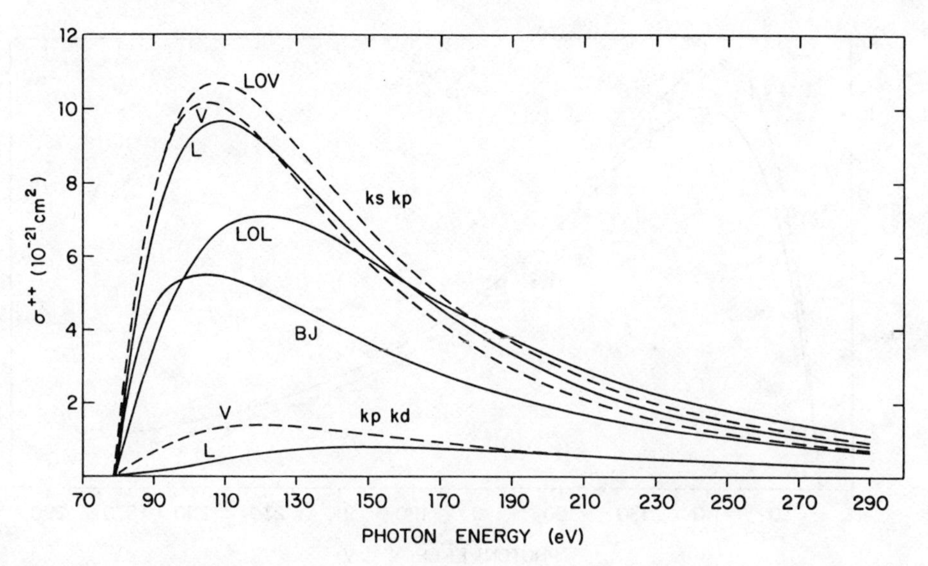

Fig. 22. Calculations of the double photoionization cross section σ^{++} for helium by Carter and Kelly, ref. 107. Curves labelled LOL(LOV) are lowest order length (velocity) results for the kskp channel. Curves labelled L(V) are length (velocity) results containing higher-order corrections for both kskp and kpkd channels. Curve labelled BJ is dipole-velocity result from Byron and Joachain, ref. 106.

PHOTOIONIZATION WITH EXCITATION

The process of photoionization with excitation, in which the remaining ion is left in a bound excited state, has thus far been calculated in only a small number of cases. However, R-matrix calculations have frequently included recouplings which leave the ion in an excited multiplet of the lowest configuration of the ion. There have been MBPT calculations[122] for Fe $3d^64s^2 \rightarrow 3d^7k$ or $3d^64pk$ which have resulted in an increase in the cross section near threshold by approximately 20%. There has also been a recent MBPT calculation by Ishihara et al [123] for neon $1s^22s^22p^6 \rightarrow 1s2s^22p^53s$. Near threshold this process was calculated to reach a maximum of 5% of the single photoionization cross section and then drop with increasing energy for neon.[123]

Early experimental work by Wuilleumier and Krause[124] showed that photoionization with excitation is an important process and constitutes approximately 20% of the total cross section for Ne

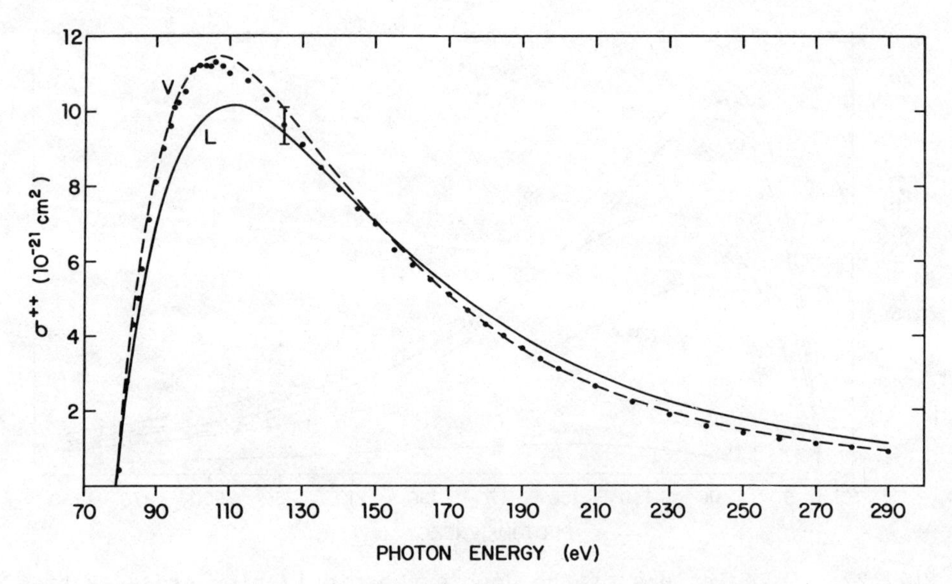

Fig. 23. Double photoionization cross section of helium. Length (solid curve) and velocity (dashed curve) calculated by Carter and Kelly, ref. 107. Experimental data (dots) from Bizau et al., ref. 120.

over a wide range of energies. Experiments using photoelectron spectrometry can measure these cross sections; but, with the exception of helium,[125-130] they have been limited so far to measuring only several points of the photoionization curve. This subject has been reviewed in articles by Krause,[3,130] Shirley,[132] Berkowitz,[133] and Wuilleumier.[4] It is now possible to use synchrotron radiation for these studies, and it is expected that many interesting measurements of photoionization with excitation will be achieved in coming years.

There has been considerable interest and controversy regarding the cross section for photoionization of helium leaving it in the n=2 excited state. Calculations for this cross section have been carried out by Jacobs and Burke[134] using a 56-term Hylleraas initial state wave function and a 1s, 2s, 2p close coupling final state wave function. Chang[135] has used the transition state many-body theory of Chang and Fano.[27] Both calculations are in rather close agreement for the total (2s + 2p) cross section with .10 Mb near threshold (Chang, velocity) and 0.11 Mb (Jacobs and Burke, velocity). However, the two calculations are in complete

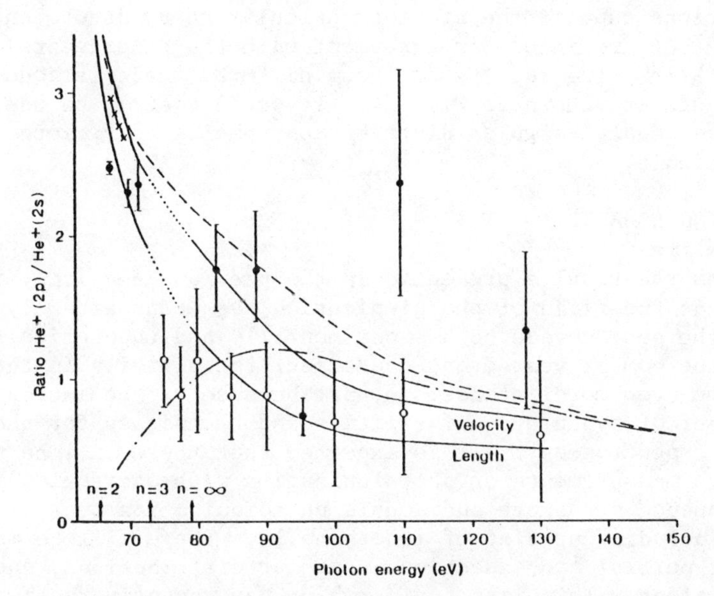

Fig. 24. Photoionization of helium with ratio R of He^+ left in 2p state to He^+ left in 2s state. ———, 6-state calculations by Berrington et al., ref. 136; ---, calculation by Jacobs and Burke (velocity), ref. 134; —.—, calculation by Chang (velocity), ref. 135; ⧳, experiment by Woodruff and Samson, ref. 129; Φ, experiment by Bizau et al., ref. 130.

disagreement for the ratio σ_{2p}/σ_{2s} near threshold, with Burke and Jacobs predicting a ratio near 3 and Chang predicting a ratio near zero. A recent experiment by Bizau et al[130] studied the angular distribution of the ejected electrons and were interpreted as favoring the calculations by Chang. Another experiment by Woodruff and Samson[129] distinguished the 2s and 2p states by observing the 304 Å fluorescent radiation with and without an electric field which quenched the 2s state. Their results favored the calculations by Jacobs and Burke.

Berrington et al[136] have recently calculated this cross section using the R-matrix method. They included 1s, 2s, and 2p eigenstates and also $\bar{3s}$, $\bar{3p}$, and $\bar{3d}$ pseudostates in an expansion of the ^1S initial and ^1P final wave functions. As a check on their calculations, they find that the position, width, and shape of the $2s2p\,^1P$ autoionizing state is accurately described. These calculations support the previous calculation by Jacobs and Burke[134] but are in better agreement with the experimental results.[129,130] The results of these different calculations and experiments are shown in Fig. 24. It would clearly be useful to have experimental angular distribution results very close to the n=2 threshold.

CONCLUDING REMARKS

From the results presented in the previous sections, it is clear that the field of photoionization has been extremely lively during the past decade both experimentally and theoretically. During the coming years, one can expect the activity in this field to become even more intense, in part because of the increase in the number of synchrotron facilities and techniques for photoelectron spectroscopy. It is expected that there will be very extensive measurements of photoionization with excitation including resonance structure and double photoionization over wide energy ranges including studies of inner shells. There will be measurements of partial cross sections, angular distributions, and spin polarizations. This last topic, which has recently received considerable attention,[137-141] was deferred to the paper by W. Johnson in this volume since it depends on relativistic effects.

Other topics which are expected to receive much attention include cross sections for atoms and ions in excited states, cross sections for negative ions,[142] and cross sections for atoms and ions in the presence of electric and magnetic fields.[142]

I am grateful to the U.S. National Science Foundation which supported this work. I am also grateful to Dr. S.L. Carter and Prof. D.J. Larson for helpful discussions.

REFERENCES

1. J.A.R. Samson, Atomic Photoionization, in: "Handbuch der Physik, Vol. 31", W. Mehlhorn, ed., Springer, Berlin (1980).
2. A.F. Starace, Theory of Atomic Photoionization, ibid.
3. M.O. Krause, Electron Spectrometry of Atoms and Molecules, in: "Synchrotron Radiation Research", H. Winnick and S. Doniach, eds., Plenum, New York (1980).
4. F.J. Wuilleumier, Atomic Physics with Synchrotron Radiation, in: "Atomic Physics 7," D. Kleppner and F.M. Pipkin, eds., Plenum, New York (1981).
5. H.P. Kelly and S.L. Carter, Phys. Scr. 21, 448 (1980).
6. P.G. Burke and K.T. Taylor, J. Phys. B8, 2620 (1975).
7. M.Ya. Amusia and N.A. Cherepkov, Case Studies in Atomic Physics 5, 47 (1975).

8. W.R. Johnson and C.D. Lin, Phys. Rev. A 14, 565 (1976).
9. U. Fano and J.W. Cooper, Rev. Mod. Phys. 40, 441 (1968).
10. "Photoionization and Other Probes of Many-Electron Interactions," F.J. Wuilleumier, ed., Plenum, New York (1976).
11. S.T. Manson, Adv. Electron. Electron Phys. 41, 73 (1976);
12. S.T. Manson and A.F. Starace, Rev. Mod. Phys. 54, 389 (1982).
13. M.Ya. Amusia, Adv. At. Mol. Phys. 17, 1 (1981).
14. W. Brandt and S. Lundqvist, Phys. Rev. 132, 2135 (1963).
15. P.L. Altick and A.E. Glassgold, Phys. Rev. 133, A632 (1964).
16. R.J.W. Henry and L. Lipsky, Phys. Rev. 153, 51 (1967).
17. F. Combet-Farnoux, reference 10, p. 407.
18. H.P.Kelly, Phys. Rev. 136, B896 (1964).
19. E.S. Chang and M.R.C. McDowell, Phys. Rev. 176, 126 (1968).
20. A. Dalgarno and G.A. Victor, Proc. Roy. Soc. A291, 291 (1966).
21. M.Ya. Amusia, N.A. Cherepkov, and L.V. Chernysheva, Sov. Phys. JETP 33, 90 (1971).
22. G. Wendin, in: "Vacuum Ultraviolet Radiation Physics," E.E. Koch, R. Haensel, C. Kunz, eds., Pergamon Vieweg, Braunschweig (1974).
23. L. Armstrong, Jr., J. Phys. B 7, 2320 (1974).
24. A.F. Starace and L. Armstrong, Jr., Phys. Rev. A13, 1850 (1976).
25. A.F. Starace and S. Shahabi, Phys. Rev. A25, 2135 (1982).
26. N.A. Cherepkov and L.V. Chernysheva, Phys. Lett. 60A, 103 (1977).
27. T.N. Chang and U. Fano, Phys. Rev. A13, 263 (1976).
28. T.N. Chang, Phys. Rev. A15, 2392 (1977); A18, 1448 (1978).
29. J.R. Swanson and L. Armstrong, Jr., Phys. Rev. A15, 661. (1977); A16, 1117 (1977).
30. L. Armstrong, Jr., and W.R. Fielder, Jr., Phys. Scr. 21, 457 (1980).
31. F.S. Ham, in:"Solid State Physics Vol. 1," F. Seitz and J. Turnbull, eds., Academic, New York (1955).
32. M.J. Seaton, Proc. Phys. Soc. 88, 801 (1966); J. Phys. B11, 4067 (1978).
33. U. Fano, J. Opt. Soc. Am. 65, 979 (1975).
34. C.H. Greene, U. Fano, and G. Strinati, Phys. Rev. A19, 1485 (1979).
35. C.H. Greene, A.R.P. Rau, and U. Fano, Phys. Rev. A, to be published.
36. C.M. Lee and W.R. Johnson, Phys. Rev. A22, 979 (1980).
37. W.R. Johnson, K.T. Cheng, K.-N. Huang, and M. LeDourneuf, Phys. Rev. A22, 989 (1980).
38. W.R. Johnson, C.D. Lin, K.T. Cheng, and C.M. Lee, Phys. Scr. 21, 409 (1980).
39. P.W. Langhoff and C.T. Corcoran, J. Chem. Phys. 61, 146 (1974).
40. W.P. Reinhardt, Comp. Phys. Commun. 17, 1 (1979).

41. A.U. Hazi and K. Reed, Phys. Rev. A24, 2269 (1981).
42. A.Z. Msezane and S.T. Manson, Phys. Rev. Lett. 48, 473 (1982).
43. J. Lahiri and S.T. Manson, Phys. Rev. Lett. 48, 614 (1982).
44. K.A. Brueckner, Phys. Rev. 97, 1353 (1955); 100, 36 (1955).
45. J. Goldstone, Proc. Roy. Soc. A239, 267 (1957).
46. S. Huzinaga and C. Arnau, Phys. Rev. A1, 1285 (1970).
47. H.P. Kelly and R.L. Simons, Phys. Rev. Lett. 30, 529 (1973).
48. R. Haensel, G. Keitel, N. Kosuch, U. Nielsen, and P. Schreiber, J. Phys. 32, Colloque C4, 236 (1971).
49. C.D. Lin, Phys. Rev. A9, 181 (1974).
50. D.J. Kennedy and S.T. Manson, Phys. Rev. A5, 227 (1972).
51. R.G. Houlgate, J.B. West, K. Codling, and G.V. Marr, J. Elect. Spectrosc. 9, 205 (1976).
52. J.A.R. Samson and J.L. Gardner, Phys. Rev. Lett. 33, 671 (1974).
53. J.B. West, P.R. Woodruff, K. Codling, and R.G. Houlgate, J. Phys. B9, 407 (1976).
54. M.Y. Adam, F. Wuilleumier, N. Sandner, S. Krummacher, V. Schmidt, and W. Mehlhorn, Jpn. J. Appl. Phys. 17, 170 (1978).
55. M.Ya. Amusia and V.K. Ivanov, Phys. Lett. 59A, 194 (1976).
56. M.J. Lynch, K. Codling, and A.B. Gardner, Phys. Lett. 43A, 213 (1973).
57. L. Torop, J. Morton, and J.B. West, J. Phys. B9, 2035 (1976).
58. G. Wendin, Phys. Lett. 51A, 291 (1975).
58a. R. Rabe, K. Radler, and H.W. Wolff, in: "VUV Radiation Physics," E.E. Koch, R. Haensel, and C. Kunz, eds., Pergamon Vieweg, Braunschweig (1974).
59. H.P. Kelly, S.L. Carter, and B.E. Norum, Phys. Rev. A25, 2052 (1982).
60. A.E. Hansen, Mol. Phys. 13, 425 (1967).
61. A.W. Fliflet, Ph.D. disseration, U. of Virginia, 1975, unpublished.
62. S.L. Carter and H.P. Kelly, J. Phys. B11, 2467 (1978).
63. M.Ya. Amusia, V.K. Ivanov and L.V. Chernysheva, Phys. Lett. 59A, 191 (1976).
64. A.W. Fliflet, R.L. Chase, and H.P. Kelly, J. Phys. B7, L443 (1974).
65. M.H. Hecht and I. Lindau, Phys. Rev. Lett. 47, 821 (1981).
66. A. Zangwill and P. Soven, Phys. Rev. Lett. 45, 204 (1980).
67. A.W. Fliflet and H.P. Kelly, Phys. Rev. A10, 508 (1974).
68. P.L. Altick, Phys. Rev. 169, 21 (1968).
69. J. Dubau and J. Wells, J. Phys. B6, 1452 (1973).
70. G.N. Bates and P.L. Altick, J. Phys. B6, 653 (1973).
71. C.H. Greene, Phys. Rev. A23, 661 (1981).

72. Z. Altun, S.L. Carter, and H.P. Kelly, J. Phys. B, to be published.
73. V.L. Carter, R.D. Hudson, and E.L. Breig, Phys. Rev. A4, 821 1971).
74. R.W. Ditchburn and R.D. Hudson, Proc. Roy. Soc. A256, 53 (1960).
75. G.H. Newsom, Proc. Phys. Soc. 87, 975 (1966).
76. T.J. McIlrath and R.J. Sandeman, J. Phys. B5, L217 (1972).
77. J.P. Connerade, M.A. Baig, W.R.S. Garton, and G.H. Newsom, Proc. Roy. Soc. A21, 1 (1980).
78. Z. Altun, S.L. Carter, and H.P. Kelly, to be published.
79. M.W.D. Mansfield and G.H. Newsom, Roc. Roy. Soc. A357, 77 (1977).
80. M. LeDourneuf, Vo Ky Lan, P.G. Burke, and K.T. Taylor, J. Phys. B8, 2640 (1975).
81. K.T. Taylor and P.G. Burke, J. Phys. B9, L353 (1976).
82. M. LeDourneuf, Vo Ky Lan, and C.J. Zeippen, J. Phys. B12, 2449 (1979).
83. S.L. Carter and H.P. Kelly, Phys. Rev. A13, 1388 (1976).
84. A.M. Cantù, M. Mazzoni, M. Pettini, and G.P. Tozzi, Phys. Rev. A23, 1223 (1981).
85. E.S. Chang and K.T. Taylor, J. Phys. B11, L507 (1978).
86. R.J.W. Henry, J. Chem. Phys. 48, 3635 (1968).
87. N.A. Cherepkov, L.V. Chernysheva, V. Radojevic, and I. Pavlin, Can. J. Phys. 52, 349 (1974).
88. F.J. Comes and A. Elzer, Z. Naturf. 23a, 133 (1968)
89. J.A.R. Samson and R.B. Cairns, J. Opt. Soc. Am. 55, 1035 (1965).
90. P.M. Dehmer, J. Berkowitz, and W.A. Chupka, J. Chem. Phys. 60, 2676 (1974).
91. E.R. Brown, S.L. Carter, and H.P. Kelly, Phys. Rev. A21, 1237 (1980).
92. M. Lamoureux and F. Combet-Farnoux, J. Phys. (Paris) 40, 545 (1979).
93. M.J. Conneely, Ph.D. Thesis, London University, 1969 (unpublished).
94. J.P. Connerade, M.W.D. Mansfield, and M.A.P. Martin, Proc. Roy. Soc. A350, 405 (1976).
95. R. Bruhn, B. Sonntag, and H.W. Wolff, Phys. Lett. 69A, 9 (1978).
96. B. Sonntag, J. de Physique, Colloque C4, 39, 9 (1978).
97. L.C. Davis and L.A. Feldkamp, Phys. Rev. A17, 2012 (1978).
98. M.Ya. Amusia, V.K. Ivanov, and L.V. Chernysheva, J. Phys. B14, L19 (1981).
99. H.P. Kelly and A. Ron, Phys. Rev. A5, 168 (1972).
100. L.J. Garvin, E.R. Brown, S.L. Carter, and H.P. Kelly, to be published.

101. R. Bruhn, E. Schmidt, H. Schröder, and B. Sonntag, to be published.
102. B. Sonntag, private communication.
103. P. Kobrin and D. Shirley, private communication.
104. F. Combet-Farnoux and M. Ben Amar, Phys. Rev. A21, 1975 (1980).
105. R.L. Brown, Phys. Rev. A1, 586 (1970).
106. F.W. Byron, Jr., and C.J. Joachain, Phys. Rev. 164, 1 (1967).
107. S.L. Carter and H.P. Kelly, Phys. Rev. A24, 170 (1981).
108. T.N. Chang, T. Ishihara, and R.T. Poe, Phys. Rev. Lett. 27, 838 (1971).
109. T.N. Chang and R.T. Poe, Phys. Rev. A12, 1432 (1975).
110. S.L. Carter, Ph.D. Dissertation, Univ. of Virginia, 1976.
111. S.L. Carter and H.P. Kelly, Phys. Rev. A16, 1525 (1977).
112. P. Winkler, J. Phys. B10, L693 (1977).
113. S.L. Carter and H.P. Kelly, J. Phys. B9, 1887 (1976).
114. J.A.R. Samson and G.N. Haddad, Phys. Rev. Lett. 33, 875 (1974).
115. D.M.P. Holland, K. Codling, J.B. West, and G.V. Marr, J. Phys. B12, 2465 (1979).
116. V. Schmidt, N. Sandner, H. Kuntzemuller, P. Dhez, F. Wuilleumier, and E. Källne, Phys. Rev. A13, 1748 (1976).
117. T.A. Carlson, Phys. Rev. 156, 142 (1967).
118. G.S. Lightner, R.J. Van Brunt, and W.D. Whitehead, Phys. Rev. A4, 602 (1971).
119. G.R. Wight and M.J. Van der Wiel, J. Phys. B9, 1319 (1976).
120. J.M. Bizau, F. Wuilleumier, D. Ederer, P. Dhez, S. Krummacher, and V. Schmidt, J. Phys. (Paris), to be published.
121. M.Ya. Amusia, E.G. Drukarev, V.G. Gorshkov, and M.P. Krazachkov, J. Phys. B8, 1248 (1975).
122. H.P. Kelly, Phys. Rev. A6, 1048 (1972).
123. T. Ishihara, J. Mizuno, and T. Watanable, Phys. Rev. A22, 1552 (1980).
124. F. Wuilleumier and M.O. Krause, in: "Proceedings of the International Conference on Electron Spectrometer, Asilomar, Calif., 1971," North-Holland, Amsterdam, 1972.
125. J.A.R. Samson, Phys. Rev. Lett. 22, 693 (1969).
126. M.O. Krause and F. Wuilleumier, J. Phys. B5, L148 (1972).
127. F. Wuilleumier, M.Y. Adam, N. Sandner, and V. Schmidt, J. Phys. (Paris), Lett. 41, L373 (1980).
128. P.R. Woodruff and J.A.R. Samson, Phys. Rev. Lett. 45, 110 (1980).
129. P.R. Woodruff and J.A.R. Samson, Phys. Rev. A25, 848 (1982).
130. J.M. Bizau, F. Wuilleumier, P. Dhez, D.L. Ederer, T.N. Chang, S. Krummacher, and V. Schmidt, Phys. Rev. Lett. 48, 588 (1982).
131. M. Krause, in ref. 10, p. 137.

132. D.A. Shirley, J. de Phys. (Paris) 39, Colloque C-4, C4-35 (1978).
133. J. Berkowitz, "Photoabsorption Photoionization, and Photoelectron Spectroscopy," Academic, New York, 1979.
134. V.L. Jacobs and P.G. Burke, J. Phys. B5, L67 (1972).
135. T.N. Chang, J. Phys. B13, L551 (1980).
136. K.A. Berrington, P.G. Burke, W.C. Fon, and K.T. Taylor, J. Phys. B, to be published.
137. N.A. Cherepkov, J. Phys. B10, L653 (1977).
138. K.T. Cheng, K.N. Huang, and W.R. Johnson, J. Phys. B13, L45 (1980).
139. K.N. Huang, Phys. Rev. A22, 223 (1980).
140. U. Heinzmann, J. Phys. B13, 4353 (1980); 13, 4367 (1980).
141. G. Schönhense, U. Heinzmann, J. Kessler, and N.A. Cherepkov, Phys. Rev. Lett. 48, 603 (1982).
142. W.A.M. Blumberg, R.M. Jopson, and D.J. Larson, Phys. Rev. Lett. 40, 1320 (1978).

A TIME-DEPENDENT LOCAL DENSITY APPROXIMATION OF ATOMIC PHOTOIONIZATION

A. Zangwill*

Department of Physics
Brookhaven National Laboratory
Upton, NY 11973 U.S.A.

INTRODUCTION

In recent years considerable effort has been expended in the calculation of accurate atomic photoionization cross sections. Interestingly, it has proved necessary to proceed far beyond the simplest Hartree-Fock approximation (HFA) to achieve this goal. The most extensive calculations to date have employed either the random phase approximation with exchange (RPAE)[1] or many-body perturbation theory (MBPT)[2]. Both of these approaches build systematically on standard Hartree-Fock theory and yield results which are generally in excellent agreement with experiment.

Concurrent with these developments, the density functional formalism has emerged as an alternative to a Hartree-Fock based description of the electronic structure of atoms, molecules and solids. In its most common form as a local density approximation (LDA) the question of atomic photoionization can again be addressed. Here too, one finds that the simplest approach fails to adequately characterize the experimental results in many cases. However, a recent straightforward generalization of density functional theory to time-dependent phenomena has been applied successfully to the problem of the optical response of atoms. In particular, highly accurate photoionization cross sections can be readily obtained. The purpose of the present article is to review this time-dependent local density approximation (TDLDA), illustrate its scope and limitations and compare it to the more familiar Hartree-Fock based methods.

*Permanent address, Dept. of Physics, Polytechnic Institute of New York, Brooklyn, New York 11201.

LOCAL DENSITY APPROXIMATION

Density functional theory is an exact formulation of the ground state properties of a many-particle system in which the particle density, n(x), plays a central role. Since the formal aspects of the theory have been reviewed elsewhere[3], only the features essential to atomic photoionization will be discussed here. In particular, the exact ground state charge density of an atom of atomic number Z is determined by solution of the following Hartree-like equations:

$$[-\nabla^2 + V_{eff}(x)]\psi_i(x) = \varepsilon_i\psi_i(x) \tag{1a}$$

$$V_{eff}(x) = -\frac{Ze^2}{|x|} + e^2\int dx' \frac{n(x')}{|x-x'|} + V_{xc}(x) \tag{1b}$$

$$n(x) = \sum_{i=1}^{Z} |\psi_i(x)|^2. \tag{1c}$$

$V_{xc}(x)$ is the exchange-correlation potential, a local function of space defined as the functional derivative of the exchange-correlatio energy functional:

$$V_{xc}(x) = \frac{\delta}{\delta n(x)} E_{xc}\{n(x)\}. \tag{2}$$

The exact solution of (1) is very simple for an atom if $E_{xc}\{n(x)\}$ is known for the system of interest. Unfortunately, this quantity is generally unknown and approximations must be introduced. By far the most commonly used choice is the local density approximation wherein one takes

$$V_{xc}(x) \to V_{LDA}(x) = \frac{\delta}{\delta n(x)} \int dx\, n(x)\varepsilon_{xc}\{n(x)\}. \tag{3}$$

Here, $\varepsilon_{xc}\{n(x)\}$ is the total exchange and correlation energy per particle of the homogeneous electron gas. Since this quantity is known essentially exactly[4], $V_{LDA}(x)$ can be evaluated and typically appears parameterized in terms of elementary functions[4,5]. Note that the LDA contains no adjustable parameters; the well-known one parameter X-α method of Slater[6] can be viewed as a phenomenological approach to the inclusion of exchange and correlation in $V_{eff}(x)$.

The physics of the LDA is that each point in the atom is considered to possess the attributes of a uniform electron gas at the density of the point in question. Clearly, the LDA is not an "approximate Hartree-Fock" scheme since HF treats exchange exactly and neglects correlation whereas the LDA treats both on an equal footing, albeit in a non-transparent fashion for an inhomogeneous system. Nonetheless, experience has shown that remarkably good results can be obtained for a wide range of ground state properties in atoms, molecules and solids[7]. However, photoionization removes an atom from its ground state and therefore is formally outside the purview of current density functional theory. Indeed, the crucial role of the excited states is made clear by the standard Golden Rule expression:

$$\sigma(\omega) = 4\pi^2 \alpha\hbar\omega \sum_{i,f} |\langle f|U^{ext}|i\rangle|^2 \, \delta(\hbar\omega - \varepsilon_f + \varepsilon_i) \quad (4)$$

Here, α is the fine structure constant and $U^{ext}(x)$ is the dipole operator. Unfortunately, the suggestively denoted quantities, ε_i and $\psi_i(x)$ in (1a) cannot be identified with a Koopman's removal energy and Slater determinant wave function as in HF theory. Nonetheless, regarding the solution of (1) as providing a basis set, one can construct an LDA to atomic photoionization using (4).

In Figure 1 the photoabsorption cross section of xenon near the 4d threshold is shown in the LDA, HF[8] and experiment[9]. Unlike HF, the length and velocity forms of the matrix elements are identical in the LDA because $V_{eff}(x)$ is completely local. This figure illustrates that neither the LDA nor the simplest HFA is sufficient to describe

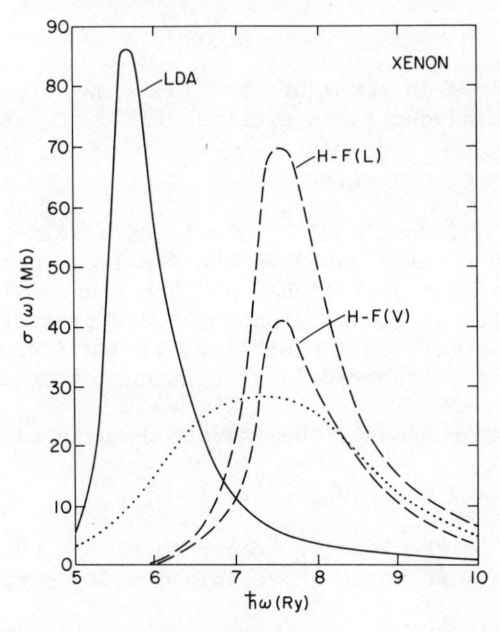

Fig. 1. Xenon 4d total photoabsorption. LDA (solid line), HFA (dashed line), experiment from Ref. 9 (dotted line).

the excitation process. In this case, as in many others, it is necessary to properly account for the electronic polarizability of th system of atomic electrons.

A particularly convenient improved approximation to this end can be obtained by use of self-consistent, first order, time-dependent perturbation theory. The essential physics to be included is that the external field distorts the atomic charge cloud (by admixture of excited orbitals) which in turn creates an electrostatic potential acting on the system. The self-consistent response of the electrons produces a mean field which reflects the atomic dielectric properties and alters the photoionization amplitudes. If this linear response approach is applied to the HFA one obtains precisely the RPAE. In what follows we consider the same approximation applied to the LDA. Given this parallelism, emphasis will be placed on direct comparisons with the RPAE.

TIME-DEPENDENT LOCAL DENSITY APPROXIMATION

Formalism

The complex, frequency dependent, dipole polarizability, $\alpha(\omega)$, is a useful quantity to study since it is simply related to the total photoabsorption cross section by

$$\sigma(\omega) = 4\pi \frac{\omega}{c} \operatorname{Im} \alpha(\omega) \tag{5}$$

and directly defined in terms of the dipole moment of the charge density disturbance induced by an external field at frequency ω:

$$\alpha(\omega) = e\int dx\ z\ \delta n(x|\omega). \tag{6}$$

In 1975, Ahlberg and Goscinski[10] formulated a self-consistent linear response approach to $\alpha(\omega)$ based on the X-α local potential. Their formulation was of the variational-perturbation variety and some approximate results at finite frequency were presented for $\alpha(i\omega)$. Here we follow the formulation of Zangwill and Soven[11] which is particularly amenable to numerical calculations of cross sections.

In LDA linear response, the induced density is given by

$$\delta n(x|\omega) = \int dx' \chi_0(x,x'|\omega)\ U^{ext}(x') \tag{7}$$

where $\chi_0(x,x'|\omega)$ is the Fourier transform of the LDA density-density response function. As noted above this density perturbation induces a potential

$$U^{ind}(x|\omega) = \int dx' \frac{\delta n(x'|\omega)}{|x-x'|} + \frac{\partial V_{xc}(x)}{\partial n(x)} \delta n(x|\omega). \tag{8}$$

The first term is the classical electrostatic response (actually this is a quasi-static approximation) and the second reflects the contribution from exchange-correlation effects. To achieve self-consistency, a new charge density is computed according to

$$\delta n(x|\omega) = \int dx' \chi_o(x,x'|\omega) \left[U^{ext}(x') + U^{ind}(x'|\omega) \right] . \tag{9}$$

Equations (8) and (9) constitute the TDLDA which requires their simultaneous solution. The resulting $\delta n(x|\omega)$ is then combined with (5) and (6) to yield the total photoabsorption. Alternatively, it can be shown[11] that the Golden Rule expression, (4), remains valid if $U^{ext}(x)$ is replaced by the complex, frequency dependent effective field

$$U^{eff}(x|\omega) = U^{ext}(x) + U^{ind}(x|\omega). \tag{10}$$

All of the atomic many-body effects of the TDLDA dielectric response are thereby built into an effective driving field for the electrons.

At zero frequency, the TDLDA equations have been solved[10-13] to yield excellent results for closed shell atomic polarizabilities. At finite frequencies, calculations are facilitated by exploiting the identity[11]:

$$\chi_o(x,x'|\omega) = \sum_{i\ occ} \Big[\psi_i^*(x)\psi_i(x')G(x,x'|\varepsilon_i + h\omega) + \psi_i(x)\psi_i^*(x')G^*(x,x'|\varepsilon_i - h\omega) \Big] . \tag{11}$$

$G(x,x'|E)$ is the LDA one-particle Green function which contains the complete sum over excited states and is obtained by solution of the LDA Schrodinger-like equation,

$$\left[-\nabla^2 + V_{eff}(x) - E \right] G(x,x'|E) = \delta(x-x'). \tag{12}$$

For closed shell atoms, all the above equations become one-dimensional and therefore are no more difficult to solve than the original LDA equations, (1).

Direct contact with the perturbed orbital approach[10,13] can be made if we consider an alternative expression for the induced density, $\delta n(x|\omega)$. Under the influence of the external field, the time-independent eigenfunctions $\psi_i(x)$ evolve according to

$$\psi_i(x|t) = \psi_i(x) + \psi_i^{+}(x|\omega)e^{-i\omega t} + \psi_i^{-}(x|\omega)e^{+i\omega t} \tag{13}$$

The induced density is then constructed as

$$\delta n(x|t) = \sum_{i=1}^{Z}\left\{|\psi_i(x|t)|^2 - |\psi_i(x)|^2\right\} \tag{14}$$

The positive frequency component of this density is then given by

$$\delta n(x|\omega) = \sum_{i=1}^{Z}\left\{\psi_i^{*}(x)\psi_i^{+}(x|\omega) + \psi_i(x)\psi_i^{-*}(x|\omega)\right\} \tag{15}$$

while the perturbed orbitals themselves satisfy the modified Sternheimer equations[13]:

$$\begin{aligned} \left[\nabla^2 - V_{eff}(x) + \varepsilon_i + \hbar\omega\right]\psi_i^{+}(x|\omega) &= U^{eff}(x|\omega)\psi_i(x) \\ \left[\nabla^2 - V_{eff}(x) + \varepsilon_i - \hbar\omega\right]\psi_i^{-}(x|\omega) &= U^{eff}(x|\omega)\psi_i(x) \end{aligned} \tag{16}$$

Simultaneous solution of (8), (15) and (16) again constitutes a self-consistent TDLDA solution for the induced density.

As a consequence of the local nature of $V_{LDA}(x)$ the TDLDA possesses several desirable qualities not available in the usual RPAE. From a purely practical point of view, the differential equations approach permits all the excited orbitals (bound and continuum) to be automatically included through the Green function regardless of the size of the atom. In the language of perturbation theory, one easily sums all the dipolar particle-hole channels which couple the atomic shells. More importantly however, the effective field, $U^{eff}(x|\omega)$, depends explicitly on space and frequency which allows a pictorial representation of atomic dynamics. In particular, one can define an effective local dielectric function,

$$\varepsilon^{-1}(x|\omega) = \frac{U^{eff}(x|\omega)}{U^{ext}(x)} \tag{17}$$

although, of course, the true atomic dielectric response is non-local according to (7).

Before passing on to specific results three characteristics of the TDLDA should be noted. First, at zero frequency it is exact within density functional theory[12] and therefore provides an explicit test of the LDA itself. Second, the cross section satisfies the f-sum rule[10] and the compressibility sum rule[11]:

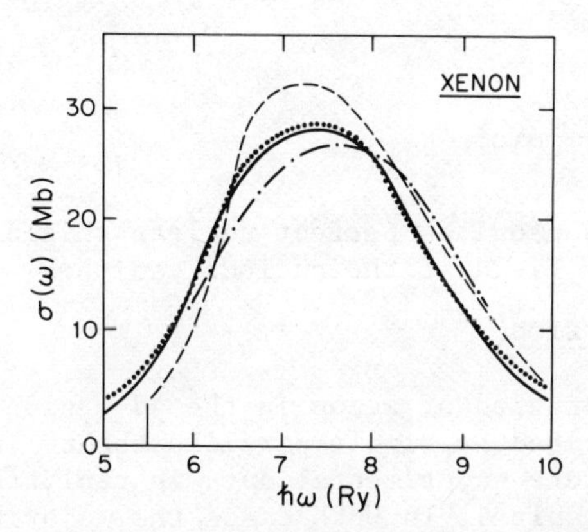

Fig. 2. Xenon 4d total photoabsorption. RPAE (dashed line), TDLDA (solid line), experiment (dotted line) and TDLDA without induced XC (dashed-dot line).

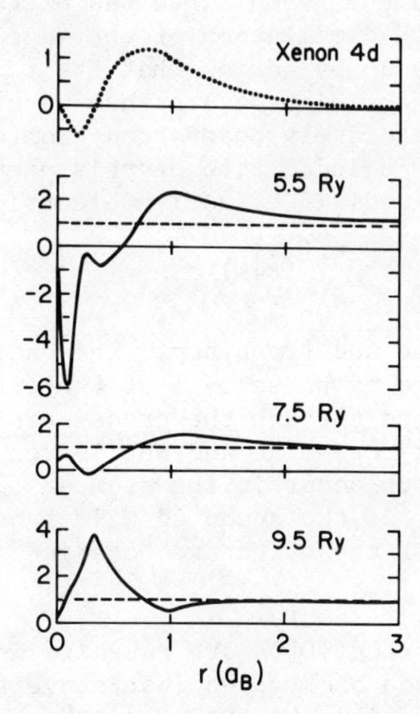

Fig. 3. Xe 4d radial wave function and effective electric field at three photon energies.

$$\int_0^\infty d\omega \; \sigma(\omega) = \frac{2\pi^2 e^2}{mc} N$$
$$\int_0^\infty d\omega \; \frac{\sigma(\omega)}{\omega^2} = \frac{2\pi^2}{c} \alpha(o) \tag{18}$$

Finally, the TDLDA provides perfect nuclear shielding[14], i.e., the effective electric field at the nucleus vanishes.

Results and Discussion

The photoabsorption of xenon in the 4d region was seen to be inadequately described by single particle theory in Figure 1. In Figure 2 the absolute experimental data is replotted along with the TDLDA and RPAE[8] results. In both cases the majority of the discrepancy with experiment is removed. Also illustrated in Figure 2 is the TDLDA result if the induced exchange-correlation potential (second term in (8)) is omitted from the calculation. This has a small but recognizable effect on the cross section and indicates that most of the physics is included in the classical induced field if the $\psi_i(x)$ (from which it is constructed) contain quantum-mechanical exchange and correlation at the level of the LDA.

The broad bump in the xenon curve has often been discussed in terms of a "collective" oscillation of the 4d electrons. Some light can be shed on this issue by noting that the redistribution of spectral oscillator strength which occurs when passing from the LDA to the TDLDA can be qualitatively understood from a simple harmonic oscillator model[15]. One indication of this phenomenon is illustrated in Figure 3 where the radial "effective electric field", defined as

$$E^{eff}(r|\omega) = -\frac{d}{dr} \text{Re } U^{eff}(x|\omega) \tag{19}$$

is plotted versus space and frequency. The oscillator analogy is reflected in the change of phase by π of the effective field as the excitation energy passes through the cross section maximum. This figure also shows that screening and anti-screening of the external field (dashed line) both occur in the atom at fixed frequency. The spatial dividing point is the xenon 4d wave function maximum which locates the position of the effective "charges" which induce a dipolar field in the atom.

Other features of the TDLDA are revealed in the total photoabsorption cross section of krypton just above threshold as illustrated in Figure 4. Unlike the RPAE result[8], the TDLDA cross section is seen to vanish at a threshold energy below the experimental onset. In the LDA, the Coulomb self-interaction of the atomic electrons is not explicitly subtracted away as in the HFA. Hence, all the electrons move in a Z-electron neutral field which decays exponentially

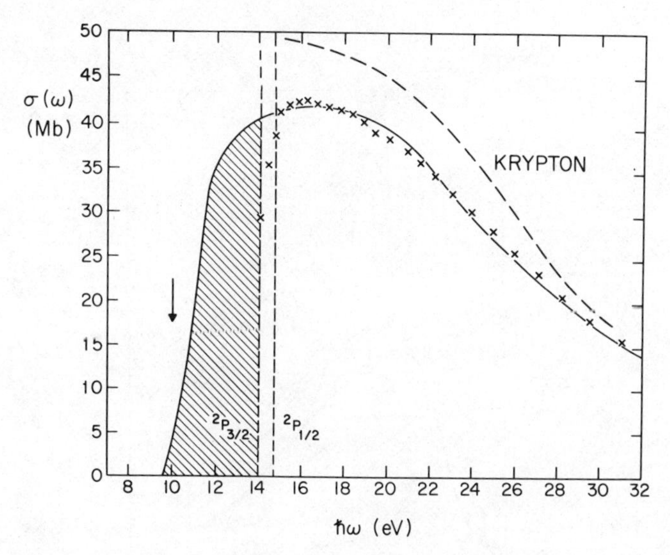

Fig. 4. Krypton photoabsorption. TDLDA (solid line), RPAE (dashed line), experiment from Ref. 18, (crosses), first I.P. denoted by arrow.

to zero at large distances and does not support Rydberg states. The lack of an asymptotic Coulomb tail leads to cross sections which vanish at threshold. The presence of the self-interaction also yields unphysically low values for the eigenvalues, ε_i, which determine the photoionization thresholds. Nonetheless, above 15 eV the TDLDA cross section is remarkably accurate. In light of the f-sum rule, (18), one expects that the calculated pseudo-continuum absorption at 10-14 eV should account for the discrete absorption strength in the real atom. Comparison with the high resolution oscillator strength measurements of Geiger[16] reveal this to be true to within 5%[17].

Returning to 4d absorption, the case of barium has attracted a tremendous amount of attention in recent years. One reason is the dramatic failure of the RPAE in the near threshold region as illustrated in Figure 5. The source of this discrepancy was identified by Wendin[20] as the effect of atomic relaxation around the photoexcited 4d core hole. An RPAE calculation which addressed this problem by using continuum orbitals appropriate to the relaxed ion[21] is also indicated in Figure 5. Great improvement is obtained by this procedure although systematic discrepancies remain. In this light, Kelly et.al. have recently noted[22] that similar results are possible using only a Tamm-Dancoff + relaxation (modified HF) scheme rather than the full RPAE + relaxation approach.

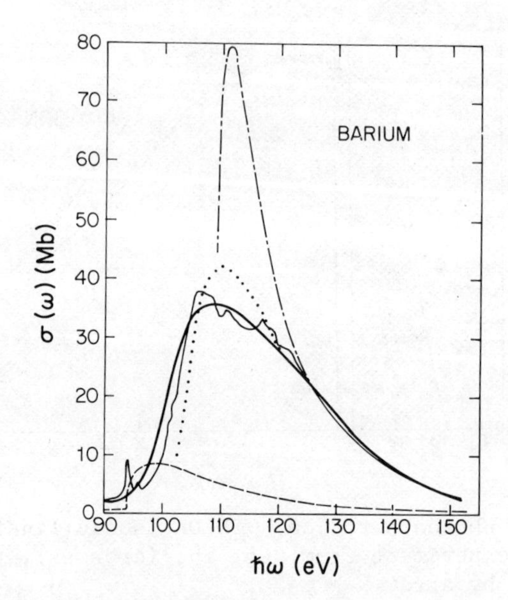

Fig. 5. Barium photoabsorption. LDA (dashed line), TDLDA (heavy solid line), RPAE (dashed-dot line), RPAE + relax (dotted line), experiment from Ref. 19 (light solid line), normalized with use of f-sum rule.

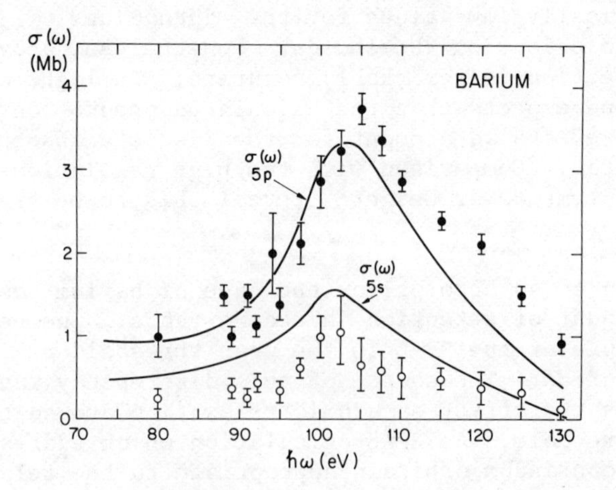

Fig. 6. Barium 5s and 5p partial cross sections. TDLDA (solid lines), data from Ref. 24 normalized to calculation for one curve only.

The TDLDA total photoabsorption cross section for barium is also illustrated in Figure 5. This calculation contains no relaxation effects; all orbitals are found in the ground state LDA potential. Clearly, relaxation (a real physical effect!) manifests itself differently in the LDA and HFA. One crude way to include some type of relaxation into a TDLDA calculation is by use of a transition state[6] potential in place of $V_{LDA}(x)$. This artifice is known to yield energy eigenvalues which closely approach experimental ionization energies and a Coulomb tail (-e/2r) develops. Unfortunately, TDLDA calculations[14] indicate that the corresponding cross sections can be substantially worse that the depicted neutral potential results. It is fair to say that the role of relaxation in the TDLDA is not yet understood. Finally, vis a vis the RPAE-TDLDA comparison, it has recently been suggested that relaxed HF potential barrier effects are somehow simulated by LDA self-interaction effects[23]. Clearly, more systematic comparisons are in order.

The success of the TDLDA for the barium total photoabsorption is mirrored in the resonant behaviour of the valence level partial photoemission cross sections near the 4d threshold (Figure 6). This

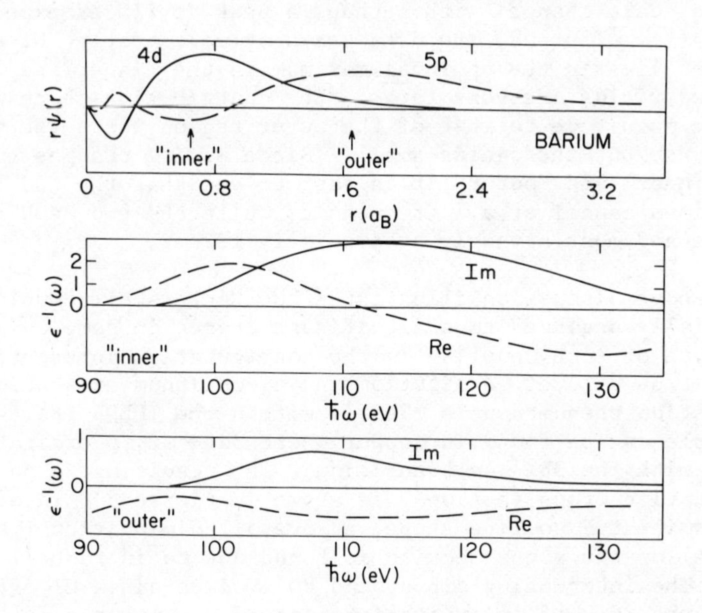

Fig. 7. Barium 4d and 5p radial wave functions and frequency dependence of effective dielectric function at two radial distances from nucleus.

effect is sometimes referred to as an autoionization phenomenon[25]. While it is true that this description is possible, a consistent physical picture emerges through the use of the space and frequency dependence of $U^{eff}(x|\omega)$ which unifies all the barium results. The top panel of Figure 7 shows the radial wave function for two relevant barium orbitals. The radii at which the 4d and 5p orbitals have maximum amplitude are denoted by "inner" and "outer", respectively. The harmonic oscillator dynamics of the highly polarizable 4d shell are best studied with the frequency dependence of the effective dielectric function, (17), evaluated at the "inner" radius (middle panel). Since the factorization

$$|\langle f|U^{eff}(\omega)|i\rangle|^2 = |\langle f|\mathrm{Re}\; U^{eff}(\omega)|i\rangle|^2 + |\langle f|\mathrm{Im}\; U^{eff}(\omega)|i\rangle|^2 \quad (20)$$

is valid in the TDLDA[11], the dramatic asymmetric barium resonance clearly reflects the response of the 4d electrons to an effective field created by the 4d shell itself.

The outer valence orbitals are dynamically inert. They simply respond to the fields created by the wildly oscillating 4d shell in accordance with their own spatial character. Hence, the dependence of $\varepsilon^{-1}(\omega)$ in the "outer" region (bottom panel) should be appropriate. However, the dielectric response in the atomic extremeties is rather weak (note scale change) with perhaps a peak at 110 eV expected for the 5p shell. Instead, the data peaks at more nearly 102 eV. The explanation lies in the 5p local maximum in the "inner" region where the induced fields are very large. The contribution there is comparable in magnitude to that of the outer region where the wave function is large but the fields small. Since $\psi_{5p}(x)$ changes sign between "inner" and "outer" it is easy to see that the two Im $U^{eff}(\omega)$ contributions cancel at all frequencies while the two Re $U^{eff}(\omega)$ contributions add most strongly at precisely 102 eV.

The above discussion illustrates how "intershell coupling" can be pictorially analyzed to a significant degree in some cases. Another type of such coupling is the configuration interaction (CI) between a true discrete excitation and a continuum excitation. This autoionization phenomenon is clearly within the TDLDA framework. A nice example can be found in copper where 3d $\rightarrow \varepsilon f, \varepsilon p$ excitations interfere with the 3p $\rightarrow$ 4s transition. The resulting 3d partial photoionization cross section[26] is shown in Figure 8. In addition to the prominent Fano line shape, an overall diminuition (relative to the LDA) of the cross section is found due to intrashell 3d polarization. The interesting dip around 80 eV is again a CI effect, but this time the 3d $\rightarrow \varepsilon f, \varepsilon p$ excitations interfere with the continuum channels, 3p $\rightarrow \varepsilon s, \varepsilon d$.

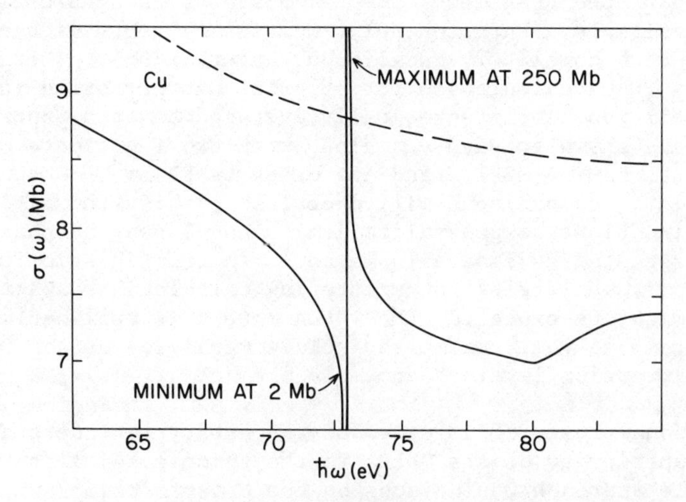

Fig. 8. Copper 3d partial photoemission cross section, TDLDA (solid line), LDA (dashed line).

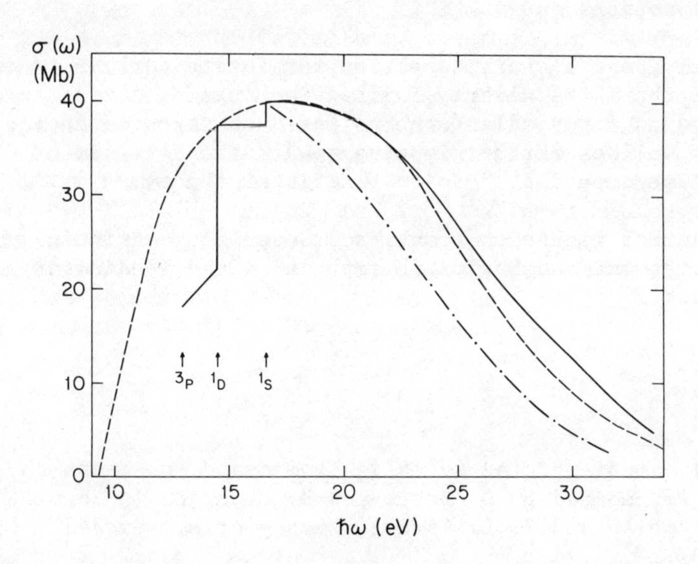

Fig. 9. Chlorine total photoabsorption. TDLDA (dashed line), MBPT (solid line) and RPAE (dashed-dot line).

Prospects

Following the RPAE developmental history, an open-shell TDLDA would be desirable. Unfortunately, multiplet theory within the LDA itself is not a completely unambiguous undertaking[7]. Therefore, an exploratory TDLDA calculation for chlorine was performed in which the $3p^5$ shell was simply spherically averaged at each opportunity. The result is shown in Figure 8 along with two legitimate open-shell calculations, one by MBPT[27] and the other by RPAE[28]. Once again, more systematic comparisons will doubtless clarify the situation. In other directions, a generalization to non-linear response applicable to laser studies is available now[17], a relativistic TDLDA is forthcoming[14], while electron scattering calculations await development. As might be expected, the TDLDA method is applicable to other finite systems as well, examples include metallic surfaces[29], semiconductor inversion layers[30] and molecules[31].

Additional research is certainly necessary to understand the apparent superiority of the TDLDA to the formally similar RPAE. One possible approach would focus on the cross section sum rules. Since the stated rules, (18), do not clearly distinguish between the two methods[11] it is possible that higher frequency moments will indicate the source of the differences. Since these higher moments do not correspond to observables, the "experimental" values will have to be provided by Monte Carlo calculations of the corresponding correlation functions[32].

Finally, a necessary direction for future work is to seek examples where the TDLDA clearly fails. Such examples would provide the testing ground for similar schemes based on improved choices for $V_{xc}(x)$[7] as well as further impetus toward the development of a rigorous time-dependent density functional theory.

The author wishes to thank Paul Soven for continuing support and encouragement and David Liberman and Göran Wendin for stimulating correspondence.

REFERENCES

1. M. Ya. Amusia, in Advances in Atomic and Molecular Physics, vol. 17, edited by D. Bates and B. Bederson (Academic, NY 1981).
2. H. P. Kelly and S. L. Carter, Phys. Scrip. (Sweden) 21, 448 (1980).
3. W. Kohn and P. Vashishta, in Theory of the Inhomogeneous Electron Gas, edited by S. Lundqvist and N. March (Plenum, NY 1982).
4. S. H. Vosko, L. Wilk and M. Nusair, Can. J. Phys. 58, 1200 (1980).
5. O. Gunnarsson and B. I. Lundqvist, Phys. Rev. B13, 4274 (1976).

6. J. C. Slater, The Self-Consistent Field for Molecules and Solids: Quantum Theory of Molecules and Solids, vol. 4, (McGraw Hill, NY, 1974).
7. A. R. Williams and U. von Barth, in Theory of the Inhomogeneous Electron Gas, edited by S. Lundqvist and N. March (Plenum, NY 1982).
8. M. Ya. Amusia and N. A. Cherepkov, Case Stud. Atom. Phys. 5, 47 (1975).
9. J. B. West and J. Morton, Atom. Data and Nuc. Tab. 22, 103 (1978).
10. R. Ahlberg and O. Goscinski, J. Phys. B 8, 2149 (1975).
11. A. Zangwill and P. Soven, Phys. Rev. A 21, 1561 (1980).
12. M. J. Stott and E. Zaremba, Phys. Rev. A 21, 12 (1980).
13. G. D. Mahan, Phys. Rev. A 22, 1780 (1980).
14. D. Liberman, LASL (unpublished).
15. A. Zangwill and P. Soven, J. Vac. Sci. Technol. 17, 159 (1980).
16. J. Geiger, Z. Phys. A 282, 129 (1977).
17. A. Zangwill, thesis (University of Pennsylvania, Philadelphia 1981).
18. G. V. Marr and J. B. West, Atom. Data Nuc. Tab. 18, 497 (1976).
19. P. Rabe, K. Radler and H. W. Wolff, in Vacuum UV Radiation Physics, edited by E. Koch et. al. (Vieweg-Pergamon, Berlin, 1974).
20. G. Wendin, in Photoionization and Other Probes of Many-Electron Interactions, edited by F. J. Wuilleumier (Plenum, NY 1976).
21. M. Ya. Amusia, in Atomic Physics 5, edited by R. Marrus et. al. (Plenum, NY 1977).
22. H. P. Kelly, S. L. Carter and B. E. Norum, Phys. Rev. A 25, 2052 (1982).
23. Z. Crljen and G. Wendin, to be published.
24. M. H. Hecht and I. Lindau, Phys. Rev. Lett. 47, 821 (1981) and private communication.
25. G. Wendin, in Vacuum UV Radiation Physics, edited by E. Koch et.al. (Vieweg-Pergamon, Berlin, 1974).
26. A. Zangwill and P. Soven, Phys. Rev. B 24, 4121 (1981).
27. E. R. Brown, S. L. Carter and H. P. Kelly, Phys. Lett. 66A, 290 (1978).
28. N. A. Cherepkov and L. V. Chernysheva, Phys. Lett. 60A, 103 (1977).
29. P. J. Feibelman, Phys. Rev. B 12, 1319 (1975).
30. T. Ando, Z. Phys. B26, 263 (1977).
31. Z. Levine and P. Soven, to be published.
32. P. C. Martin, in Many-Body Physics, edited by C. DeWitt and R. Balian (Gordon and Breach, NY 1968).

SHAPE RESONANCES IN THE PHOTOIONIZATION SPECTRA OF FREE AND CHEMISORBED MOLECULES

Torgny Gustafsson

Department of Physics and the Laboratory for Research on
the Structure of Matter, University of Pennsylvania
Philadelphia, PA 19104-3859

INTRODUCTION

With the emergence of synchrotron radiation as a powerful spectroscopic tool, it has become possible to study in detail the photoionization cross sections immediately above the various ionization potentials, not only of core levels (which can also be done by x-ray absorption) but also of all the different valence levels (though the use of photoelectron spectroscopy). The most prominent features in these cross sections are in molecular physics known as shape resonances and constitute major modifications of a simple hydrogenic photoabsorption cross section. Essentially the same phenomenon is sometimes in the condensed matter literature referred to as XANES (x-ray absorption near edge structure), to contrast it with EXAFS (extended x-ray absorption fine structure). This very rich field has over the last five years developed into several different directions. The purpose of the present paper is to summarize some fairly recent experiments concerned with shape resonances in free diatomic molecules and to show how this knowledge can be used in surface physics to extract information otherwise difficult to obtain. Many of the properties of free molecules are retained after chemisorption. It is possible to perform much more accurate theoretical calculations for free molecules than for the molecule-surface complex. By cautiously extrapolating these calculations to the chemisorbed phase much new insight has been gained.

EXPERIMENTAL CONSIDERATIONS

The first evidence for shape resonances in molecules was obtained by measuring the x-ray absorption coefficient, using Bremsstrahlung, around core levels[1,2]. As the photoexcitation cross sections of valence levels are very low at these (keV) energies, the total photoabsorption cross section is essentially identical to the photoabsorption cross section of the core level under study. With closely overlapping core levels, or with valence levels, one has to be able to separate overlapping continua, i.e. using photoelectron spectroscopy instead. Monochromatized Bremsstrahlung is too weak for reasonable counting rates. Some of the very first data showing evidence for shape resonances (although not interpreted as such at the time) were obtained using line sources[3] and using (e,2e) spectroscopy[4]. The latter technique has been utilized very effectively by Brion and coworkers[5]. It is based on a simple idea. A high energy E (several keV) electron is incident on the sample. It loses an energy ΔE to an electron bound with an energy E_B in the target. This electron may be emitted and emerges with a kinetic energy $\Delta E-E_B$. By using two electron energy analyzers and tuning them to an energy of $E-\Delta E$ and $\Delta E-E_B$ respectively, the two emerging electrons are detected in coincidence. By sweeping ΔE, the energy dependence of the probability to excite the bound electron can be measured. It can be shown that (in the limit of zero momentum transfer) these probabilities are simply related to the partial photoionization probability. Among the many advantages of this technique is the fact that in a single instrument, by just adjusting a few voltages, one can change the excitation energy (ΔE). This should be contrasted with photon based techniques, where change of excitation (photon) energy may involve tedious change-overs of gratings. Among the disadvantages of the present generation of (e,2e) spectrometers are moderate energy resolution and long data accumulation times. Also, attempts to use this spectroscopy as a competitive tool for surface studies have so far been unsuccessful.

Synchrotron radiation has been used by many groups for shape resonance studies. Its photon energy spectrum is continuous and the intensity generally very high compared to other continuum sources. Resolution can in most monochromators be traded for intensity or vice versa. The field of monochromator technology is presently in a state of very rapid development. Novel designs with much improved performances are being suggested every year. As an example, let us consider the toroidal grating monochromator (TGM) now being constructed by the Penn-Oak Ridge collaboration at the National Synchrotron Light Source (NSLS) at Brookhaven. It uses four different gratings, one prefocussing mirror, and for low photon energies, also a postfocussing mirror. By using a movable exit slit, much improved resolution is obtained. Due to the slow divergence of the beam (the entire instrument is well over 10 m long) the sample position can be fixed nevertheless. The instrument will cover the entire photon energy range from 5 to 1000 eV. Realistic ray tracing calculations show a (calculated) energy resolution of better than 30 meV up to 1000 eV at 1 mrad horizontal acceptance (for most of this range, this value is actually less than 10 meV and at the carbon K-edge $\lesssim 4$ meV). At large horizontal acceptance, the transmitted intensity increases as do the imaging errors. However, at 10 mrad acceptance better than 100 meV resolution should be achieved both at the carbon and oxygen K-edges.

SHAPE RESONANCES IN FREE DIATOMIC MOLECULES

The core level (nitrogen 1s) photoabsorption spectrum of N_2 in Fig. 1 demonstrates many of the aspects of shape resonances we will discuss later[6]. The spectrum shows lots of structure both above and below the nitrogen K-edge (409.9 eV). Peaks B-E are

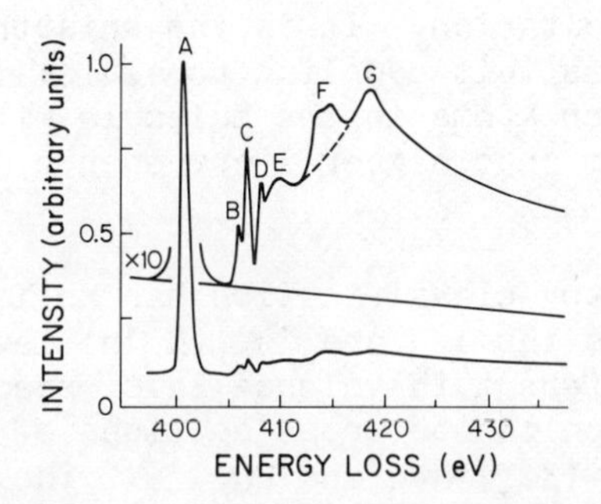

Fig. 1. Electron energy loss spectrum of N_2 in the region around the N 1s threshold (after ref. 6). The measured quantity is qualitatively very similar to the photoabsorption cross section.

(weak) Rydberg states. Peak F is a two electron excitation and lies beyond the scope of the present discussion. Peak A, which falls below the K-edge, is extremely strong. The region above the edge is dominated by a weaker, but broader, peak, G. Neither of these features have any correspondence in the atomic nitrogen photoabsorption spectrum. Instead, they are manifestations of the interaction between the outgoing photoelectron and the highly anisotropic molecular potential. Theoretical calculations have shown that A is a shape resonance of π_g symmetry, whereas G has σ_u symmetry[7-9]. The origin of these resonances can be understood in two complementary ways. In a multiple scattering picture, one would argue that at some particular photoelectron kinetic energy, the deBroglie wavelength would be right for strong Bragg scattering between the two nuclei. This would lead to a quasistationary state. Excitation to these states leads to an enhanced cross section. Alternatively, one may view these states as due to competition between the centrifugal (repulsive) and nuclear (attractive) part of the total electrostatic potential. The centrifugal part varies as $l(l+1)/r^2$, where r is measured <u>from the molecular centre</u>. The nuclear attraction varies as $1/r$, where r here is measured <u>from either of the two nuclei</u>. This result is a total potential with a ridge. The photoelectron is created in the attractive region around the nuclei, and has to overcome this potential barrier in order to escape. For a suitable set of parameters, it is possible to support quasibound states in this potential. Theoretical work has shown that the resonance A in Fig. 1 is due to the l=2 component whereas G is mainly due to l=3. In an atom, no mechanism exists for this $s \rightarrow d(f)$ excitation. It is the anisotropy of the molecular field that makes this coupling possible. The wave created on one of the nitrogen atoms in the molecule will be scattered by the spectator atom into higher l-components, making a transition possible[7-9].

The $u \leftrightarrow g$ symmetry classification has no importance in core level spectroscopy as the $1\sigma_g$ and $1\sigma_u$ (N 1s) levels are degenerate. The situation is different in valence level spectroscopy. In Fig. 2 we show the photoionization cross sections of the three outermost valence levels in N_2 ($3\sigma_g$, $1\pi_u$ and $2\sigma_u$)[10]. The σ_u shape resonance should be observed in this energy range. We identify the peak at $\hbar\omega \sim 26$ eV in the $3\sigma_g$ cross section as this resonance. It is here located 12.5 eV above threshold, which compares to ~11.5 eV in the core level spectrum (Fig. 1). The $3\sigma_g$ cross section shows also another peak at low photon energies ($\hbar\omega \sim 23$ eV). This is the same feature as F in Fig. 1 and is due to a two-electron excitation ($1\pi_u^4\ 3\sigma_g^2\ 1\pi_g^0 \rightarrow 1\pi_u^3\ 3\sigma_g^1\ 1\pi_g^1\ (ns\sigma)^1 \rightarrow 1\pi_u^4\ 3\sigma_g^1\ 1\pi_g^0\ (ns\sigma)^0$ + free electron where $ns\sigma$ is an intermediate Rydberg state. The most interesting cross section in Fig. 2 is however that of the $2\sigma_u$ level. This cross section is quite flat and shows no evidence for the σ_u resonance. This result is expected based on dipole selection rules. $\sigma_u \rightarrow \sigma_u$ transitions are dipole forbidden; the flat cross section gives us confidence in the theoretical arguments.

The same two shape resonances are also observed in electron-molecule scattering. Here, they are shifted to higher kinetic energy. This is due to the added electron, which produces more repulsive potential, in fact pushing the π resonance well into the continuum[11] and making the σ resonance very broad and weak[12].

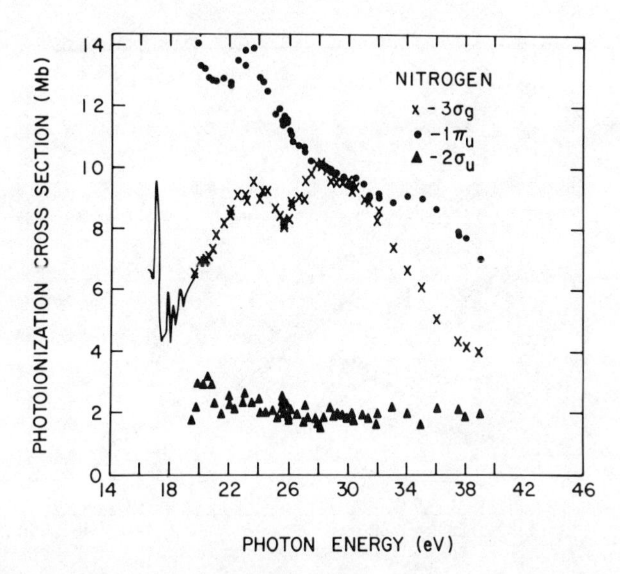

Fig. 2. Partial photoionization cross sections of the three outermost valence orbitals of free N_2[10].

The two resonances in Fig. 1 are observed in almost all diatomics studied to date. In Fig. 3 we show data for the valence levels of CO[10]. The 5σ level shows the same features as its counterpart in N_2, $3\sigma_g$: A strong resonance around $\hbar\omega = 24$ eV, and on the low energy side a two electron excitation. The CO case is in one important aspect different from N_2: The 4σ level (corresponding to $2\sigma_u$ in N_2) shows a strong resonance at ~ 32 eV. As CO is heteronuclear, a classification according to odd-even symmetry breaks down and all σ initial states may be dipole coupled to the resonance. The 1π cross section is monotonically decreasing with photon energy. This level can couple to the resonance, but the resonance enhancement is obscured by the much stronger $\pi \rightarrow \delta$ transitions[8,13].

The kinetic energy of the photoelectron on resonance is 9.6 eV for CO 5σ and 12.3 eV for CO 4σ[10]. This difference has been attributed to differences in the initial state wave function. This suggests that the discussion above which focussed solely on the behavior in the final state is somewhat simplified. It is

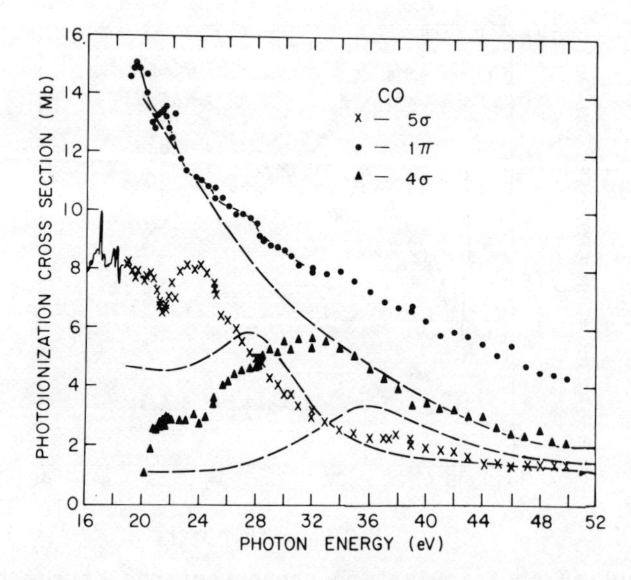

Fig. 3. Partial photoionization cross sections of the three outermost valence orbitals of free CO[10]. The dashed curves are the results of a theoretical calculation[9].

nevertheless interesting to try to study the dependence of the resonance energy on a single variable in order to understand shifts in resonance energy from molecule to molecule. In a multiple scattering picture, one expects that the photoelectron wave-vector k on resonance should fulfill a condition of the type $k \cdot d$ = constant (d is the internuclear separation). The kinetic energy of the photoelectron on resonance, E_R, would then be proportional to d^{-2}. In Fig. 4 we show data on E_R from valence excitation studies of several different diatomics[14]. The data follow the naive prediction quite nicely and show correctly that in F_2 (d = 1.417 Å) the σ resonance should be pulled into the discrete part of the spectrum.

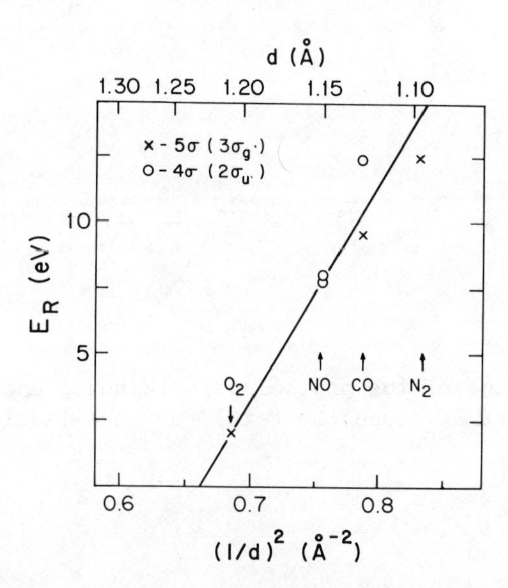

Fig. 4. Dependence of the photoelectron kinetic energy on resonance (E_R) on d^{-2}, where d is the ground state internuclear separation. The data are for valence orbital excitations[14].

A different viewpoint is due to Hitchcock and Brion[15], who instead focus on the nuclear charges of the constituent atoms (Z). The attractive part of the potential will be stronger as Z increases (for a given l quantum number). To first order, we would therefore expect E_R to scale with Z. The quite comprehensive studies of Hitchcock and Brion give convincing support for this correlation (Fig. 5). Data on small hydrocarbons can be incorporated in this trend, if the nuclear charges on the hydrogen atoms are neglected. The scaling parameter is then called ΣZ_{HA}, where the subscript stands for "heavy atom". It is particularly satisfying that the correlation describes the F_2 data so well. A similar good correlation has been obtained between the l=2 (π) resonances and ΣZ_{HA}. Finally, we may note that for O_2 the resonance falls in the continuum ($E_R > 0$) for the valence levels (Fig. 4) but in the discrete (Fig. 5) for the core levels.

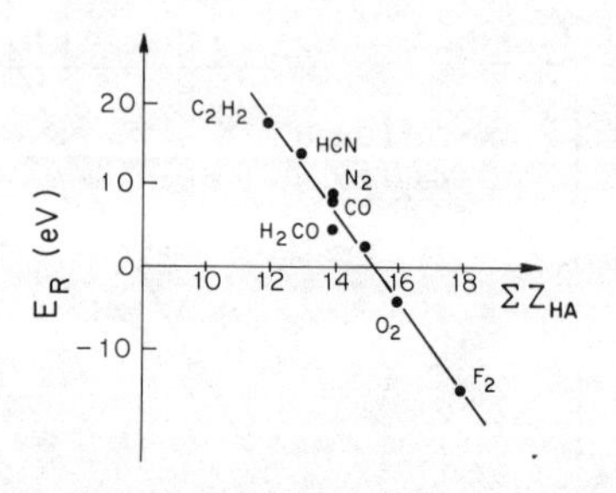

Fig. 5. Dependence of the photoelectron kinetic energy on resonance (E_R) on ΣZ_{HA} (see the text) for core level excitations[15].

It should be obvious from Figs. 4 and 5 that there is convincing evidence both for a dependence on internuclear separation and on the direct influence of the charge distribution. Dehmer and coworkers have studied the appearance of shape resonances in the various vibrational sublevels in free CO and N_2[16]. They have found that the resonance energy shifts from sublevel to sublevel in a way that can be accounted for with a model that focusses on the dependence on d. As we will see below, the shifts between free and adsorbed molecules are hard to account for in terms of d-dependence and are instead attributed to changes in the screening upon adsorption.

SHAPE RESONANCES IN CHEMISORBED MOLECULES

If the orientation of a molecule could be kept fixed in space relative to the exciting photon beam the photoionization event can be made very specific. Selection rules, more powerful than those based on odd-even symmetry, can be used. One way of experimentally establishing such conditions is to let the molecules adsorb onto a solid surface. On a surface, the orientation of the molecule is presumably fixed, or, at the very least, has just a few fairly well defined values. This provides us with a powerful tool to actually determine the orientation of adsorbed molecules. Consider, for example, photoexcitation from an initial state i that has σ symmetry[13]: The (angle-resolved) photoexcitation cross section for excitation to the final state f will be:

$$\frac{d\sigma}{d\Omega} \propto G_\sigma(\Omega)\cos^2\theta_A + G_\pi(\Omega)\sin^2\theta_A\sin^2\phi_A + G_{\sigma\pi}(\Omega)\cos\theta_A\sin\theta_A\sin\phi_A$$

where θ_A is the angle between the A vector and the molecular axis and ϕ_A is an angle in the xy plane that defines the orientation of the electric vector.

The first two terms describe the excitation strength from the σ initial state to $\sigma(\pi)$ final states. The third term is an interference term, which describes the way the $\sigma \rightarrow \sigma$ and $\sigma \rightarrow \pi$ transition amplitudes beat against each other. In gas phase photoemission (randomly oriented molecules), one has to integrate over all possible θ_A and ϕ_A so the interference term integrates out.

If $\theta_A = 0^\circ$, i.e. the electric vector is parallel to the molecular axis, the last two terms drop out and only σ final states are probed. Transitions to the π shape resonance are consequently forbidden in this geometry. Conversely, for $\theta_A = 90^\circ$ (the electric vector perpendicular to the axis) only the second term remains and only π final states are probed. Transitions to the $\pi(\sigma)$ resonances are strongly allowed (forbidden). These rules are in principle independent of the angular acceptance of the electron energy analyzer. These selection rules were first applied by Allyn et al[17] in order to study the controversial case of the c(2x2) phase of CO on Ni(100). The results are shown in Fig. 6. The photon energy range is such that transitions from the molecular 4σ level are observable. When A is parallel with the surface a smoothly increasing cross section is observed. With the light incident at an angle of 45°, a huge resonance is observed, which we identify with the σ resonance. We conclude that the molecular axis is perpendicular to the surface, a conclusion supported by later work with other techniques.

The arguments above were based on the symmetries of the free molecule. In the c(2x2)CO phase on Ni(100) it is known that the molecule adsorbs on top of a Ni atom[18]. The correct symmetry classification is then C_{4v}. The way the free CO wavefunctions mix upon adsorption are known and it can be shown that our arguments remain valid.

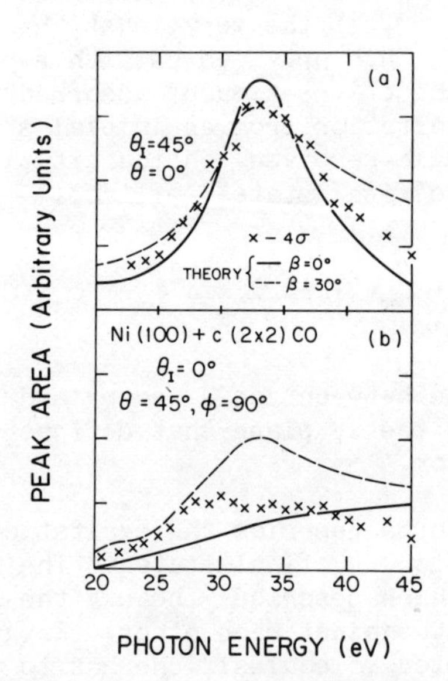

Fig. 6. Photoionization cross sections of the CO 4σ level in the c(2×2)/Ni(100) system[17]. In panel (a), the light is incident at an angle of 45°, and in (b) along the surface normal. The full (dotted) curves are theoretical calculations for a molecule standing straight up (tilted 30° away from the surface normal)[9].

The resonance energy has shifted from $\hbar\omega$ = 32 eV in free CO (Fig. 3) to $\hbar\omega$ = 36 eV in CO on Ni(100). The same shift ($\pm$ 1 eV) is observed on all other surfaces where CO adsorbs molecularly and for which experiments have been performed to date (Cu(100), Pt(111), Ni(111), Ni(110), Co(0001), etc.). This shift is even more dramatic if one considers that the binding energy of the 4σ level is lower in the chemisorbed phase, due to improved hole screening from the metal electrons. E_R shifts actually from ~12 eV in free CO to ~20 eV in the chemisorbed case. Within the fairly crude estimates possible[19], there are no indications that the d-spacing has changed sufficiently to solely account for this effect.

The σ resonance is, not surprisingly, also present in $Cr(CO)_6$, where the geometric structure of course is well known[20]. The resonance here has shifted to $E_R \sim 18$ eV, within a few volts of the chemisorbed value. This effect has been attributed to the screening induced by the substrate.

In Fig. 7 we show data on the photoionization cross sections of the $2\sigma_u$ level in N_2 on Ni(110)[21]. Strikingly a strong resonance

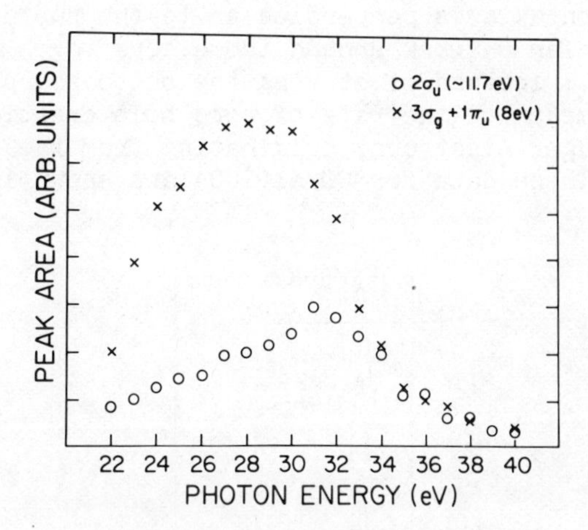

Fig. 7. Partial photoionization cross sections for the valence levels in the N_2/Ni(110) system[21].

is observed here, whereas the gas phase cross section (Fig. 2) is quite flat. Obviously, the (g,u) symmetry must be broken so that one can excite to the resonance. This strength of this symmetry breaking leads one to consider bonding configurations where the two nitrogen atoms are inequivalent. The most drastic such configuration would be one where the nitrogen axis is perpendicular to the surface. In fact more detailed work using photoelectron spectroscopy has shown this to be the preferred bonding configuration[21].

One of the most exciting new possibilities for studies of this kind is the frequency range around the carbon and oxygen K-edges. An advantage in using core states as initial states are that they are simple and fairly well understood. Ambiguities with band assignments in valence band spectra can be avoided. In particular, studies of chemisorbed small hydrocarbons promise to become very interesting. These molecules undergo complex arrangements as a function of substrate temperature and coverage. Any input that could limit the number of possible structural models would be most welcome. For example, it is known that C_2H_4 chemisorbs on Pt(111) at liquid nitrogen temperatures with its axis parallel to the surface. Gentle heating to room temperature leads to a loss of one hydrogen atom and conversion to a C-CH_3 specie, where the carbon-carbon axis is perpendicular to the surface[22]. The best example so far on work around these core edges is due to Stohr and coworkers[23]. Instead of stuying the photoabsorption coefficient directly, they measured the rate of core hole creation by measuring the number of Auger electrons, originating from the filling of the C1s hole. These data for CO/Ni(100) are shown in Fig. 8.

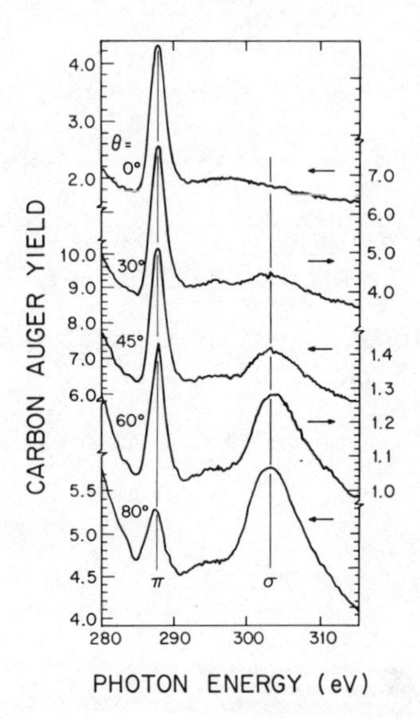

Fig. 8. Carbon Auger yield (≈ the Cls photoabsorption cross section) for the c(2x2)/Ni(100) system. θ is the angle between the surface normal and the light incidence direction[23].

For normally incident light a strong π resonance is observed below the vacuum level with no trace of the σ resonance in agreement with the data in Fig. 6. As the angle of incidence is changed, the strength of the π resonance decreases and the σ resonance increases. These observations can be quantified along the lines outlined above and the conclusion is, again, that the CO molecule is adsorbed perpendicular to the surface. A similar conclusion has been reached by the same group for the NO/Ni(100) system[23].

CONCLUSIONS

Studies of shape resonances in free and chemisorbed molecules have already yielded several substantial results. The existence of such resonances have been documented in a large range of systems and we have an empirical knowledge of their properties. The structural results that have been obtained have proved to be of great value in surface science. The simiplicity of the arguments used is quite appealing and allows one to draw conclusions about structural parameters without detailed numerical work.

A second generation of experiments on shape resonances have already appeared. These include studies of the angular distribution of the photoelectrons both for free molecule[24] (β-parameters) and for molecules on surfaces. Another very elegant variant is the study of the fluorescent light emitted by decaying excited ionic states[25]. By determining the degree of polarization of the fluorescence, one can in principle experimentally directly distinguish between, say, the $\sigma \rightarrow \sigma$ and $\sigma \rightarrow \pi$ transition channels in the total cross section. Detailed studies of the decay of the resonance excitation have been reported[26]. Finally, resonance excitations have been observed in the vibrational spectra in electron-molecule scattering of weakly bound (physisorbed) species. Such data may provide independent information to derive both adsorbate orientation and structure[27]. The first data on photoexcitation to shape resonances in physisorbed molecules have also been reported, but the experimental situation here is less clear[28].

ACKNOWLEDGEMENTS

This research has been supported by the National Science Foundation MRL program under Grant No. DMR-79-23647. My collaborators in these experiments include C.L. Allyn, M. Copel, D.E. Eastman, F. Greuter, W. Gudat, L.I. Johansson, H. Levinson, G. Loubriel, S.J. Oh and E.W. Plummer.

REFERENCES

1. T.M. Zimkina and V.A. Fomichev, Dokl. Akad. Nauk. SSSR 169, 1304 (1966) (Sov. Phys. Dokl. 11, 726 (1966)).
2. R.E. LaVilla and R.D. Deslattes, J. Chem. Phys. 44, 4399 (1966).
3. J.A.R. Samson and J.L. Gardner, J. Electron Spectrosc. 8, 35 (1976).
4. M.J. van der Wiel and C.E. Brion, J. Electron Spectrosc. 1, 309 (1972/1973).
5. C.E. Brion and A. Hamnett in: "Advances in Chemical Physics", Vol. 45, p. 1, J. Wm. McGowan, ed., J. Wiley and Sons, New York (1981).
6. G.R. Wight, C.E. Brion and M.J. van der Wiel, J. Electron Spectrosc. 1, 457 (1972/1973).

7. J.L. Dehmer and D. Dill, J. Chem. Phys. 65, 5327 (1976).
8. N. Padial, G. Csanak, B.V. McKoy and P.W. Langhoff, J. Chem. Phys. 69, 2992 (1978).
9. J.W. Davenport, Phys. Rev. Lett. 36, 945 (1976).
10. E.W. Plummer, T. Gustafsson, W. Gudat and D.E. Eastman, Phys. Rev. A15, 2339 (1977).
11. M. Krauss and F.H. Mies, Phys. Rev. A1, 1592 (1970).
12. Z. Pavlovic, M.J.W. Bones, A. Herzenberg and G.J. Schulz, Phys. Rev. A6, 676 (1972).
13. T. Gustafsson, Surf. Sci. 94, 593 (1980).
14. T. Gustafsson and H. Levinson, Chem. Phys. Lett. 78, 28 (1981).
15. A.P. Hitchcock and C.E. Brion, J. Phys. B14, 4399 (1981).
16. See e.g. Roger Stockbauer, B.E. Cole, D.L. Ederer, John B. West, Albert C. Parr and J.L. Dehmer, Phys. Rev. Lett. 43, 757 (1979).
17. C.L. Allyn, T. Gustafsson and E.W. Plummer, Chem. Phys. Lett. 47, 127 (1977) and Solid State Comm. 28, 85 (1978).
18. S. Andersson, Solid State Comm. 21, 75 (1977).
19. S. Andersson and J.B. Pendry, Phys. Rev. Lett. 43, 363 (1979); M. Passler, A. Ignatiev, F. Jona, D.W. Jepsen and P.M. Marcus, Phys. Rev. Lett. 43, 360 (1979) and S.Y. Tong, A. Maldondo, C.H. Li and M.A. van Hove, Surf. Sci. 94, 73 (1980).
20. G. Loubriel and E.W. Plummer, Chem. Phys. Lett. 64, 234 (1979).
21. K. Horn et al, Surf. Sci. (in press).
22. See e.g. M. Albert, L.G. Sneddon, W. Eberhardt, F. Greuter, T. Gustafsson and E.W. Plummer, Surf. Sci. 120, 19 (1982).
23. J. Stohr, K. Babeschke, R. Jaeger, R. Treichler and S. Brennan, Phys. Rev. Lett. 47, 381 (1981).
24. B.E. Cole, D.L. Ederer, R. Stockbauer, K. Codling, A.C. Parr, J.B. West, E.W. Poliakoff and J.L. Dehmer, J. Chem. Phys. 72, 6308 (1980).
25. E.D. Poliakoff, J.L. Dehmer, Dan Dill, A.C. Parr, K.H. Jackson and R.N. Zare, Phys. Rev. Lett. 46, 907 (1981).
26. G. Loubriel, T. Gustafsson, L.I. Johansson and S.J. Oh, Phys. Rev. Lett. 49, 571 (1982).
27. J.E. Demuth, D. Schmeisser and Ph. Avouris, Phys. Rev. Lett. 47, 1166 (1981).
28. H.-J. Lau, J.-H. Fock and E.E. Koch, Chem. Phys. Lett. (in press); M. Copel, F. Greuter and T. Gustafsson (to be published).

ATOMIC COLLISIONS IN THE HIGH ENERGY REGIME

Sheldon Datz

Oak Ridge National Laboratory
Oak Ridge, Tennessee 37830

INTRODUCTION

An understanding of the physics of electronic and atomic collisions draws heavily on our knowledge of atomic structure. However the converse is also true, i.e., experiments in collision physics often shed light on aspects of atomic structure. The prime example here is the "high energy atomic collision physics" experiment of Rutherford which established the basic structure of the nuclear atom. The fundamental contributions of Bohr to the theory of atomic structure are matched by his early contributions to the theory of charge changing collisions. Prof. Fano, in the opening paper of this conference, pointed out the equivalence of the structural and collisional views of a scattering resonance, i.e., should one treat it as a perturbation on a scattering trajectory or, as in his view, should it be treated as a structural entity undergoing a time evolution in the center of mass system?

The study of atomic collisions in the high energy regime has two functions, one is to test the velocity and/or energy dependence of phenomena which are observable at low energies and the other is to study channels which are not readily observable at lower energies.

Thus, the rubric "high energy atomic physics" covers a very large number of phenomena ranging from the formation of transient super-heavy atoms and multiple ionization events in very violent collisions to charge transfer in rather more delicate ones. The increasing availability of high energy accelerators and the development of sources which produce multicharged ions with large stored energy, coupled with greater sophistication in experimental techniques has brought about a considerable increase in our understanding of these processes in recent years and periodic reports of progress in some of these areas have been made to this conference. It is clear that a comprehensive review is well beyond the scope of this paper; instead we will focus on some areas in which recent studies have made some long standing problems more tractable, and on some unanticipated observations which give additional information and pose new questions.

The first part of paper will deal with the relation of Coulomb ionization to the molecular orbital (MO) picture of the united atom and the application of these results to the structure of K shell in super-heavy atomic systems.

In the second part we will discuss recent results on electron transfer into discrete states and charge transfer events which involve more than one electron, e.g., capture to continuum and "transfer ionization".

In the final section we will describe the recent observation of radiation arising from bound state transitions in one and two dimensional "atoms" and "molecules" formed when high energy electrons are injected into a crystal along directions close to atomic rows and planes.

K SHELL IONIZATION IN HIGH Z SYSTEMS

Inner shell ionization of electrons to the continuum in ion-atom collisions can occur by two different processes.[1] For low Z_1 (projectile) particles on high Z_2 (target) atoms the only available process is Coulomb excitation which is variously treated by plane wave Born approximation (PWBA), the binary encounter approximation (BEA), and the semiclassical approximation (SCA). When Z_1 becomes comparable to Z_2 and the ion velocity v_i is lower than the velocity of the bound electron in question, v_e, the electrons adjust adiabatically to the approach of the two nuclei and enter molecular orbitals (MO) which in the limit of fused nuclei approach the atomic orbitals of the united atom $Z_{ua} = Z_1 + Z_2$. This stacking of electrons can lead to a promotion of an innershell electron to the continuum or to a vacant outer orbital by direct curve crossing, rotational coupling, or radial coupling between molecular levels when such channels are available.

In asymmetric collisions the lowest orbital, the 1sσ, connects the 1s orbital of Z_2 with the 1s of Z_{ua}. This is illustrated in Fig. 1 for the $_{35}$Br-$_{40}$Zr system.[2] When the system is sufficiently asymmetric the 2pσ is far removed from 1sσ radial coupling between the two is weak and the only way to form a vacancy in the 1s shell of the high Z collision partner is by Coulomb ionization from the 1sσ. Since Coulomb ionization cross sections are dependent on the electron binding energy, measurements of the impact parameter depen-

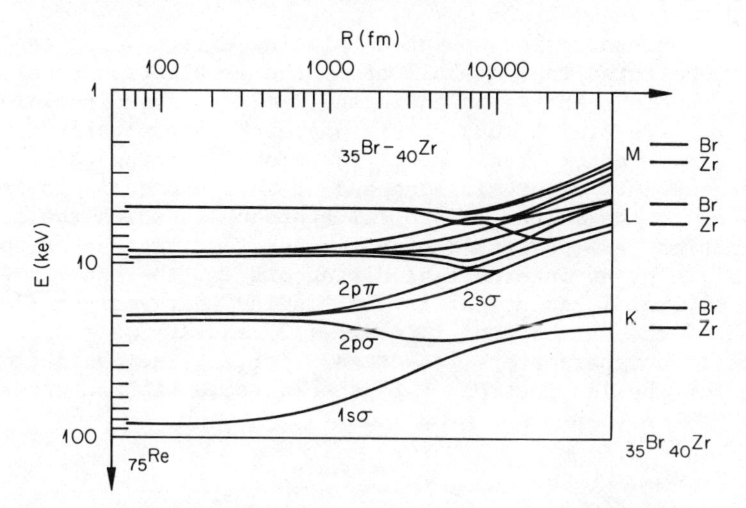

Fig. 1. Molecular orbital diagram for the system Br + Zr.

dence of Coulomb ionization can yield information on the structure of the 1sσ orbital.

Both the PWBA and BEA assume straight line trajectories for the projectile ion. However, at low ion velocities the nuclear repulsion will deflect the ion leading to hyperbolic trajectories, a larger distance of closest approach for a given impact parameter. This problem was first treated by Bang and Hansteen[3] in a semiclassical approximation (SCA) wherein the ion path is treated classically and the electron-ion momentum exchange is treated by first order perturbation theory. Although the results of this treatment do not lend themselves easily to a simple scaling law, as do the PWBA and BEA the determination of the impact parameter dependence of the cross section is a natural consequence of this approach. They utilize the concept of "adiabatic distance" given in terms of the ion velocity v_i and the ionization potential of the electron Is

$$r_{ad} = hv_i/I_s \tag{1}$$

which is physically related to maximum distance of closest approach which can cause ionization, i.e., beyond this distance the electron is able to adjust adiabatically to the Coulomb impulse imparted by the projectile. As long as the $r_{ad} \leqslant r_s$, the electron shell radius, the SCA may be used directly to predict impact parameter dependence. The magnitude of the cross section should scale simply with $Z_1{}^2$ but the cross sections fall below the calculated ones for small impact parameter collisions especially for higher Z. This comes about because of the so called enhanced "binding effect". As the projectile Z_1 penetrates the ls shell of Z_2 the ls electron sees a higher effective unclear charge which in the limit of fused nuclei is equal to $Z_1 + Z_2$ (see Fig. 2 top). This approach is basically a modification of a one center atomic model in which a binding effect can be incorporated in a smooth fashion and it is phenomenologically equivalent to the molecular orbital approach in which the electronic wave functions are described with respect to a two center Coulomb potential; e.g. at internuclear distance R = 0 the binding energy of the lsσ orbital is identical to the binding energy of the ls electron of the united atom $Z_{ua} = Z_1 + Z_2$ and $\langle r_{1s\sigma}\rangle = \langle r_{1s}\rangle_{ua}$. Thus for impact parameters less than ~ $\langle r_{1s}\rangle_{ua}$ the ionization cross section for the ls electron of Z_2 should scale with $Z_1{}^2$ and be a

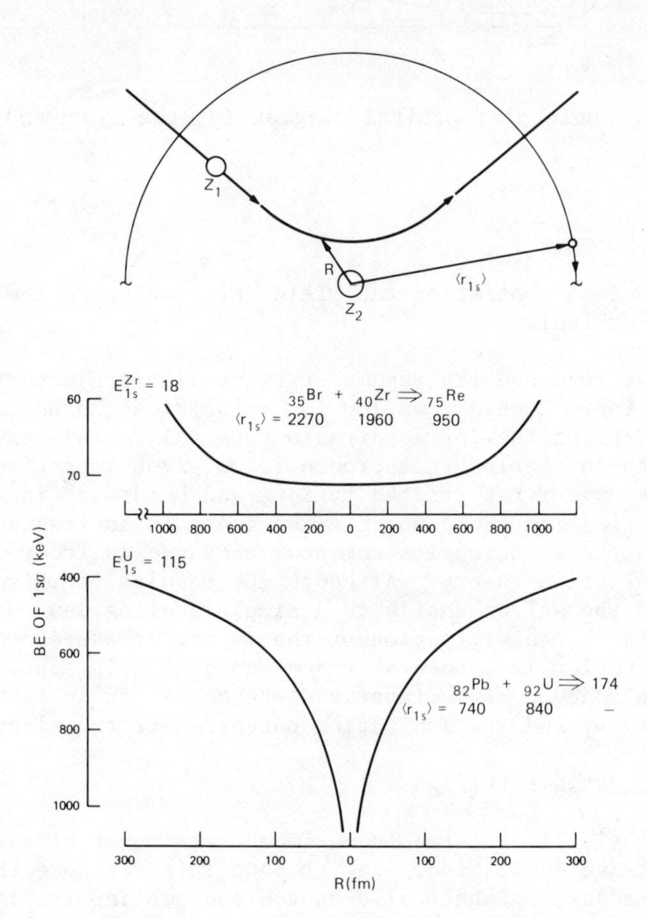

Fig. 2. (Top) Trajectory of projectile Z_1 passing through K shell of target atom Z_2. (Middle) Binding energy of lsσ orbital for the system Br + Zr as a function of R. (Bottom) Binding energy of the lsσ orbital for the system Pb + U as a function of R.

function of Z_{ua} independent of the particular $Z_1 + Z_2$ combination at a given v_i.

This scaling has been nicely demonstrated by Gaukler[4] for $Z_{ua} = 67-68$ and is shown in Fig. 3 together with a comparison with a RHFS SCA calculation by Trautmann et al.[5] for the Cl-Sn system in which the ua wave function is centered on the target nucleus. (Improvements upon this approximation and other aspects of this section are discussed by Schuch.[6] Here the measured cross sections for ionization of 1s electron of Z_2 are plotted as a function of impact parameter for various $Z_1 + Z_2$ combinations all of which give ~ the same Z_{ua}. The range is impressive i.e., the apparent success of this relatively simple approach can be understood from Fig. 1 and Fig. 2b where it is shown that the 1sσ binding energy for systems with $Z_{ua} \sim 70$ is relatively constant over the region of $R \leqslant 1000$ fm.

The situation typified in Fig. 2 (middle) is quite different for very large Z_{ua} where relativistic effects are strong and become ever stronger as the united atom core is approached. This situation is illustrated in Fig. 2 (bottom) for the Pb+U system.[2] In this case one would not expect the flat impact parameter dependence, P(b), for 1s ionization of Z_2 displayed in Fig. 3. However by properly inverting the procedure used above in which the 1sσ binding energy is used to predict P(b), the measured P(b) can be used to determine binding energy as a function of R. This has been done by Liessen et al.[7] and the results are shown in Fig. 4. Here the binding energies for the 1sσ for $R_o = 100$ fm and 50 fm derived from the measured 1s ionization cross sections of Z_2 are plotted as a function of Z_{ua}. The dashed curves represent the calculated values and demonstrate the increasingly steep R dependence of the 1sσ binding energy with increasing Z_{ua}. The relatively good agreement of binding energy down to 50 fm for $Z_{ua} = 178$ lends credence to the calculated united atom binding energy (solid line) which lies in the lower continuum.[8]

CHARGE TRANSFER

In the previous section we have considered only processes which lead to electron loss to the continuum. Electron transfer is, in a sense, a more delicate process involving discrete states in both the initial and final configuration. Even so charge transfer cross sections can become very large in near resonant collisions and can completely dominate inner shell ionization total cross sections.[9]

The theoretical treatment of this problem began more than 50 years ago with Thomas who proposed that transfer took place in two steps; one in which the electron detached from the target and accelerated to the speed of the projectile and second is deflected from the target nucleus to motion parallel to the projectile. This

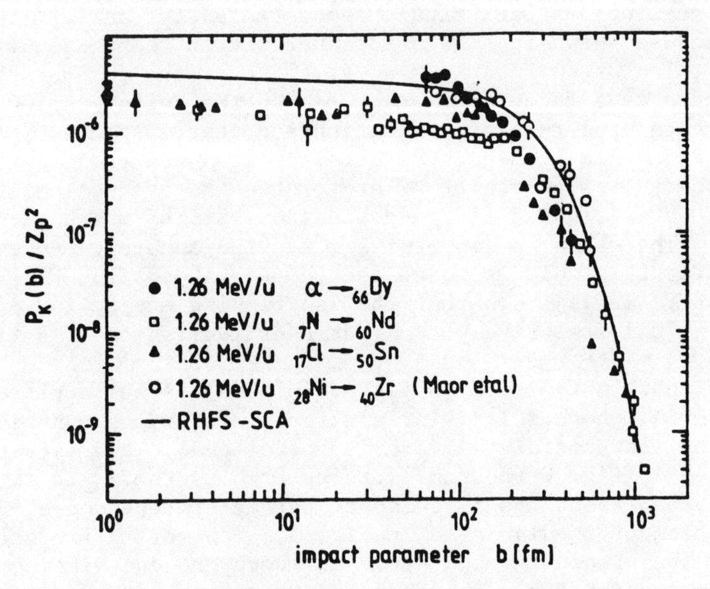

Fig. 3. Target K- vacancy production probabilities $P_k(b)/Z_p^2$, ($Z_p \equiv Z_1$) as a function of impact parameter, b.

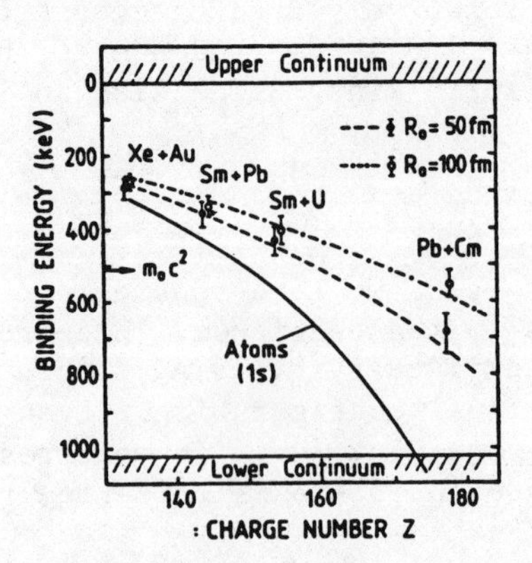

Fig. 4. 1sσ binding energies as a function of ($Z_1 + Z_2$). Experimental points are from Liessen et al.[7]

classical approach is not without intrinsic merit and has been refined and quantified over the years.

At intermediate velocities ($v_i \lesssim v_e$) non-classical models are used in which the electron is scattered into a bound state of the projectile. An approximation to the result can be obtained using first order perturbation and this was done in the Oppenheimer-Brinkman-Kramers treatment. This approach, which has been dubbed OBK, suffers because of the non-orthoganality of the initial and final states.

Numerous improvements have been carried out over the years and these are covered in recent review papers by e.g., Shakeshaft[10] and Horsedal-Pedersen.[11] In this paper we discuss recent experiments for testing various theoretical models. These include dependence on impact parameter, state-to-state transfer cross sections and some new effects which have been observed that involve more than single electron interactions.

Impact Parameter Dependence of K to K Charge Transfer

The case of non-resonant K shell to K shell charge transfer at intermediate velocities ($v_i/v_k \sim 0.5$) has been investigated by Horsdal-Pedersen[11] who measured the impact parameter dependence for this process in a number of systems. In this work Auger spectra from relaxation of the target atom were recorded in coincidence with projectile ions which had captured a single electron and had been scattered into given angle. The scattering angle defined the collisional impact parameter and the coincidence of a KLL Auger electron from the target with a capture by the projectile was taken as evidence for a K to K transfer.

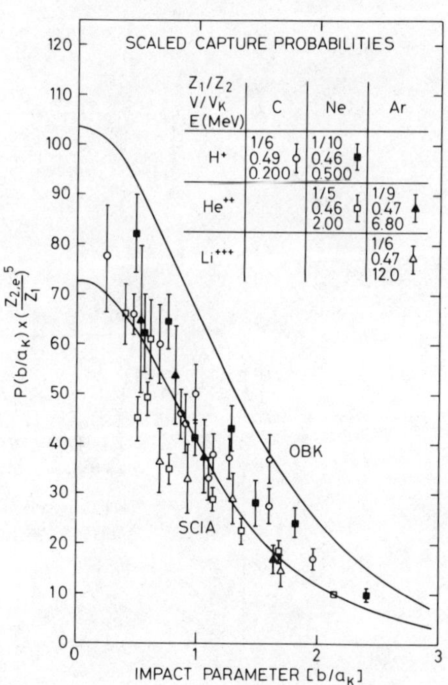

Fig. 5. Scaled K to K transfer probabilities as a function of reduced impact parameter.

The scaled capture probabilities for the asymmetric H^+ + C, H^+ + Ne, He^{++} + Ne, He^{++} + Ar and Li^{+++} + Ar systems are shown in Fig. 5. Both the OBK and the semi-classical impulse approximation calculations (SCIA) predict simple scaling with

$(Z_1/Z_2)^5$ for impact parameter dependent charge transfer probability P(b) in terms of scaled impact parameter $bZ_2 = b/r_k$ and velocity $v/Z_2 = v/v_k$. All data were taken for $v/v_k \sim 0.5$ but Z_1/Z_2 ranged from 1/6 to 1/10. The OBK does predict the form of P(b) but overestimates the cross sections while the SCIA does rather well in both aspects.

The results shown in Fig. 6 for the impact parameter dependence of K to K transfer in the nearly symmetric F^{8+} + Ne system,[12] however show strong deviations from the monotonic behavior shown in Fig. 5. Here the reduced velocity is somewhat lower $v/v_k = 0.31$. The structure in this angular dependence is interpreted in terms of an interference effect between the amplitudes for charge transfer and for elastic scattering which add coherently. The two potential curves in question here are the 2pσ and 1s. shown in Fig. 7 in the F^{8+} + Ne diabatic correlation diagram. We enter with a vacancy in the 2pσ; an odd number of transitions between the 2pσ and the 1sσ in the total trajectory leads to a net charge transfer while an even number leads to elastic scattering. At $v/v_k = 0.31$ only 0, 1 or 2 transfers are possible. At higher velocities where only a single transfer is possible the interference effects disappear.[12]

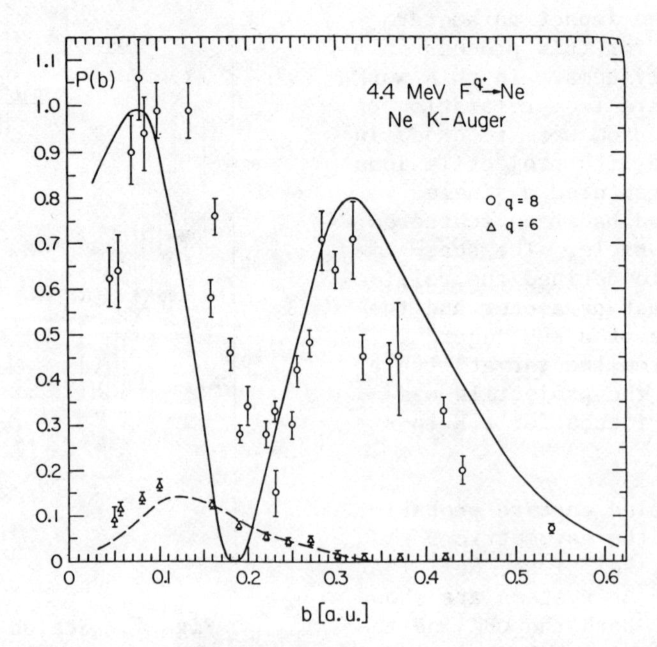

Fig. 6. Impact parameter dependence of K to K charge transfer for 4.4 MeV F^{8+} + Ne. Note that $\langle r_k \rangle$ of Ne is ~ 0.104 au and compare with Fig. 5.

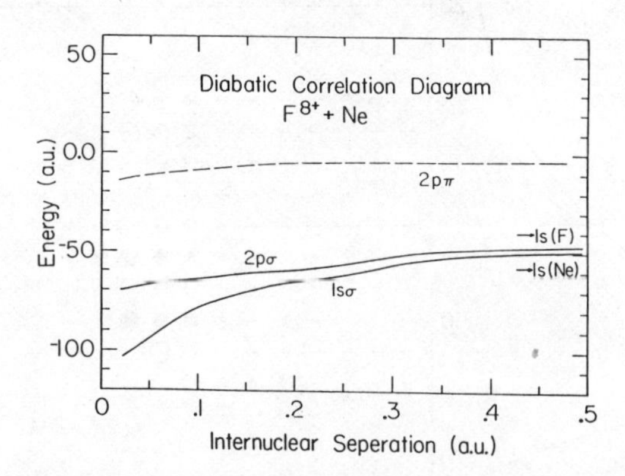

Fig. 7. MO correlation diagram for the F^{8+} + Ne system.

The solid curve shown in Fig. 6 was derived by Lin and coworkers[13] who used a two center expansion method and a Hartree-Fock-Slater potential. The agreement is reasonably good.

Charge Transfer to Specified n,ℓ states

In the case of very highly charged ions colliding with target atoms with relatively loosely bound electrons transfer takes place to high n states of the projectile. In first order this can be understood simply on the basis of energy defect for the transition. If n is large enough (Rydberg states) the core configuration becomes relatively unimportant and the charge tranfer cross sections to given nℓ states depend only on the projectile ion charge state and velocity. The results of a classical trajectory Monte Carlo (CTMC) calculation by Olson[14] for capture of a 1s electron from atomic hydrogen on to an ion of charge 14^+ of a velocity of 100 keV/amu (~ 2 v_o) are shown in Fig. 8. The left hand portion of the figure displays the total cross section for capture into a given n state. The right hand portion shows the weighting of ℓ states within a given n state. The black circles show the calculations by Olson and the open ones represent extrapolations and interpolations by Knudsen et al.[15] At the lower n values there is a strong tendency to populate the maximum ℓ (Yrast) levels which is mitigated at n values above the maximum in the left hand curve. The n dependence of the charge transfer cross section for this process has been tested by Knudsen et al.[15] who measured the light emission from excited Au^{13+} formed by charge capture of 20 MeV Au^{14+} (~ 100 keV/amu) from an H_2

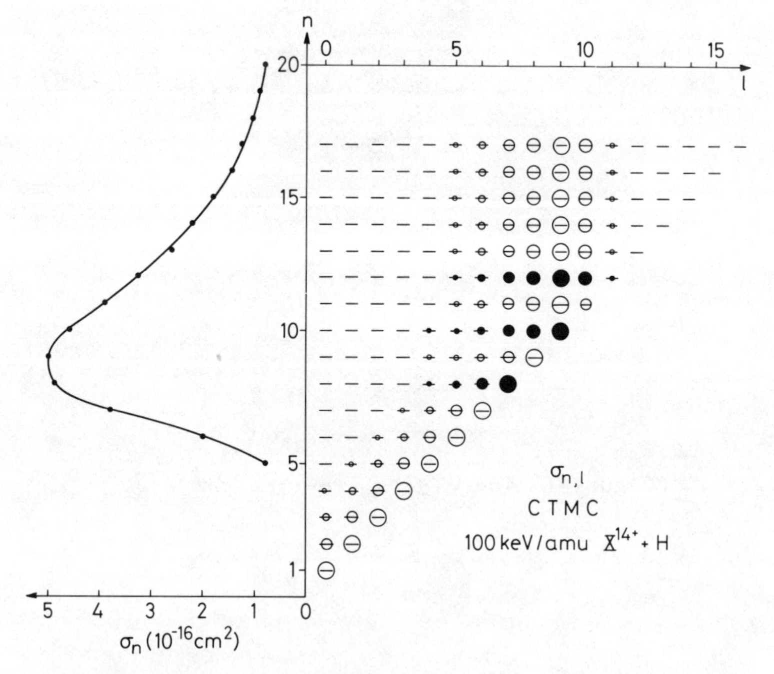

Fig. 8. Calculated cross sections $\sigma_{n,\ell}$ for 100 keV/amu 14^+ bare ions (Si^{14+}) colliding with atomic hydrogen.

target. The observed spectrum is shown in Fig. 9 and a comparison of the derived partial cross section with various theories for Si^{14+} is shown in Fig. 10. The uppermost curve in Fig. 10 is 1/3 the value given by a simple OBK calculation it is clearly too high and predicts a maximum at $n \simeq q$, the ionic charge, the Eikonal calculation of Eichler and Chan[16] lowers the value somewhat but retains the $n \simeq q$ prediction. The CTMC calculation of Olson[14] gives a lower maximum n value at $n \simeq q^{2/3}$ and the closest fit is obtained with the unitarized distorted wave approximation of Ryufuku and Watanabe.[17]

Although to my knowledge, the above is the first reported work on the distribution of n states in charge transfer to highly ionized species, two other groups, Prof. Cocke at the Kansas State University and the group at the University of Nagoya are making great strides in this area of study. They both work with low velocity highly charged ions. At Kansas State these are recoil ions formed by collision of high velocity ions obtained from a Van de Graaff accelerator while at Nagoya the highly charged ions are obtained from a high powered electron bombardment ion source (EBIS). The multiply charged ions are passed through a target gas cell and the loss or gain of energy accompanying electron capture is measured to determine the final $n\ell$ state. An example of such a measurement obtained

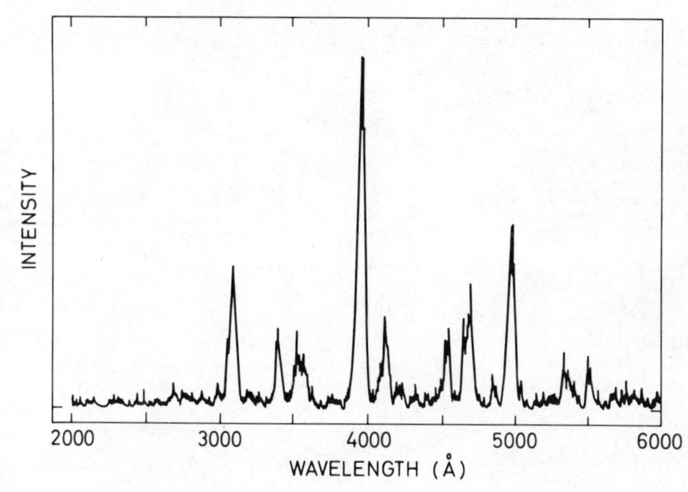

Fig. 9. Emission spectrum from 20 MeV Au^{14+} colliding with H_2 to give Au^{13+}, (n,ℓ). The three most prominent lines are the 12→11 transition at 3060 Å and the 14→13 at 4969 Å.

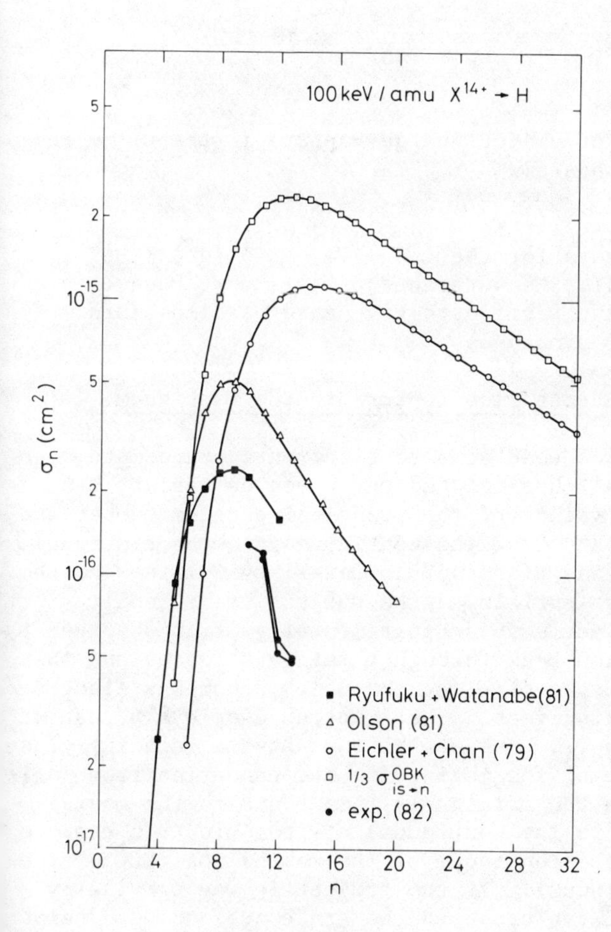

Fig. 10. Measured cross sections for populating some n states, σ_n, in charge transfer collisions between 20 MeV (~ 100 keV/amu) Au^{14+} with H_2 in units of cm^2 per molecule (●). Also shown in units of cm^2 per atom are calculations for Si^{14+} on H at 100 keV/amu by the UDWA theory (ref. 17) (■); CTMC calculations (Δ) (ref. 14) and the EIK theory (o) (ref. 16). Also shown is 1/3 of the calculated OBK cross section (□).

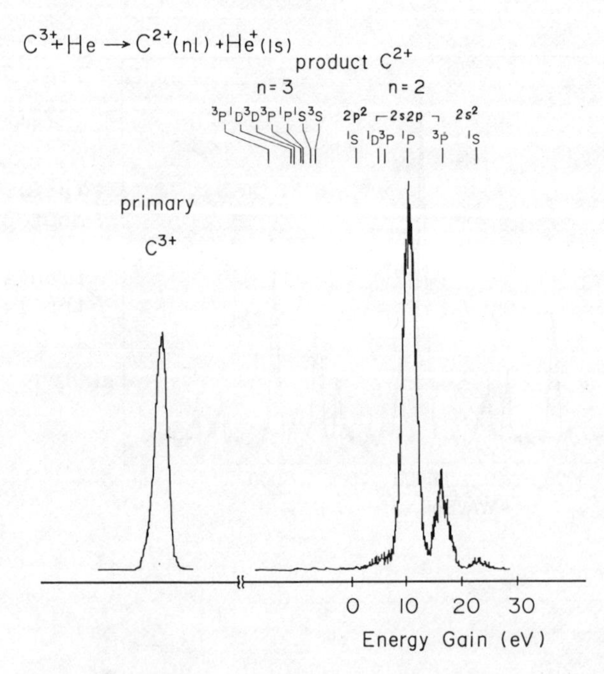

Fig. 11. Energy gain spectrum for charge capture in the $C^{3+} + He \rightarrow C^{2+}(n,\ell) + He^{+}$ system.

by Ohtani et al.[18] at Nagoya for the $C^{3+} + He \rightarrow C^{2+}(n\ell) + He^{+}$ system is shown in Fig. 11. Measurements of this sort have been made for ions up to O^{8+} and promise to give much detailed information on charge transfer processes.

Charge Capture to Bound States Plus Capture to the Continuum

In the previous section we discussed electron capture into high Rydberg states. As Macek[19] has pointed out electron capture can occur to ever increasing values of the principal quantum number and continue smoothly from just below the continuum into the continuum itself. This phenomenon was discussed in detail by Sellin[20] at the last ICAP and an extensive article on the subject has recently appeared in the Physical Review.[21] Experimentally it is observed by passing a high velocity ion beam through a thin gas target and then measuring the energy spectrum of electrons which accompany electrons which accompany the existing ions. The spectrum displays a cusp at an electron velocity $v_e = v_i$. These electrons have a velocity close to zero in the rest frame of the moving ion and can arise from ionization of electrons which had originally been bound to the moving ion (ELC ≡ electron loss to the continuum) or from electron capture from the target atom to the continuum of the moving ion (ECC ≡ electron capture to the continuum). If the projectile was originally totally stripped only ECC can occur and we are concerned here only with a comparison of the cross section for ECC with that for capture to bound states. Such a comparison can be made using the data of Vane et al.[22] who determined the yield of ECC electrons for the process:

$$O^{8+}\ (20\ \text{MeV}) + \text{Ar} \rightarrow O^{[8-(n-1)]+} + \text{ECC} + \text{Ar}^{(n+m)+} + me^- \qquad (1)$$

by measuring the number of electrons contained in the ECC cusp which occurred in coincidence with the capture of a given number, (n-1), of electrons to bound states on the oxygen ion. In this case the Ar target must have lost n electrons to the rest frame of the O ion (and perhaps in more electrons to its continuum). The measured cross sections for ECC, as a function of n, are listed in column 2 of Table I. We now wish to compare these numbers with the cross section $\sigma_B(n)$ for capture of an equal number, n, electrons to bound states, i.e.,

$$O^{8+}\ (20\ \text{MeV}) + \text{Ar} \rightarrow O^{(8-n)+} + \text{Ar}^{(n+m)+} + me^- \qquad (2)$$

These cross sections can be derived from the single and multiple charge capture cross sections measured by McDonald and Martin[23] for the same system; these are listed in column 1 of Table I. However since the measured total bound state cross sections do not include ECC and are really the sum of $\sigma_B(n) + \sigma_{ECC}(n+1)$ one must correct column 1 to yield $\sigma_B(n)$ which is listed in column 3 of Table I.

A comparison of $\sigma_{ECC}(n)$ with $\sigma_B(n)$ yields the interesting result that the probabilities of capture of the nth electron into bound or continuum states are comparable in magnitude, the ratio ranging from 2 at n=1 to 0.74 at n=3.

Table I. Cross sections (in cm^2) for capture to continuum and capture to bound states for 20 MeV O^{8+} + Ar.

	$\sigma_B(n)+\sigma_{ECC}(n+1)^a$	$\sigma_{ECC}(n)^b$	$\sigma_B(n)$
n=1	2.2×10^{-17}	9.3×10^{-18}	1.9×10^{-17}
n=2	6.0×10^{-18}	3.0×10^{-18}	5.0×10^{-18}
n=3	8.5×10^{-19}	1×10^{-18}	7.4×10^{-19}
n=4		9×10^{-20}	

a. from ref. 23
b. from ref. 22

Charge Transfer plus Loss to Continuum ("Transfer Ionization")

Another process involving charge transfer to multiply charged ions has recently come under study. In general the process dubbed "transfer ionization" (TI), involves the loss from the target atom of more electrons than are captured by projectile. (Thus in this broad sense capture plus ECC falls into this category!) In a more restricted sense one of the ways by which this process is thought to occur by the transfer of two electrons to high n states on the projectile followed by autoionization of the doubly excited state so formed. Here we will discuss a few examples of this phenomenon.

Consider first the experiments of Cocke et al.[24] Beams of low velocity Ar^{q+}, $q = 2 \rightarrow 9$ eV collide with neutral inert gas atoms. Using coincidence techniques, the charge states of both the Ar ion and the target atom following the collision are measured and the cross sections determined for such processes as e.g.

$$Ar^{q+}(E = 500\ q\ eV) + Xe$$

$$\rightarrow Ar^{(q-1)+} + Xe^{+},\ \text{(single capture)} \tag{3a}$$

$$\rightarrow Ar^{(q-1)+} + Xe^{2+} + e^{-}\ \text{(TI)} \tag{3b}$$

$$\rightarrow Ar^{(q-2)+} + Xe^{2+},\ \text{(double capture)} \tag{3c}$$

As can be seen from Fig. 12, following a threshold at Ar^{3+}, transfer ionization becomes comparable in magnitude to simple charge transfer. This threshold, comes about because the cross sections for forming doubly excited states by two electron transfer should be large when the process becomes exoergic since curve crossings at large separations now become possible i.e., the condition for exoergicity in transfer ionization of Ar^{q+} + B is given by

$$RE[Ar^{q+}] + Re[Ar^{(q-1)+}] - Ex^{**}[Ar^{(q-2)+}] > IP(B) + IP(B^{+}) \tag{4}$$

where RE is the recombination energy, IP is the ionization potential and Ex** is the double excitation energy of $Ar^{**(q-2)+}$. This threshold condition was found to hold true for all of the rare gas target atoms (He→Xe) studied.

Another example of target ionization accompanying charge transfer is shown in Fig. 13 where cross sections for various charge changing processes in the system 20 MeV Au^{q+} + He are shown as a function of q (q = 5-21).[15] The superscripts refer to the initial and final charge states of He and the subscripts to the initial and final Au charge states. As one would anticipate simple charge transfer of one electron $\sigma^{02}_{q,q-1}$ increases with increasing q, the cross sections for direct ionization of the He, $\sigma^{01}_{q,q}$ and σ^{02}_{qq} generally decrease with increasing q. Most noteworthy is the increase in

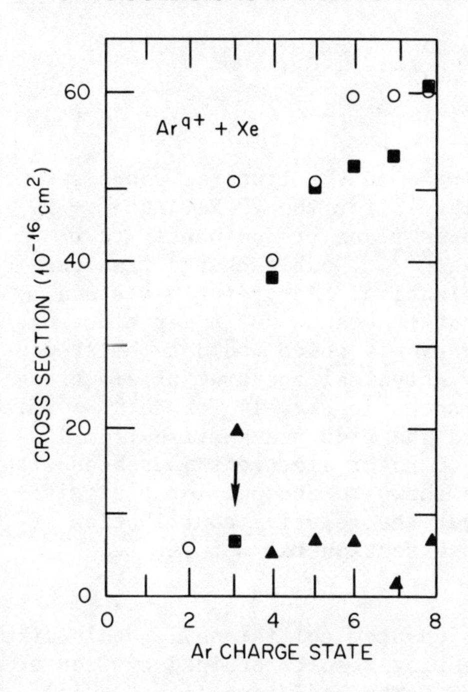

Fig. 12. For the collision system Ar^{q+} + Xe; (E = 500 q eV) single capture cross sections Eq. 3a (o); transfer ionization, Eq. 3b (■) and double capture cross sections Eq. 3c (▲) as a function of q. The arrow indicates the threshold for exothermic transfer ionization.

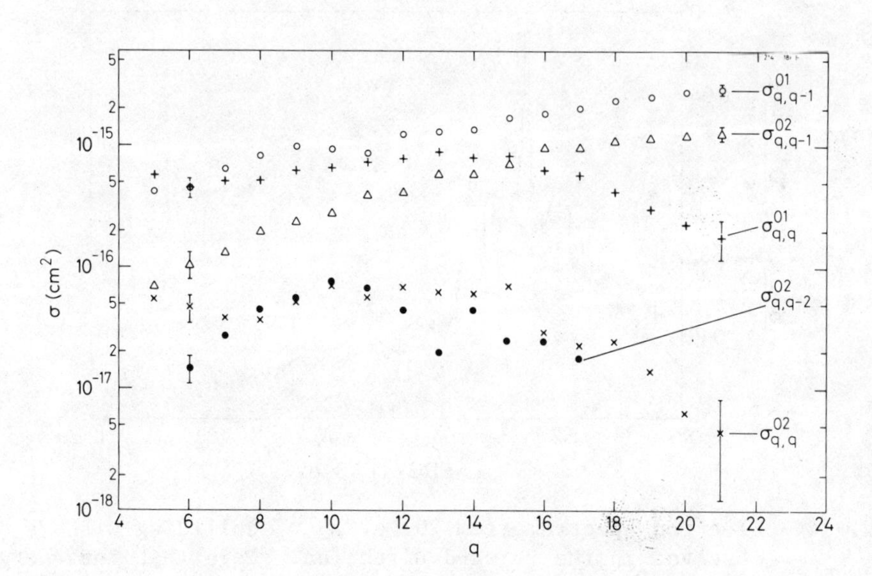

Fig. 13. Partial cross sections for 20 MeV Au^{q+} (q = 5-21) colliding with He.

the "transfer ionization" channel, $\sigma^{02}_{q,q-1}$, i.e.,

$$Au^{q+} + He \rightarrow Au^{(q-1)+} + He^{++} + e^{-} \quad (5)$$

which rises to almost 50% of the single charge transfer cross section at q = 21. As was shown in Fig. 10 for the 20 MeV $Au^{14+} + H_2$ system, single electron transfer takes place predominantly to n = 5 to 8 states and theory predicts n α $q^{2/3}$. Double charge transfer to states could easily lead to autoionization. Accordingly a study by the Aarhus group has been carried out to search for Auger electrons originating from these autoionizing events which would be emitted from the rest frame of the Au ion. A typical spectrum of electrons moving in the Au beam direction is shown in Fig. 14 for Au^{18+} + He. The central peak is the ECC cusp and the side peaks arise from low energy Auger electrons. The yield of Auger electrons increases with the charge on the Au ion and can be shown to account for a significant fraction of the $\sigma_{q,q-1}$. However the relative contribution of charge capture plus ECC to this cross section has not yet been determined.

As a coda to these sections on ion-atom collisions I should like to mention a very recent and interesting result obtained by Groh et al.[25] on the collisions of Xe^{q+} + Xe. The Xe^{q+} ions (q = 3 → 15) were obtained from an EBIS source at energies ~ 10 q keV and

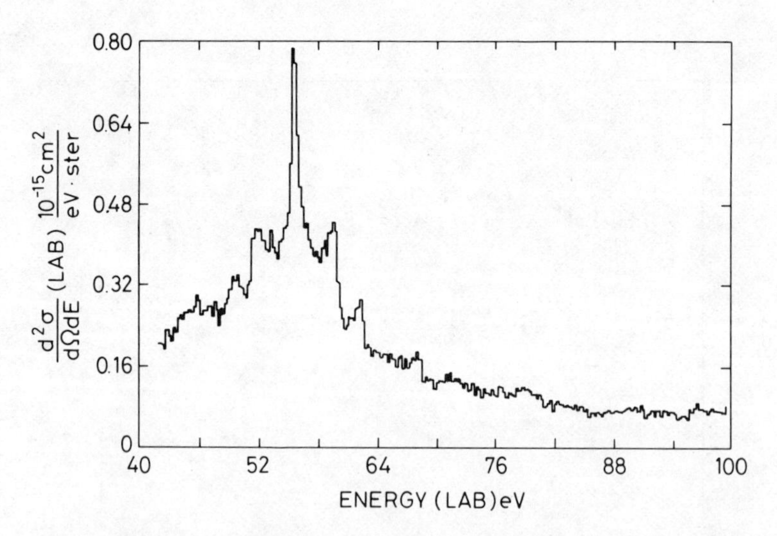

Fig. 14. Electron spectrum from 20 MeV Au^{18+} colliding with He observed in the forward direction. Note that the energy transformation from the Au rest frame to the lab system greatly magnifies the Auger electron energy. For example an Auger electron with 1 eV forward recoil would appear at ~ 70 eV lab.

collided with neutral Xe. As in the experiments described above the charge states of the recoil ions in coincidence with a given charge capture event were determined. The result for target Xe atom ionization accompanying double charge transfer

$$Xe^{q+} + Xe \rightarrow Xe^{(q-2)+} + Xe^{i+} + (i-2)\,e \qquad (6)$$

as a function of q is shown in Fig. 15. Here the ordinate represents the fraction of recoils which have lost i electrons. Thus i=2 represents two electron transfer to bound states of $Xe^{(q-2)+}$, i=3 two electron transfer plus a single ionization ... i=7 two electron transfer plus ionization of 5 additional electrons! This plot of multiple ionization probability as a function of input ion charge is reminiscent of the plots of multiple ionization as a function of scattering angle reported for e.g., 50 keV Ar^+ + Ar collision by Everhart group[26] in the 1950s. In this case multiple ionization was treated by Russek[27] as a statistical basis i.e., the number of electrons "boiled off" in the collision was statistically determined by the total energy deposited in the outer shell in the collision process. For the situation described in Fig. 15 the energy is supplied by the stored potential energy in the incident multiply charged ion. Treatments involving the Russek approach have been applied to this case and initial results look encouraging.[28]

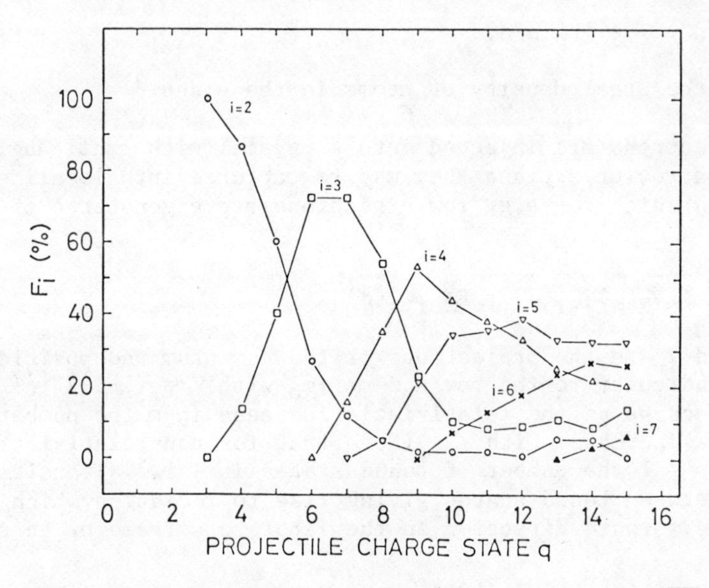

Fig. 15. For the collision system Xe^{q+} ($\sim$ 10 q keV) + $\overline{Xe}$, the charge state distribution of $\overline{Xe}$ obtained in coincidence with $X^{(q-2)+}$ as a function of q.

CHANNELING RADIATION: ATOMIC AND MOLECULAR PHYSICS IN ONE AND TWO DIMENSIONS

Many experiments in atomic physics are designed to test the fundamental concepts of quantum mechanics. In this section I will describe some recent experiments which demonstrate the applicability of these concepts to one and two dimensional systems. Although the first experimental result on channeling radiation were just published in 1979[29] a considerable body of information has been obtained and we can only give a brief outline of some aspects of the work in this paper.

In 1965 Lindhard[30] demonstrated that a charged particle moving closely parallel to an atomic row in a crystal experiences a continuum ("string") potential made up from the atomic potentials in the row. The two dimensional potential, $\overline{V}_R(\rho)$ is cylindrically symmetric and is function only of ρ, the distance from the row. For any atomic screened Coulomb potential

$$\overline{V}_R(\rho) = (2Z_1 Z_2 \; e^2/d)K_R(\rho/a) \tag{7}$$

where Z_1 and Z_2 are the atomic numbers of the projectile and lattice atom, d is the distance between atoms in a row and a is the screening radius. Similarly if we are dealing with atomic planes (sheets of atoms) the one dimensional continuum potential is a function of ρ the distance from the plane and is given by

$$\overline{V}_p = 2\pi n Z_1 Z_2 \; e^2 \; K_p(\rho/a) \tag{8}$$

where n is the areal density of atoms in the plane.[31]

When electrons are injected into a crystal with small angle with respect to a row or a plane they may be captured into localized bound states which for e.g. row directions are eigenstates of the Hamiltonian

$$[\vec{p}_\perp{}^2/2m + \overline{V}_R(\vec{\rho})]\psi_i(\vec{\rho}) = E_{\perp,i}\psi_i(\vec{\rho}) \tag{9}$$

where $p_\perp$ and ρ are the projections of the momentum and position on a plane perpendicular to the row and $m \approx \gamma m_o$ with $\gamma = (1-\beta^2)^{-1/2}$ and $\beta = v/c$. Because of the relativistic increase in m the number of bound states increases with γ. (Note that for non-relativistic electrons, $\gamma \simeq 1$ the number of bound states is $\leqslant 1$.) Transitions can occur between bound states giving rise to radiation which for $\beta \simeq 1$ in the forward direction in the laboratory frame of energy

$$\hbar\omega_L \simeq 2\gamma^2(E_{\perp,j} - E_{\perp,i}) \tag{10}$$

The factor of $2\gamma^2$ comes from transformation to the rest frame, γ, and from the Doppler transformation, $(1+\beta)\gamma$.

Table II. Experimental and calculated line energies for 54 MeV (γ=107) electrons in bound states transitions in {110} of silicon.

Transition $i \to j$	Photon Energy (keV) Calc.	Photon Energy (keV) Est.	Transition $i \to j$	Photon Energy (keV) Calc.	Photon Energy (keV) Est.
$1 \to 0$	125.3	122.5	$4 \to 3$	49.0	49.1
$2 \to 1$	89.0	88.8	$5 \to 4$	38.4	38.3
$3 \to 2$	64.5	64.3			

Planar Channeling Radiation

In this case we have a one dimensional potential and only one quantum number, n, and dipole selection rules dictate Δn odd. An example of the photon spectrum observed in the forward direction for the injection of 54 MeV electrons along the (110) planar direction in Si is shown in Fig. 16.[32] The bound state $\Delta n = 1$ transitions are evident up to $n = 5 \to 4$ and at higher energies the $\Delta n = 3$ transitions are also evident. One can now compare the observed spectrum with calculations based on e.g. Hartree-Fock descriptions of the Si atom. This can be done directly through the solution for the one-dimensional Schrödinger equation or one may work in momentum space and use the many-beam formulation of the Schrödinger equation for the transverse motion.[33] The results of the many-beam calculations which use Doyle-Turner scattering factors derived from Hartree-Fock wave functions are compared with experimental results in Table II. It is evident that the agreement is excellent.

Axial Channeling Radiation

In this case we have a two dimensional potential with quantum numbers n and ℓ. We use as an example the channeling radiation spectrum from 16.88 MeV electrons injected along a <100> direction in a diamond crystal which is shown in Fig. 17.[34] Again, as can be seen from Table III, the agreement is satisfactory.

Table III. Experimental and calculated line energies for 16.88 MeV electrons for bound state transitions in <100> diamond.

Transition	Experiment	Theory
2p-1s	58.3	58.0
3d-2p	34.4	34.0
2s-2p*	21.5	22.8
3p-1s	100.6	100.8

*Identification of this line is not unambiguous because transitions from higher states also lie in this energy region.

In both the planar and axial case the "nucleus" is extended in space and can be aligned with the incoming electron beam direction. The initial population of states depends upon impact parameter and the transverse momentum. Hence by tilting the crystal with respect

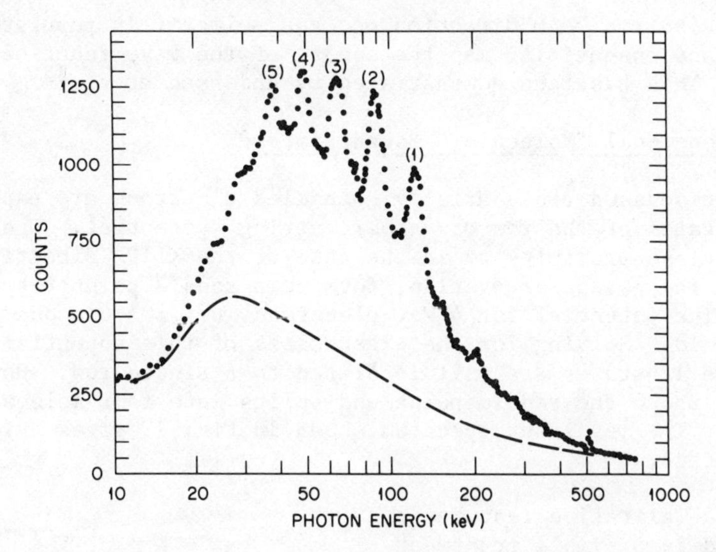

Fig. 16. Photon spectrum from channeling of 54 MeV electrons in the (110) plane of Si. The peaks labeled 1→5 correspond to transitions n=1→0, n=2→1, ... respectively. The lower spectrum is that obtained from a nonchanneled beam for the same fluence.

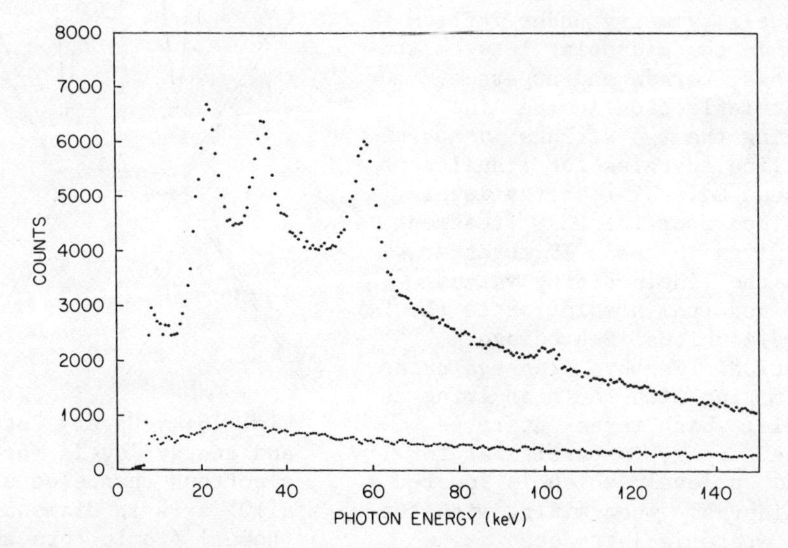

Fig. 17. Photon spectrum from channeling of 16.88 MeV electrons in <100> axial direction. The lower spectrum is that from a non-channeled beam for the same fluence.

to the electron beam direction one can selectively populate certain states and essentially map the square of the wave function for that state. This has been demonstrated by Andersen et al.[35]

Two-dimensional "Molecular" Bound States[36]

As discussed above axially channeled electrons are captured into bound states of the row or "atomic string" potential. When two rows lie in close proximity as in the case of the <110> direction in diamond the potential overlap, forming a saddle point between the rows. The potential for 4 MeV electrons in <110> diamond is shown in Fig. 18. Solving for the eigenstates of this potential one finds that the 1s states are well localized to a single row. However, the 2p lies above the saddle point and splits into four molecular type levels. The resultant spectrum shown in Fig. 19 gives evidence for this effect.

The qualitative features are obtained in a simple treatment analogous to the linear-combination-of-atomic-orbitals (LCAO) method in chemistry. When the transverse Hamiltonian is diagonalized in the subspace spanned by the four single-string 2p states for a pair of strings, four eigenstates are obtained which may be classified according to their symmetry under reflection in the mid-point between the strings, gerade and ungerade, and under reflection in the line connecting the two strings, σ and π. The line energies for transitions between molecular energy levels obtained from the LCAO treatment are given in Table IV together with the single-string values obtained from a solution to the two-dimensional Schrodinger equation. We have also evaluated the mixing with the near-lying 2s levels, which turns out to be quite strong, in particular for the σ_g2p level, which is lowered considerably when mixing with the σ_g2s state is introduced. The splitting between the lines is seen to increase by ~ 150 eV so that it is now ~ 100 eV larger

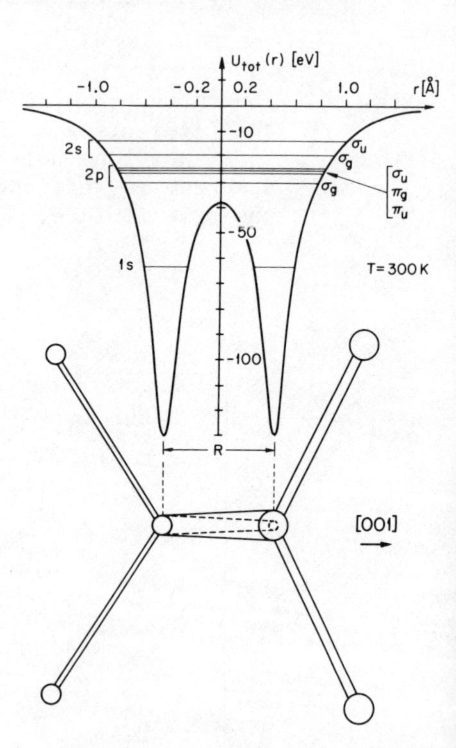

Fig. 18. (Upper) Potential and energy levels for 4 MeV electrons channeled along <110> axis in diamond. (Lower) Atomic core and bond configuration in diamond viewed along the <110> direction.

than the observed splitting. Also, a dipole transition 2s-1s becomes allowed, and, in fact, a corresponding weak line is visible in the spectrum (see Fig. 19). However, while the LCAO-type model is very instructive for qualitative purposes, its accuracy is limited and difficult to assess.

Table IV. Photon energies (in electron volts) for 2s-1s and 2p-1s transitions for 4-MeV electrons in <110> diamond.

Transition	Single string	1st order LCAO	With 2s-2p mixing	Many beam	Experiment
$\sigma_g 2s-\sigma_u 1s$			7253	7103	6933
2p-1s	5801				
$\sigma_u-\sigma_g$		6025	6015	6019	5897
$\pi_g-\sigma_u$		5888		5887	5751
$\pi_u-\sigma_g$		5771		5742	5624
$\sigma_g-\sigma_u$		5279	5115	5090	5084

The values obtained for channeling radiation energies from the many-beam formulation are listed in Table IV. For the 2p-1s transitions, there is reasonable agreement with the result of the LCAO-type calculations. The upper three lines and the more accurate

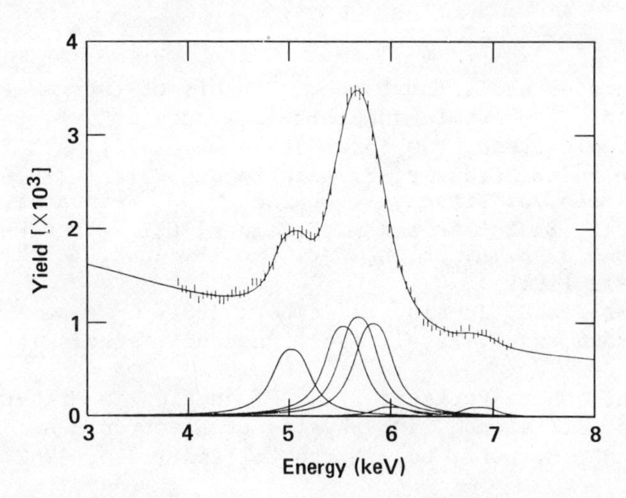

Fig. 19. Decomposition of the <110> 2p-1s and 2s-1s lines (see Table IV). See ref. 36 for fitting procedure.

2s-1s line are higher than the measured values by ~ 120 eV. However, the most significant deviation is for the separation of the σ_g-σ_u line which is larger by ~ 130 eV than deduced from the experiment. This we attribute to the accumulation of charge in the tetrahedral bonds in diamond.

The effect on the <110> potential of a charge accumulation in the bond regions may be understoood qualitatively from Fig. 18. Each atom in a <110> row is bonded to two atoms in a neighboring row and to one atom in each of the next-nearest neighboring rows. Hence accumulation of charge in the bonds increases the electron density between close-lying pairs of <110> rows and increases the potential energy of channeled electrons in this region. Since the σ_g2p state has a high density here, this level will increase in energy relative to that of other 2p levels, i.e., the splitting will be reduced, as required by experiment. With use of the many-beam method and the Gaussian parameters obtained from the x-ray analysis, the energy levels have been recalculated as a function of the electron density enhancement and the closest match with the observed splitting is obtained to an electron density of 1.7/Å^3 in the center of bonds. Although this determination is probably less accurate than that obtained from the x-ray data, it does demonstrate that information on charge distributions and potentials in crystals can be obtained from channeling radiation.

This work was supported by the U.S. Department of Energy, Division of Basic Energy Sciences, under Contract No. W-7405-eng-26 with Union Carbide Corporation.

REFERENCES

1. D. H. Madison and E. Merzbacher, "Theory of Charged-Particle Excitation", in "Atomic Inner Shell Processes", B. Crasemann, ed. (Academic Press, New York, 1975) pp. 1-72.
2. B. Müller and W. Greiner, Z. Naturforsch. 31a 1 (1976) and G. Soff, W. Betz, J. Kirsch, V. Oberacker, J. Reinhardt, K. Wietschorke, B. Müller and W. Grieiner, GSI Report M-6-78.
3. J. Bang and J. M. Hansteen, Kgl. Dan. Vidensk. Selsk. Mat. Fys. Med. 31 #13 (1959).
4. G. Gaukler, Ph.D. Thesis, Heidelberg 1981.
5. D. Trautmann and F. Rösel, Nucl. Instrum. Methods 169, 259 (1980).
6. R. Schuch, "Characteristic X Ray Production in High-Energy Heavy-Ion Collisions", in "Physics of Electronic and Atomic Collisions", S. Datz, ed. (North Holland Press, 1982) pp. 151-167.
7. D. Liessen, P. Armbruster, F. Bosch, S. Hagmann, P. H. Mokler, S. Schmidt-Böcking, R. Schuch, J. B. Wilhelmy and H. J. Wollersheim, Phys. Rev. Lett. 44, 983 (1980).

8. B. Fricke and G. Soff, Atomic Data and Nuclear Data Tables 19, 83 (1977).
9. See e.g., J. Barrett, B. M. Johnson, K. W. Jones, R. Schuch, I. Tserruya and T. H. Kruse, Abstr. XII Int. Conf. on Phys. of Elect. and Atomic Coll., S. Datz, ed. (North Holland Press, 1981) pp. 716-717.
10. R. Shakeshaft, "Atomic Rearrangement Collisions at Assymtotically High Impact Velocities" in "Physics of Electronic and Atomic Collisions", S. Datz, ed. (North Holland Press, 1982) pp. 123-138.
11. E. Horsdal-Pedersen, "Electron Capture from K Shells by Light Ions", in "Physics of Electronic and Atomic Collisions", S. Datz, ed. (North Holland Press, 1982) pp. 139-149.
12. S. Hagman, C. L. Cocke, J. R. McDonald, P. Richard, H. Schmidt-Böcking and R. Schuch, Phys. Rev. A 25, 1918 (1982).
13. C. D. Lin, unpublished.
14. R. E. Olson, Phys. Rev. A 24, 1726 (1981).
15. H. Knudsen, P. Hvelplund, L. H. Andersen, S. Bjørnelund, M. Frost, H. K. Haugen and E. Samsø, Physica Scripta (in press).
16. J. Eichler and F. T. Chan, Phys. Rev. A 20, 104 (1979).
17. H. Ruyfuku and T. Watanabe, Phys. Rev. A 21, 745 (1980) and H. Ruyfuku, JAERI-M-82-031 (1982).
18. S. Ohtani, Physica Scripta (in press)
19. M. E. Rudd and J. Macek, Case Studies in Atomic Physics 3, 125 (1972).
20. I. A. Sellin, in Atomic Physics 7, D. Kleppner and F. M. Pipkin, eds. (Plenum Press, 1981) pp. 455-489.
21. M. Breinig, S. B. Elston, S. Huldt, L. Liljeby, C. R. Vane, S. D. Berry, G. A. Glass, M. Schauer, I. A. Sellin, G. D. Alton, S. Datz, S. Overbury, R. Laubert and M. Suter, Phys. Rev. A 25, 3015 (1982).
22. C. R. Vane, I. A. Sellin, S. B. Elston, M. Suter, R. S. Thoe, G. D. Alton, S. D. Berry and G. A. Glass, Phys. Rev. Lett. 43, 1388 (1979).
23. J. R. McDonald and F. W. Martin, Phys. Rev. A 14, 1965 (1971).
24. C. L. Cocke, R. DuBois, T. J. Gray, E. Justiniano and C. Can, Phys. Rev. Lett. 46, 1671 (1981).
25. W. Groh, A. Muller, C. Achenbach, A. S. Schlachter and E. Salzborn, Phys. Lett. 85A, 77 (1981).
26. See e.g., E. N. Fuls, P. R. Jones, E. P. Ziemba and E. Everhart, Phys. Rev. 107, 704 (1957).
27. See e.g., A. Russek and M. T. Thomas, Phys. Rev. 114, 1538 (1959).
28. E. Salzborn, private communication.
29. M. J. Alguard, R. S. Swent, R. L. Pantell, B. L. Berman, S. D. Bloom and S. Datz, Phys. Rev. Lett. 42, 1148 (1979); R. L. Swent, R. H. Pantell, M. J. Alguard, B. L. Berman, S. D. Bloom and S. Datz, Phys. Rev. Lett. 43, 1723 (1979); J. U. Andersen and E. Laegsgaard, Phys. Rev. Lett. 44, 1079 (1980).
30. J. Lindhard, K. Dansk. Vidensk. Selsk. Mat. Fys. Medd. 34 No. 14 (1965).

31. For a general review of channeling see: D. S. Gemmell, Rev. Mod. Phys. 46, 129 (1974).
32. R. L. Swent, R. H. Pantell, S. Datz, M. J. Alguard, B. L. Berman, R. Alverez and D. C. Hamilton, in Physics of Electronic and Atomic Collisions, S. Datz, ed. (Plenum Press, 1981) p. 861.
33. J. U. Andersen, E. Bonderup, E. Laegsgaard, B. B. Marsh and A. H. Sørensen, Nucl. Instrum. Methods 194, 209 (1982).
34. S. Datz, R. Fearick, B. L. Berman, J. Kephart, R. H. Pantell and H. Park, unpublished.
35. J. U. Andersen, K. R. Eriksen and E. Laegsgaard, Physica Scripta 24, 588 (1981).
36. J. U. Andersen, S. Datz, E. Laegsgaard, J. F. P. Sellschop and A. H. Sørensen, Phys. Rev. Lett. 49, 215 (1982).

HEAVY PARTICLE COLLISIONS

Larry Spruch* and Robin Shakeshaft†

*Physics Department, New York University
New York, NY 10003 U.S.A.

†Physics Department, University of Southern California
Los Angeles, CA 90007 U.S.A.

I. INTRODUCTION

The speaker (LS) is on record as believing that occasionally, for all its obvious disadvantages, talks should be given by those who had not been deeply involved, if at all, in the subject to be covered. Possible advantages include great freedom in the choice of topics, the lack of a compulsion to cover every detail of the topics chosen, and an objectivity (if some ignorance) in accreditation. Neither of us had been knowledgeable in two of the three topics we have chosen. To avoid pitfalls in our treatment of the third, asymmetric charge transfer, we have given a completely qualitative discussion of developments in that area.

We have obviously chosen areas we believe to be of particular significance, but having picked three areas, we cannot begin to give the details necessary for a thorough understanding. Our primary purpose is to interest those who have not read the original papers in doing so.

II. NUCLEAR RESONANCES AND THE PROBABILITY OF K-SHELL IONIZATION

If atomic physics has been enjoying a renaissance because of new developments within the field, it is also true that there are

This research was supported in part by the Office of Naval Research under Contracts N00014-76-C-0317 and N00014-77-C-0553 and by the National Science Foundation under Grants PHY-7910413 and PHY81-19010.

areas of atomic physics which are so well understood that agreement between theory and experiment exceeds that in any other field of physics. It is for the latter reason that an area of interplay between atomic physics and another branch of physics can be extremely useful in studying that other branch. Thus, for example, relatively long nuclear half-lives, longer than perhaps picoseconds, can be measured directly, while relatively short nuclear half-lives, shorter than perhaps 10^{-21} seconds, can be determined by measuring the associated half-width, but intermediate half-lives can be difficult to measure by either approach. Now the orbital period of a K-shell electron in a nucleus of atomic number Z is of order $(2\pi a_o/Z)/(Ze^2/\hbar)$ or roughly $10^{-16}/Z^2$ seconds. One can then hope to determine intermediate nuclear half-lives if one can find a measurable effect of the nuclear half-life on a K-shell electron. Consider, for example, the scattering through an angle θ in the center of mass frame of a nuclear projectile P and a target nucleus T, with atomic numbers Z_P and Z_T, respectively. The incident relative kinetic energy $E_i = \frac{1}{2}Mv_i^2$, with M the reduced nuclear mass, is assumed to be in the neighborhood of a resonance with a half-life Δt. (We will talk of ionization, but the argument would be the same for excitation.) We let P(ion) = $P_{fi}(E_i,\theta)$ be the ratio of the probability of the ionization of the K-shell electron in the course of the nuclear scattering process to the probability that there is no ionization during the nuclear scattering. Ionization is from the initial (1s) state i, with normalized wave function $\phi_i(\vec{r})$, to the final continuum state with normalized wave function $\phi_f(\vec{r})$. We will find that P(ion) will depend upon Δt. We expect a significant effect if the nuclear width $\Gamma \approx \hbar/\Delta t$ is at most of the order of the binding energy U_K of the K-shell electron, or, equivalently, if the half-life or time-delay is at least of the order of the orbital period of the K-shell electron.

In section II we use capital letters for nuclear energies, momenta, coordinates (but not velocities), wave functions, and scattering amplitudes, and small letters for the corresponding electronic properties. We have $\vec{K}_i\hbar = M\vec{v}_i$ and $\vec{K}_f\hbar$ for the incident and emergent relative momenta of the nuclei, E_i and E_f for the associated energies, $\vec{R}$ for the P-T separation and V for the P-T interaction, ε_i for the initial K-shell energy of the electron (e^-) and ε_f for the e^- energy in its final ionized state, and $\vec{r}$ for the T-e^- separation. We introduce ω via $E_i-E_f = \hbar\omega = \varepsilon_f-\varepsilon_i$. We will study the problem both in the semi-classical approximation (SCA) and, in the context of quantum theory, in the distorted wave Born approximation (DWBA). We now list a number of approximations which we will make in both the SCA and the DWBA. We will later list additional approximations to be made separately in the SCA and DWBA approximations. The approximations which are made more for convenience of discussion than out of dire necessity are indicated by a star. With the approximations we will make, the mathematics is trivial; in the discussion below the most difficult step is the integration of an exponential. The hard part is of course the justification of the approximations.

i) *The recoil of T can be neglected.

ii) The P-e^- interaction can be treated as a perturbation; a necessary condition for the validity of (ii) is that $Z_P << Z_T$.

iii) The e^--e^- interactions in the neutral target atom can be neglected.

We note that the ejected K-shell electrons have a continuous energy spectrum and therefore provide a poor signature for ionization; a good signature is provided by the X-ray radiation or by one of the Auger electron lines which follow the ionization. (Since the X-ray is emitted by an electron whose initial state is any of a number of p-states --- primarily the 2p state --- with equally populated projections of the angular momentum, and whose final state is an s-state, the X-ray radiation will be spherically symmetric.)

A. The Semi-Classical Approximation

iv') The electron will be treated quantum mechanically, but we assume that the motions of both P and T can be described classically. Indeed, in the lab frame, it follows from assumption (i) that T is always at rest.

v') We assume further that P has an impact parameter of zero with respect to T and moves with constant momentum $\vec{K}_i\hbar$ from time $t = -\infty$ to $t = -\frac{1}{2}\Delta t$, with $\vec{R}(t) = \hat{K}_i v_i(t + \frac{1}{2}\Delta t)$ in this interval, that P then collides with T, the two forming a composity system (with $\vec{R}(t) = 0$) for a time interval Δt, that is, until $t = + \frac{1}{2}\Delta t$, and that P then departs with constant momentum $\vec{K}_f\hbar$, with $\vec{R}(t) = \hat{K}_f v_i(t-\frac{1}{2}\Delta t)$. (For inelastic nuclear scattering, one would simply replace v_i by v_f, the outgoing relative velocity, in this last expression.) The angle θ between $\vec{K}_f$ and $\vec{K}_i$ is fixed by the location of the detector.

The electron is subject to a perturbation H' which for the moment we write simply as $H'(\vec{r},\vec{R}(t))$. The amplitude a_{fi} that the electron will be in a final state f at $t = + \infty$ if it was in an initial state i at $t = -\infty$ is then given, in first order time-dependent perturbation theory, by

$$i\hbar a_{fi} = \int_{-\infty}^{\infty} e^{i\omega t} H'_{fi}[\vec{R}(t)]dt \quad , \tag{2.1}$$

where

$$H'_{fi}[\vec{R}(t)] = \int d^3r\phi_f^*(\vec{r})H'[\vec{r},\vec{R}(t)]\phi_i(\vec{r}) \ . \tag{2.2}$$

The dependence of a_{fi} on θ and E_i (through the presence of $\vec{R}(t)$) is often suppressed. With the changes of variables $t \to t \pm \frac{1}{2}\Delta t$, we can rewrite Eq. (2.1) as

$$i\hbar a_{fi} = \exp(-\tfrac{1}{2}i\omega\Delta t)[\,i\omega^{-1}H'_{fi}(0) + I(-\infty,0;\hat{K}_i)]$$

$$+ \exp(\tfrac{1}{2}i\omega\Delta t)[-i\omega^{-1}H'_{fi}(0) + I(0,\infty;\hat{K}_f)] \quad , \tag{2.3}$$

where

$$I(\alpha,\beta;\hat{K}) \equiv \int_\alpha^\beta \exp(i\omega t)H'_{fi}(\hat{K}v_i t)dt \ . \tag{2.4}$$

We then have

$$P(\text{ion}) \equiv P_{fi}(E_i,\theta) = |a_{fi}(E_i,\theta)|^2 \ . \tag{2.5}$$

To proceed further, we would use

$$H'[\vec{r},\vec{R}(t)] = -Z_p e^2/|\vec{r}-\vec{R}(t)| \ . \tag{2.6}$$

Since P(ion) depends upon a_{fi} which in turn depends upon Δt (through the interference of the two terms in Eq. (2.3) for a_{fi}), a comparison of a theoretical estimate and experimental determination of P(ion) gives an estimate of Δt.

B. A Quantum Approach, in the DWBA

In addition to assuming the validity of the DWBA, we make the following assumptions.

iv") *The nuclear scattering process is an elastic one.

v") *The question of the P-T interaction V did not explicitly arise in the SCA. In the DWBA we need not specify V but we will assume that it is spin-independent and spherically symmetric, that is, that $V = V(R)$. The effects of $V(R)$ are contained in the exact elastic nuclear scattering amplitudes which will appear; these will be denoted in general by $F(\hat{R} \leftarrow \vec{K})$, with $\hat{R}$ denoting an arbitrary emergent direction, and by $F(\hat{K}_f \leftarrow \vec{K}_i) \equiv F_{e\ell}(E,\theta)$ for the scattering process of interest.

vi") θ, the angle of scattering of P, is not small.

vii") The dimension over which the effects of V are significant is very small compared to the dimension a_o/Z_T of the K-shell. This is eminently reasonable with regard to the nuclear component of V, and it is not unreasonable even with regard to the Coulomb component of V, since large-angle nuclear scattering is determined largely by V(R) at small R.

viii") We often neglect the difference between K_i and K_f (but never when either appears in an exponent).

Note that, as opposed to the SCA, the DWBA preserves conservation of linear and angular momentum and of energy.

We now invoke the DWBA, as discussed in Taylor (1972), for example, to write for the ionization amplitude $f(\text{ion}) \equiv f_{fi}(E_i,\theta)$

$$f(\text{ion}) = -(M/2\pi\hbar^2)\langle\Psi^{(-)}_{\vec{K}_f}(\vec{R})|H'_{fi}(\vec{R})|\Psi^{(+)}_{\vec{K}_i}(\vec{R})\rangle \quad , \tag{2.7}$$

where, in our present time-independent formalism, $H'_{fi}(\vec{R})$ differs from the $H'_{fi}[\vec{R}(t)]$ of Eq. (2.2) only in the replacement of $H'[\vec{r},\vec{R}(t)]$ by the time-independent form $H'(\vec{r},\vec{R})$. As $R \sim \infty$, the exact nuclear scattering wave function $\Psi^{(+)}_{\vec{K}_i}$ behaves as

$$\Psi^{(+)}_{\vec{K}_i}(\vec{R}) \sim e^{i\vec{K}_i\cdot\vec{R}} + F(\hat{R} \leftarrow \vec{K}_i)(e^{iK_iR}/R) \; . \tag{2.8}$$

Similarly, since $\Psi^{(-)*}_{\vec{K}} = \Psi^{(+)}_{-\vec{K}}$, we have

$$\Psi^{(-)*}_{\vec{K}_f}(\vec{R}) \sim e^{-i\vec{K}_f\cdot\vec{R}} + F(\hat{R} \leftarrow -\vec{K}_f)(e^{iK_fR}/R) \; . \tag{2.9}$$

In line with approximation vii"), only the asymptotic forms of the nuclear wave functions are relevant. (Note that these forms contain the exact elastic scattering amplitudes.) The insertion of (2.8) and (2.9) into Eq. (2.7) gives four terms. We now make a peaking approximation:

ix")* We drop all contributions involving exponentials which contain $\vec{K}_i$ or K_i and $\vec{K}_f$ or K_f unless the exponent can almost vanish, and we approximate any factor $g(\hat{K}_i,\hat{K}_f)$ of such an exponential by its value for the direction(s) $\hat{K}_i$ and/or $\hat{K}_f$ at which the exponent almost vanishes. We therefore drop the term proportional to $\exp i(\vec{K}_i-\vec{K}_f)\cdot\vec{R}$ since the exponent is negligible only for θ very small, a region we have excluded, and we drop the term proportional to $R^{-2}\exp i(K_i+K_f)R$. In the coefficients of the terms with $\exp i(\vec{K}_i\cdot\vec{R} + K_fR)$ and $\exp i(-\vec{K}_f\cdot\vec{R} + K_iR)$, we approximate $\hat{R}$ by $-\hat{K}_i$ and by $\hat{K}_f$, respectively. Further, we use

$$K_i-K_f = (K_i^2-K_f^2)/(K_i + K_f) \approx (2M/\hbar^2)\hbar\omega/(2K_i) = \omega/v_i \ .$$

Finally, since the angle of scattering is the same, we use rotational invariance to give $F(-\hat{K}_i \leftarrow -\vec{K}_f) = F(\hat{K}_i \leftarrow \vec{K}_f) = F_{e\ell}(E_f,\theta)$, the last step following by definition, we drop terms in $\exp(\pm i)(K_i + K_f)R$ after having performed the angular integration over $d\hat{R}$, and we approximate $1/K_f$ by $1/K_i$ to arrive at

$$i\hbar f(\text{ion}) = F_{e\ell}(E_i,\theta)\int_0^\infty H'_{fi}(\hat{K}_fR)e^{i\omega R/v_i}\,dR/v_i$$

$$+ F_{e\ell}(E_f,\theta)\int_0^\infty H'_{fi}(-\hat{K}_iR)e^{-i\omega R/v_i}\,dR/v_i \ .$$

Setting $t = -R/v_i$ in the second integral and $t = R/v_i$ in the first, and introducing

$$Q \equiv F_{e\ell}(E_f,\theta)/F_{e\ell}(E_i,\theta) \ , \tag{2.10}$$

we have

$$P(\text{ion}) = |f(\text{ion})/F_{e\ell}(E_i,\theta)|^2 =$$

$$|I(0,\infty;\hat{K}_f) + QI(-\infty,0;\hat{K}_i)|^2/\hbar^2 \tag{2.11}$$

where the I's, defined by Eq. (2.4), are precisely those which appeared in the SCA.

We will be concerned with the case $\hbar\omega \ll E_i$, for the probability of the electron picking up an energy comparable to E_i is negligible. If we are off resonance, it follows that we can approximate K_f by K_i and therefore Q by 1, so that P(ion) is independent of $F_{e\ell}(E_i,\theta)$, as is to be expected; with H' treated in first order perturbation theory, for a non-resonant nuclear reaction, the back reaction of the electron on P can be neglected, f(ion) is proportional to $F_{e\ell}$, and P(ion) is independent of $F_{e\ell}$. For a resonant nuclear reaction, on the other hand, the slight change from E_i to E_f can be very important. In other words, in the time independent DWBA, with the further approximation of using the asymptotic forms of the nuclear scattering wave functions, the dependence of P(ion) on the half-life of the compound nucleus originates in the strong E dependence of $F(E,\theta)$ near a resonance. More precisely, on physical grounds we might expect the energy $\hbar\omega$ given up to the K electron to be of order U_K. (Formally, this follows from the quantum matrix element $H'_{fi}[\vec{R}]$ for the electron defined by Eq. (2.2) with $H'[\vec{r},R(t)]$ replaced by $H'[\vec{r},\vec{R}]$. Thus, with a factor $\exp(-Z_T r/a_o)$ from ϕ_i, and in the approximation in which ϕ_f is proportional to $\exp(i\vec{k}_f\cdot\vec{r})$, we expect $H'_{fi}[\vec{R}]$ to fall off rapidly for k_f larger than Z_T/a_o.) If $F(E,\theta)$ is to vary significantly as E varies by an amount U_K, one must have $\Gamma \lesssim U_K$.

C. Comparison of the SCA and the DWBA

Even though energy, momentum and angular momentum are conserved in the DWBA but not in the SCA, there is a close relation between the two approaches, and indeed we have already seen that they involve the identical I integrals. The analogy can be pursued further if $\omega\Delta t \ll 1$, and if we can approximate $F(E_f,\theta) = F(E_i-\hbar\omega,\theta)$ by $F(E,\theta) - \hbar\omega\partial F(E,\theta)/\partial E$, evaluated at $E = E_i$, so that, using the quantum mechanical time delay τ defined by

$$\tau = \tau(E,\theta) = -i\hbar\partial \ell n F(E,\theta)/\partial E \quad ,$$

we can write $Q = 1-i\omega\tau$. In the SCA, we then have from Eqs. (2.5) and (2.3), writing the latter with an over-all factor $\exp(\frac{1}{2}i\omega\Delta t)$ and then using $\exp(-i\omega\Delta t) \approx 1-i\omega\Delta t$,

$$\hbar^2 P(\text{ion}) = \left|(1-i\omega\Delta t)I(-\infty,0;\hat{K}_i) + I(0,\infty;\hat{K}_f) + \Delta t H'_{fi}(0)\right|^2 . \quad (2.12)$$

To obtain P(ion) in the DWBA from Eq. (2.12) for P(ion) in the SCA, we must drop the last term and replace Δt by τ in the first term. The SCA and DWBA estimates of P(ion) may not be quite as close as they seem to be. Firstly, Δt is real while $\Delta\tau$ can be complex. Secondly, the DWBA P(ion) has no $H'_{fi}(0)$ term. Indeed, in the SCA, the $H'_{fi}(0)$ term originates in the P-e^- interaction during the time P is at R = 0, while in the DWBA we used the asymptotic forms of $\Psi_i^{(+)}$ and $\Psi_f^{(-)}$, which are surely incorrect for small R and in particular at R = 0. To obtain the DWBA analog of the $H'_{fi}(0)$ term, one must expand the nuclear wave functions (<u>not</u> their asymptotic forms) in partial waves, and study the monopole component of H', originating in the region R < r. The point is that the possibility of P penetrating the classically forbidden region is rather larger for the resonant than for the non-resonant case. Nevertheless, when all is said and done, the contribution from the region $R < r$ will normally be quite small; the monopole contribution can be significant, but its inclusion does not change the form of Eq. (2.12), though a slight redefinition is necessary.

We close this section with a few comments on the literature.

Similar processes had been considered earlier, but the present process was first considered, in the SCA, by Ciochetti and Molinari (1965). Blair et al. (1978) recorded, without proof, the DWBA theoretical result. The analysis presented above follows very closely that of Feagin and Kocbach (1981). A proof of the DWBA result, somewhat of a tour de force, has been given by Blair and Anholt (1982). (The Appendix of this paper contains a study of various SCA analyses.) See also McVoy and Weidenmüller (1982). The first experiment showing the effect of the nuclear half-life on the ionization probability was described in the paper by Blair et al. referred to just above. In that experiment on ^{58}Ni and in a second on ^{88}Sr by Chemin et al. (1981), protons were elastically scattered across a resonance for which Γ was comparable in magnitude to U_K, and the effect was both expected and seen. In an experiment by Duinker et al. (1980) on the scattering of protons by ^{12}C across a resonance, the effect was seen even though it is not expected, since here Γ is much larger than U_K; this problem has not yet been resolved. A short but very nice review of a number of experiments which deeply involve the interplay of atomic and nuclear physics, not just the effect of a nuclear resonance on K-shell ionization considered here, has been given by Merzbacher (1982). See also "Two Notes on Sec. II" just before the list of references.

III. EFFECT OF COLLISIONS ON THE EMISSION OF RADIATION

A. Introduction

Consider a medium containing N two-level one-electron atoms. The n-th atom has the normalized wave function

$$\Psi^n(\vec{r}_n,\vec{R}_n,t) = A_1^n(\vec{R}_n,t)\psi_1(\vec{r}_n) + A_2^n(\vec{R}_n,t)\psi_2(\vec{r}_n) \quad , \tag{3.1}$$

where $\vec{R}_n$ locates the center of mass of the atom and where $\vec{r}_n$ is the electron coordinate relative to the atom's center of mass. The polarization (dipole moment density) of the medium at position $\vec{X}$ and time t is

$$\vec{P}(\vec{X},t) = Ne\vec{r}_{12}\rho_{12}(\vec{X},t) + \text{c.c.} \quad , \tag{3.2}$$

where

$$\vec{r}_{12} = \int d\vec{r}\psi_1(\vec{r})\vec{r}\psi_2(\vec{r})^* \quad , \tag{3.3}$$

and where, assuming no correlations between atoms, ρ_{12} is an off-diagonal element of the ensemble average density matrix:

$$\rho_{ij}(\vec{X},t) = \frac{1}{N} \sum_{n=1}^{N} A_i^n(\vec{X},t)A_j^n(\vec{X},t)^* \quad . \tag{3.4}$$

The polarization of the medium or, equivalently, the "coherence" $\rho_{12}(\vec{X},t)$, governs the response of the medium to both an applied field and (as in spontaneous emission) to the vacuum field.

The population densities of states 1 and 2 in a macroscopically small region centered at $\vec{X}$ are $\rho_{11}(\vec{X},t)$ and $\rho_{22}(\vec{X},t)$, respectively. The total fractions of atoms in states 1 and 2 are $\int\rho_{11}(\vec{X},t)d\vec{X}$ and $\int\rho_{22}(\vec{X},t)d\vec{X}$. The populations change due to pumping and spontaneous emission (natural decay). Changes in populations due to collision induced transitions are neglected here. Changes in the coherence ρ_{12} occur because of pumping, natural decay, dipole dephasing between atoms of different velocities (Doppler broadening), and collisional decay (pressure broadening). We consider two-level atoms which do not interact with one another, and which can emit and absorb radiation and scatter from stationary (infinitely massive structureless) perturbers. Recently, new insight has been gained into the understanding of collisional effects on the coherence. We describe here the modern view of collisional effects on ρ_{12}, following closely the work of Berman (1975) and Berman et al. (1982).

B. Qualitative Discussion

Traditionally, the destruction of ρ_{12} by collisions is attributed to a loss of coherence of the phases for different values of n (phase interruption) of the products $A_1^n(\vec{R}_n,t)A_2^n(\vec{R}_n,t)^*$. (The traditional theory is described by, among others, Sobel'man (1972).) However, an alternative explanation, applicable in the classical regime, is that collisions destroy ρ_{12} by reducing the overlap of $A_1^n(\vec{R}_n,t)$ and $A_2^n(\vec{R}_n,t)^*$. To see this, we assume for simplicity that the collision is impulsive and that the scattering angle is small compared to unity. The effective potential between an active atom in state i and a perturber is, to a <u>first approximation</u>,

$$V_i(\vec{R}) = \int\psi_i(\vec{r})^* V(\vec{r},\vec{R})\psi_i(\vec{r})d\vec{r} \quad , \tag{3.5}$$

where $\vec{R}$ is the atom-perturber separation and $V(\vec{r},\vec{R})$ is their interaction. An atom entering a collision with momentum $\hbar\vec{k}$ in state i will scatter through an angle θ_i given approximately by

$$\theta_i = -\frac{1}{\hbar k}\int_{-\infty}^{\infty} \frac{\partial V_i(\vec{R})}{\partial b}\, dt \quad , \tag{3.6}$$

where in the integrand we may write $\vec{R} = \vec{b} + \vec{v}t$, with $\vec{b}$ and $\vec{v}$ the impact parameter and velocity of the incident atom. Assuming the diffraction angle $1/(kb)$ to be small, the scattering is classical if

$$\theta_i >> 1/(kb) \ . \tag{3.7}$$

If this condition is satisfied, the population $\rho_{ii}(\vec{R},t)$ follows a classical trajectory. The trajectories for the two populations are classically distinguishable if

$$|\theta_2 - \theta_1| >> 1/(kb) \quad , \tag{3.8}$$

in which case the overlap of $A_1^n(\vec{R}_n,t)$ and $A_2^n(\vec{R}_n,t)$ is effectively zero after the collision. Therefore, assuming that the above inequalities hold, if an atom enters a collision in a superposition state, so that initially $\rho_{12} \neq 0$, the separation of population trajectories results in ρ_{12} vanishing after the collision. While this view differs from the traditional (phase interruption) view, the difference is not so great when one considers the manner in which the overlap $A_1^n(\vec{R}_n,t)A_2^n(\vec{R}_n,t)^*$ vanishes in the classical limit; the quantum-mechanical overlap acquires a large phase which varies rapidly with $\vec{R}_n$ and the overlap vanishes when averaged over slight variations of $\vec{R}_n$. It is interesting to combine Eqs. (3.6) and (3.8); approximating $\partial V_i/\partial b$ by V_i/b, Eq. (3.8) becomes

$$\hbar^{-1}\left|\int[V_2(R) - V_1(R)]dt\right| >> 1 \ . \tag{3.9}$$

The value of b for which the left-hand-side equals unity is denoted as b_W, the "Weisskopf radius". For $b < b_W$ collisions are classical and destroy ρ_{12} through trajectory separation (or phase interruption in the traditional view). For $b > b_W$ the scattering is nonclassical and does not destroy ρ_{12}. Thus ρ_{12} survives collisions only in the diffractive zone, that is, for $b > b_W$ and therefore in a narrow forward scattering cone.

C. Quantitative Discussion: Transport Equations

A more quantitative discussion requires the use of transport equations for the density matrix. To begin, we consider a beam of one-level atoms, each of mass m and incident velocity $\vec{v}$, scattering from a collection of perturbers with volume density N. The transport equation, in the form of Eq. (3.14b) below, could be written down without derivation simply on the basis of physical principles. However, we sketch a derivation here since it facilitates the derivation of the somewhat more complicated transport equations for two-level atoms. The atoms in the incident beam are assumed to have wavepackets of similar form but random impact parameters. Thus the normalized incoming wavepacket of a typical atom is

$$\phi_{in}(\vec{k}) = \exp(-i\vec{b}\cdot\vec{k})\phi_o(\vec{k})$$

where $\phi_o(\vec{k})$ is independent of the impact parameter $\vec{b}$ and is sharply peaked when $\vec{k}$ equals $m\vec{v}/\hbar$. After a collision, an atom is represented by the outgoing wavepacket $\phi_{out}(\vec{k})$, where (Taylor, 1972)

$$\phi_{out}(\vec{k}) = \phi_{in}(\vec{k}) + (i/\pi)\int d\vec{k}'\delta(k^2-k'^2)f(\vec{k}' \rightarrow \vec{k})\phi_{in}(\vec{k}'); \quad (3.10)$$

$f(\vec{k}' \rightarrow \vec{k})$ is the scattering amplitude. The ensemble average probability densities before and after the collision are $\rho_{in}(\vec{k})$ and $\rho_{out}(\vec{k})$, respectively, where the domain of integration is the cross sectional area A of the incident beam,

$$\rho_\alpha(\vec{k}) = (1/A)\int d^2b|\phi_\alpha(\vec{k})|^2 \;, \quad [\therefore \rho_{in}(\vec{k}) = |\phi_o(\vec{k})|^2] \;, \quad (3.11)$$

with α in or out. Now on average the beam encounters one perturber in the time interval $\tau = 1/(NvA)$. The collision rate of change of the probability density $\rho_{11}(\vec{k},t) \equiv \rho(\vec{k},t)$ is therefore

$$\left.\frac{\partial\rho(\vec{k},t)}{\partial t}\right|_{coll} = \frac{\rho_{out}(\vec{k})-\rho_{in}(\vec{k})}{\tau}$$

$$= Nv\int d^2b\,[|\phi_{out}(\vec{k})|^2 - |\phi_{in}(\vec{k})|^2] \,. \quad (3.12)$$

If Eq. (3.10) is used to substitute for $\phi_{out}(\vec{k})$ in Eq. (3.12), the integration over b may be done by assuming A to be sufficiently large that we can use, for $\vec{k} \neq \vec{k}'$

$$\int d^2b\phi^*_{in}(\vec{k}')\phi_{in}(\vec{k}) = (2\pi)^2\delta^2(\vec{k}'_\perp - \vec{k}_\perp)\phi^*_o(\vec{k}')\phi_o(\vec{k}) \;, \quad (3.13)$$

where $\vec{k}_\perp$ denotes the component of $\vec{k}$ perpendicular to $\vec{v}$. An integration over the variable $\vec{k}'$ may then be done using the fact that $\phi_o(\vec{k}')$ is highly localized --- see the analogous discussion of Taylor (1972), pages 49-51. Setting $\rho_{in}(\vec{k}) = \rho(\vec{k},t)$, the following transport equation is obtained:

$$\left.\frac{\partial\rho(\vec{k},t)}{\partial t}\right|_{coll} = -\Gamma(\vec{k})\rho(\vec{k},t) + \int d\vec{k}'W(\vec{k}' \to \vec{k})\rho(\vec{k}',t) \quad , \qquad (3.14a)$$

where, with $\sigma(\vec{k})$ denoting the total cross section,

$$W(\vec{k}' \to \vec{k}) = Nv|f(\vec{k}' \to \vec{k})|^2 \, k^{-2}\delta(k-k') \quad , \qquad (3.15)$$

$$\Gamma(\vec{k}) = 4\pi(N\hbar/m)\mathrm{Im}f(\vec{k} \to \vec{k}) = Nv\sigma(\vec{k}) = \int d\vec{k}'W(\vec{k} \to \vec{k}') \quad . \qquad (3.16)$$

Note that $k = k'$ in Eq. (3.14a) since the perturbers do not recoil. Further, $k = mv/\hbar$ since $\phi_o(\vec{k})$ is highly localized. However, since Eq. (3.14a) is linear, it applies when $\rho_{in}(\vec{k})$ is a superposition of densities localized in different regions of $\vec{k}$-space; hence $\rho(\vec{k},t)$ may represent a broad distribution in $\vec{k}$-space. From Eqs. (3.15) and (3.16), Eq. (3.14a) can be written in the more transparent form

$$\left.\frac{\partial\rho(\vec{k},t)}{\partial t}\right|_{coll} = -Nv\sigma(\vec{k})\rho(\vec{k},t)+Nv\int d\hat{k}'|f(\vec{k}'\to\vec{k})|^2\rho(\vec{k}',t). \qquad (3.14b)$$

The transport equations for the density matrix $\rho_{ij}(\vec{k},t)$, with $\rho_{ij}(\vec{k},t)$ the momentum space analog of $\rho_{ij}(\vec{R},t)$ of Eq. (3.4), of two-level atoms can be derived similarly if collision induced transitions between the two levels are neglected. We have (Berman, 1975)

$$\left.\frac{\partial\rho_{ii}(\vec{k},t)}{\partial t}\right|_{coll} = -\Gamma_{ii}^{vc}(\vec{k})\rho_{ii}(\vec{k},t) + \int d\vec{k}'W_{ii}(\vec{k}'\to\vec{k})\rho_{ii}(\vec{k}',t), \qquad (3.17a)$$

$$\left.\frac{\partial\rho_{12}(\vec{k},t)}{\partial t}\right|_{coll} = -(\Gamma_{12}^{ph}(\vec{k}) + \Gamma_{12}^{vc}(\vec{k}))\rho_{12}(\vec{k},t)$$

$$+ \int d\vec{k}'W_{12}(\vec{k}' \to \vec{k})\rho_{12}(\vec{k}',t) \quad , \qquad (3.17b)$$

where, if $f_i(\vec{k} \to \vec{k}')$ is the scattering amplitude for an atom in state i,

$$W_{ij}(\vec{k}' \to \vec{k}) = Nvf_i(\vec{k}' \to \vec{k})f_j^*(\vec{k}' \to \vec{k})k^{-2}\delta(k-k') \quad , \qquad (3.18)$$

$$\Gamma_{ij}^{vc}(\vec{k}) = \int d\vec{k}' W_{ij}(\vec{k} \to \vec{k}') \quad , \tag{3.19}$$

$$\Gamma_{12}^{ph}(\vec{k}) = -2\pi i N(\hbar/m)[f_1(\vec{k} \to \vec{k}) - f_2^*(\vec{k} \to \vec{k})] - \Gamma_{12}^{vc}(\vec{k}) \ . \tag{3.20}$$

The superscript vc denotes velocity changing. If velocity changing collisions are neglected, that is, if $W_{ij}(\vec{k}\to\vec{k}') = \bar{W}_{ij}(\vec{k})\delta^3(\vec{k}'-\vec{k})$ so that $\Gamma_{ij}^{vc}(\vec{k}) = \bar{W}_{ij}(\vec{k})$, the integral term and the term in Γ_{ij}^{vc} cancel in Eqs. (3.17), and these equations reduce to the much simpler equations of the traditional pressure broadening theory:

$$\left.\frac{\partial\rho_{ii}(\vec{k},t)}{\partial t}\right|_{coll} = 0 \quad , \tag{3.21a}$$

$$\left.\frac{\partial\rho_{12}(\vec{k},t)}{\partial t}\right|_{coll} = -\Gamma_{12}^{ph}(\vec{k})\rho_{12}(\vec{k},t) \ . \tag{3.21b}$$

With velocity changing collisions neglected, Γ_{12}^{ph} is, in the traditional view, the collision decay rate due to phase interruption of the atomic dipole. If velocity changing collisions are allowed but if the scattering amplitudes f_1 and f_2 are equal, it follows from Eqs. (3.18)-(3.20) and the optical theorem that $\Gamma_{12}^{ph}(\vec{k}) = 0$; thus there is no phase interruption of the atomic dipole during the collision if the atom scatters as a structureless entity.

The qualitative analysis of the previous subsection indicates that, in general, $W_{12}(\vec{k} \to \vec{k}')$ vanishes in the classical scattering regime due to trajectory separation. (More accurately, W_{12} oscillates rapidly and the integral over the classical region vanishes.) W_{12} gives a nonvanishing contribution only in the diffractive scattering region. Thus any departure from a transport equation for ρ_{12} of the form of Eq. (3.21b) arises from diffractive scattering. Such a departure may be observed by creating a photon echo, as we now briefly discuss.

D. Laser Spectroscopy

Suppose that a single-mode laser of frequency Ω interacts with the atoms. Assume the laser field $\vec{E}$ propagates in the z direction, i.e., $\vec{E} = \vec{E}_0\cos(Kz-\Omega t)$. In the absence of collisions and under appropriate initial conditions $\rho_{12}(\vec{k},t)$ factors into

$$\rho_{12}(\vec{k},t) = \rho_{12}(\vec{k}_T)\rho_{12}(k_z,t) \tag{3.22}$$

where, following convention, we use $\rho_{12}(x)$ to denote different functions for different arguments x, and where $\vec{k}_T$ is the component of $\vec{k}$ perpendicular to the z-axis. This factorization is assumed to hold, to a first approximation, in the presence of collisions. Substituting into Eq. (3.17b), integrating over $\vec{k}_T$, and defining the normalization of $\rho_{12}(\vec{k}_T)$ --- it is not defined by Eq. (3.22) --- to be $\int\rho_{12}(\vec{k}_T)d^2k_T = 1$, we find

$$\left.\frac{\partial\rho_{12}(k_z,t)}{\partial t}\right|_{coll} = -[\Gamma^{vc}_{12}(k_z) + \Gamma^{ph}_{12}(k_z)]\rho_{12}(k_z,t)$$

$$+ \int_{-\infty}^{\infty} dk_z' W_{12}(k_z' \to k_z)\rho_{12}(k_z',t) \quad , \tag{3.23}$$

$$W_{12}(k_z' \to k_z) = \int d^2k_T \int d^2k_T' W_{12}(\vec{k}' \to \vec{k})\rho_{12}(\vec{k}_T') \quad , \tag{3.24}$$

$$\Gamma^{\beta}_{12}(k_z) = \int d^2k_T \Gamma^{\beta}_{12}(\vec{k})\rho_{12}(\vec{k}_T) \quad , \tag{3.25}$$

where β is vc or ph.

The field frequency seen in the rest frame of an atom is $\bar{\Omega} = \Omega - Kv_z$ (Doppler shift) where $K = \Omega/c$ and $v_z = \hbar k_z/m$. If $|\bar{\Omega} - \omega| << |\bar{\Omega} + \omega|$, where ω is the transition frequency for the two atomic levels, $\rho_{12}(k_z,t)$ will oscillate in time with the field as $\rho_{12}(k_z,t) = \tilde{\rho}_{12}(k_z,t)\exp(i\bar{\Omega}t)$ where $\tilde{\rho}_{12}(k_z,t)$ varies slowly with time (rotating wave approximation). From Eq. (3.23), we then have (Berman et al., 1982)

$$\left.\frac{\partial\tilde{\rho}_{12}(k_z,t)}{\partial t}\right|_{coll} = -[\Gamma^{vc}_{12}(k_z) + \Gamma^{ph}_{12}(k_z)]\tilde{\rho}_{12}(k_z,t)$$

$$+ \int_{-\infty}^{\infty} dk_z' W_{12}(k_z' \to k_z)\exp[iK(v_z - v_z')t]\tilde{\rho}_{12}(k_z',t) \ . \tag{3.26}$$

(Note that since we are concerned here with collisional effects, we have not included a term originating in $(\partial/\partial t)\exp(i\bar{\Omega}t)$.) This equation governs the collision rate of change of ρ_{12} in the presence of a single-mode laser. To obtain the full rate of change of ρ_{12} with time, the impact approximation is assumed. In this approximation a collision is regarded as instantaneous compared to all other relevant time scales. Then $\partial\rho_{12}/\partial t|_{coll}$ can be simply added to the time derivative $\partial\rho_{12}/\partial t|_{rad}$ due to coupling with the radiation field to give the full time derivative.

Only the diffractive scattering contributes to the integral of

Eq. (3.26). Now diffractive scattering occurs in a very narrow forward cone (assuming the atoms are not moving too slowly) and so the accompanying velocity changes are small. Let δv be the characteristic value of the velocity change $v_z' - v_z$ in the diffractive region. A coherence $\tilde{\rho}_{12}(k_z,0)$ which is prepared at time $t = 0$ will subsequently decay; let τ_c be the coherence lifetime. If $K\delta v\tau_c << 1$ the exponential in the integrand of Eq. (3.26) may be set equal to unity. Further, $\tilde{\rho}_{12}(k_z',t)$ is expected to vary slowly over the diffractive region so that it may be taken out of the integral of Eq. (3.26) at the value $\tilde{\rho}_{12}(k_z,t)$. In this case, the term in Γ_{12}^{vc} cancels the integral term and Eq. (3.26) reduces to the traditional equation

$$\left.\frac{\partial\tilde{\rho}_{12}(k_z,t)}{\partial t}\right|_{coll} = -\Gamma_{12}^{ph}(k_z)\tilde{\rho}_{12}(k_z,t) \quad , \tag{3.27}$$

leading to the prediction that $\tilde{\rho}_{12}$ has the decay rate $\mathrm{Re}\Gamma_{12}^{ph}(k_z)+\gamma_{12}$, where γ_{12} is the natural decay rate. (If the experimental situation involves a distribution of k_z, we must average over k_z and include the free induction decay rate due to a relative dephasing of atomic dipoles with different velocities.) However, suppose instead that $K\delta v\tau_c >> 1$. For $t > 1/(K\delta v)$ the exponential oscillates rapidly over the diffractive region and the integral in Eq. (3.26) vanishes so we obtain (Berman et al., 1982)

$$\left.\frac{\partial\tilde{\rho}_{12}(k_z,t)}{\partial t}\right|_{coll} = -[\Gamma_{12}^{vc}(k_z) + \Gamma_{12}^{ph}(k_z)]\tilde{\rho}_{12}(k_z,t) \quad , \tag{3.28}$$

leading to the prediction of the larger decay rate $\mathrm{Re}[\Gamma_{12}^{vc}(k_z) + \Gamma_{12}^{ph}(k_z)] + \gamma_{12}$. Now τ_c, the effective coherence lifetime of ρ_{12}, depends on the experimental situation. Without going into the details of a photon echo --- a lucid discussion is given in Sargent et al., 1974 --- suffice it to say that in a photon echo experiment τ_c can be made large; the condition $K\delta v\tau_c >> 1$ can therefore be attained and the larger decay rate confirmed. This was done recently (Mossberg et al., 1980), establishing for the first time the influence of diffractive scattering on the emission of radiation.

IV. ASYMMETRIC CHARGE TRANSFER

A measure of the great practical importance of the charge transfer process is the very considerable experimental and theoretical effort devoted to that process. The areas in which significant progress has very recently been recorded include atom capture as well as electron capture, the eikonal approximation, and versions of the Glauber approximation. Unfortunately, space permits only one topic, an important step in our understanding of asymmetric charge transfer.

To appreciate this step, it will be useful to consider earlier developments in asymmetric and symmetric charge transfer. There have been a number of relatively recent reviews in these areas (Basu et al., 1978; Belkić et al., 1979; Shakeshaft and Spruch, 1979; Shakeshaft, 1982), and we limit ourselves to some brief comments.

We wish to consider electron capture by a projectile P (a bare nucleus of charge $Z_p e$) incident with a high velocity $\vec{v}$ on a neutral atom. The target nucleus T has a charge $Z_T e$, and the process is

$$P^{(Z_p)} + \text{Atom} \longrightarrow P^{(Z_p - 1)} + \text{Ion}^{(+1)} \; .$$

One must and can do better (Briggs and Taulbjerg, 1979) but we assume that all electron-electron interactions are negligible. In a Born expansion in a non-relativistic context, the n-th term represents the contribution associated with n scatterings. We make one remark on potential scattering (scattering by a target with no internal degrees of freedom) before considering charge transfer. The Born expansion, for sufficiently large incident energy E and for many potentials, is a convergent expansion in powers of $\bar{V}/E$, where $\bar{V}$ is a characteristic value of V. The first Born term therefore dominates for sufficiently large E for potential scattering.

We begin our consideration of charge transfer by studying the symmetric case, for which $Z_p = Z_T \equiv Z$. In the first Born term, the main contribution originates in components of the target and final bound state wave functions for which the velocity of the electron is comparable with $\vec{v}$, and those amplitudes are very small for v large --- more precisely, for $v \gg Ze^2/\hbar$, a characteristic electron velocity in the initial and final state --- even if, as we do for simplicity, we consider capture from and to ground states. The second Born term can be described roughly as follows. The electron can initially have a small speed, for it is given a speed close to v in a close collision with P. The electron then moves, in this intermediate state, almost as a free particle. (The uncertainty in its energy is $\Delta E \approx p\Delta p/m \approx \hbar v/a$, with a an atomic dimension, so that $\Delta E/E$ falls off as $1/v$, but the off-the-energy shell component gives a significant contribution). The electron is then scattered elastically by T, emerging with velocity close to $\vec{v}$, and is captured. The second Born term dominates over the first even though it involves an additional collision (and therefore an additional factor, proportional to e^2, which often suggests that the term involved is of higher order) because the second Born term does not require high speed components in the initial and final bound states. It is widely believed, though it has not been proved, that higher order Born terms are dominated by the second, for they suffer from having still further factors of e^2, and they have no compensating advantages since the second Born term already allows low velocities in the initial and final states.

For many applications of great current interest, one has Z_P small, say unity, but $Z_T \gtrsim 5$, and incident energies E such that v is rather large compared to $Z_P e^2/\hbar$ but not compared to $Z_T e^2/\hbar$. It is then inappropriate to ignore multiple e^--T collisions; rather, all e^--T collisions must be included. However, we can continue to ignore multiple e^--P collisions. In the present asymmetric analog of the second Born term in the symmetric case, the electron in the intermediate state is described by a Coulomb wave rather than by a plane wave. A natural starting point is to assume that the Coulomb wave is on the energy shell. This amounts to the impulse approximation, developed largely in this context by Briggs (1977) and also by Kocbach (1980) and Amundsen and Jakubassa (1980). While this approach gives good results at larger incident energies, theory and experiment begin to disagree at energies rather above the value at which the disagreement had been expected. The point is that the off-the-energy-shell component of the intermediate state wave function --- now a Coulomb wave --- must be retained. The analysis is tricky, and requires further approximations. It is a major achievement that the final result is obtained in tractable form: the predicted asymmetric charge transfer cross section differs from the impulse approximation prediction by a rather simple factor, one which gives considerably better agreement with the data at lower energies. We note incidentally that this work not only provides a theoretical foundation for asymmetric charge transfer but also provides much deeper insight into a number of earlier approaches, placing them in a hierarchy of successive approximations. See Macek and Taulbjerg (1981), Briggs, Macek and Taulbjerg, to be published, and Macek and Alston, to be published.

Two Notes on Sec. II:

1) Many intermediate nuclear half-lives can be determined by means independent of the measurement of P(ion) --- most generally by matching scattering data to the Breit-Wigner formula, but also by using special techniques, such as channeling. One time interval which might be determined most easily by a measurement of P(ion) is the time interval during which two heavy ions remain in one another's neighborhood in the course of a scattering process.

2) The argument of the first paragraph of Sec. II can be reversed; one can use a detailed knowledge of the properties of a nuclear resonance to determine an atomic property. Thus, let us rewrite the equation above Eq. (2.10) as

$$f(\text{ion}) = F_{e\ell}(E_i,\theta)A + F_{e\ell}(E_f,\theta)B$$

where A and B depend only upon atomic properties. This form remains valid even if one includes contributions from small values of R.

Now assume, for example, that the nuclear resonant state is an $s_{\frac{1}{2}}$ state, and choose θ to be 90° so that the only relevant interference term arises from the monopole term, with both P and e^{-} emerging in spherically symmetric distributions. One can then show that $B = A^{*}$. If one were not at a resonance, one would have $F_{e\ell}(E_i,\theta) \approx F_{e\ell}(E_f,\theta)$, and therefore $f(\text{ion}) \approx 2\, F_{e\ell}(E_i,\theta)\text{ReA}$. At resonance, however, one can also, at least in principle, determine the imaginary component of A. It is not clear however if theory and experiment are now good enough to determine ImA. (See Blair et al., 1978, and references therein.)

References

Amundsen, P.A. and D. Jakubassa, 1980, Charge transfer in asymmetric heavy-ion collisions, J. Phys. B, 13:L467.

Basu, D., S.C. Mukherjee and D.P. Sural, 1978, Electron capture processes in ion-atom collisions, Physics Reports, 42:145.

Belkić, Dz., R. Gayet and A. Salin, 1979, Electron capture in high-energy ion-atom collisions, Physics Reports, 56:279.

Berman, P.R., T.W. Mossberg and S.R. Hartman, 1982, Collision kernels and laser spectroscopy, Phys. Rev. A, 25:2550.

Blair, J.S., P. Dyer, K.A. Snover, and T.A. Trainor, 1978, Nuclear "time-delay" and X-ray-proton coincidences near a nuclear scattering resonance, Phys. Rev. Lett., 41:1712.

Blair, J. and R. Anholt, 1982, Theory of K-shell ionization during nuclear resonance scattering, Phys. Rev. A, 25:907.

Briggs, J.S., 1977, Impact-parameter formulation of the impulse approximation for charge exchange, J. Phys. B, 10:3075.

Briggs, J.S. and K. Taulbjerg, 1979, Charge transfer by a double-scattering mechanism involving target electrons, J. Phys. B, 12:2565.

Briggs, J., J. Macek, and K. Taulbjerg, Theory of asymmetric charge transfer, Comment. At. Mol. Phys., in press.

Chemin, J.F., R. Anholt, Ch. Stoller, W.E. Meyerhoff, and P.A. Amundsen, 1981, Measurement of ^{88}Sr K-shell ionization probability across the nuclear elastic-scattering resonance at 5060 keV, Phys. Rev. A, 24:1218.

Ciochetti, G. and A. Molinari, 1965, K electron shell ionization and nuclear reactions, Nuovo Cimento, 40B:69.

Duinker, W., J. van Eck, and A. Niehaus, 1980, Experimental evidence for the influence of inner-shell ionization on resonant nuclear scattering, Phys. Rev. Lett., 45:2102.

Feagin, J.M. and L. Kocbach, 1981, Inner-shell excitation in central collisions of light ions with atoms: an interplay between atomic and nuclear processes, J. Phys. B: At. Mol. Phys., 14:4349.

Kocbach, L., 1980, Impulse approximation calculations for capture of K-shell electrons by fast light nuclei, J. Phys. B, 13:L665.

Macek, J. and K. Taulbjerg, 1981, Correction to Z_P/Z_T expansions for electron capture, Phys. Rev. Lett., 46:170.
Macek, J. and S. Alston, Theory of electron capture from a hydrogen-like ion by a bare ion, Phys. Rev. A, in press.
McVoy, K.W. and H.A. Weidenmüller, 1982, Analysis of K-shell ionization accompanying nuclear scattering, Phys. Rev. A, 25:1462.
Merzbacher, E., 1982, The interplay between nuclear and atomic phenomena in ion-atom collisions, in Proceedings of the Divisional Conference of the European Physical Society on "Nuclear and Atomic Physics with Heavy Ions", Bucharest, Romania, June 1981, to be published.
Mossberg, T.W., R. Kachru and S.R. Hartmann, 1980, Observation of collisional velocity changes associated with atoms in a superposition of dissimilar electronic states, Phys. Rev. Lett., 44:73.
Sargent, M., M.O. Scully and W.E. Lamb, 1974, Chap. 13, in "Laser Physics", Addison-Wesley, Reading.
Shakeshaft, R., 1982, Atomic rearrangement collisions at asymptotically high impact velocities, in "Invited Papers of the XII International Conference on the Physics of Electronic and Atomic Collisions", Gatlinberg, Tennessee, July 1981, S. Datz (editor), North Holland, Amsterdam.
Shakeshaft, R. and L. Spruch, 1979, Mechanism for charge transfer (or for the capture of any light particle) at asymptotically high impact velocities, Rev. Mod. Phys., 51:369.
Sobel'man, I.I., 1972, Chap. 10, in "Introduction to the Theory of Atomic Spectra", Pergamon, New York.
Taylor, J.R., 1972, in "Scattering Theory", Wiley, New York.

LIGHT SCATTERING AS A PROBE FOR ATOMIC INTERACTIONS

Keith Burnett

Department of Physics, University of Colorado and Joint Institute for Laboratory Astrophysics, University of Colorado and National Bureau of Standards, Boulder, Colorado 80309 U.S.A.

INTRODUCTION

In this talk we should like to describe some recent work on the scattering of light by atomic and molecular collisions.[1-8] This work, in our opinion, offers the prospect of a rather direct technique for studying the mechanisms of heavy particle collisions and reactions. We should also like to show that knowledge of such mechanisms, as well as being interesting in its own right, has important consequences for the field of laser assisted and modified collisions. Let us consider to begin with a conventional atomic or molecular crossed beam experiment where, for the sake of discussion, we suppose that we can specify all the relevant ingoing states of the colliding partners: translational and internal energies, spin, etc. Let us suppose that we can also measure all the corresponding quantities for the final or product states. If we have a ground state to ground state scattering problem with a single open channel there are well-established inversion procedures.[9] For a multichannel problem, i.e. one where several asymptotic states are accessible, inversion is rarely possible. It may, in some cases, be possible to calculate an "ab initio potential" and do full quantal calculations; more often, however, we are forced to consider models of the interaction between the collision partners and consider which model fits the data best. It would, of course, be most advantageous if we could obtain some information more directly about the collision complex rather than have to be content with asymptotic information which may not tie down the form of the potential very well.

What we should like to have is the equivalent of the bound-bound spectroscopy for molecules that enables such accurate information to be obtained for the interatomic potentials that bring about bound molecular states. This is precisely what we are going to discuss: the free-free spectroscopy of the transient species formed during a collision.

To see how we should be able to study the evolution of a collision let us consider first how intermolecular potentials between atoms bound together are studied. This is done, of course, via spectroscopy. One starts with the Born-Oppenheimer approximation for the total molecular wave function: this enables one to describe the motion of the nuclei in a potential that depends on the separation between them. This result, the existence of a specific adiabatic potential, rests on there being no appreciable mixing between electronic states. One of its corollaries, the Franck-Condon Principle, enables one to interpret and invert (e.g. using the R.K.R.[10] method) the vibrational spectra in terms of the interatomic potentials in different electronic states. To what extent can we extend such a technique to free-free spectra, in other words, to absorption in the middle of a transient molecule — a collision complex — and deduce information about the potentials between atoms as they collide?

It has been shown that this is indeed a profitable exercise. In particular, the careful work of many groups has shown how absorption and emission by collisions in a gas may be analyzed to obtain rather accurate interatomic potentials. To see absorption in the middle of a collision we can look at absorption in the far wings of a collision broadened spectral line. By 'far wings' we mean detunings large compared to the inverse of a collision duration. The absorption profile in the far wings can often be analyzed using the quasi-static theory[11] of Kuhn,[12] and Jablonski[13] which has been refined for studying interatomic potentials. Looking at pure absorption, however, restricts the type of physics we can observe, as we shall see below.

SCATTERING EXPERIMENTS

Collisional Redistribution of Radiation: Collision Induced Fluorescence

Let us consider what we can learn from the following type of experiment. We have a dilute atomic vapor (say a metal vapor) in the presence of a few torr of buffer gas in a cell. A laser beam, whose frequency is close to, but detuned from, the resonance transition of the metal, passes through the cell; the detuning from resonance is greater than the inverse of the duration of a metal atom-rare gas perturber collision. If we observe the fluorescence, induced by the absorption of light in the middle of collisions, what more can we learn from this experiment than we could from a pure absorption experiment?

We can envisage the absorption process as occurring in the manner shown in Fig. 1. If the excited state of the atom-perturber quasi-molecule has a single adiabatic potential surface then we learn little more from the experiment than we would from a pure absorption experiment (the signal does not have exactly the same frequency dependence as the absorption experiment since some of the quasi-molecules excited are bound and do not contribute to the fluorescence around the atomic line). This does not mean that this experiment is not worth doing since one may be able to study systems more conveniently using laser excitation followed by fluorescence (or other detection methods), rather than by a classical absorption or emission experiment.[14-16] For example, laser excitation followed by degenerate four-wave mixing detection of excited states can be much more sensitive than absorption experiments. If the excited state of the molecule is not a single adiabatic channel then the door is open for us to study the dynamics of excited state mixing during the collision. In order to show how one goes about this we shall discuss an experiment we have performed.[4] A

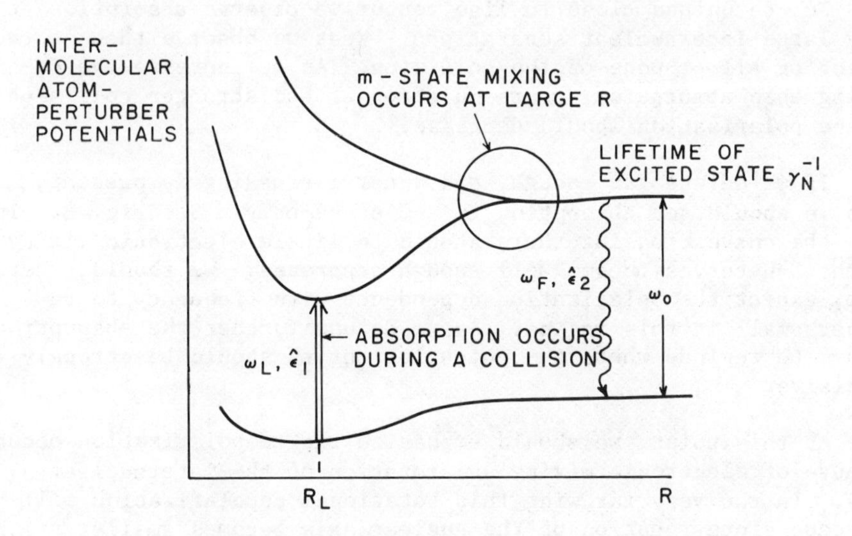

Fig. 1. Absorption in the middle of a collision.

schematic of the experiment is shown in Fig. 2. In this experiment the atom is strontium and we are near resonant with the λ460.7 nm transition. The laser is a broadband cw dye laser operating with Stilbene 3 pumped with a uv argon ion laser. With this system we observe the polarization of the collisionally redistributed fluorescence: the incident laser is, of course, polarized.

If we want to observe single collision dynamics we need to operate near to where the collisional and spontaneous emission damping rates are comparable. Measurements are made over a range of pressures and theory developed by Burnett and Cooper[1] used to extrapolate to zero pressure; and thus extract the single collision quantities of interest (see the theory section below). The other condition we need is that multiple scattering should be absent. This is easy to achieve for the incoming path since absorption in the far wings will be very small. The fluorescence, around line center, will, of course, be heavily trapped at any reasonable densities; say $\sim 10^{12}$ cm^{-3} -- even for rather small physical lengths to a window. It is possible to partially eliminate these problems by the use of a re-entrant window system and that is what we did. In the experiment the polarization emitted is observed as a function of incoming laser frequency: for convenience we rotate the laser polarization, rather than changing the outgoing polarization detected. As we vary the laser frequency the region where absorption takes place varies with it. What should we expect to see for red detunings?

For detunings close to line center we observe absorption at very large internuclear separations. What we observe then is the effect of all or none of the collision. As we increase the detuning then absorption occurs in stronger and stronger collisions, so the polarization should decrease.

If we detune far enough, and a curve crossing is present, then we should get absorption to a Π state only. See Fig. 3. Inside the curve crossing there should be little electronic mixing if the Σ-Π curves become well enough separated. We should, therefore, expect the polarization dependence with frequency to be rather small in this region. As we detune further the absorption occurs to regions where the potential curves should be strongly repulsive.

At this point, we should emphasize that depolarization occurs because of electronic mixing <u>and</u> rotation of the internuclear axis. In the very far wing this rotational depolarization will decrease since rotation of the nuclear axis becomes smaller as the trajectories become more and more curved.

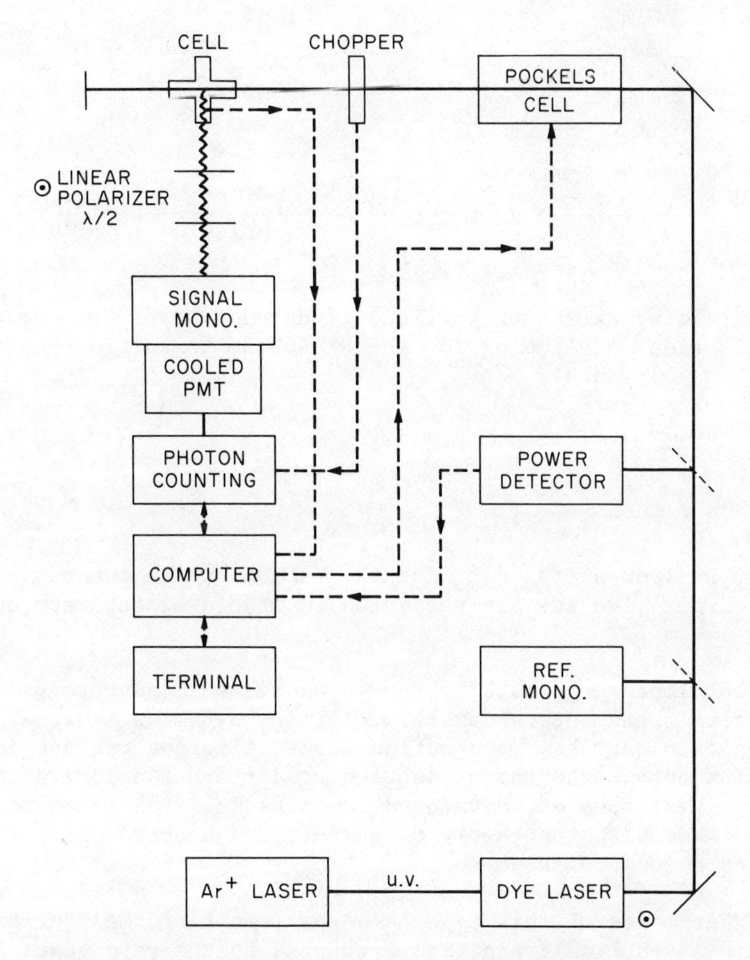

Fig. 2. Schematic of experiment on strontium-rare gas depolarization.

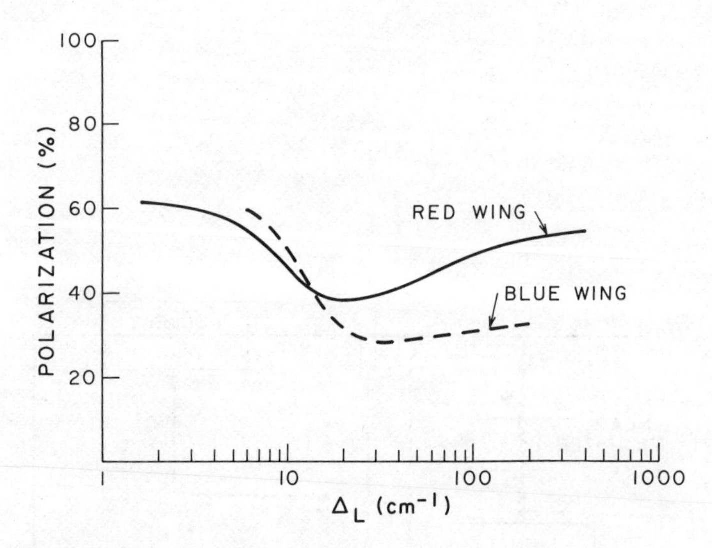

Fig. 3. Polarization of collision induced fluorescence as a function of detuning for the Sr 460 nm transition perturbed by argon.

If we look at the data for the case in which the perturber is argon, Fig. 4, we see the variation we should expect from our discussion.

The simple arguments we have used are not meant to produce definitive statements about the collision dynamics. Rather, they are meant to show how information about different regions in the collision enter into the production of different polarizations. Cooper[18] has, however, developed semiclassical collision models that (linked with the theory of Burnett and Cooper) point to the dynamics discussed above.

Experiments of this type have now been performed for the strontium λ460.7 nm transition perturbed by the rare gases (He, Ne, Ar, Kr, Xe), and a full analysis of how the different dynamical features vary with the rare gases is now under way.

We should mention that fully quantal calculations by Julienne at NBS in Washington are also showing how very strong a constraint on the non-adiabatic part of the potentials this type of data can be.[19]

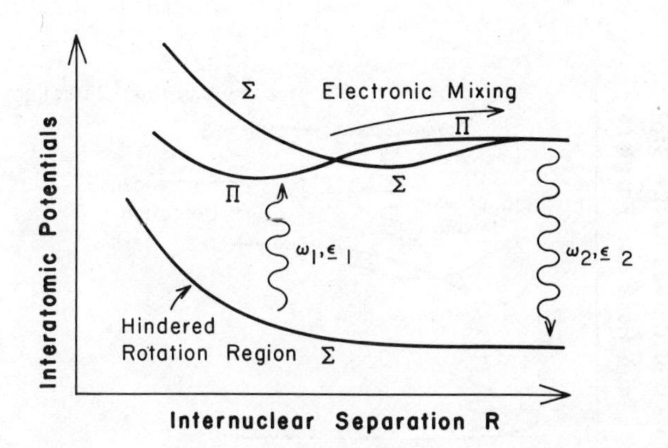

Fig. 4. Schematic of strontium-rare gas potential curves.

Collision-Induced Two-Photon Absorption

A second type of experiment that can give complementary information to that described above is collision-induced two-photon absorption. When we observe collision-induced absorption followed by fluorescence we can study the evolution of the system from the absorption event to the completion of the collision. Re-emission of a photon before completion of the collision is possible, but unlikely, due to the large difference, for a heavy particle collision, between a spontaneous emission decay time, γ_N^{-1}; and a collisional duration, τ_C: typically $\gamma_N \tau_C \sim 10^{-4}$. This type of process is easier to observe in absorption to higher lying states. We could as shown in Fig. 5 look for the absorption of a second laser photon before the completion of the collision.[20] This type experiment probes propagation between two regions in a collision complex. This is just about the maximum amount of information one can hope to obtain about a collision complex. Experiments of this type may be the best way to further constrain and test any conclusions from collision dynamics obtained from the single photon experiments.

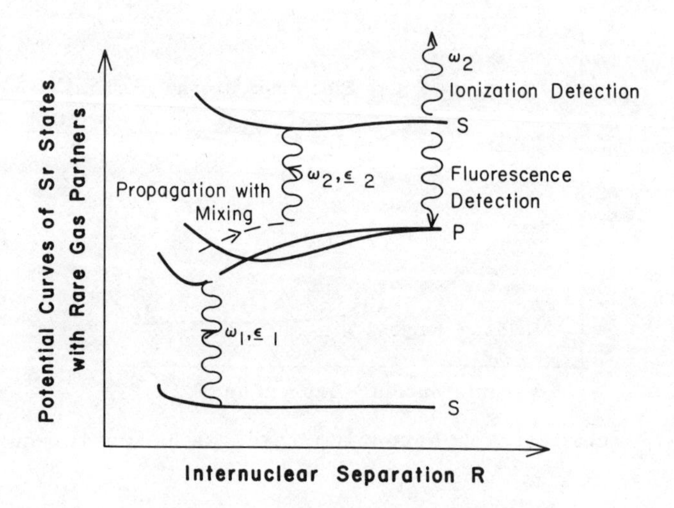

Fig. 5. Collision-induced two-photon absorption.

SCATTERING FROM ATOM-MOLECULE AND REACTIVE ENCOUNTERS

Some work on collision-induced absorption in atom-molecule collisions has been done,[21] however, results for this case are not very extensive, as yet. Moreover, the best manner for analyzing the light scattering problem in which, as in this case, other inelastic channels are open (vibrational and rotational excitation) is far from clear.

I think it is fair to say that many physical chemists see the type of experiments we have discussed as holding an exciting prospect for the study of reaction dynamics. In this case, we are talking about electronic absorption, corresponding to transitions between the ground and excited states of the reactive complex. First experiments to study the adiabatic potential surfaces -- corresponding to the far wing absorption pressure broadening experiments -- have been performed and the extension to other cases is under way.[23] It is our feeling that the main obstacle in the way of real progress in these experiments is a practical analysis scheme for the frequency and polarization dependent cross sections of the type

$$AB + C + h\nu \rightarrow A + (BC)^* \quad .$$

The formal extension of the techniques used on atom-atom collisions to this case is not too difficult, but exactly how much can be learned from such experiments is an open question. Personally we feel the development of practical analysis schemes for such experiments is most important and worth the expenditure of considerable effort.

IMPLICATIONS FOR LASER-MODIFIED COLLISION STUDIES

What we have discussed above is how the dynamics of collisions can be studied using light scattering. Exactly the same techniques can be applied to "radiative"[20] collisions. (Some use the terminology, "laser-induced collisions" to describe optical and radiative collisions.) For example, if we are interested in the efficient population of a given state using a radiative collision it may well be worthwhile to study the frequency and polarization dependence of the cross section. In this way, one could obtain a detailed picture of the mechanism of a collision and the possibilities for optimization. One case in which it may be particularly useful to do such studies is for laser-induced or enhanced catalysis on a surface.[8] We believe it may be possible to develop a scheme for characterizing the dynamics of the complex formed in a collision of an atom or molecule with a surface in the manner sketched above for atom-atom collisions. The information gained should bear directly on the planning of laser-induced or enhanced catalysis on surfaces, as well as being a way of studying the molecule-surface interaction potential.

We should also like to discuss the role the scattering experiments described above can have in the planning of strong field experiments. By strong field we mean a field intense enough to saturate the collision complex. To see how saturation enters the collision problem let us return to our discussion of collision-induced absorption and introduce the concept of a dressed state of the atom + radiation field. If we think of the radiation field purely quantum-mechanically, then the ground state of the atom in the presence of a radiation field is really a ground state plus N photons. In like manner the excited state with one photon having been absorbed is an excited state plus (N-1) photons (see Fig. 6). During a collision these two states can become degenerate and be coupled strongly -- in our usual way of thinking this means absorption can occur. It is, however, most useful to think in terms of the dressed state since it is not limited in any way to

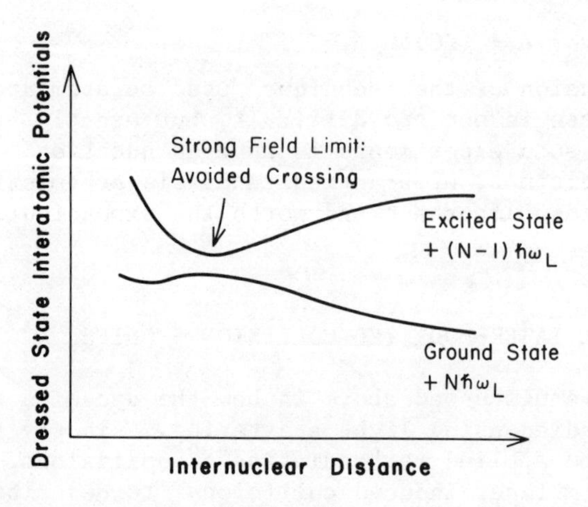

Fig. 6. Absorption in the middle of a collision: dressed state picture.

weak fields -- in fact, it is in the strong field limit that the dressed state method becomes simplest. We can, to put it succinctly, treat the dressed atom of atom + radiation field as a new atom with new Einstein A and B rates <u>and</u> a new interatomic potential between it and a perturber. The whole problem of strong field collision-induced processes then becomes a simple one of collision rates between dressed atom states.[24-26]

Let us consider what occurs in the strong field limit using our new picture. If the coupling $\underline{E}\cdot\underline{d}$ is strong enough it will cause the crossing of dressed state curves during the collision to become an avoided crossing. We can give an estimate of when this occurs: it will occur when the coupling $\underline{E}\cdot\underline{d}$ is large compared to the energy uncertainty due to the finite time spent in the crossing region, $\sim\hbar\Delta$. So we get the strong field or saturation condition to be $\underline{E}\cdot\underline{d}/\hbar\Delta \gtrsim 1$. This is, of course, the same condition we would get from the atomic picture if we thought in terms of Rabi flopping, although this dressed state picture shows clearly why one would expect the crossed section for the overall process to decrease at high field strengths.

This switching off of the cross section at high field strengths is, as we see (in Fig. 6), accompanied by a depression of a region of the dressed state potential surface; this may be a way in which, e.g., chemical reaction barriers may be lowered.

The calculation of dressed state collision rates presents a considerable challenge to practitioners of molecular scattering theory[2,3] and offers the potential for studying collision dynamics; though perhaps not as clearly as the weak field experiments do. The quantitative analysis of an experiment of this type involves solving a set of rate equations in the dressed frame -- extra time dependence may be present in pulsed laser experiments (the commonest method for producing intense fields) and can also be taken into account.

We have made a detailed analysis of this type[17] in the experiment[18] we performed on the J=0 to J=1, λ = 460.7 nm transition in strontium perturbed by argon; a case where dressed state collision rates are available from the work of Light and Szöke.[19] In this case we find that the polarization of the dressed atom fluorescence varies rapidly with applied field strength (as Light and Szöke predicted). Although we do not want to go into detail there is one practical consequence of this result that we should like to mention. Schemes for using intense fields to switch collision and reaction schemes have often ignored the consequences of the symmetry (or lack of symmetry) of the collision complex. The rapid dependence of polarization on field strength in this case showed clearly how important such effects can be.

One of the ways to study the structure of the collision complex is, of course, through the type of spectroscopy of collisions we have described. The results of such studies can then be used to anticipate what effects the symmetry etc. of the collision will have on the strong field saturation of the relevant cross sections.

We should briefly mention two important features we have not discussed so far: the effect of resonances and satellites in the absorption cross sections.[30] Obviously if we wish to saturate a collision complex we should like to know how to access the longest lived collision complexes. For the same reason, of ease of saturation, the positions of satellites are important things to know. As of yet, little use has been made[31] of resonances and satellites in laser modified collision processes and yet they offer the possibility of larger cross sections and easier saturation. Again more work is needed on the dynamics of collision complexes near regions where satellites or resonances can be involved.

A SUMMARY OF THE THEORY

In this section we shall discuss a few of the technical aspects of the theory needed to analyze the results of the experiments described above. Our discussion will be very brief and the reader is referred to the literature for all of the details.[1-8]

From the point of view of scattering theory the type of processes we have discussed are examples of off-shell[32] events. The objects we, therefore, need to deal with are off-shell T-operators. The reason why we need off-shell T-operators is that we are really dealing, even in the case of light scattering by an atom-atom collision, with a three-body problem -- the third body being an incoming or outgoing photon. The problem is more complicated than the normal one of scattering theory since we also need to discuss the creation and decay of coherences in the atoms undergoing collisions. What is necessary then is a theory that includes these off-shell events in a scattering theory for density operators,[1-3,5,6] and this is what has been developed. We shall briefly describe here the applications of the theory to different experiments.

Collisionally Redistributed Fluorescence

This experiment looks at "half off-shell" events: the incoming photon is detuned from resonance and the collision has to take up that energy defect. The polarization and intensity of the fluorescence as a function of laser detuning may be specified by the following sum over the multipoles present in the excited state[1]

$$I_F = I_L \sum_K M_{23}^K \left[\frac{2f^K(\Delta)}{\gamma_c^K + \gamma_N} - \frac{f(\Delta_L)}{\gamma^1(\Delta_L)}\right] .$$

Here γ_N is the natural width, γ_c^K is the relaxation rate for the Kth rank-excited multipole (K = 0 population, K = 2 alignment), and M_{23}^K contains dipole moment and polarization properties.

Here $f(\Delta_L)$ is the usual "unified" absorption profile

$$f(\Delta) = \frac{1}{\pi} \frac{\gamma'(\Delta_L)}{[\omega_o - \omega_L + \Delta_c^1(\Delta_L)]^2 + \gamma^1(\Delta_L)^2}$$

with

$$\gamma^1(\Delta_L) = \frac{1}{2}\gamma_N + \gamma_c^1(\Delta_L)$$

whose $\gamma_c^1(\Delta_L)$ and $\Delta_c^1(\Delta_L)$ are the usual frequency dependent width and shift. The "generalized" absorption profile $f^K(\Delta)$ contains all the effects of the absorption in the middle of a collision followed by propagation to the end of the collision.

If we want to calculate the full spectrum of scattered light then we find that the inclusion of these off-shell events prevents us from using the quantum fluctuation regression theory. To calculate spectra the dipole autocorrelation must be calculated directly without the use of any regression assumptions.[1] This is an interesting example of where a microscopic calculation of a correlation function can be calculated for a system driven from thermal equilibrium.

Collision-Induced Two-Photon Absorption

This process is, as far as the atom-perturber collision is concerned, fully off-shell. Fully off-shell T-operators are more difficult to calculate than their half-off-shell versions. The semiclassical propagator approach,[33] however, can be used to estimate them[34] and work is in progress to determine exactly how useful experimental studies for the multichannel case would be.

Localization of Off-Shell Events

We have, in our discussion above, always talked as if the "off-shell event," i.e. absorption of a photon, is localized. Let us now sketch our justification for this. First let us define when we are off-shell. The natural definition of when an energy is off shell comes from the duration of a heavy particle collision τ_c, the quantity $\hbar/\tau_c$, being an energy defect that the collision will "notice." τ_c is defined via the cross section πb^2 for the collision and the relative collision velocity, u thus:

$$\tau_c \sim \frac{b}{u} \sim \frac{\sqrt{\sigma}}{u} \quad .$$

If we use a semiclassical propagator and evaluate the T-matrix elements using the method of stationary phase we find the absorption of a photon detuned from the atomic resonance an amount Δ_L is localized within

$$\tau_{\text{localization}} \sim \tau_c \left(\frac{1}{\Delta_L \tau_c}\right)^{1/2} \quad .$$

This shows why we can say when we are detuned by an amount much greater than the inverse of a collisional duration, i.e.

$$\left(\frac{1}{\Delta_L \tau_c}\right)^{1/2} << 1 \quad ,$$

then we can speak of a localized absorption since $\tau_{localization} \ll \tau_c$.

This discussion applies to the localization along a given trajectory and in a given electronic state. For the case where electronic mixing is present the question of localization has been discussed by Cooper.[18]

ACKNOWLEDGMENTS

I should like to thank J. Cooper, P. D. Kleiber, A. Gallagher, W. J. Alford, R. J. Ballagh, A. Ben-Reuven, P. Julienne M. G. Raymer, and A. Szöke for many helpful discussions. This work was supported by the National Science Foundation (Grant No. PHY79-04928) and the Research Corporation.

REFERENCES

1. K. Burnett, J. Cooper, R. J. Ballagh and E. W. Smith, Phys. Rev. A 22:2205 (1980); K. Burnett and J. Cooper, Phys. Rev. A 22:2027 (1980); K. Burnett and J. Cooper, Phys. Rev. A 22:2044 (1980).
2. S. Mukamel, J. Chem. Phys. 71:2884 (1979); Y. Rabin, D. Grimbert and S. Mukamel, Phys. Rev. A 26:271 (1982).
3. G. Nienhuis, Physica 93c:393-407 (1978); J. Phys. B 15:535-550 (1982).
4. P. Thomann, K. Burnett and J. Cooper, Phys. Rev. Lett. 45:1326 (1980).
5. D. Voslamber and J. B. Yelnik, Phys. Rev. Lett. 41:1233 (1978).
6. A. Ben-Reuven and Y. Rabin, Phys. Rev. A 19:2056 (1979).
7. F. H. Mies, Quantum theory of atomic collisions in intense laser fields, in: "Theoretical Chemistry: Advances and Perspectives," Vol. 8, D. Henderson, ed., Academic Press, New York (1981); T. F. George, J. Phys. Chem. 86:10 (1982).
8. Kai-shue Lam and T. F. George, J. Chem. Phys. 76:3396 (1982); T. F. George, et al., Theory of molecular rate processes in the presence of intense laser fields, in: "Chemical and Biochemical Applications of Lasers," Vol. IV, C. B. Moore, ed., Academic Press, New York (1979).
9. H. Pauly, Elastic scattering cross sections. I: Spherical potentials, in: "Atom-Molecule Collision Theory, A Guide for the Experimentalist," R. B. Bernstein, ed., Plenum, New York (1980).

10. R. J. LeRoy, Applications of Bohr quantization in diatomic molecule spectroscopy, in: "Semiclassical Methods in Molecular Scattering and Spectroscopy, M. S. Child, ed., N.A.T.O. Advanced Study Institutes Series (Series C), Reidel, New York, (19).
11. R. E. M. Hedges, D. L. Drummond and A. Gallagher, Phys. Rev. A 6:1519 (1972).
12. H. G. Kuhn, Phil. Mag. 18:987 (1934); Proc. Roy. Soc. A 158:212 (1937).
13. A. Jablonski, Acta Phys. Polon. 6:371 (1937); 7:196 (1938); Phys. Rev. 68:78 (1945).
14. J. L. Carlsten, A. Szöke and M. G. Raymer, Phys. Rev. A 15:1029 (1977).
15. J. V. McGinley, Thesis, Oxford University (1981).
16. M. G. Raymer, J. L. Carlsten and G. Pichler, J. Phys. B 12:L119 (1979).
17. P. Ewart, A. I. Ferguson S. V. O'Leary, Optics Commun. 40:147 (1981); P. Ewart and S. V. O'Leary (1982), to be published; M. Dagenais, Phys. Rev. A 24:1404 (1981); Y. Prior, A. R. Bogdan, M. Dagenais and N. Bloembergen, Phys. Rev. Lett. 46:111 (1981).
18. J. Cooper, "Why half a collision is better than a whole one," (Invited talk), in: Sixth International Conference on Spectral Line Shapes, K. Burnett, ed., de Gruyter, Berlin (in press).
19. P. Julienne, "Non-adiabatic effects in line broadening," (Invited talk), in: Sixth International Conference on Spectral Line Shapes, K. Burnett, ed., de Gruyter, Berlin (in press).
20. J. C. White, Opt. Lett. 242 (1981); M. H. Nayfeh and G. B. Hillard, Phys. Rev. A 24:1409 (1981); L. I. Gudzenko and S. I. Yakovlenko, Zh. Eksp. Teor. Fiz. 62:1686 (1972) [Sov. Phys.-JETP 35:877 (1972)]; S. E. Harris et al., in: "Tunable Lasers and Applications," S. Mooradian, T. Jaeger and P. Stoketh, eds., Springer, New York (1976), p. 193; S. E. Harris et al., Laser induced collisional energy transfer, in: "Atomic Physics 7," D. Kleppner, ed., Plenum, New York (1981); P. L. deVries, C. Chang, T. F. George, B. Laskowski, J. R. Stallcop, Phys. Rev. A 22:545 (1980).
21. A. Gallagher, The absorption and emission of radiation by the collision complex, in: "Physics of Electronic and Atomic Collisions," S. Datz, ed., North Holland, Amsterdam (1982), pp. 403-411.
22. Y. Rabin and P. Hering, contributed paper European Conference on Atomic Physics, April 6-10, 1981, Ruprecht-Karls-Universität, Heidelberg, EPS Conference Abstracts Volume 5A, Part 11; T. Lukasik and S. C. Wallace, Phys. Rev. Lett. 47:240 (1981).

23. P. Arrowsmith, F. E. Bartoszek, S. H. Bly, T. Carrington, Jr., P. E. Charters and J. C. Polanyi, J. Chem. Phys. 73:11,4895 (1980); P. Hering, P. R. Brooks, R. F. Curl, Jr., R. S. Judson and R. S. Lowe, Phys. Rev. Lett. 44:687 (1980).
24. K. Burnett, J. Cooper, P. D. Kleiber and A. Ben-Reuven, Phys. Rev. A 25:1345-1357 (1982).
25. Y. Rabin and A. Ben-Reuven, J. Phys. B 13:2011 (1980).
26. S. Reynaud and C. C. Cohen-Tannoudji, "Dressed atom approach to collisional redistribution," J. Physique (in press).
27. P. D. Kleiber, J. Cooper, K. Burnett, C. V. Kunasz and M. G. Raymer, "Theory of time dependent intense field collisional fluorescence," Phys. Rev. A (submitted).
28. P. D. Kleiber, K. Burnett and J. Cooper, Phys. Rev. Lett. 47:22,1595 (1981).
29. J. C. Light and A. Szöke, Phys. Rev. A 1363 (1978).
30. J. Szudy and W. E. Baylis, J. Quant. Spectrosc. Radiat. Transfer 15:641 (1975).
31. A. M. Bonch-Bruevich, T. A. Vartanyan and V. V. Khromov, Zh. Eksp. Teor. Fiz. 78:538 (1980) [Sov. Phys.-JETP 51 (1980)]; T. A. Vartanyan, Yu. N. Maksimu, S. G. Przhibelskii and V. V. Khromov, Pis'ma Zh. Eksp. Teor. Fiz. 29:281 (1979).
32. E. W. Schmid and H. Ziegelmann, "The Quantum-Mechanical Three Body Problem," Vieweg, Braunschweig (1974).
33. H. J. Korsch and R. Möhlenkamp, J. Phys. B 10:3451 (1977); H. J. Korsch, Phys. Rev. A 14:1645 (1976); B. J. B. Crowley and B. Buck, J. Phys. G 4:9 (1978).
34. K. Burnett, unpublished work.

ELECTRON-PHOTON CORRELATION STUDIES OF SPIN EXCHANGE, SPIN ORBIT AND QUANTUM BEATS

H. Kleinpoppen and I. McGregor

Atomic Physics Laboratory,
University of Stirling
Stirling, FK9 4LA

INTRODUCTION

Electron impact studies, both theoretical and experimental, have played an important part in our understanding in the field of atomic physics. The electron provides a useful probe into the region of the electron cloud surrounding the atom. Amongst the most notable examples in electron collision studies which can be thought of are the Franck-Hertz experiment proving the quantised nature of energy loss in collisions between electrons and atoms, the Ramsauer-Townsend effect in the structure of scattering cross-sections and electron spin effects in Mott scattering.

Other studies have included the measurement of both optical and electron excitation functions where the intensity is measured as a function of impact energy. These have been extended to include investigations of structure (resonances) in these excitation functions caused by compound atomic states being produced during the excitation process. A considerable effort has also been devoted to measurements of line polarisations following electron impact. These studies provide an insight into the magnetic sublevels which are involved in the excitation process. The electron impact work has provided much of the groundwork which is now accepted but was "discovered" with considerable difficulty.

The experimental work has seen considerable development in the ability to produce electron beams with both a well defined energy and a narrow energy spread using electron monochromators of various types and also the ability to detect individual particles or photons. Today's experiments fully exploit these facilities and allow much detailed information regarding the actual scattering information to be obtained. This experimental advance allows stringent tests to be applied to theoretical models for scattering processes, enabling a fuller understanding of collision processes to be achieved.

ELECTRON PHOTON COINCIDENCE TECHNIQUES

One area in which electron impact processes have, in recent years, taken a significant step forward is in the use of coincidence techniques applied to electrons and photons. By this method, electrons having lost a known amount of energy and scattered through a known angle are detected simultaneously with photons emitted in a given direction from the particular excited atomic state in question. Only when there is an electron signal in coincidence with a delayed photon signal is an actual measurement made. The important feature regarding this method is that the observation is thus restricted to the radiation emitted by only those atoms which have been excited by electrons scattered into a given direction defined by the detector. Thus a certain sub-ensemble of excited atoms are selected in the experiment.

This method has been employed in many areas of atomic physics. Imhoff and Read[1] have used electron-photon coincidences to measure helium lifetimes thus ensuring the complete absence of cascade processes from affecting the measurement. Pochat et al[2] have measured differential cross-sections for electron impact excitation of $n = 4$ and 5 states of helium using the decay photons of appropriate wavelengths to uniquely specify the coincident scattered electrons. In addition to several other similar examples, it has also been employed for particles other than photons and electrons e.g. between two electrons as in the (e, 2e) experiments and between ions and photons in ion-atom collision experiments.

Another important area in which the electron-photon coincidence method has been applied is in evaluating differential cross-sections for the excitation of magnetic sublevels. Using knowledge of the magnetic sublevel cross-sections, of the relevant lifetime and of the hyperfine structure allowed line polarisation values to be calculated. A calculation of Flower and Seaton,[3] using the theory of Percival and Seaton,[4] accurately reproduced the experimental values[5] of the threshold polarisation values for the resonance lines of ^{6}Li, ^{7}Li and ^{23}Na. King et al[6] showed that "threshold polarisations" can be measured in electron atom excitations at energies in excess of threshold by detecting, in coincidence with the photons, those electrons scattered in the forward direction only. Axial symmetry requires that the threshold condition $\Delta M_L = 0$ applies in this situation.

ELECTRON PHOTON ANGULAR CORRELATIONS

In order to excite atomic magnetic sub-levels with $M_L \neq 0$ it is necessary to detect electrons outside of the forward direction. Information about those magnetic sub-levels can be obtained by investigating the angular correlations between electrons scattered inelastically from an atom and the photon emitted in the scattering plane. Such a study commenced in 1968[7] to investigate the excitation of the 2^1P state of helium and the first results were reported by Eminyan et al[8] in 1973.

The 2^1P state of helium excited from the ground state by electron bombardment, in which the electron is scattered in some direction (θ_e, ϕ_e), can be described as a coherent superposition of magnetic sublevels as

$$\psi = a_{-1}|1-1> + a_0|10> + a_1|11> \qquad (1)$$

where a_m describe the excitation to a sublevel $|LM>$. Mirror symmetry in the scattering plane results in $a_1 = a_{-1}$ and since a_0 may be assumed real and positive, a_1 can be defined as $a_1 = |a_1|e^{i\chi}$. The total cross-section is then $\sigma = a_0^2 + 2a_1^2$ and the partial cross-section for M = 0 may be written as $\lambda = a_0^2/\sigma$. These parameters, λ and χ, could then be determined by measuring an angular correlation function and fitting the data to the following expression obtained from the theory formulated by Macek and Jaceks[9]

$$N \propto \lambda \sin^2\theta + (1 - \lambda)\cos^2\theta - \sqrt{\lambda(1 - \lambda)} \sin\theta \cos\theta \cos\chi \qquad (2)$$

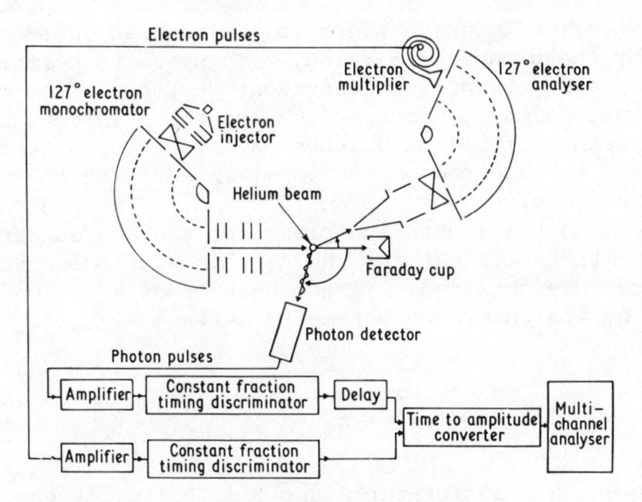

Fig. 1. A typical electron-photon coincidence apparatus as used for helium in reference 8.

An outline of the apparatus is given in figure 1. The apparatus consists of an electron gun and a 127° cylindrical electrostatic monochromator to produce the incident electron beam, a further 127° analyser and channeltron at a fixed electron scattering angle of θ_e, to select electrons having lost an appropriate amount of energy, and an interference filter and photomultiplier to record photons of the correct wavelength at a photon engle of θ_γ. The photon detector lies in the electron scattering plane ($\phi_\gamma = \pi$). The signals from the detectors are fed to the coincidence circuitry and the resulting time spectrum recorded in a multichannel analyser. The photon detector could be rotated in the scattering plane allowing the coincidence rate as a function of scattering angle to be obtained.

By fitting the data to equation (2) the values for the parameters λ and $|\chi|$ could be obtained and values for the orientation and alignment parameters[10] of the target atom determined. The orientation parameters indicates the degree of atomic orientation, i.e. the amount of angular momentum imparted to the atom in the collision. It should be noted that all these parameters require no further normalisation before being compared with theoretical calculations.[11] Also such comparisons can be made without the usual averaging over electron scattering angles as in line polarisation measurements or summation over degenerate magnetic sublevels as in measurements of differential cross-sections. This early work on helium has been considerably extended[12] and has also included such atomic systems as hydrogen,[13] neon,[14] argon,[15] krypton and mercury.

ELECTRON-POLARISED PHOTON CORRELATIONS

One extension in which there is now renewed interest was performed by the group at Stirling. This was to perform a full polarisation analysis of the coincident photons[16] from which the Stokes parameters may be obtained. The state which was studied was the 3^1P state of helium which decays to the 2^1S state by emission of a visible photon (501.6nm) in addition to the decay to the ground state with a uv photon (53.7nm). The photons are viewed normal to the scattering plane and the Stokes parameters associated with the linear (η_1 and η_3) circular polarisation (η_2) measurements were obtained. The coherence of a photon beam can be denoted by the coherence parameter defined as

$$\mu = \frac{\eta_1 + i\eta_2}{\sqrt{1 - \eta_3^2}}$$

For the $3^1P - 2^1S$ coincident photons from helium it was proven that $\mu = 1$ implying that each and every photon is in the same polarisation state. This result is a direct consequence that the excitation process was coherent thus validating the earlier assumption of equation (1). The polarisations can be related to the λ and χ parameters by

$$\eta_1 = 2\lambda - 1 \quad \eta_2 = 2[\lambda(1-\lambda)]^{\frac{1}{2}}\sin\chi \quad \eta_3 = -2[\lambda(1-\lambda)]^{\frac{1}{2}}\cos\chi \ .$$

They may also be related to the orientation and alignment parameters as

$$O_{1-} = -\eta_2/2 \quad A_0 = -(1+3\eta_1)/4 \quad A_{1+} = -\eta_3/2 \quad A_{2+} = (\eta_1 - 1)/4$$

It may be seen that the circular polarisation specifies the orientation whilst the alignment is given by the linear polarisation components. Additionally from this measurement the sign of the orientation (and χ) can be obtained.

CLASSICAL GRAZING MODEL

The recent interest[17] lies in the application of a classical grazing model to the problem of atomic orientation produced by electron scattering. Kohmoto and Fano have shown that a classical grazing model can be used to relate the attractive or repulsive nature of the interaction to the sign of the orientation produced. Theoretical calculations[18] predict that in the excitation of the 2^1P state of helium by electron impact that the sign of the orientation changes in going from small to large scattering angles. The implication is that an attractive force is present at small scattering angles whilst a repulsive force is present at large scattering angles.

The polarisation measurement of the 3^1P excitation is in agreement with this model since a positive orientation was measured for small angles implying an attractive force there. This measurement did not go to large enough scattering angles for a change in sign of the orientation to be confirmed. Angular correlation data for the 3^1P excitations,[19] whilst indicating a possible sign reversal for the orientation in going from small to large angles, do not positively confirm it. At present we are preparing to extend the circular polarisation measurements for the 3^1P state so as to directly measure such a sign change.

We shall pursue some of the consequences of a classical grazing model in which a repulsive force, due to electron-electron interactions, acts at small impact parameters and an attractive force, due to atomic polarisability, acts at large impact parameters. Four such classical paths are illustrated in figure 2 using a right handed coordinate system. Electrons scattered through θ to the left by either a repulsive or attractive forces are detected by the electron detector at $+\theta$.

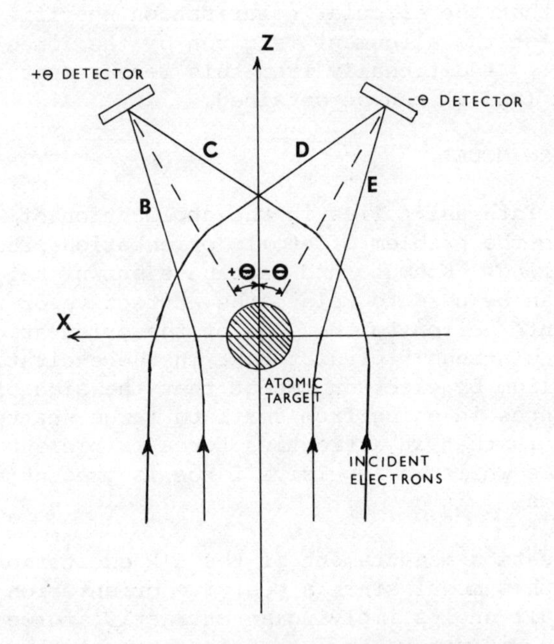

Figure 2: Classical electron trajectories of scattering angles ±θ for attractive and repulsive forces. Intensities B, C, D and E are explained in text.

Similarly, those electrons scattered to the right are incident on a detector at $-\theta$. Those electrons passing on the left side will produce a negative orientation of the excited atom, $\langle L_y \rangle < 0$, and those that pass on the right side a positive orientation, $\langle L_y \rangle > 0$. The sign of the orientation can be experimentally determined by observing the scattered electrons in coincidence with either left-handed or right-handed circularly polarised light emitted in the y-direction (positive or negative orientation respectively).

Let C and E represent probabilities of scattering an electron through $+\theta$ and $-\theta$, respectively, with the emission of a left-hand circularly polarised photon in the y-direction. Correspondingly B and D are probabilities of electron scattered through $+\theta$ and $-\theta$, respectively, with a right hand circularily polarised photon in the y-direction. Four possible experiments can be considered.

Defining the circular polarisation as $\eta_2 = (I_{RHC} - I_{LHC})/(I_{RHC} + I_{LHC})$ then measurements of the circular polarisation of photons emitted in the y-direction in coincidence with electrons scattered through $+\theta$ $(-\theta)$ can be written as

$$\eta_2(+\theta) = \frac{B - C}{B + C} \qquad \eta_2(-\theta) = \frac{D - E}{D + E} \qquad (3)$$

Measurements of the right-left electron scattering asymmetry in coincidence with right- (or left-) hand circularly polarised photons gives

$$A_{RHC} = \frac{D - B}{D + B} \qquad A_{LHC} = \frac{E - C}{E + C} \qquad (4)$$

Because of parity invariance with respect to reflection in the scattering plane B = E and C = D. This results in

$$\eta_2(+\theta) = -\eta_2(-\theta) \qquad A_{RHC} = -\eta_2(+\theta) \qquad A_{LHC} = +\eta_2(+\theta) \qquad (5)$$

Whilst the first relationship, equation (5), has been confirmed experimentally the equality of the electron scattering asymmetry with the circular polarisation of the coincident photons has so far not been tested. However, Hermann and Hertel[20] have performed a related experiment in which an electron scattering asymmetry was measured for super-elastic collisions with sodium atoms excited by circularly polarised laser light.

With this model the scattering process can be described by a repulsive amplitude (related to B) and an attractive amplitude (related to C). When, integrating over all impact parameters, the repulsive forces are balanced by the attractive forces there is no orientation produced in the atom. When these forces are out of balance then there is a net attractive (or repulsive) force acting and the atom is left in an orientated state.

SPIN-ORBIT EFFECTS IN ELECTRON-PHOTON COINCIDENCE EXPERIMENTS

The electron-photon coincidence technique has been extended to the study of electron collisions with heavy atoms which are no longer adequately described by a LS coupling scheme. In our group at Stirling we have investigated the excitation of the 6^3P_1 intercombination state of mercury using the polarisation correlation method of Standage and Kleinpoppen.[16] The presence of spin orbit interactions was confirmed when the excitation process was shown to no longer be completely coherent.[21] Blum et al[22] and da Paixao[23] have introduced new parameters which are required to describe the electron-photon coincidence function for heavy atoms. Four parameters (λ, $\bar{\chi}$, Δ and ε) instead of the usual two (λ and χ) are necessary to analyse correlations from P-state excitations. These parameters may be defined as

$$\lambda = \sigma_0/(\sigma_0 + 2\sigma_1) \qquad \cos\bar{\chi} = \mathrm{Re}\langle a_0 a_1\rangle/|a_0 a_1|$$

$$\cos\Delta = |\langle a_0 a_1\rangle|/(\sigma_0\sigma_1)^{\frac{1}{2}} \qquad \cos\varepsilon = -\langle a_{-1}a_1\rangle/\sigma_1 \tag{6}$$

These new parameters Δ and ε can be interpreted as a measure of the degree of coherence in the excitation between magnetic sub-levels with $M_L = 0$ and $M_L = \pm 1$ and between $M_L = 1$ and $M_L = -1$, respectively.

Since a photon polarisation correlation can determine only three independent quantities it is necessary to perform polarisation correlations with several differing geometries in order to fully extract these parameters above. At Stirling a polarisation correlation measurement at a photon polar and azimuthal angle of $\pi/2$ ($\theta_\gamma = \phi_\gamma = \pi/2$) was combined with a polarisation correlation experiment at a photon polar angle of $\pi/2$ and a photon azimuthal angle of $3\pi/4$ ($\theta_\gamma = \pi/2$, $\phi_\gamma = 3\pi/4$). In the former measurement the two linear polarisations and the circular polarisation of the coincident photons were obtained, whilst in the latter, one of the linear polarisations was determined. It is worth noting that it is necessary to perform at least one circular polarisation measurement in order to determine separately $\bar{\chi}$ and Δ.

A schematic diagram of the apparatus can be seen in figure 3. An electron beam from a gun-lens system was focused onto the interaction region. The scattered electrons were energy selected by means of a 127° cylindrical analyser and detected in a channeltron. The overall energy resolution of the system was 400 meV, which is sufficient to resolve the 6^3P_1 state from the 6^1P state. The timing circuitry ensures that due to the long lifetimes of the metastable $6^3P_{0,2}$ states, no significant contributions from

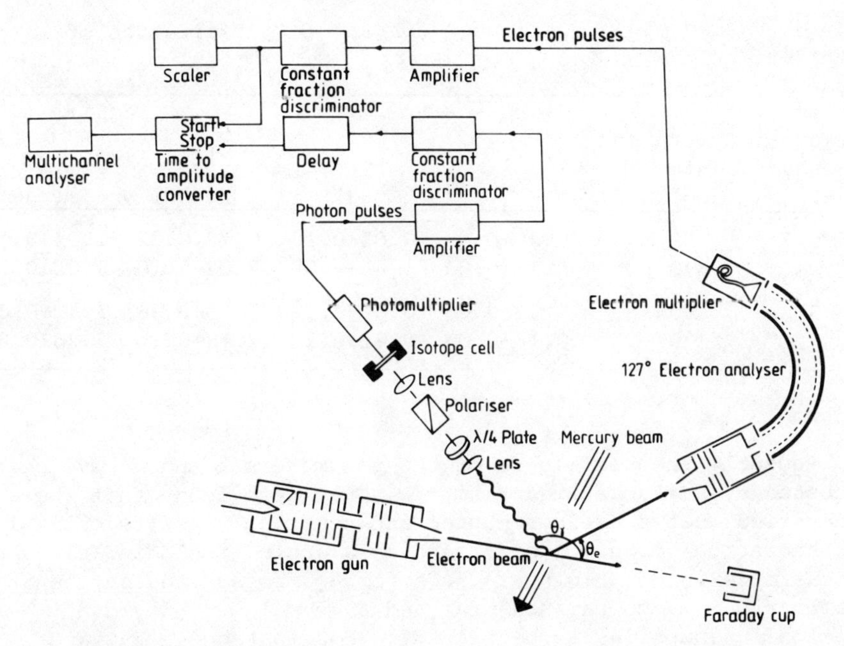

Fig. 3. A schematic diagram of the apparatus used in mercury studies.

these states are present. Photons from the interaction region are detected by a photomultiplier after passing through a polariser and an isotope cell. An isotope cell[24] was employed in order that the effects of hyperfine structure were eliminated. This filter absorbs the radiation from the odd isotopes of mercury whilst transmitting that from the even isotopes.

The normalised Stokes parameters (polarisation components) are related to the parameters defined above by

$$\eta_1(\phi_\gamma = \pi/2) = \frac{-4[\lambda(1-\lambda)]^{\frac{1}{2}} \cos\Delta \cos\bar{\chi}}{(1+\cos\varepsilon) + (1-\cos\varepsilon)\lambda}$$

$$\eta_2(\phi_\gamma = \pi/2) = \frac{-4[\lambda(1-\lambda)] \cos\Delta \sin\bar{\chi}}{(1+\cos\varepsilon) + (1-\cos\varepsilon)\lambda}$$

$$\eta_3(\phi_\gamma = \pi/2) = \frac{(3\lambda-1) - (1-\lambda)\cos\varepsilon}{(1+\cos\varepsilon) + (1-\cos\varepsilon)\lambda}$$

$$\eta_2(\phi_\gamma = 3\pi/4) = (3\lambda-1)/(\lambda+1)$$

The values obtained from these measurements are presented in Table 1. An important feature to remember about the parameters ε and Δ is that for the case where LS coupling applies both $\cos\varepsilon = 1$ and $\cos\Delta = 1$.

Table 1 Values of λ, $\bar{\chi}$, ε and Δ for the 6^3P_1 excitation of mercury.

Incident electron energy (eV)	Scattering angle (deg)	λ	χ (rad)	$\lvert\varepsilon\rvert$ (rad)	$\lvert\Delta\rvert$ (rad)
5.5	50	0.26 ± 0.03	0.83 ± 0.15	0.77 ± 0.62	1.04 ± 0.11
	70	0.18 ± 0.03	0.48 ± 0.08	2.02 ± 0.16	1.25 ± 0.06
6.5	50	0.23 ± 0.03	0.85 ± 0.12	1.55 ± 0.26	1.12 ± 0.08
	70	0.16 ± 0.03	0.75 ± 0.13	2.10 ± 0.15	1.26 ± 0.06

Recently the usefulness of the parameters ε and Δ have been discussed by Bartschat and Blum.[25] They have found that whereas a comparison between experimental and numerical results only allows the theoretical model to be tested as a whole, including all approximations made in describing the scattering process and on the atomic wavefunctions, a comparison of ε and Δ may allow individual dynamical assumptions to be tested. For instance a P state excitation process can be considered as a two step process where initially an elastic scattering occurs, in the presence of spin-orbit interactions, followed by an excitation in which spin-dependent forces may be neglected. In this instance the states $M_L = 1$ and $M_L = -1$ are coherently excited and a definite phase relation exists between a_1 and a_{-1}. This results in $\cos\varepsilon = 1$ whereas nothing can be stated regarding Δ. In fact recently such results have been obtained by Register[26], for superelastic scattering of electrons on the 6^1P_1 state of barium, who measured a value of $\cos\varepsilon$ close to unity and $\cos\Delta$ almost zero for 100eV electrons at an angle of 5°.

Blum[27] has also obtained a further relationship between ε and λ assuming that a 6^3P_1 state of a heavy atom may be represented by adding a 6^1P_1 contribution to the 6^3P_1 wavefunction

$$\phi(6^3P_1) = \alpha\phi'(6^3P_1) + \beta\phi'(6^1P_1) \tag{8}$$

where $\phi'(6^3P_1)$ and $\phi'(6^1P_1)$ are pure Russel-Saunders states. This relation is given as

$$\frac{-\lambda}{1-\lambda} \le \cos\varepsilon \le 1 \tag{9}$$

Applying this to the results in table 1 it is found that both values for 50° violate expression (9) whereas those values for 70° are consistent with it. The measurement at 5.5eV is in the presence of a strong negative ion resonance whilst that at 6.5eV is free from such effects. This violation of expression (9) suggests that additional spin dependent effects must be considered when dealing with the excitation of this state (at a scattering angle of 50°).

Before leaving this discussion on mercury one should also mention a coincidence experiment reported by Hanne et al.[28] In this experiment they used a technique similar to that of King et al[6] to measure the "threshold polarisation". The state which was investigated was excitation from the 6^1S_0 ground state to the 6^3P_1 state. By detecting the photons in coincidence with the electrons scattered in the forward direction the selection rule $\Delta M_L = 0$ applies. Consequently excitation from the 6^1S_0 ground state of the $M = \pm 1$ sublevels can only occur through transfer of spin angular momentum.

According to McConnell and Moiseiwitsch[29] the polarisation for a natural mixture of isotopes, will be -0.77 if electron exchange with spin transfer is the appropriate excitation mechanism.

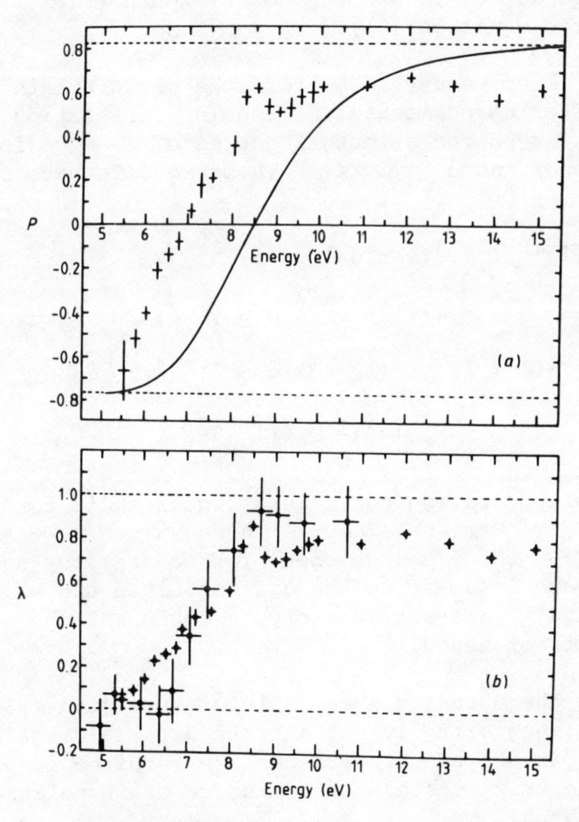

Fig. 4. Results obtained for exchange versus direct excitation mechanisms.

For mercury intermediate coupling applies (equation (8)) and if excitation exists solely via the 6^1P_1 admixture polarisation should be +0.82. The polarisation measurement thus provides detailed information about the relative contributions of the two mechanisms.

The results obtained are shown in figure 4 and it may be seen that for electrons scattered in the forward direction the excitation mechanism changes dramatically near the threshold of the excitation. The mechanism at low incident energies is primarily one of exchange whereas at incident energies greater than about 9eV direct excitation is by way of the Russel Saunders 6^1P_1 state.

The work on heavy atoms has not been restricted to mercury. Studies have been carried out on the inert gases including the $5\,^1P_1$ and $5\,^3P_1$ states of krypton[30] and on the $6\,^3P_1$ state of xenon. Because of the difficulty of carrying out polarisation measurements for the UV photons in the decay of these states the angular correlation technique has been used. This restriction reduces the number of parameters which can be obtained since Δ can no longer be explicitly evaluated. The three parameters which can be determined are χ', defined as $\cos\chi' = \cos\bar{\chi}\cos\Delta$, λ and ε (with the usual definitions from equation (6)).

It is still necessary to perform the measurements at two different scattering geometries. Angular correlation functions were measured for photon azimuthal angles of $\phi_\gamma = \pi$ (in the electron scattering plane) and $\phi_\gamma = 3\pi/4$. These results were fitted to the functions

$$N(\phi_\gamma = \pi) \propto \tfrac{1}{2}(1+\lambda) + \tfrac{1}{2}(1-3\lambda)\cos^2\theta - \tfrac{1}{2}(1-\lambda)\cos\varepsilon\,\sin^2\theta - [\lambda(1-\lambda)]^{\frac{1}{2}}\cos\chi'\sin 2\theta \qquad (10)$$

$$N(\phi_\gamma = 3\pi/4) \propto \tfrac{1}{2}(1+\lambda) + (1-3\lambda)\cos^2\theta - [\tfrac{1}{2}\lambda(1-\lambda)]^{\frac{1}{2}}\cos\chi'\sin 2\theta$$

By initially carrying out a least squares fit for the data for $\phi_\gamma = 3\pi/4$ λ and χ' were obtained. These values were then inserted into the function for $\phi_\gamma = \pi$ to check for consistency and to obtain ε. Tables 2 and 3 summarise the data obtained and whilst the values for λ and χ' are satisfactory, ε could only be extracted for the 30° data of xenon.

Inserting the data for λ and ε into expression (9) it is again found that the inequality is not satisfied. This again indicates that further spin dependent processes are required in addition to considering the triplet state as a mixture of Russell Saunders $^3P_1$ and $^1P_1$ states.

Table 2 Values of the parameters λ and $\bar{\chi}$ for the 5^3P_1 and 5^1P_1 states of krypton

E_i (eV)		θ_e (deg)	λ	$\bar{\chi}$(rad)
36	$^3P_1$	20	0.584 ± 0.124	1.633 ± 0.084
	$^3P_1$	30	0.517 ± 0.081	1.716 ± 0.079
	$^1P_1$	20	0.510 ± 0.073	1.960 ± 0.050
	$^1P_1$	30	0.449 ± 0.073	1.782 ± 0.080
60	$^3P_1$	20	0.829 ± 0.289	1.988 ± 0.249
	$^3P_1$	30	0.782 ± 0.157	1.451 ± 0.178
	$^1P_1$	20	0.717 ± 0.217	1.336 ± 0.130
	$^1P_1$	30	0.583 ± 0.098	1.790 ± 0.107

Table 3 Values of the parameters λ, $\bar{\chi}$ and ε for the 6^3P_1 state of xenon (E_i, the incident electron energy, is 80eV)

θ_e (deg)	λ	$\bar{\chi}$ (rad)	ε(rad)
20	0.499 ± 0.033	1.239 ± 0.019	
30	0.403 ± 0.034	1.504 ± 0.035	$1.214 + 0.076$

QUANTUM BEATS

Quantum beats have been extensively studied in beam foil spectroscopy,[31] but only recently have results been reported for their observation following electron impact excitation.[32] For beats to be observed the excitation to the participating levels must be coherent and the modulation period of the decay must be greater than the instrumental resolution of about 10^{-9}s obtainable in electron photon coincidence experiments. Since the internal relaxation times from fine or hyperfine interactions of coherently excited states are $T_{fs} \sim 10^{-11}$s and $T_{hfs} \sim 10^{-8}$s only beats between suitable hyperfine structure levels may be observed. The reported measurement is in the 3^2P decay of sodium where beats were measured from the hyperfine structure of the J = 3/2 levels.

A 100eV electron beam intersected a modulated sodium beam in a field free interaction region. Electrons scattered through 3° having lost 2.1eV were detected in coincidence with sodium D-line photons using standard coincidence electronics. The results oscillations about an exponential decay are shown in figure 5 plotted as $I(t)e^{\Gamma t}$ where

$$\Gamma t = -\tfrac{1}{2}s^2\gamma^2 + \gamma(t - t_0)$$

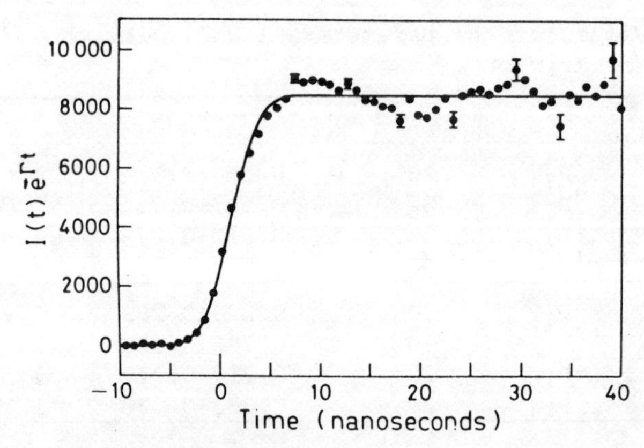

Fig. 5. Quantum beats as observed in reference 32.

with γ the reciprocal of the lifetime of the excited state, s is the Gaussian standard deviation and t_0 is the prompt response time. A detailed analysis of the beat structure based on Macek and Jaecks[9] provides a function of the form

$$I(t) = G[A + \sum_i B_i \cos(\omega_{FF'}t + \psi_{FF'})]$$

where $\omega_{FF'}$ and $\psi_{FF'}$ are the beat frequencies and phase factors. Due to the complexity of this function, and the experimental time resolution information regarding the orientation and alignment parameters, contained in B_i, could not be extracted.

CONCLUSION

While new applications of the electron-photon coincidence method, such as zero field quantum beats, will continue the full analysis of the experimental results is complex. The determination of parameters, such as the spin orbit phase parameters ε and Δ, which allow specific dynamical features of theoretical models to be tested will take on a greater significance. For heavy atoms this is of particular significance since there is only limited theoretical results available and effects, such as spin orbit interaction, must be taken into account. Perhaps also the "ultimate" electron impact experiments are starting[33] where polarised electrons are used as the projectile in coincidence experiments.

REFERENCES

1. R.E. Imhoff and F.H. Read, J. Phys B 4 (1971) 450.
2. A. Pochat, D. Rozvel and J. Peresse, J. de Physique 34 (1973) 701.
3. D.R. Flower and M.J. Seaton, Proc. Phys. Soc. 91 (1967) 59.
4. I.C. Percival and M.J. Seaton, Philos. Trans. Roy. Soc. London Series A 251 (1958) 113.
5. H. Hafner and H. Kleinpoppen, Z Physik 198 (1967) 315.
6. G.C.M. King, A. Adams and F.H. Read, J. Phys B 5 (1972) 254.
7. H. Kleinpoppen, Columbia University, unpublished report (1968).
8. M. Eminyan, K. MacAdam, J. Slevin and H. Kleinpoppen, Phys. Rev. Lett. 31 (1973) 576; and J. Phys. B 7 (1974) 1519.
9. J.H. Macek and D.H. Jaecks, Phys. Rev. A4 (1971) 2288.
10. U. Fano and J.H. Macek, Rev. Mod. Phys. 45 (1973) 553.
11. D.H. Madison and W.N. Shelton, Phys Rev A7 (1973) 449.
12. K. Blum and H. Kleinpoppen, Physics Reports 52 (1979) 205; and J. Slevin, H.Q. Porter, M. Eminyan, A. Defrance and G. Vassilev, J. Phys B 13 (1980) 3009.
13. S.T. Hood, E. Weigold and A.J. Dixon, J. Phys B 12 (1979) 631; and J. Slevin, M. Eminyan, J.M. Woolsey, G. Vassilev and H.Q. Porter, J. Phys B 13 (1980) L341.
14. A. Ugabe, P.J.O. Teubner, E. Weigold and A. Arriola, J Phys B 10 (1977) 71.
15. H. Arriola, P.J.O. Teubner, A. Ugabe and E. Weigold, J. Phys B8 (1975) 1275; and I.C. Malcolm and J.W. McConkey, J. Phys B 12 (1979) 511.
16. M.C. Standage and H. Kleinpoppen, Phys Rev Lett 36 (1975) 577.
17. M. Kohmoto and U. Fano, J Phys B 14 (1981) L477; and H.J. Beyer, H. Kleinpoppen, I. McGregor and L.C. McIntyre Jr. J Phys B 15 (1982) in press.
18. W.C. Fon, K.A. Berrington and A.E. Kingston, J Phys B 13 (1980) 2309; and D.H. Madison and K.H. Winters, Phys Rev Lett 47 (1981) 1885.
19. R. McAdams and J.F. Williams, J Phys B 15 (1982) L247.
20. R.W. Hermann and I.V. Hertel, Coherence and Correlation in Atomic Collisions Eds. H. Kleinpoppen and J.F. Williams (1980: New York: Plenum) 625: and J Phys B 13 (1980) 4285.
21. A.A. Zaidi, I. McGregor and H. Kleinpoppen, Phys Rev Lett 45 (1980) 1168.
22. K. Blum, F.J. de Paixao and Gy. Csanak, J Phys B 13 (1980) L257.
23. F.J. da Paixao, N.T. Padial, Gy. Csanak and K. Blum, Phys. Rev. Lett. 45 (1980) 1164.
24. A.A. Zaidi, I. McGregor and H. Kleinpoppen, J Phys B 11 (1978) L151.
25. K. Bartschat and K. Blum, J. Phys B 15 (1982) in press.
26. D. Register as quoted in ref. 25.
27. K. Blum, private communication.

28. G.F. Hanne, K. Wemhoff, A. Wolcke and J. Kessler, J Phys B 14 (1981) L507.
29. J.C. McConnell and B.L. Moiseiwitsch, J Phys B 1 (1968) 406.
30. I. McGregor, D. Hils, R. Hippler, N.A. Malik, J.F. Williams, A.A. Zaidi and H. Kleinpoppen, J. Phys B 15 (1982) L411.
31. H.J. Andra in Progress in Atomic Spectroscopy part B p829, Plenum Press, New York, 1969.
32. P.J.O. Teubner, J.E. Furst, M.C. Tonkin and S.J. Buckman, Phys. Rev. Lett. 46 (1981) 1569.
33. K. Bartschat, K. Blum, G.F. Hanne and J. Kessler, J.Phys. B 14 (1981) 3761.

HIGH RESOLUTION LASER SPECTROSCOPY OF SMALL MOLECULES

W. Demtröder, D. Eisel, H.J. Foth, M. Raab,
H.J. Vedder, H. Weickenmeier

Fachbereich Physik, Universität Kaiserslautern
6750 Kaiserslautern, W.-Germany

INTRODUCTION

For the investigation of atomic spectra in most cases Doppler-limited resolution is sufficient to separate spectral lines. Sub-Doppler spectroscopy is only required if the substructure of lines, caused by narrow fine structure or by hyperfine structure is to be resolved, or if very accurate line positions, line profiles or line shifts are to be measured.

In contrast to this situation for atoms, molecular spectra exhibit a much more complex structure with a density of lines which may exceed that of atomic spectra by several orders of magnitude. This is, first of all, due to the manifold of rotational and vibrational levels within each electronic state and furthermore to a larger variety of angular momentum coupling, such as spin-rotation interaction, Λ-type doubling, fine and hyperfine structure. In addition different kinds of perturbations may further increase the line density and the complexity of the spectrum. Even for small molecules, such as diatomic or triatomic molecules, the spacings between rotational lines of an electronic transition may become much smaller than the Doppler-width. This implies that single rotational lines often cannot be resolved with "classical" Doppler-limited techniques.

There are, however, several possible ways to circumvent these difficulties and to gain detailed information on molecular structures in spite of the complex structure of the spectrum: The first method relies on a reduction of the line density in absorption spectra by extreme cooling of the molecules down to temperatures of a few Kelvin. At these low temperatures only the lowest vibrational level and very few rotational levels are thermally populated and therefore only transitions from these few levels appear as lines in the absorption spectrum. The cooling can be achieved either through adiabatic expansion in supersonic molecular beams[1] or by "freezing" the molecules in a matrix of noble gases[2]. A disadvantage of the matrix isolation technique is the influence of the host lattice on the level energies of the guest molecule, which results in broadening and shift of the molecular lines. The adiabatic expansion method, on the other hand, uses seeded beam techniques in order to reach these low temperatures and therefore needs large vacuum chambers and high pumping speeds.

The second, most frequently used method reduces the linewidth of molecular transitions instead of the density of lines. This is achieved by applying Doppler-free spectroscopic techniques such as saturation spectroscopy[3], polarization spectroscopy[4] or Doppler-free two photon spectroscopy[5]. Another way of reducing the Doppler-width uses a collimated molecular beam which is crossed perpendicular by a laser beam[6]. All these methods require single mode tunable lasers to fully utilize the Doppler-free resolution.

A third method, which comes more and more into use, is based on various optical double resonance techniques and requires generally at least two tunable lasers. The essential point of this method is the labelling of a selected molecular level by optical pumping with a "pump laser". A second "probe laser" is tuned through the spectral range of interest, but only those molecular transitions are monitored, which start from the "labelled" level, selected by the pump transition. The method works with pulsed and with cw lasers. Combined with one of the sub-Doppler techniques a drastic reduction of the line density as well as a reduction of the line width can be achieved. This makes double resonance techniques very attractive.

We will now discuss the second and third method in more detail and illustrate them by several examples.

SUB-DOPPLER LASER SPECTROSCOPY

From the variety of different techniques which recently have been developed to outwit the Doppler-width of spectral lines in gas phase spectroscopy[7] only two examples are selected here. They shall illustrate the resolution achieved so far and the gain in information about molecular structures obtainable from resolved features in sub-Doppler spectra, which are completely masked in Doppler-limited spectroscopy.

Laser Spectroscopy in Collimated Molecular Beams

When the beam from a tunable single mode laser is crossed perpendicularly with a collimated molecular beam, the distribution of molecular velocity components parallel to the laser beam is reduced through the collimating apertures by a factor ε which equals the collimation ratio of the molecular beam. In practice collimation ratios of 1/100 are readily realised which implies that the Doppler-width of a molecular absorption line is reduced from a typical value of $\Delta\nu_D \cong 1000$ MHz in a gas cell at room temperature to about 10 MHz. For many transitions this reduced Doppler-width comes close to the natural linewidth.

Since the intersection path of the laser beam and the molecular beam is typically less than 1 cm and the density of absorbing molecules is low ($\leqq 10^{12}/cm^3$) the relative attenuation of the laser intensity may be below 10^{-10} for molecular transitions with small absorption cross section. Sensitive detection schemes are therefore demanded to obtain a sufficiently good signal to noise ratio.

One sensitive method monitors the fluorescence photons emitted from the laser excited molecules. Since under collision free conditions each absorbed photon generates a fluorescence photon, the "excitation spectrum" $N_{Fl}(\lambda_L)$ which represents the number of fluorescence photons N_{Fl} counted as a function of laser wavelength λ_L, reflects the true absorption spectrum, if the fluorescence detector has a sensitivity independent of the fluorescence wavelengths λ_{Fl}. A brief estimation may illustrate the sensitivity: If 1% of the fluorescence emitted into the solid angle 4π can be imaged by the collimating lens onto the cathode of a photomultiplier with 20% quantum efficiency and a thermal dark current of 10 photoelectrons per second, we need $5\cdot10^3$ absorbed photons/s in order to achieve a signal to noise ratio S/N = 1. If the incoming laser has a power of 1 W, which corresponds to $3\cdot10^{18}$ photons/s at $h\cdot\nu = 2$ eV, the relative attenuation, just detectable, is $2\cdot10^{-15}$!

Another even more sensitive detection method utilizes a second laser with sufficiently small wavelength to ionize those molecules which had been excited by the first laser. If the intensity of the ionizing laser is so high that each excited molecule is ionized before it is deactivated by emission of a fluorescence photon, each absorbed photon of the first laser is converted into an ion which can be detected with an efficiency close to 100%. In this ideal case single absorbed photons can be monitored, which also implies that single absorbing molecules are detectable[8]. In practice, the sensitivity may be somewhat lower. However, with cw argon lasers as ionizing source, the sensitivity has proved to be at least 1 - 2 orders of magnitude higher than with fluorescence detection.

For the infrared region where low vibrational-rotational levels in the electronic ground state are excited by a tunable infrared laser, the two detection methods, discussed above, lose their sensitivity. The spontaneous lifetimes of these levels are typical in the millisecond range and the excited molecules in the beam travel several meters before they radiate. The efficiency of collecting the

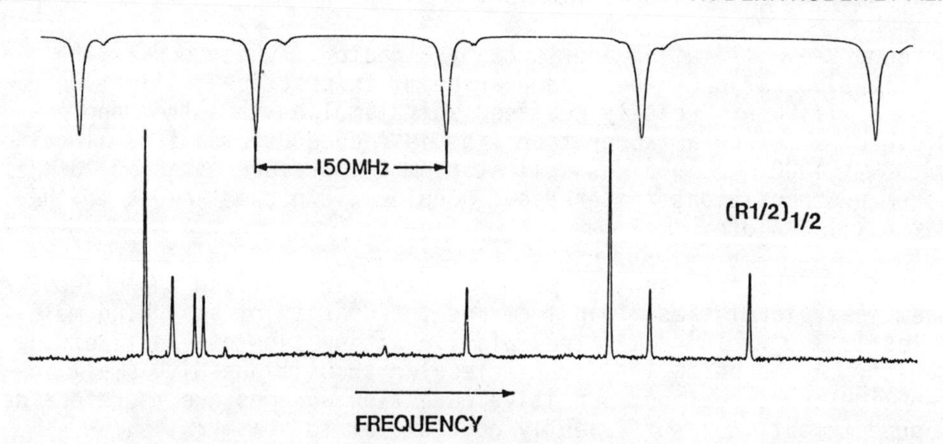

Fig. 1. Optothermal detection of the J = 1/2 → 3/2, v = 0 → 2 overtone transition in NO with completely resolved hyperfine structure (from ref. 10).

fluorescence onto a detector is therefore poor and furthermore the quantum efficiency of photodetectors is much lower in the infrared than in the visible range. The photoionization method fails because the second ionizing laser will generally ionize both states of the infrared transition induced by the first laser.

Scoles and coworkers[9] have developed an optothermal detection scheme, where the vibrational energy of molecules excited by an infrared laser is monitored by a cooled bolometer which is placed in the molecular beam downstreams from the interaction region with the laser. This detector measures the total energy (translational + internal energy) of molecules transferred to the bolometer when the molecules impinge and stick on the cold surface. The sensitivity of the system is high, and changes of the internal energy of 10^{-13} W induced by the absorption of the laser can still be detected. Figure 1 which shows the completely resolved hyperfine structure of the J = 1/2 → 3/2; v = 0 → 2 overtone transition in the NO molecule, excited by a color center laser, demonstrates the sensitivity of the system[10]. The line-width of about 1.5 MHz is due to the residual Doppler-width caused by the finite acceptance angle of the detector.

Figure 2 illustrates for the complex NO_2 spectrum how much more information can be obtained with sub-Doppler resolution. The upper, Doppler-limited excitation spectrum was taken with a single mode dye laser traversing a cell which contained NO_2 gas at a pressure of 0.2 torr. The lower spectrum represents a small section of the upper one, indicated by the bar, recorded with sub-Doppler resolution of about 12 MHz in a collimated supersonic NO_2 beam. The triplet structure of the lines reflects the hyperfine structure due to the nuclear

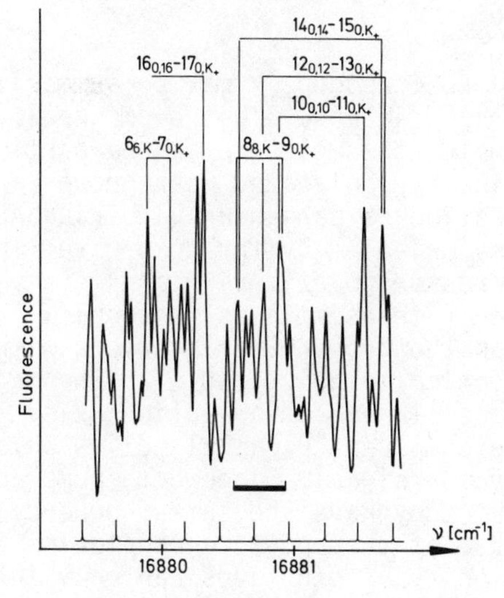

Fig. 2. Doppler-limited excitation spectrum of NO_2 (left spectrum) and a section of this spectrum, marked by the bar, recorded with sub-Doppler resolution in a collimated NO_2 beam (from ref. 11).

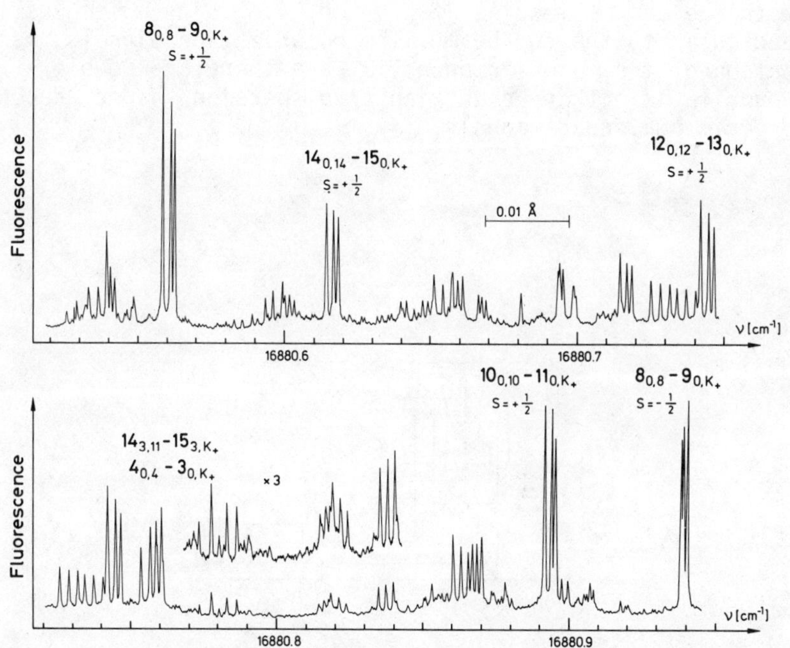

spin I = 1 of the nitrogen nucleus. The fine structure splittings due to the interaction of the electron spin S = 1/2 with the angular momentum of molecular rotation, appear as coarse features at this high resolution. In order to demonstrate, that even small peaks in the spectrum are real lines and not noise, part of the spectrum is shown in the insert with three times increased sensitivity[11].

Molecular Polarization Spectroscopy

The basic idea of polarization spectroscopy[12] may be summarized as follows: When a circularly polarized intense electro- magnetic wave passes through a molecular sample, the molecule can be partly oriented by optical pumping[13]. A linearly polarized weak probe wave, passing the sample in opposite direction, experiences birefringence when it interacts with the molecules oriented by the pump wave. The sample cell is placed between two crossed polarizers P and A (see Fig. 3), which block the probe wave if the sample is isotropic, i.e. without the pump wave. However, when the probe wave interacts with oriented molecules, it becomes elliptically polarized with the major axis being turned against the original linear polarization axis. The second polarizer then transmits the component parallel to its polarization axis and the detector receives a signal. Since both pump and probe wave come from the same laser, they have the same frequency ω but opposite wave vectors $\pm\vec{k}$. Because of the opposite Doppler-shift $\pm\vec{k}\cdot\vec{v}$ for a molecule moving with velocity $\vec{v}$ both waves can only interact with the same molecules if $\vec{k}\cdot\vec{v} = 0$, i.e. with molecules moving perpendicular to the two beams. The polarization signals, recorded as a function of the laser frequency ω, are therefore Doppler-free. The line profile is either Lorentzian or dispersion shaped, depending on the experimental conditions[14].

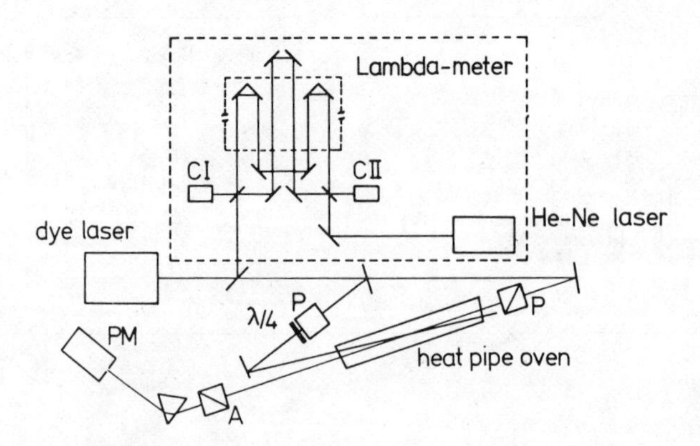

Fig. 3. Experimental arrangement for polarization spectroscopy with absolute wavelength measurements with a travelling Michelson wavemeter.

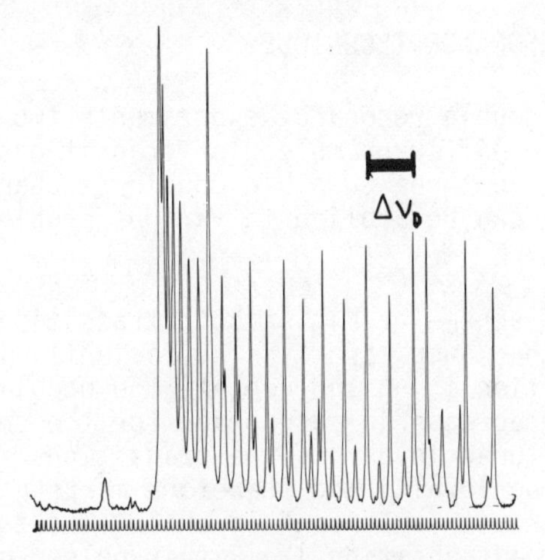

Fig. 4. Polarization spectrum of the bandhead of the v' = 9 ← v" = 14 band in the $C^1\Pi_u \leftarrow X^1\Sigma_g$ transition of Cs_2.

The sensitivity of polarization spectroscopy is about two to three orders of magnitude higher than that of saturation spectroscopy. One reason is the suppression of the background by the crossed polarizers, another the intensity of the probe laser. If only 10^{-6} of an incoming probe power of 1 mW is transmitted through A, the photomultiplier receives 10^{-9} W, which is far above the detector noise limit. A further advantage for molecular spectroscopy is the different response to Q-lines ($\Delta J = 0$) and to P- and R-lines ($\Delta J = \pm 1$), which can be attributed to different optical pumping of the M-sublevels for both kind of transitions[15]. The wavelengths of molecular transitions can be determined at least within 10^{-5} nm when the laser is stabilized onto the center of the Doppler-free lines and its wavelength is measured with a travelling Michelson wavemeter[16].

So far polarization spectroscopy has been applied to a number of small molecules[17,18]. Figure 4 illustrates a small section of a Doppler-free polarization spectrum of Cs_2, which shows rotational lines of the $C^1\Pi_u \leftarrow X^1\Sigma_g$ transition around the bandhead of the v' = 9 ← v" = 14. Note that many rotational lines fall within the Doppler-width of about 600 MHz. The frequency marks, which are 60 MHz apart, are generated with a 120 cm confocal Fabry-Perot interferometer.

OPTICAL DOUBLE RESONANCE TECHNIQUES

For optical double resonance experiments two lasers are required which are tuned to different molecular transitions sharing a common level. In Fig. 5 a-c three possible double resonance schemes are illustrated which can be applied to various problems in molecular spectroscopy.

In the first scheme of Fig.5a both transitions share a common lower level. If the "pump laser" is sufficiently intense, it will saturate the transition i → k and deplete the population N_i. This depletion causes a decrease in absorption for the second laser, if this "probe laser" is tuned to any of the transitions i → m. The pump laser "labels" the common level i and therefore marks a few transitions i → m of the probe laser with a "known" lower level out of a manifold of other allowed probe transitions[19]. Either pulsed or cw lasers can be used. In case of cw lasers the pump intensity is chopped at a frequency f where the period 1/f is long compared to the relaxation time. The population N_i and with it the absorption of the probe laser will be therefore modulated at the same frequency f. When the probe absorption is monitored through a lock in amplifier only those probe transitions are selected which start from the labelled level i. When using polarized lasers, the molecules in level i will be partly oriented and the polarization state of a linearly polarized probe laser will be altered, when interacting with these oriented molecules. This optical double resonance polarization technique[20] represents a very powerful Doppler-free method to assign complex and irregular molecular spectra. Figure 6 shows as example two double resonance signals in the polarization spectrum of Cs_2. The pump laser was stabilized onto a R transition J" = 72 → J' = 73 while the probe laser was tuned through the spectrum. The two signals correspond to the schemes of Fig. 5a and 5c. The additional small signals are caused by collisions which partly transfer the orientation of the pumped level to neighbouring rotational levels.

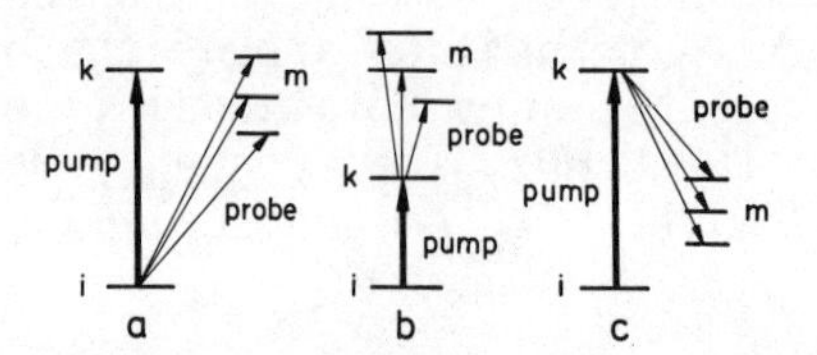

Fig. 5. Possible optical-optical double resonance schemes.

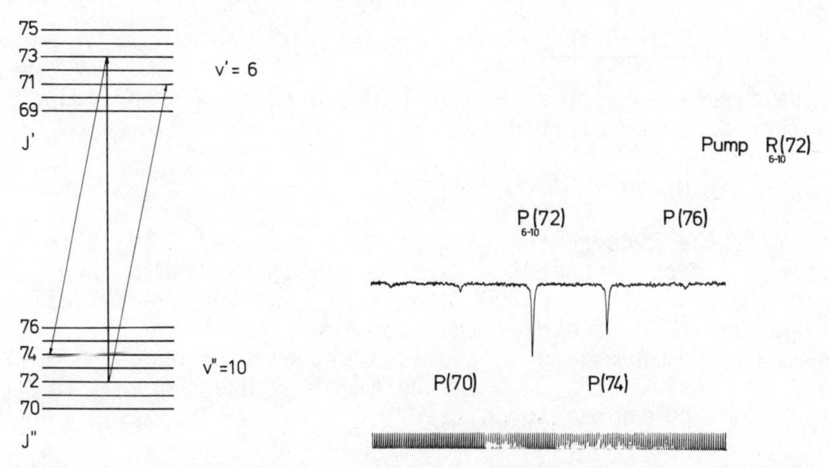

Fig. 6. Optical-optical double resonance signals in Cs_2.

The second double resonance scheme of Fig. 5b, often called "stepwise excitation" allows one to reach highly excited molecular states with visible lasers. In contrast to direct excitation by ultraviolet one photon transitions, where the parity of the upper states is opposite to that of the initial ground state, levels of the same parity are excited by this resonant two photon transitions. Using single mode lasers, sub-Doppler double resonance spectra are obtained. This has been demonstrated by R.W. Field and his coworkers[21] who measured several perturbed electronic transitions of the BaO molecule and determined spectroscopic constants of electronic states, unknown so far.

The advantages of optical double resonance methods become obvious when perturbed molecular spectra are investigated where the assignment of irregularly spaced lines is extremely difficult when conventional methods are used. Since the labelled level is known, the assignment of the final levels m of the probe transitions is much easier because of selection rules which leave only few choices for possible assignments. Bernheim et al.[22] applied this method to the detailed studies of highly excited states of the Li_2 molecule.

Schawlow and coworkers[23] investigated two step excitation of the Na_2 molecule by using an elegant polarization labelling technique. A narrow band pulsed pump laser induces an anisotropy for molecules in a selected level k = (v',J') in the excited B state of Na_2. A second, broad band linearly polarized pulsed probe laser passes through the cell with oriented molecules. All those wavelengths within the spectral profile of the probe pulse, which correspond to transitions from the labelled level k to higher excited levels m, are simultaneously transmitted through the crossed polarizer A of Fig. 3 and are photographed behind a spectograph. This technique permitted the assignment of a number of Rydberg series to high lying molecular Rydberg levels[24].

Auto-Ionizing Molecular Rydberg States

The term energy of an atomic Rydberg state with principal quantum number n can be represented by

$$E_n = I - R_A/(n - \delta)^2 \tag{1}$$

where R_A is the Rydberg constant of this atom and $\delta(l)$ is the quantum defect which depends on the angular momentum quantum number l of the Rydberg electron. The quantum defect δ allows for the deviation of E_n in the actual force field of the atom from the energy of a level with the same quantum number n in atomic hydrogen (for which $\delta = 0$). Since without external fields $E_n < I$, the atomic Rydberg levels of singly excited atoms can not autoionize.

The situation is different for molecules. An electronic Rydberg state of a diatomic molecule AB can be represented, within the Born-Oppenheimer approximation, by a potential curve $E_n(R)$ which dissociates for internuclear separations $R \to \infty$ into an excited Rydberg atom A* and a ground state atom B. For sufficiently large values of n the excited electron is, on the average, far away from the molecular ion core because $\langle r \rangle \propto n^2$! It can therefore not contribute significantly to the binding of the molecular ion core. This implies that the Rydberg potential curves $E_n(R)$ are essentially the same as the potential curve $E_0^+(R)$ of the ionic ground state, apart from a vertical shift $\Delta E = C_M \cdot R/(n - \delta)^2$ on the energy scale, where C_M is a slowly varying function of R (see Fig. 7).

The essential point is now that the energy $E_n(v^*, J^*)$ of a molecular Rydberg level with vibrational quantum number v* and rotational quantum number J* lies above a level $E(v^+, J^+)$ in the ion ground state, if its vibrational and rotational energy is sufficiently high, i.e. if

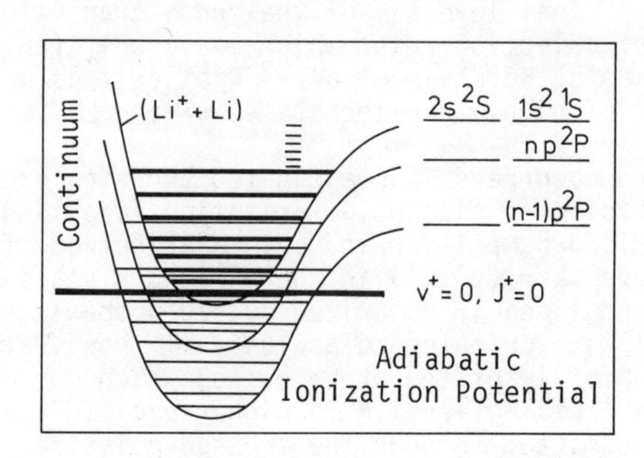

Fig. 7. Potential curves of molecular Rydberg states and of the molecular ion ground state.

$$E^*_{vib}(v^*) + E^*_{rot}(J^*) > \Delta E + E^+_{vib}(v^+) + E^+_{rot}(J^+) \quad . \tag{2}$$

When a sufficiently strong interaction exists between the nuclear motion and the electronic motion (which implies a breakdown of the BO-approximation) the energy $E^*_{vib} + E^*_{rot}$ may be partly converted to electronic energy and the electron can gain sufficient energy to leave the core.

This auto-ionization process competes with the radiative decay of the Rydberg level, often also with possible predissociation. However, the spontaneous transition probability is approximately proportional to n^{-3}. For large quantum numbers n the radiative decay therefore becomes very slow and even small auto-ionization probabilities may already result in complete auto-ionization of a Rydberg level if predissociation can be neglected. This means that the excitation of these molecular Rydberg levels above the ionization limit can be very effectively monitored through the resulting ion.

The relevance of studying molecular Rydberg levels is not only justified by the investigation of the auto-ionization process and its causes but also by the fact that each Rydberg level $E_n(v^*,J^*)$ converges for $n \to \infty$ against a well defined rotational-vibrational level $E^+(v^+,J^+)$ of the ion ground state. Thus the molecular constants of this state can be obtained from measuring Rydberg series of the neutral molecule.

Unfortunately, the situation in real molecules is more complex than considered so far. This is due to different kinds of perturbations which affect the energies of Rydberg levels and shift them from their expected values. For large quantum numbers n, for instance, the energy separation of adjacent Rydberg states becomes comparable to the spacings between rotational levels. In a classical picture this means that the revolution period of the Rydberg electron becomes comparable to the rotation period of the molecular ion frame. This leads to a decoupling of the electron angular momentum from the internuclear axis and the molecule has to be described by an intermediate case between Hund's coupling cases a, respectively b and d[25].

Fortunately, a theoretical approach to the description of high lying auto-ionizing Rydberg levels has been developed, named multichannel quantum defect theory[26,27]. This theory has been applied very successfully by Jungen et al.[28] to a quantitative description of the auto-ionization in H_2. The agreement between theory and experiment[29] is striking. It may be expected that this theory may well be adapted to other Rydberg molecules, in particular to "hydrogen-like" molecules, such as Li_2.

We have therefore started in Kaiserslautern experimental investigations of auto-ionizing Rydberg states of Li_2. The experimental procedure is illustrated in Fig. 8. The Li_2 molecules are prepared in a

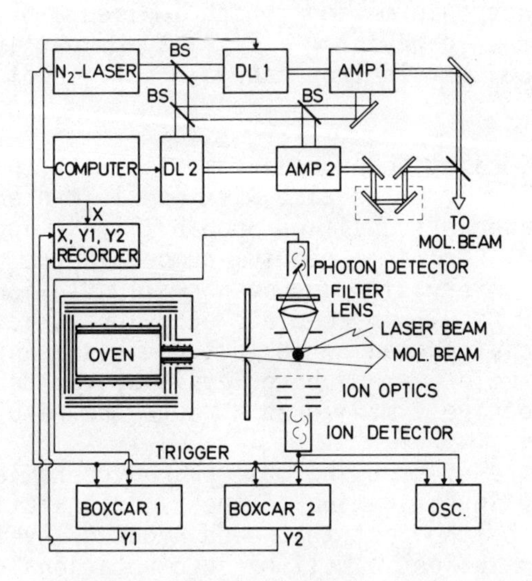

Fig. 8. Experimental arrangement for observing autoionization in Li_2.

supersonic molecular beam which is crossed perpendicularly by two collinear laser beams from two pulsed dye lasers L1 and L2, pumped by the same nitrogen laser. The first laser pumps the Li_2 molecules from a level (v_i'',J_i'') in the $X^1\Sigma_g$ ground state into a selected level (v_k',J_k') in the $B^1\Pi_u$ state. Since the molecular constants of both states are known[30], the transition $(v_i'',J_i'') \rightarrow (v_k',J_k')$ can be assigned by observing the laser induced fluorescence. When the pump laser L1 is stabilized onto this transition, the second laser L2 is switched on and induces transitions from the intermediate level (v_k',J_k') into higher Rydberg levels (v^*,J^*), which can auto-ionize. The number of ions is monitored as a function of the wavelength λ_2 of L2, at first with L1 off and then with L1 on. The difference between the two readings gives those ions which are produced by L2 on a transition $(v_k',J_k') \rightarrow (v^*,J^*)$ or by direct photoionization

$$Li_2(v_k',J_k') + h\nu_2 \rightarrow Li_2^+ + e^- \qquad (3)$$

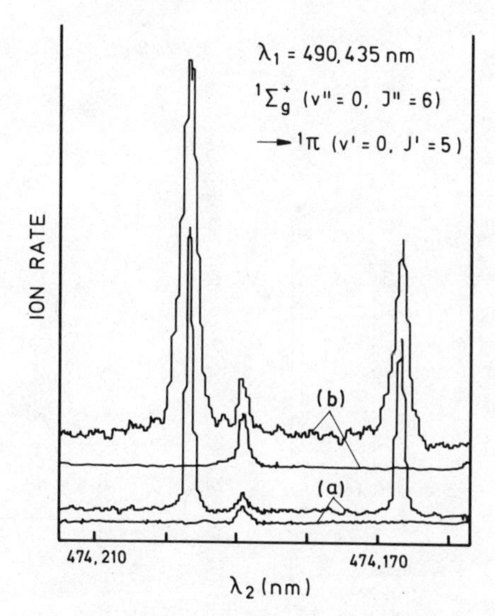

Fig. 9. Linewidth of autoionization lines at different electric fields a) E = 20 V/cm, b) E = 100 V/cm

In our experiment $2h\nu_2$ was larger than the ionization limit of $Li_2(X^1\Sigma_g)$. The second laser alone can therefore also produce ions by two photon ionization. The probability for this process is, however, sufficiently large only for resonant two photon ionization when the photon energy $h\nu_2$ fits to a molecular transition

$$Li_2(X^1\Sigma_g(v'',J'')) + h\nu_2 \rightarrow Li_2\ B^1\Pi_u(v_2',J_2') \quad . \tag{4}$$

The lower trace in Fig. 10A shows the number of ions $N_{ion}(2\nu_2)$ produced with L1 off. The peaks correspond to resonant transitions (4), as could be proved by observation of the corresponding fluorescence (lowest trace in Fig. 10B). When the pump laser L1 is switched on, the upper trace in Fig. 10A is recorded. The difference between the two curves (note that the upper curve is vertically shifted by an amount indicated by the arrow in order to separate the two recordings) gives the number of ions produced by one photon $h\nu_2$ on a transition from a labelled level (v_k',J_k') in the intermediate state. The auto-ionization lines can be seen superimposed on a continuum which increases with decreasing wavelength λ_2. The continuum is attributed to the direct photoionization process (3) and from its onset at

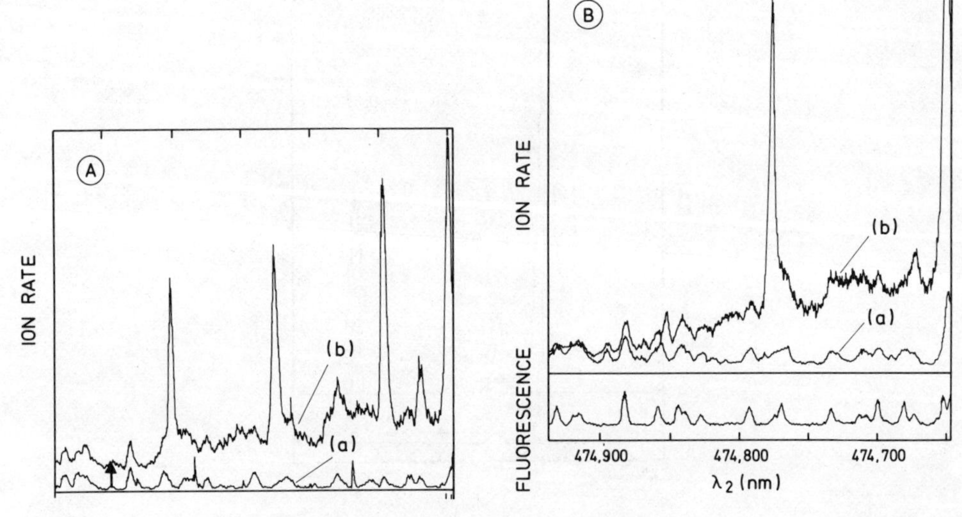

Fig. 10. Ion rate as a function of λ_2. a) Laser 1 off, b) L1 on. The intermediate level ($v_k' = 0$, $J_k' = 6$) was excited by L1. The two spectra A and B were taken at E = 50 V/cm and 20 V/cm. The lowest trace in B is the fluorescence spectrum induced by L2.

λ_2 = 474.9 nm, measured at different electric field strengths, the adiabatic ionization potential

$$IP = E(X^2\Sigma_g^+, v^+ = 0, J^+ = 0) - E(X^1\Sigma_g^+, v'' = 0, J'' = 0) \qquad (5)$$

is determined[31] to be IP(Li_2) = 41475 ± 8 cm^{-1}. A comparison between Figs. 10A and 10B shows that with increased strength of the electric field used to extract the ions, some new auto-ionization lines appear close above the ionization threshold. They can be attributed to very high Rydberg levels with n ≅ 50 and $v^* = 0$ which auto-ionize by rotational coupling into $v^+ = 0$. The molecules in these levels have a large polarizibility and the induced dipole moment may enhance the coupling efficiency.

The linewidth of the auto-ionization lines is very narrow and mainly determined by the laser bandwidth of about 3 GHz. In order to investigate the linewidth in more detail, the bandwidth of L2 was reduced to 1.5 GHz by inserting an extra etalon into the laser cavity. The laser linewidth could be determined from the width of the lines $I_{Fl}(\lambda_2)$ induced by L2. Figure 9, which shows two auto-ionization lines excited at two different field strengths, illustrates that the linewidth increases with increasing electric field. The steps in the line profiles are due to the stepwise scanning of the laser wavelength. From a comparison of these line profiles with the fluorescence line profiles excited by the same laser, an upper limit $\Delta\nu \leqq 300$ MHz for the auto-ionization linewidth at low fields can be obtained. This implies that the lifetime of the auto-ionizing state is at least 0.5 ns, probably longer[32].

Compared to atomic auto-ionizing states, where lifetimes below 10^{-12} s are common, the molecular levels exhibit a much weaker coupling to the ionization continuum. The reason is that the molecular ionization process takes place through a coupling between nuclear motion and electronic configurations, whereas in atoms a direct electronic coupling of the excited electron to the continuum makes the auto-ionization probability by far larger.

Similar experiments on auto-ionization of molecular Rydberg levels have been performed on Na_2 and K_2 by Martin et al.[33] and by Leutwyler et al.[34,35]. A complete analysis of the auto-ionization spectra of all these alkali molecules is, however, still pending since a full understanding of all perturbations requires very detailed studies of the many molecular states closely below the ionization limit. These will not only be Rydberg states but also doubly excited states which can interact with the Rydberg states. For Li_2 such investigations are under way in our laboratory.

Accurate Determination of Dissociation Energies and Long Range Potentials of Molecular Ground States

Until recently most of our knowledge about the long range part of the interatomic interaction potential came from scattering experiments and from measurements of line broadenings and shifts. The application of narrow band tunable lasers to the excitation of fluorescence spectra terminating on high vibrational levels has added another, very accurate method for detailed studies of potential curves and dissociation energies[36].

Assume a high vibrational level v' is excited in an upper electronic state which is less tightly bound than the ground state. The potential curve of this state will then be shifted towards larger internuclear separations R compared to the ground state (see Fig. 11). If the optical excitation starts from a thermally populated low vibrational level v" of the ground state, it will reach the upper level v' around its inner turning point if the Franck Condon factor is maximum for this case. Since the excited molecule makes many vibrations during its lifetime, the fluorescence will be emitted preferentially from the outer turning point where the molecule spends a longer time. This fluorescence terminates either on high vibrational levels below the dissociation limit (bound-bound transitions) or in the continuum above the dissociation limit (bound-free transitions).

From the rotational and vibrational spacings of the fluorescence lines terminating on bound levels below the dissociation limit an accurate determination of the potential V(R) is possible[36]. From the structure in the continuous fluorescence spectrum of bound-free transitions the difference between upper and lower potential can be determined[37]. Figure 12 illustrates both kinds of fluorescence spectra. It shows the long wavelength section of the fluorescence spectrum of Cs_2 excited on a transition $X^1\Sigma_g^+(v''=0, J'=49) \rightarrow D^1\Sigma_u^+(v'=50, J'=48)$ which terminates on high vibrational levels of the ground state up to about v" = 150. The structural continuum to the long wavelength side of the lines is due to bound-free transitions. From this spectrum the dissociation energy of the Cs_2 $X^1\Sigma_g$-state could be accurately determined[38] as $D_e = 3648 \pm 8$ cm^{-1}. The limitation of the accuracy is caused by the insufficient resolution of the monochromator used to disperse the fluorescence. The small rotational spacings of the highest vibrational levels could not be resolved. Here the optical-optical double resonance scheme of Fig. 5c can improve the situation. If the fluorescence transitions are replaced by stimulated emission, the line width of the double resonance signal is only limited by the width of the upper level, labelled by the pump transition. This allows to resolve even level spacings close to the dissociation limit, which are smaller than the Doppler width of the optical transition.

Molecular Spectroscopy below the Natural Linewidth

When the two beams of pump laser and probe laser in the double resonance scheme of Fig. 5c travel collinearly instead of antiparallel a particular situation arises. The simultaneous interaction of both waves with the molecule may be regarded as a resonant stimulated Raman scattering process. It can be shown[39] that in such a case the width γ_{DR} of the double resonance signal is given by

$$\gamma_{DR} = \gamma_i + \gamma_m + (1 - \lambda_{pump}/\lambda_{probe})\, \gamma_k \tag{5}$$

when the levels i and m are vibrational-rotational levels of the electronic ground state they have long radiative lifetimes and their widths are small compared to γ_k. The width of the double resonance signal is then approximately equal to the fraction $(1 - \lambda_{pump}/\lambda_{probe})$ of the upper level width γ_k.

With this technique therefore line widths can be achieved[40] which are below the natural linewidth of the optical transitions $i \rightarrow k$ or $k \rightarrow m$. This is demonstrated in Fig. 13 which shows two double resonance signals in the polarization spectrum of Cs_2, which are generated with a fixed pump laser and a tunable probe laser according to the level scheme of Fig. 6. The left signal corresponds to the double resonance scheme of Fig. 5a, the left, narrower signal to Fig. 5c.

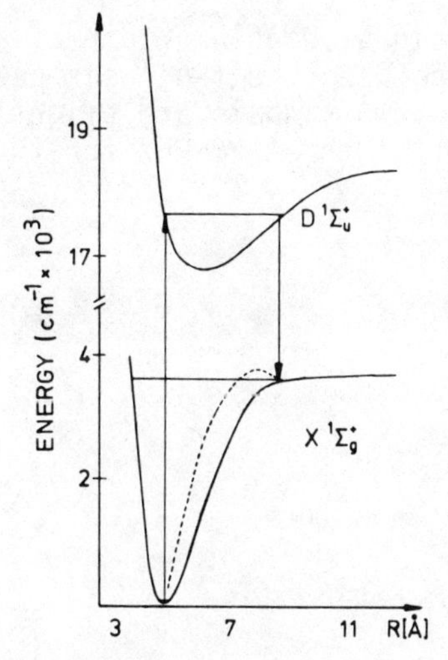

Fig. 11. Schematic potential diagram for the excitation of high vibrational levels with subsequent fluorescence to levels closely below the dissociation limit.

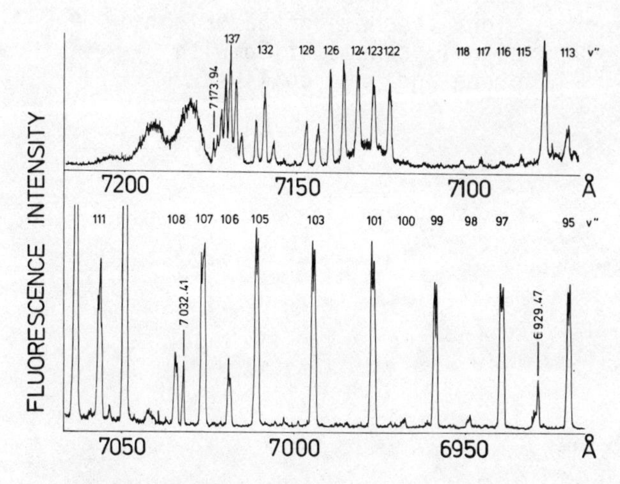

Fig. 12. Long wavelength section of the fluorescence spectrum of Cs_2 excited on the transition $X^1\Sigma_g(v'' = 0, J'' = 49) \rightarrow D^1\Sigma_u^+(v' = 50, J' = 48)$.

Both signals have a residual Doppler-width of about 6 MHz due to the finite crossing angle between the two laser beams. The natural line width is 20 MHz and the two signals are slightly power-broadened. This results in a total width of 28 MHz for the double resonance sig-

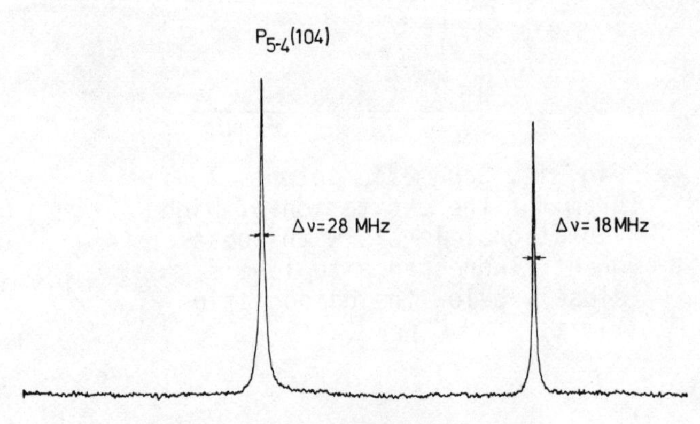

Fig. 13. Two double resonance signals of Cs_2. Left signal: scheme of 5a, right signal: scheme of 5b with probe and pump collinear.

nal of scheme 5a, while the width of the signal due to scheme 5c is only 18 MHz. This "subnatural linewidth" spectroscopy can be used to resolve for example very narrowly spaced levels in shallow potential minima, close to the dissociation limit.

REFERENCES

1. D.H. Levy, L. Wharton, R.E. Smalley, Laser Spectroscopy in Supersonic Jets, in: Chemical and Biochemical Applications of Lasers, Vol. II, C.B. Moore, ed., Academic Press, New York (1977).
2. H.E. Hallam, Vibrational Spectroscopy of Trapped Species, Wiley & Sons, London (1973).
3. V.S. Letokhov, V.P. Chebotayev, Nonlinear Laser Spectroscopy, Springer, Berlin, Heidelberg, New York (1977).
4. C. Wieman and T.W. Hänsch, Doppler-free Polarization Spectroscopy, Phys. Rev. Lett. 36:1170 (1976).
5. G. Grynberg, B. Cagnac, Doppler-free multiphoton spectroscopy, Rep. Progr. Phys. 40:791 (1977).
6. R. Abjean, M. Leriche, On the shapes of absorption lines in a divergent atomic beam, Opt. Commun. 15:121 (1975).
7. W. Demtröder, Laser Spectroscopy, Springer, Berlin, Heidelberg, New York (1981)
8. G.S. Hurst, M.G. Payne, S.P. Kramer, and J.P. Young, Resonance ionization spectroscopy and one atom detection, Rev. Mod. Phys. 51:767 (1979).
9. T.E. Gough, R.E. Miller, G. Scoles, Appl. Phys. Lett. 30:338 (1977).
10. R.E. Miller, PhD-Thesis, University of Waterloo, Ontario, Canada (1980).
11. H.J. Foth, H.J. Vedder, and W. Demtröder, Sub-Doppler Laser Spectroscopy of NO_2 in the λ = 592.5 nm Region, J. Mol. Spectrosc. 88:109 (1981).
12. R.E. Teets, F.V. Kowalski, W.T. Hill, N. Carlson, T.W. Hänsch, Laser Polarization Spectroscopy, Proc. Soc. Photo-Opt. Inst. Engin. 113:80, San Diego (1977).
13. R.E. Drullinger and R.N. Zare, Optical Pumping of Molecules, J. Chem. Phys. 51:5532 (1969).
14. M. Raab, G. Höning, W. Demtröder, C.R. Vidal, High Resolution Laser Spectroscopy of Cs_2: II. Doppler-free Polarization Spectroscopy of the $C^1\Pi_u \leftarrow X^1\Sigma_g^+$ System, J. Chem. Phys. 76:4370 (1982).
15. G. Höning, PhD Thesis, University of Kaiserslautern, Germany (1980).
16. F.V. Kowalski, R.E. Teets, W. Demtröder, and A.L. Schawlow, An improved wavemeter for cw lasers, J. Opt. Soc. Am. 68:1611 (1978).
17. V. Stert and R. Fischer, Doppler-Free Polarization Spectroscopy Using Linear Polarized Light, Appl. Phys. 17:151 (1978).
18. R.E. Teets, N.W. Carlson, A.L. Schawlow, Polarization Labelling Spectroscopy of NO_2, J. Mol. Spectrosc. 415 (1979).
19. M.E. Kaminsky, R.T. Hawkins, F.V. Kowalski, and A.L. Schawlow, Identification of Absorption Lines by Modulated Lower-Level Population, Phys. Rev. Lett. 36:671 (1976).
20. R. Teets, R. Feinberg, T.W. Hänsch, and A.L. Schawlow, Simplification of Spectra by Polarization Labelling, Phys. Rev. Lett. 37:683 (1976).
21. R.A. Gottscho, P.S. Weiss, R.W. Field, Sub-Doppler Optical-Optical Double Resonance Spectroscopy of BaO, J. Mol. Spectrosc. 82:283 (1980).

22. R.A. Bernheim, L.P. Gold, P.B. Kelley, C. Tomczyk, A spectroscopic study of the $E^1\Sigma_g^+$ and $F^1\Sigma_g^+$ states of $^7Li_2$ by pulsed optical-optical double resonance, J. Chem. Phys. 74:3249 (1981).
23. N.W. Carlson, A.J. Taylor, and A.L. Schawlow, Identification of Rydberg States in Na_2 by Two-Step Polarization Labelling, Phys. Rev. Lett. 45:18 (1980).
24. N.W. Carlson, A.J. Taylor, K.M. Jones, and A.L. Schawlow, Two-Step Polarization Labelling Spectroscopy of Excited States of Na_2, Phys. Rev. A24:822 (1981).
25. G. Herzberg, Molecular Spectra and Molecular Structure, Vol. I, p. 218 ff., Van Nostrand Reinhold Comp., New York (1950).
26. M.J. Seaton, Quantum defect theory, Proc. Phys. Soc. 88:801 (1966).
27. U. Fano, Quantum defect theory of l-uncoupling in H_2 as an example of channel interaction treatment, Phys. Rev. A2:353 (1970).
28. M. Raoult and Ch. Jungen, Calculation of vibrational preionization by multichannel quantum defect theory, J. Chem. Phys. 74: 3383 (1981).
29. P.M. Dehmer, W.A. Chupka, Very high resolution study of photoabsorption, photoionization and predissociation in H_2^+, J. Chem. Phys. 65:2243 (1976).
30. M.M. Hessel, C.R. Vidal, The $B^1\Pi_u$-$X^1\Sigma_g^+$ band system of the $^7Li_2$ molecule, J. Chem. Phys. 70:4439 (1979).
31. D. Eisel, W. Demtröder, Accurate ionization potential of Li_2 from resonant two-photon ionization, Chem. Phys. Lett. 88:481 (1982).
32. D. Eisel, PhD-thesis, University of Kaiserslautern (1982).
33. S. Martin, J. Chevaleyre, S. Valignant, J.P. Perrot, M. Broyer, B. Cabaud, A. Hoareau, Autoionizing Rydberg States of the Na_2 Molecule, Chem. Phys. Lett. 87:235 (1982).
34. S. Leutwyler, A. Hermann, L. Wöste, E. Schuhmacher, Isotope selective two-step photoionization studies of K_2 in a supersonic molecular beam, Chem. Phys. 48:253 (1980).
S. Leutwyler, M. Hofmann, H.P. Härri, E. Schuhmacher, The adiabatic Ionization Potentials of the Alkali Dimers Na_2, NaK and K_2, Chem. Phys. Lett. 77:257 (1981).
35. S. Leutwyler, T. Heinis, and M. Jungen, Auto-ionizing Rydberg States in Na_2, Chem. Phys. Lett. 87 (1982).
36. F. Engelke, Studies of Molecular Interaction by Laser Excitation, Comments Atom. Mol. Phys. 11:13 (1981).
37. J. Tellinghusen and M.B. Moeller, Chem. Phys. 50:301 (1980).
38. M. Raab, H. Weickenmeier, W. Demtröder, The Dissociation Energy of the Cesium Dimer, Chem. Phys. Lett. 88:377 (1982).
39. V.P. Chebotayev, in: V.S. Lethokov, V.P. Chebotayev, Non Linear Laser Spectroscopy (Springer Series in Opt. Sciences Vol. 4, Springer, Berlin (1979).
40. R.P. Hackel, S. Ezekiel, Observation of subnatural linewidths by two-step resonant scattering in I_2-vapor, Phys. Rev. Lett 42:1736 (1979).
41. M. Raab, PhD-Thesis, University of Kaiserslautern (1981).

FAST ION BEAM LASER SPECTROSCOPY (FIBLAS) :

A CASE STUDY : N_2O^+

Michel L. Gaillard

Laboratoire de Spectrométrie Ionique et Moléculaire
(associé au CNRS)., Université de Lyon I,
Campus de La Doua, 69622 Villeurbanne Cedex (France)

INTRODUCTION

To a number of atomic and molecular physicists in the early seventies, ion spectroscopy has appeared as a "major challenge" area. From the experimental point of view, the field could be considered as still reasonably fresh : note for example that ten years ago, Herzberg[1] could identify only seven molecular ions whose absorption spectra were known. It was generally agreed at that time that ion physics was of high applied interest in gas discharges, plasmas and, thus, astrophysics. The relevance of molecular ion physics in diluted plasma chemistry soon turned out to be crucial for ionospheric and interstellar reactions. However vague and sometimes poorly justified, those were the general assertions which were invoked whenever rationalization was needed (like in drafting proposals).

By all account however, one must admit that the main incentive was something which is probably best described as "experimentalist discomfort". Physics teaches us that there is formally very little difference between the way we theoretically describe an atomic (or, for that matter, molecular) ion and its neighboring isoelectronic neutral counterpart. Experimentally however, the difference was enormous. Orders of magnitude seemed to distinguish the level of sophistication achieved in the two cases. Somehow, this lack of symmetry was felt as a disgrace. Such was at least the feeling when people began to search for ways to bring experimental investigations in ion physics up to a comparable level of precision with neutral physics. This proved to be a major endeavour : it took several years before anyone could claim some measure of progress at least in so far as spectroscopy was concerned.

The history of the entire process is by now rather complex. The general direction of research which is indicated in the preceding paragraph was never clearly spelled out during the course of the work, probably repressed in favor of arguments more palatable to funding agencies. For that reason, many different groups took part in the effort often unwittingly and almost always independently from each others. As a result and however interesting it may look to former participants, the historical approach is certainly not best suited for an introduction to the present state of the art. Despite the fact that the author could not refrain from a few general remarks, of mainly historical interest, the rest of the current presentation steers away from some of the reviewer's temptations and tries to avoid the credit awarding attitude which poorly befitted the author's strong bias. No attempt has been made to cover the field in a comprehensive fashion. We choose instead the "case study" approach and based our report on a detailed examination of a test example of current interest in our laboratory. In that context, note that use of the plural carries no majestic pretence but refers to a number of collaborators (and nevertheless friendly coworkers) involved with the FIBLAS project in Lyon and Orsay.[2-5] Without their help, work would have been much slower and certainly much less enjoyable over the past few years. During the course of the presentation, efforts will be made to spell out the underlying general principles illustrated by the example at hand. We shall not try to hide however that successful experimental research often benefits from a set of fortunate circumstances which may be difficult to gather in coincidence again in all generality.

FAST ION BEAM LASER SPECTROSCOPY OF SMALL MOLECULAR IONS

The development of fast ion beam laser spectroscopy techniques (for short : FIBLAS) is not so unusual a case of simultaneous but independent technical evolution both in atomic and molecular physics. Although the concepts involved in both cases were quite similar, the apparatus used in the pioneering experiments were widely different, ranging from the table top mass spectrometer for the early molecular physics work[6] to the largest tandem Van de Graaff accelerators for some of the atomic physics experiments.[7]

Whereas the interest of the atomic physicists concentrated immediately on high resolution spectroscopy (either by time resolved techniques[8] or by more laser specific approaches[9]), the essential effort in molecular physics was initially devoted to the study of photodissociation phenomena in diatomic molecular ions.[10] It is only recently that high resolution potential of the FIBLAS approach has been fully applied for the spectroscopic investigation of various molecular ions.

Part of that delay is due to the fact that straightforward extension of high resolution techniques was hindered by a dramatic loss of sensitivity of traditional detection schemes in the case of molecular ion beams. Indeed, the detection of the absorption or emission of electromagnetic radiation by molecular ion beams poses a difficult problem. The density of a typical accelerated ion beam (10 μA of singly charged ions accelerated to one thousandth of the speed of light) is only of the order of 10^8 particules cm^{-3}. The density problem is further enhanced by the spread of the beam population over the various rovibrational energy levels of the ground and metastable states which are initially prepared in the ion source. Direct detection of the absorption of such a smaller number of ions is clearly not feasible. The discussion by Carrington et al.[11] indicates that monitoring of the total fluorescence does not provide a sufficient increase in sensitivity unless the fluorescence probability is larger than 10^{+7} s^{-1}. This criterion is easily verified for the optical excitation of singly charged atomic ions but obviously fails for vibrational excitation, and even for a large class of electronic excitation, of molecular ions. In the very favorable case of N_2^+, Holt et al.[12] recently proved that fluorescence could be used successfully as a resonance detection channel. Nevertheless, it is clear that for a more general application of the method, the development of indirect detection schemes was crucial.

Several solutions have been found and successfully applied.[13] They fall into two categories. In the first group are detection schemes which apply to excited upper states of high fluorescence yield but low fluorescence probability. In that case the excited state has a long lifetime (typically in the microsecond range) and a long flight path in the accelerated beam. It thus becomes experimentally feasible to probe the light induced change in beam state population by a second interaction, resulting that time in a change of ion concentration. This final change is then easily measured electrically on the ion beam current either directly or by synchronuous lock-in amplification with (or without) mass (alternatively charge or energy) analysis. The sensitivity of such a detection scheme based on charge exchange reaction in a differentially pumped gas cell was clearly demonstrated by Wing et al.[14] and illustrated by the work of Carrington et al.[11] Similar in concept, although widely different in experimental realization, is the detection scheme recently devised by Carrington et al.,[15] based on beam deceleration followed by a gas phase chemical reaction with mass analysis of the reaction products. This certainly does not exhaust the possible choice for the second interaction which may include other collision channels,[16] externally applied fields[17] or even a second laser interaction leading to photoionization or dissociation.[18]

The second group of indirect detection methods concerns the excited states of low fluorescence yield. In that case, higher sensitivity is expected from observation of the non radiative decay channels. In diatomic molecular ions two main channels are to be considered, either direct dissociation or predissociation. Both lead to a change of beam composition which can be easily detected by the usual combination of beam analysis (either magnetic or electrostatic) and beam current measurement. From the point of view of high resolution spectroscopy however, repulsive states are of little interest and only the case of relatively slow predissociation has to be considered. This has been successfully applied to several systems including CH^+, NO^+, O_2^+, O_3^+, N_2O^+, CH_3I^+ and H_2S^+. It yields spectroscopic results which usefully complement, above the predissociation limit, the results of classical emission spectroscopy below that limit. Fast ion beam photofragment spectroscopy thus emerges as a technique of its own and has been the subject of recent reviews by two of its cofounders.[21,22] Its main advantage, whenever it applies, is experimental simplicity. The required hardware is basically a tandem mass spectrometer with provision for a colinear beam laser interaction between the spectrometer. Photofragment ions produced by the interaction can then be detected with nearly unit efficiency. Nature was so kind as to provide a nearly ideal test system for the method : O_2^+ in the $b^4\Sigma_g^-$ excited state. FIBLAS studies of O_2^+, pioneered by Durup and Moseley[21,22] lead to useful determinations of transition energies and molecular constants with record breaking precision in ionic physics. Lifetime measurements and fragment velocity analysis provided an amazingly detailed understanding of the predissociation mechanism. Remained to be seen however how easily one could extend the method to other eventually more complex molecular ions.

THE CASE FOR N_2O^+

Hunting for a new molecular ion that could be easily investigated by fast ion beam photofragment laser spectroscopy, we found N_2O^+ to be best suited to match the idiosyncrasies of the available experimental set up. The accelerator based device which we inherited from previous fast ion beam work can hardly be considered as representative of main stream FIBLAS instrumentation. Under somewhat extreme conditions, it does illustrate however the principles involved rather clearly (Fig. 1). Production of copious beams (microamperes) of singly charged ions is obtained effortlessly from gazeous compound fed into a radiofrequency discharge source of standard design. Stable operating conditions require however that the deposit rate on the source body of condensate arising from molecular dissociation should be kept as low as possible. This rules out prolongated operation with carbon or sulfur rich gases. On the contrary, N_2O^+ is obtained from N_2O a stable non polluting gas. Discharge chemistry was further found to provide for direct synthesis of N_2O^+ starting from a mixture of nitrogen and oxygen. This proved to be an inexpensive way of producing isotopically substituted molecules from readily available isotopically enriched nitrogen gas.

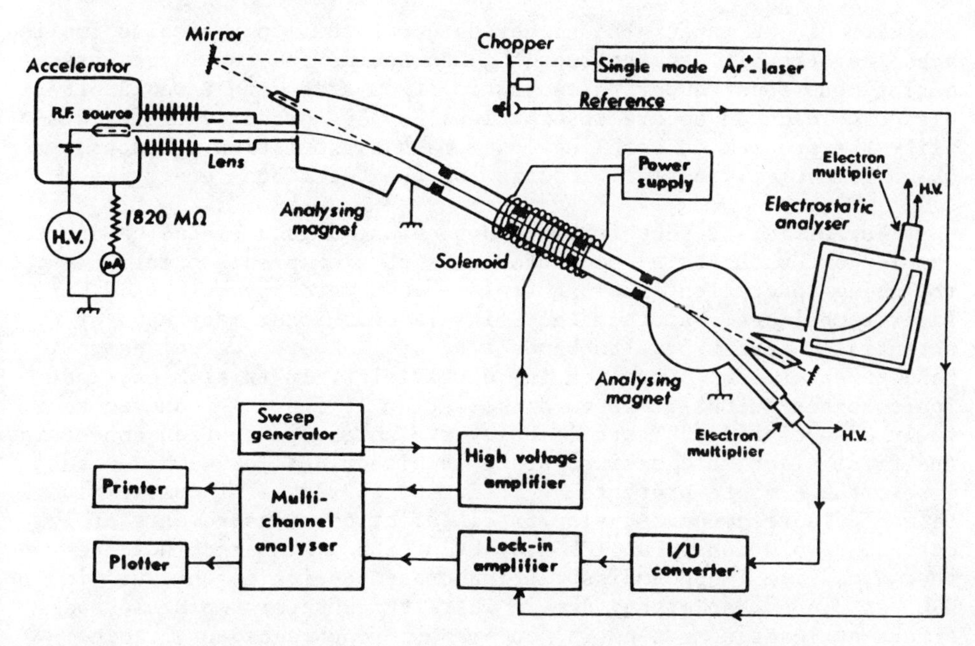

Fig. 1. FIBLAS experimental set up based on a 150 keV Sames accelerator and tandem magnetic mass analysis, eventually followed by electrostatic photofragment energy analysis.

Discharge sources suffer from many drawbacks, including a well known lack of stability, but their most basic flaw is the rather large energy spread of the extracted ion beam due to fluctuations in source voltage and plasma heating phenomena. High energy acceleration is able to offset the difficulty via kinematic compression of the source velocity spread along the beam direction.[23] With a RF discharge source, the energy to be used is however much higher than otherwise required for example when working with an electron bombardment source. This puts rather stringent requirements on the stability of the high voltage power supply since kinematic compression occurs only under the assumption that the accelerator fluctuations can be kept negligibly small with respect to the source spread. Our previous experience with O_2^+ [2] established that a compromise can be achieved with a terminal voltage in the 100 keV range and a 10^{-4}-10^{-5} stability.

Several side benefits can be gathered from the resulting use of high energy acceleration. Obviously, the Doppler shift (and thus the sweepable wavelength range in the ion reference frame, for a fixed laser wavelength) increases with the beam velocity for a colinear ion-laser beam geometry. As a result a substantial amount of spectroscopic information can be reaped using a single laser line, an experimental convenience which becomes crucial in the ultraviolet range. The higher speed of the ion beam also implies a shorter transit time between the ion source and the laser interaction zone thus favouring excitation from states that would otherwise have decayed before interaction. Finally higher energies simplify the problem of handling lower mass dissociation products which remove only a small fraction of the total beam energy.

For efficient photofragment detection, a good FIBLAS candidate should correspond to a documented case of slow predissociation following absorption in a spectral range covered by current C.W. laser techniques. Unfortunately, in diatomic ions, many studies of potential interest have to be delayed by lack of powerful narrow band tunable light source in the ultraviolet. In triatomics, most ionic states which lie above dissociation limits are found to be fully dissociated.[24] There is however a notable exception concerning the first electronic excited state of either COS^+ or N_2O^+. In both cases, the $\tilde{A}$ state presents an interesting example of competition between fluorescence and slow predissociation. Measurements at low optical resolution in an ICR trap by Orth and Dunbar[25] indicate that a maximum in photodissociation cross section is observed in N_2O^+ at 3.6 eV, an energy domain where the Doppler tuning method offers at least some spectral coverage through accidental coïncidences with the discrete ultraviolet lines of high power ion gas lasers.

A final requirement has to be satisfied by a good FIBLAS candidate in relation with the fact that high resolution of laser techniques is always obtained at the expense of broad spectral coverage. As pointed out by Saykally and Woods,[26] methods able to reveal the intimate details of the spectrum are quite impotent without the existence of data from previous optical spectroscopy work providing the initial broad overview and general understanding. The volume of information which has already been gathered both experimentally and theoretically on N_2O^+ indicates that the analysis of its structure raises issues of more than pure academic interest. Indeed N_2O^+ is an important intermediary in the reaction :

$$O^+(^4S) + N_2(X^1\Sigma_g^+) \rightarrow NO^+(X^1\Sigma^+) + N(^4S) + 1.1 \text{ eV}$$

which has been referred to as "... the principal ionospheric ion molecule reaction".[27] Information of highest precision on the bound states of N_2O^+ was obtained so far through optical emission. The analysis of Callomon and Creutzberg[28] based on photographic recordings of hollow cathod emission has been the reference for all later studies ; they successfully identified the spectrum of the transition $\tilde{A}^2\Sigma^+ \rightarrow \tilde{X}^2\Pi$ between the two lowest states of the ion. Despite lower energy resolution, photoionization also provided large amount of data. Focusing our interest on the first electronic excited state $\tilde{A}$, the situation was aptly summarized in a recent paper by Nenner et al.[29] The quantum yield of the lowest vibrational level (000) has been found equal to one. For all the vibrationally excited states, however a competition between fluorescence toward the ground state and predissociation toward the NO^+ + N limit was observed. This trend is well confirmed by lifetime measurements of selected vibrational levels. A sharp decrease of lifetime is observed between the (000) level (224±10 ns [30]) and higher vibrationally excited levels (166±15 ns [31] for the level identified with one quantum of excitation on the symmetric stretching mode (100)). This is interpreted as further evidence in favor of an onset of predissociation just above the (000) level of the $\tilde{A}$ state and agrees well with the result of early photodissociation experiments.[25,32]

From all previously known evidence, FIBLAS of N_2O^+ could thus be expected to bring new high resolution information on a molecular ion of high practical interest. Although we are still far from having exhausted all the experimental opportunities, this program has been partially fulfilled. The predissociation of N_2O^+ has been very easy to detect, easier even than the famed O_2^+ example if it was'nt for the fact that best results are obtained with the U.V. lines of the krypton laser, a device which is comparatively more awkward and expensive to operate in that domain. Obviously, photofragmentation of N_2O^+ can be observed with a more conventional set up,[33] using the low intensity beams extracted from an electron bombardment source. Such a source has a much finer ion velocity distribution. As a result, low energy acceleration under a few thousand volts provides enough velocity distribution compression to achieve comparable Doppler absorption width along the beam direction. Difficulties would however arise from the narrow domain of Doppler tuning at such low beam energies. It is clear that little usable molecular information can be extracted from the observation of only a few isolated lines however well resolved (see for example, the unfortunate situation in H_2S^+[34]).

FIBLAS OF N_2O^+ IN THE NEAR ULTRAVIOLET : HIGHLIGHTS

Spectroscopy

Photofragment spectroscopy had been applied so far to ions with very low fluorescence yield. Thus until now FIBLAS never entered a true competition with traditional emission spectroscopy. N_2O^+ offered such an opportunity for the first time since the relaxation of vibrationally excited levels of the $\tilde{A}$ state can be observed in both channels : fluorescence and photofragmentation. A clear comparison was given in Ref. 4 and is reproduced in Figs. 2 and 3. The emission spectrum of a pure N_2O DC discharge was obtained by Horani and Velghe and is used in Fig. 2 for identification purposes of the FIBLAS spectrum recorded with the 3374 Å line of a multimode krypton laser. The fast beam advantage becomes obvious when the same spectral region is recorded with a monomode laser (Fig. 3). An improvement of the resolution by more than an order

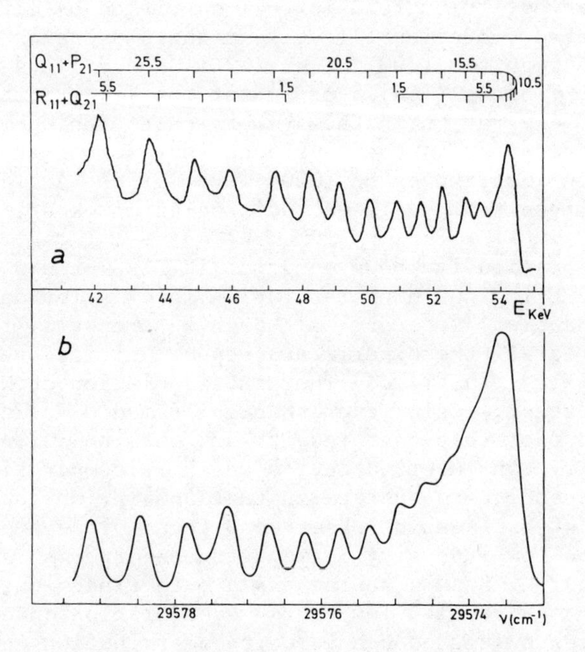

Fig. 2. a) Upper part : Experimental photopredissociation spectrum : Recorded NO^+ photofragment current as a function of the ion beam energy. The laser was operated multimode on the 3374 Å line.
b) Lower part : Microdensitomer recording of the emission spectrum of N_2O^+ in a DC discharge.

of magnitude is clearly to be seen. Within the fifty inverse centimeters which can be scrutinized with our set up by Doppler tuning the same laser line in colinear copropagating beam geometry (Fig. 4), previous analysis[28] of the emission spectrum predicted that two branch heads would be observed, corresponding to rotational progressions of the $\tilde{A}^2\Sigma^+(100) \rightarrow \tilde{X}^2\Pi_{3/2}(000)$ subband. N_2O^+ is known to remain linear in both upper and lower states. Only the ν_1 stretching vibrational mode is active in the transition. As a result, the observed band structure is clearly no more complicated than in

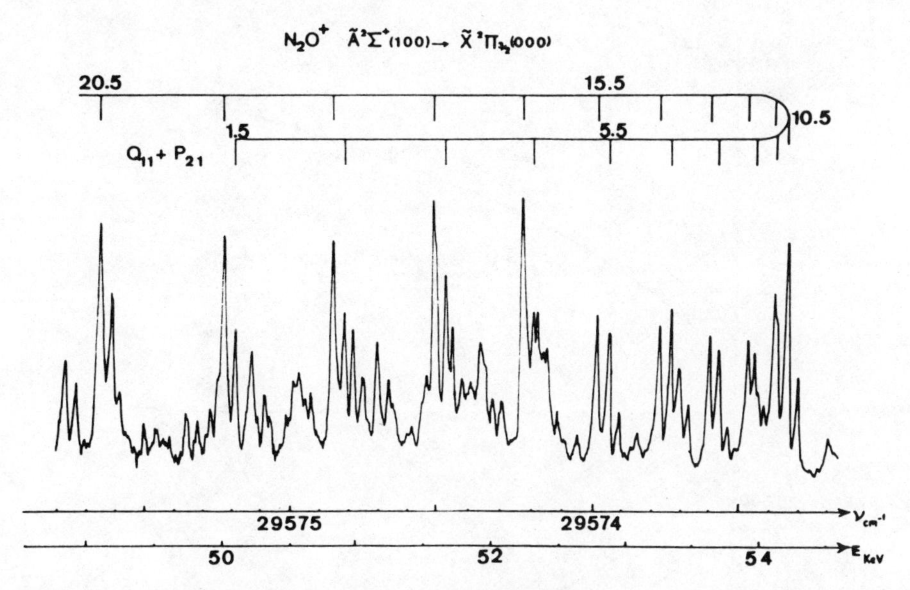

Fig. 3. FIBLAS spectrum of N_2O^+ obtained with a monomode Kr^+ laser operating on the 3374 Å line.

a diatomic molecule. This may provide some comfort to those atomic physicists who are already afraid at that point of getting too deeply concerned with the complex inner working of a polyatomic. In order to further spare such sensitivities, we shall not emphasize here the improvement in molecular constants, both rotational and vibrational, which can be obtained from the FIBLAS results. A forthcoming report[35] summarizes the current state of the analysis for the interested reader. Let us instead focus the attention on a property of the molecule which went so far unnoticed and which is brought to central light by the sudden increase in resolution, namely : hyperfine structure. Under close scrutiny (Fig. 5) it becomes obvious that the detailed structure of any given rotational line across the FIBLAS spectrum is significantly more complicated than a traditional $^2\Sigma^+ - ^2\Pi_{3/2}$ transition in a spinless molecule. Hyperfine interactions have seldom been studied in the optical spectra of triatomic molecules, let alone molecular ions. The N_2O^+ case is a priori further complicated by the fact that two nitrogen nuclei are likely to contribute. It is well known however in atomic physics that FIBLAS is ideally suited for isotopic shift and hyperfine interaction studies. This has been in fact the main application of the method in atomic spectroscopy.[36] With a fast beam set up, isotope separation and positive isotope identification is automatically carried out for example via magnetic analysis before the laser interaction. This advantage carries over to the molecular case and opens the possibility of stable isotope labelling. This was a crucial step in the study of hyperfine interactions in free N_2O^+

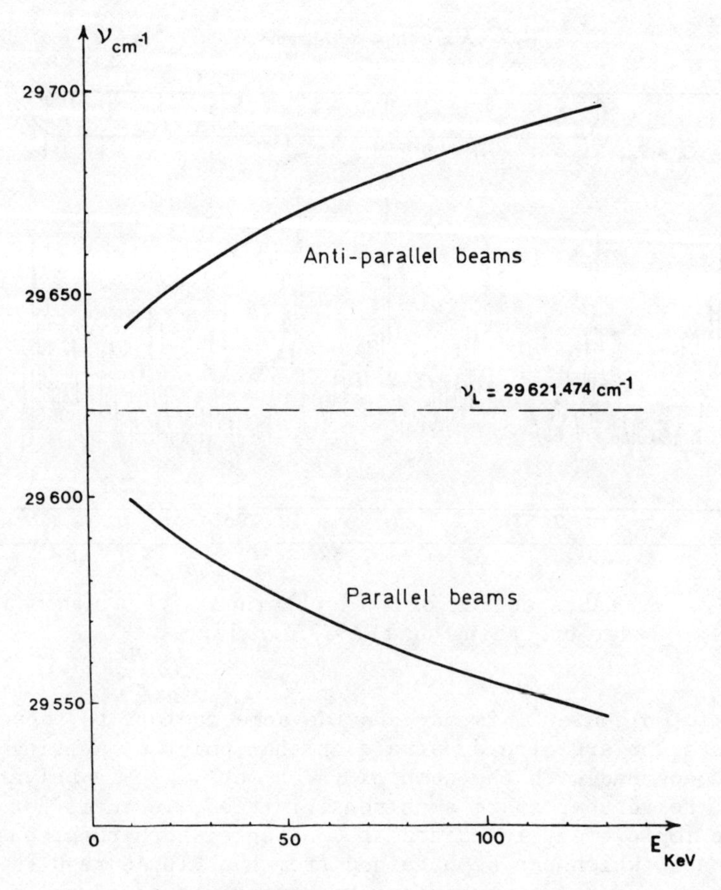

Fig. 4. Doppler tuning of a fast N_2O^+ ion beam near the 3374 Å line of the Kr^+ laser.

as reported recently in Ref. 5. Two main conclusions arise from our study of isotopically substituted $^{15}N_2O^+$, $^{14}N^{15}NO^+$ and $^{15}N^{14}NO^+$: i) gross features of the hyperfine structure are shown to arise from a single nucleus ii) the active nucleus is identified as the outer nitrogen. Without entering into the details of the analytical description which can be found in Refs. 5,35, the general picture emerges from inspection of Figs. 6 and 7. The lower state $^2\Pi_{3/2}$ hyperfine splittings are inversely proportional to J and thus become negligible as soon as J>>1. In the upper state, a single nuclear spin I is first coupled to the spin angular momentum S to produce a resultant G which, in turn, couples to the rotational

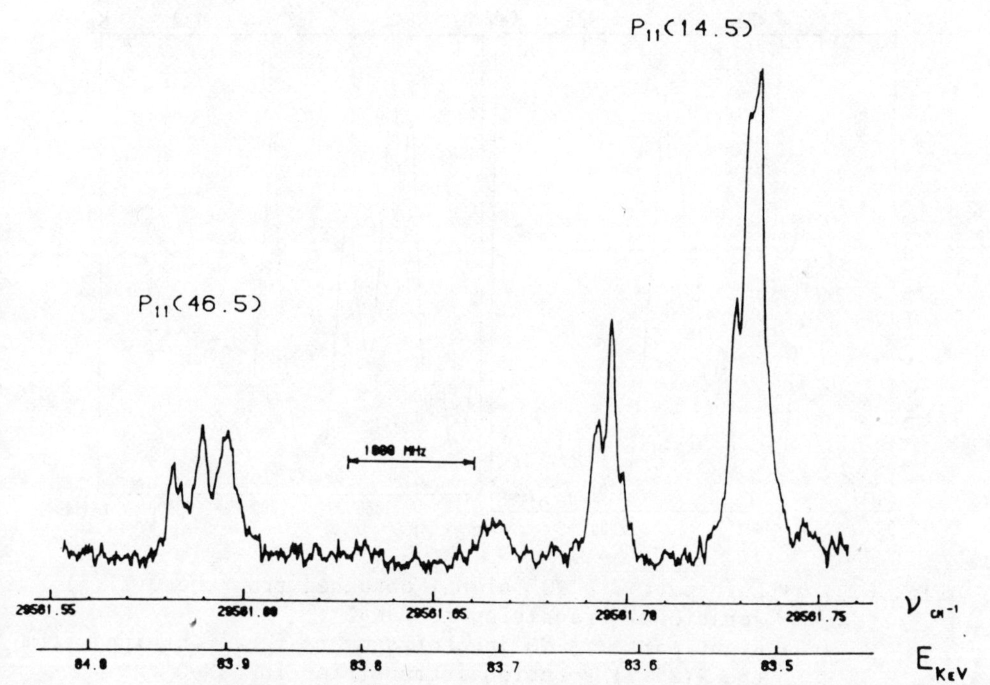

Fig. 5. Enlargement of 6 GHz wide portion of the spectrum showing the structure of the P_{11} rotational components observed with accelerating voltage of circa 84 keV. Voltage step size was 1 Volt, integration time constant : 3 sec/channel.

angular momentum N to yield F. In $^{14}N_2O^+$ the nuclear spin is I=1 with a positive nuclear moment which leads to the six F' components labelled on Fig. 6a, in two groups corresponding respectively to G'=1/2 and G'=3/2. In $^{15}N_2O^+$, I=1/2 and the nuclear moment changes sign. One observes four F' components only and a complete reversal of the pattern of Fig. 6b. Fig. 7 provides the experimental evidence pointing out to the outer nitrogen as the active nucleus. The parent molecules are identified by the isotopic shift of their caracteristic lines. The hyperfine patterns observed in isotopically selective fragment detection channels label unambiguously the nucleus responsible for the hyperfine splitting. Our description is in good agreement with general theoretical considerations based on the known molecular wavefunction of the $^2\Sigma$ state which contains a single unpaired 7σ molecular orbital. According to the tables of linear molecule wavefunctions of Mc Lean and Yoshimine,[37] the 7σ orbital of N_2O is not a strongly bounding orbital. Expended as a linear combination of atomic orbitals, it shows a dominant contribution from the 2s orbital localized on the outer nitrogen. This is the origin of the large Fermi contact interaction observed on the outer nitrogen. Our interpretation is well confirmed quantitatively by recent SCF LCAO calculations carried out specifically for the N_2O^+

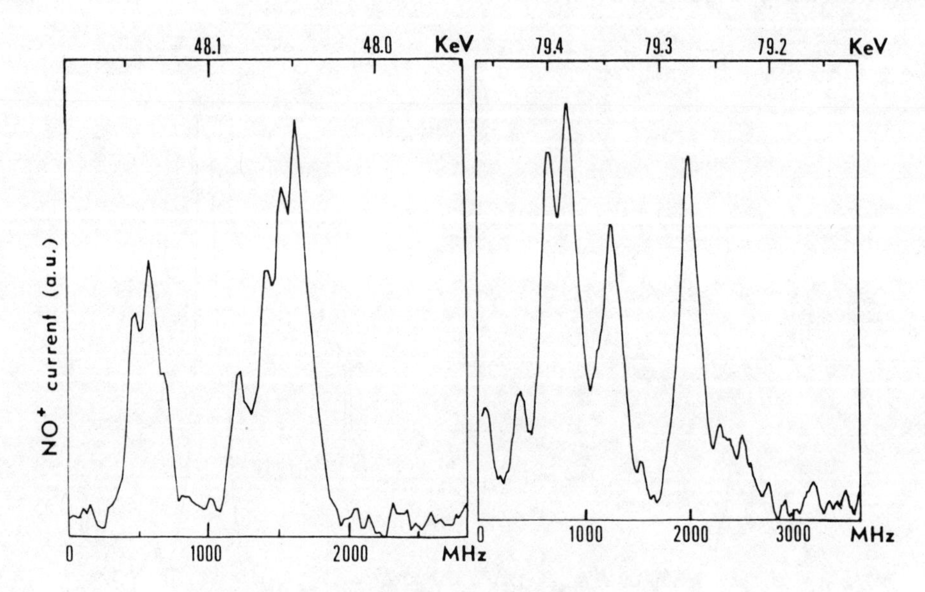

Fig. 6. a) Left part : $^{14}NO^+$ signal obtained from the Q(21.5) rotational transition in $^{14}N_2O^+$.
b) Right part : $^{15}NO^+$ photofragment signal obtained from the R(25.5) rotational transition in $^{15}N_2O^+$.

ion.[38] More generally, study of the hyperfine interaction in the excited state of N_2O^+ provides a good example on how the nuclear spins can be used to probe the electronic wavefunction at a given nuclear site, in a polyatomic molecule.

Bound breakage mechanism

In the early planning stage of the experiment, FIBLAS was not expected to contribute significantly to the understanding of the predissociation mechanism of the $\tilde{A}$ state of N_2O^+. Indeed, predissociation was already known to be extremely slow and thus hardly able to modify the natural lifetime of the excited state. The resulting line broadening is well below the FIBLAS instrumental limit set at several tens of megahertz. A marked improvement of the instrumental linewidth could conceivably be obtained by "in flight" velocity selective optical pumping.[9,39] the cost in signal to noise ratio and laser power is however prohibitive at the moment. Under such circumstances, it has been impossible to come anywhere close to reproduce via pure linewidth measurements, the lifetime study which contributed so much to the understanding of the predissociation mechanism in the O_2^+ case.[2,20-22] In that sense, the N_2O^+ example certainly fails as the perfect demonstration experiment for FIBLAS.

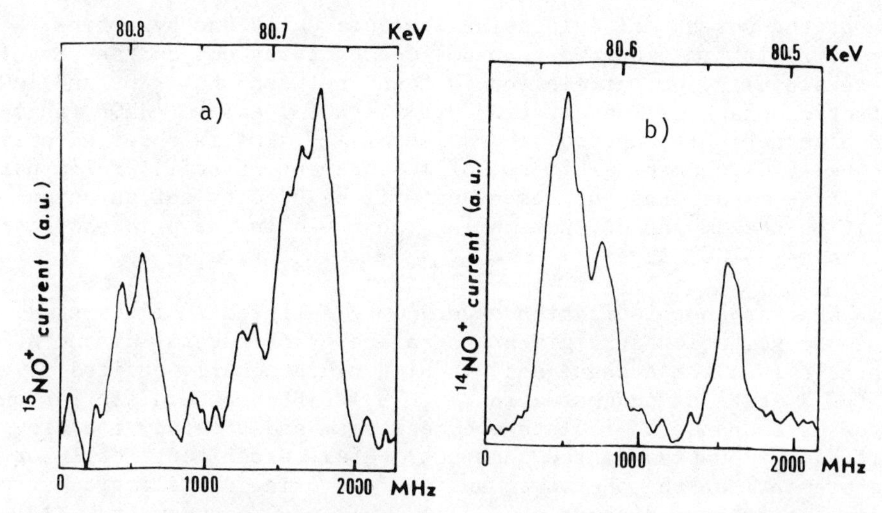

Fig. 7. Individual rotational components obtained for
a) Left part : $^{14}N^{15}NO^+$ observed on the $^{15}NO^+$ photofragment.
b) Right part : $^{15}N^{14}NO^+$ observed on the $^{14}NO^+$ photofragment.

Let us point out however that the situation is far from desperate. The good fluorescence yield of the $\tilde{A}$ state makes it a likely candidate for lifetime measurement via anyone of the various time resolved techniques which have already been successfully tested in atomic physics.[40-42] Furthermore, results of a completely different nature have been obtained which already contribute significantly to the analysis of the bound breakage mechanism during predissociation. We refer here to an other aspect of the isotope labelling experiments which gave us a chance to distinguish between two production mechanisms for the NO^+ fragment : either ejection of the outer nitrogen (direct N-N bound breakage) or loss of the central nitrogen ("scrambling"[43]). This is an approach which has been pioneered by Berkowitz and Eland[43] in their photoionization experiment. Utilizing a sample of $^{14}N^{15}NO$, they obtained the photoion yield curve of $^{15}NO^+$ and $^{14}NO^+$ individually. The $^{15}NO^+$ curve displayed a prominent increase at the $\tilde{A}^2\ ^+(100)$ threshold whereas the $^{14}NO^+$ curve did not. Thus, according to their measurements, "scrambling" appeared more probable at energies below the $\tilde{A}^2\ ^+$ state than above this state.

In our photodissociation experiments, beams of $^{14}N^{15}NO^+$ and $^{15}N^{14}NO^+$ cannot be analyzed magnetically. Individual lines on the FIBLAS spectrum can however be unambiguously attributed to either one of the parent molecule using isotopic shift and hyperfine structure information. We then noticed that, at our current level of sensitivity, no correlation is found between the $^{15}NO^+$ and $^{14}NO^+$ detection channels : excitation of $^{14}N^{15}NO^+$ leads to $^{15}NO^+$ fragments and alternatively, photodissociation of $^{15}N^{14}NO^+$ is observed only in the $^{14}NO^+$ channel. This set of measurement is still preliminary but it is clear that the technique will be used to set an upper limit on the amount of "scrambling" observed in the predissociation of the $\tilde{A}^2$ +(100) level.

Whereas photoionization experiments suffered from the simultaneous excitation at a given photon energy of several dissociative channels (via the $\tilde{A}$ state and via high vibrationally excited levels of the $\tilde{X}$ state as suggested in Ref. 29,) FIBLAS photodissociation gives us a chance to isolate one mechanism and to study its properties with unprecedented accuracy. Future experiments call for measurement of the photofragment energy[3] using the electrostatic energy analyzer indicated on Fig. 1. This new measurement will provide important information on the kinetic energy release distribution for NO^+ + N formation starting from sharply defined excited levels of the $\tilde{A}$ state. This should give us a chance to study the influence of the symmetry of the excited vibrational mode (for example the role of molecular bending) and thus confirm the early results of Thomas et al.[32] which indicate that the kinetic energy release distribution is sensitive to the identity of the exact vibrational level of $\tilde{A}$ initially populated.

CONCLUSION

The still undergoing study of N_2O^+ has been presented as an example likely to illustrate the considerable amount of progress which has been achieved in ion spectroscopy (whether atomic or molecular) during the past decade. Main improvements resulted of course from the conjunction of two independent technical developments : lasers and fast ion beams. However contrived and difficult it may have looked in its first implementations, the combination became a nearly perfect match as soon as the power of the colinear beam approach was fully understood. It is a well publicized fact that in several domain of spectroscopy (lifetime measurements, to mention only the first historical achievement,) record of accuracy nowadays belongs to the field of singly charged ions. Recent development of ion cooling in electrostatic traps[44] is likely to bring further progress in that direction. Rather than a drawback, the charge of the ion provides the spectroscopist with a much needed control over the ion motion in beams as well as in traps.

For all their useful properties, lasers are known to suffer from drastic limitations in power and/or wavelength range. A a result, their introduction in fast atomic ion beam spectroscopy has often been considered to result in a considerable reduction in scope of ion physics, focusing the interest on singly charged ions at the expense of the much broader and potentially more exciting field of multiply charged ions. This is a trend which is unfortunately not likely to be reversed in the near future. Furthermore, it should be obvious that FIBLAS of highly charge ions (if it is ever to come to age) will depart considerably from neutral atom spectroscopy if only because of the radiative width of the resonance transition. It should however come as a comforting thought to the atomic physicists to realize that whatever has been lost on one side was in fact more than compensated by the wide applicability of FIBLAS in the molecular ion field. With its emphasis on new methodological developments, atomic ion beam laser spectroscopy has gathered a reserve of experimental techniques which can, and probably soon will be carried over to molecular ions physics. As potential candidates, we already mentioned state selective time resolved spectroscopy. Obviously Zeeman effect study[45] fast beam radiofrequency[46] and multiphoton[47] methods have a strong appeal in the molecular physics context.

Perhaps even more than other fields of application in laser spectroscopy, ion physics is likely to benefit rapidly from progress in the U.V. range. Frequency doubled CW gas and dye lasers are now able to provide on a routine basis the few milliwatts of monomode laser power which have been necessary for the N_2O^+ photodissociation experiment. Further in the U.V., the Doppler tuning method could take advantage with much benefit of the high spectral brightness[48] of injected excimer amplifiers operated at high repetition rate.

With the extension of FIBLAS to polyatomic molecular ions, it becomes obvious that the combination of mass spectrometry and "in flight" optical diagnostic (eventually at very high resolution) offers unprecedented opportunities in molecular structural analysis via isotope labelling and optical isotope (or isomer) shift measurements. Infrared photodissociation which has already been observed in several polyatomic ion beams[49-50] may have the power to turn the entire scheme into a much needed very high sensitivity analytical tool.

REFERENCES

1. G. Herzberg, Rev. Chem. Soc., London 25:201 (1971).
2. M. Carré, M. Druetta, M.L. Gaillard, H.H. Bukow, M. Horani, A.L. Roche and M. Velghe, Mol. Phys. 40:1453 (1980).
3. M. Broyer, M. Larzillière, M. Carré, M.L. Gaillard, M. Velghe and J.B. Ozenne, Chem. Phys. 63:445 (1981).
4. M. Larzillière, M. Carré, M.L. Gaillard, J. Rostas, M. Horani and M. Velghe, J. Chimie Phys. 77:689 (1980).
5. S. Abed, M. Broyer, M. Carré, M.L. Gaillard and M. Larzillière, Phys. Rev. Lett. 49:120 (1982).
6. J.B. Ozenne, D. Pham and J. Durup, Chem. Phys. Lett. 17:422 (1972).

N.P.F.B. Van Asselt, J.G. Maas and J. Los, Chem. Phys. Lett. 24:555 (1974).
7. J.D. Silver, N.A. Jelley and L.C. Mc Intyre, Appl. Phys. Lett. 31:278 (1977).
8. H.J. Andra, in: "Beam Foil Spectroscopy," I.A. Sellin and D.J. Pegg, ed., Plenum, New-York, p.835 (1975).
9. M. Dufay and M.L. Gaillard, in: "Laser Spectroscopy III," J.L. Hall and J.L. Calsten, ed., Springer, Berlin, p.231 (1977).
10. J. Durup, "Etats Atomiques et Moléculaires couplés à un continuum, Atomes et Molécules hautement excités," éditions du C.N.R.S., Paris, p.107 (1977).
11. A. Carrington, D.R.J. Milverton and P.J. Sarre, Mol. Phys. 35:1505 (1978).
12. R.A. Holt, T.D. Gaily and S.D. Rosner, private communication.
13. A. Carrington and P.J. Sarre, J. Phys., Paris, C.I. 54 (1979).
14. W.H. Wing, G.A. Ruff, W.E. Lamb Jr. and J.J. Spezeski, Phys. Rev. Lett. 36:1488 (1976).
15. A. Carrington, P.G. Roberts and P.J. Sarre, J. Chem. Phys. 68:5659 (1978).
16. F. Beguin, M.L. Gaillard, H. Winter and G. Meunier, J. Phys. Paris 38:1185 (1977).
17. J.R. Hiskes, Phys. Rev. 122:1207 (1961).
18. A. Carrington, Proc. Roy. Soc. 367:433 (1979).
A. Carrington, J. Buttenshaw and P.G. Roberts, Mol. Phys. 38:1711 (1979).
19. A. Carrington, P.G. Roberts and P.J. Sarre, Mol. Phys. 35:1523 (1978).
20. P.C. Cosby, J.B. Ozenne, J.T. Moseley and D.L. Albritton, J. Molec. Spectrosc. 79:203 (1980).
21. J.T. Moseley and J. Durup, J. Chim. Phys. 77:673 (1980).
22. J.T. Moseley and J. Durup, Ann. Rev. Phys. Chem. 32:53 (1981).
23. S.L. Kaufman, Opt. Comm. 17:309 (1976).
24. J.H.D. Eland, "Theory, Techniques and Applications," Academic Press, New-York, 3:231 (1978).
25. R.G. Orth and R.C. Dunbar, J. Chem. Phys. 66:1616 (1977).
26. R.J. Saykally and R.C. Woods, Ann. Rev. Phys. Chem. 32: (1981).
27. D.G. Hopper, Chem. Phys. Lett. 31:446 (1975).
28. J.H. Callomon and F. Creutzberg, Phil. Trans. Roy. Soc. A 277:20 (1974).
29. I. Nenner, P.M. Guyon, T. Baer and T.R. Govers, J. Chem. Phys. 72:6587 (1980).
30. J.H.D. Eland, M. Devoret and S. Leach, Chem. Phys. Lett. 43:97 (1976).
31. R. Frey, B. Gotchev, W.B. Peatman, H. Pollak and E.W. Schlag, Chem. Phys. Lett. 54:411 (1978).
32. T.F. Thomas, F. Dale and J.F. Paulson, J. Chem. Phys. 67:793 (1977).
33. A. Carrington, D.R.J. Milverton and P.J. Sarre, Mol. Phys. 32:297 (1976).

34. C.P. Edwards, C.S. McLean and P.J. Sarre, Chem. Phys. Lett. 87:11 (1982).
35. S. Abed, M. Broyer, M. Carré, M.L. Gaillard and M. Larzillière, Chem. Phys. (submitted for publication).
36. See for example : O. Poulsen, these proceedings.
37. A.D. McLean and M. Yoshimine, "Tables of linear molecule wave-functions," I.B.M., San José, p.196 (1967).
38. Ph. Millié, private communication.
39. M. Dufay, M. Carré, M.L. Gaillard, G. Meunier, H. Winter and A. Zgainski, Phys. Rev. Lett. 37:1678 (1976).
40. H. Winter and M.L. Gaillard, Z. Phys. A 281:311 (1977).
41. P. Ceyzeriat, D.J. Pegg, M. Carré, M. Druetta and M.L. Gaillard, J. Opt. Soc. Am. 70:901 (1980).
42. M.L. Gaillard, D.J. Pegg, C.R. Dingham, H.K. Carter, R.L. Mlekodaj and J.D. Cole (submitted to Phys. Rev. A).
43. J. Berkowitz and J.H.D. Eland, J. Chem. Phys. 67:2740 (1977).
44. W. Neuhauser, M. Hohenstatt, P. Toschek and H. Dehmelt, Phys. Rev. Lett. 41:233 (1978).
45. M. Larzillière, M. Carré, M.L. Gaillard and F. Stoeckel, Opt. Comm. 37:27 (1981).
46. S.D. Rosner, T.D. Gaily and R.A. Holt, Phys. Rev. Lett. 40:851 (1978).
47. O. Poulsen and N.I. Winstrup, Phys. Rev. Lett. 47:1522 (1981).
48. H. Pummer, T. Srinivasan, H. Egger, K. Boyer, T.S. Luk and C.K. Rhodes, Opt. Lett. 7:93 (1982).
49. A. Von Hellfeld, D. Feldmann, K.H. Welge, A.P. Fournier, Opt. Comm. 30:193 (1979).
50. H.J. Coggiola, P.C. Cosby, J.R. Peterson, J. Chem. Phys. 72:6507 (1980).

RESONANT FAST-BEAM/LASER INTERACTIONS:

SATURATED ABSORPTION AND TWO-PHOTON ABSORPTION

Ove Poulsen

Institute of Physics
University of Aarhus
DK 8000 Aarhus C, Denmark

INTRODUCTION

The advent of lasers in spectroscopy has made possible highly precise measurements of spectroscopic as well as of fundamental interest. Particular emphasis has been put onto the elimination of the Doppler effect, which was one of the main obstacles in classical spectroscopy. This can be achieved using well collimated atomic beams or non-linear field/atom interactions, which, combined with quantum interference methods, are capable of yielding a resolution beyond the natural linewidth. In historical perspective, these methods were developed because of the problems associated with the Doppler effect, the possibilities offered by the high intensity and narrow spectral band width of lasers and, most important, an ever persistent wish to obtain very high optical resolution.

In that perspective it was early recognized, that lasers could interact with fast accelerated atom or ion beams. The advantages of using fast beam laser interactions are numerous, the most important being high time resolution and time control, complete velocity control, velocity compression resulting in narrow Doppler profiles, and a clean environment free of collisions. Furthermore many species, not easily stored in cells, are available in ion sources, most notably single ionized atoms and molecules, the extreme being on-line produced short-lived nuclei. A comprehensive review of fast beam laser spectroscopy using <u>linear</u> atom/field interactions has been given by Andrä[1] and before going into details with some recent developments in fast beam laser spectroscopy, a short up-dated review of the rapid developement of this subfield of laser spectroscopy will be given.

Lifetimes

The early impact of lasers in fast beam spectroscopy was in cascade-free measurements of atomic lifetimes[2]. These early experiments were carried out in a crossed beam geometry, resulting in a high spatial time resolution along the fast accelerated particle beam. This method has since been developed further[3] and is now able of yielding highly precise reference lifetimes in several elements. Going to a collinear geometry, lifetimes can be obtained either by Doppler switching[4] the ions into resonance over a short time interval or, more general, by pulse modulating the laser field[5,6].

High-Resolution Spectroscopy

Besides lifetimes, fine- and hyperfine structure, isotope shifts and g values can be measured in fast-beam laser-spectroscopy. The first demonstration used a crossed-beam geometry, the sharp time definition making it possible to measure hyperfine structures[7] using quantum beats and g values using time-differential level crossings[8]. Simultaneously, the collinear geometry, where the laser field is superimposed onto the fast accelerated particle beam, was shown to yield very narrow resonances, due to kinematic velocity compression. The infrared spectrum of HD^+ was observed[9], using this high optical resolution technique, which was later applied to the study of on-line produced radioactive isotopes[11]. Fundamental Lamb shift and fine-structure measurements have been carried out in one- and two-electron atoms[12,13,14], again demonstrating the wide applicability of fast beam laser spectroscopy.

Non-Linear Methods

None of these methods, except for the quantum interferences, are Doppler free and the obtainable optical resolution is dependent on the quality of the accelerator used. The first successful experiment demonstrating a first-order Doppler free saturated absorption in a fast metastable neon beam[15], using a perpendicular fast beam laser interaction, initiated 'in-flight' saturation spectroscopy[16], using a collinear geometry and taking full advantage of the Doppler labelling capability in the fast ionic beam. The hole burning is only limited by the homogeneous linewidth, but its absolute position is not Doppler free. Even higher optical resolution can be obtained by observing Ramsey fringes in saturated absorption experiments with separated laser fields[17] or transient behaviour of a coupled two-level system studied in a collinear geometry using fast Doppler switching[18].

In this work we will describe the next generation of first-order Doppler-free saturated absorption experiments, carried out in a three-level system[19] in fast atom/ion beams as well as resonant stimulated Raman processes[19] and resonantly enhanced two-photon absorption[20], in all three cases taking full advantage of the velocity

(Doppler) tuning capability, making it possible resonantly to interact on two optical transitions using only one laser field, retroreflected along the fast accelerated particle beam. It will be shown, that high optical resolution, Doppler free to first order, precise velocity control and high time resolution can be obtained in three-level fast beam laser spectroscopy.

EXPERIMENTAL

The apparatus used in Fast Beam Laser Spectroscopy is shown in Fig.1.

The accelerator consists of a universal ion source which can produce single-ionized species of all elements. These ions are accelerated to energies ranging from 30 to 300 keV and then mass separated in a high-resolution magnet. Thereafter the ions enter a drift region of length three meters, where provision is made for charge exchange in a metal vapour. This charge exchange populates many metastable levels, thus enhancing the usefulness of this apparatus. The same holds true for the ion source, which populates ionic metastable levels very efficiently. The main parameters of the accelerator is a beam divergence of $\theta \simeq 10^{-3}$ and an energy stability of $\delta E/E \simeq 7 \cdot 10^{-5}$. This energy is achieved by active correction of the High-Voltage supply. Using a slit stabilizing servo loop to long term stabilize the beam position, the low frequency voltage fluctu-

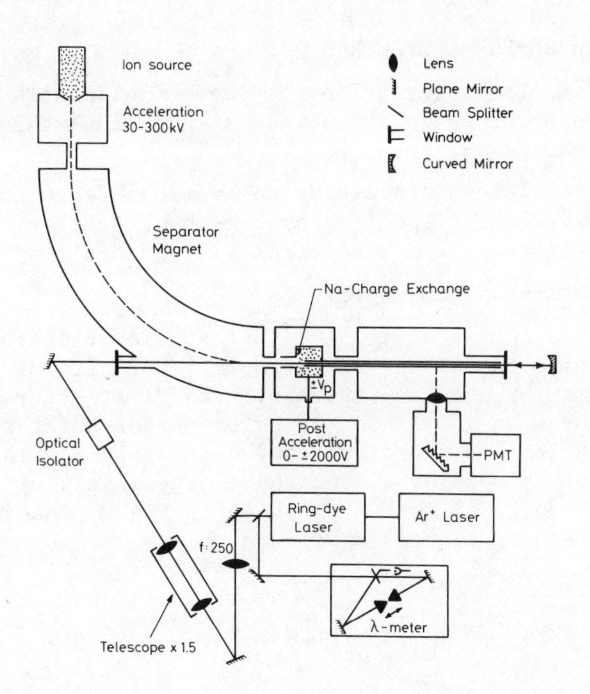

Fig.1: Experimental apparatus, consisting of 300 keV isotope separator, CW ring dye laser, and λ meter.

ations, which converted into angular noise after mass separation, can be compensated for by applying a correction voltage either (i) directly back to the HV terminal (slow), (ii) the metal-vapour charge exchange cell (fast), where it is added to a post-acceleration voltage used for precise velocity control of a neutral particle beam or (iii) added to a Faraday cage (fast) used for precise velocity control of a ionic beam. Only saturated absorption in ^{238}UII, later to be discussed, has been performed on this improved accelerator. All other results discussed have been obtained with an energy stability $\delta E/E \simeq 2\cdot10^{-4}$.

The laser system consists of an actively stabilized CW dye ring laser, with an rms spectral bandwidth of $\simeq$0.7 MHz. The laser beam is mode matched to the particle beam, with a beam waist of $\omega_0 \simeq 1.5$ mm and a far-field diffraction angle of $\theta \simeq 1.3\cdot10^{-4}$rad, whereafter it is superimposed onto the particle beam. A curved mirror retroreflects the laser field back onto itself, with a TGG Faraday isolator, providing isolation of the dye laser. The wavelength of the dye laser is measured using a fringe counting interferometer (λ-meter) absolutely calibrated against the I_2-absorption spectrum[22]. A stability and relative reproducibility of 10^{-8} is observed with an absolute reproducibility of 10^{-7}, only limited by the I_2-reference absorptions.

The return field is chopped and dual spectra, with both fields, respectively only the forward field present, may be recorded using a computer controlled digital lock-in multiscaling system, which also sweeps the laser frequency and/or particle energy.

THREE-LEVEL SPECTROSCOPY

The interactions of laser fields with a three-level atom have been intensively studied using thermal absorbers and atomic beams[23,24].

Before discussing the various features observable in three-level laser spectroscopy, with particular emphasis on fast beam experiments, the kinematics will be dealt with.

Fast-Beam Kinematics

The velocity of a fast accelerated atom is typically 1 mm/ns compared to an average thermal velocity of 10^{-4} mm/s. Thus the kinematics becomes important: the first-order Doppler shift can be as large as 100 Å and the second-order shift several GHz. Thus resonant three-level spectroscopy, using only one laser field can be carried out by Doppler tuning the energy levels appropriate. A simplified three-level system is shown in Fig.2. The excited velocity classes $\beta_\alpha = (v_\alpha/c)$ are determined by

$$\begin{aligned} \sigma_1 &= \sigma_L\, \gamma(\beta_a)(1-\beta_a) \\ \sigma_2 &= \sigma_L\, \gamma(\beta_b)(1+\beta_b) \end{aligned} \qquad (1)$$

where σ_L is the laser wave number, σ_i the atomic rest-frame energies, and $\gamma(\beta) = (1-\beta^2)^{-\frac{1}{2}}$. For a particular particle velocity $\beta_a = \beta_b = \beta$, given by

$$\beta = \frac{\sigma_2-\sigma_1}{\sigma_2+\sigma_1} , \tag{2}$$

the laser field is resonant on both atomic transitions. This velocity β is solely given by (2) and subsequently determines the resonant laser frequency

$$\sigma_0 = \sigma_1 / \gamma(\beta)(1-\beta) . \tag{3}$$

Thus the three levels are effectively coupled using one laser field.

The next important feature in fast beam laser spectroscopy concerns the velocity distribution. Due to kinematic velocity compression[10], the initial thermal distribution of velocities in the ion source is reduced by a factor $R = \frac{1}{2}\sqrt{kT/eV}$, where T is the ion source temperature and V the acceleration voltage. Including the voltage spread δV, the Doppler width in our accelerator is 20 - 200 MHz, depending on mass and acceleration voltage. The transverse velocity distribution (particle wave front curvature) not affected during acceleration, contributes $\sigma\beta_0\delta\theta$ to the linewidth. σ is the wave number, β_0 the transverse particle velocity, and $\delta\theta$ the particle beam divergence. As will be seen later, this constitutes the principal limitation in obtainable resolution in incoherent hole burnings in three-level atoms. The length of the fast beam/laser interaction is $\simeq$1 m, yielding longitudinal transit times T ranging from 0.5 µsec to 7.5 µsec.

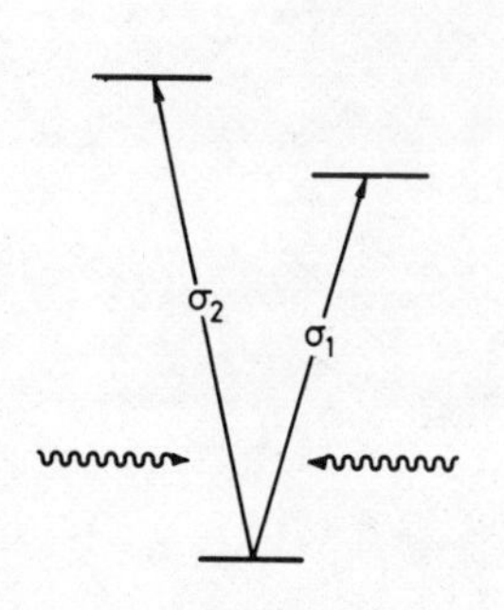

Fig. 2: Three-level atom subjected to two opposite running laser fields.

Three-Level-Atoms : Phenomena

The three different configurations available in a three-level atom, the V, the inverted V and the cascaded configuration have been studied using the resonant condition given by Eqs.(2) and (3).

In Fig.3a,b are shown the V and the inverted V configurations. Levels ℓ,m are coupled by the laser field running antiparallel to the fast beam and levels m,n are coupled by the forward running laser field. The dominant process in case of the V-configuration (fig.3a) is hole burning in the velocity distribution of level m, taking place at velocities β_a and β_b (see eq.(1)). On resonance, given by Eq.(3), the two holes collapse, and a decrease in emission is observed from levels ℓ,n, assuming the velocity β falls within the Doppler profile. This hole burning Doppler free to first order and only limited by the homogeneous linewidth $\gamma_{\ell m} + \gamma_{nm}$, is insensitive to power shifts and with a depth strongly dependent on the branching ratios Γ_i.

In case of the inverted V - configuration (fig.3b) hole/peak creation is also present, together with resonantly enhanced stimula-

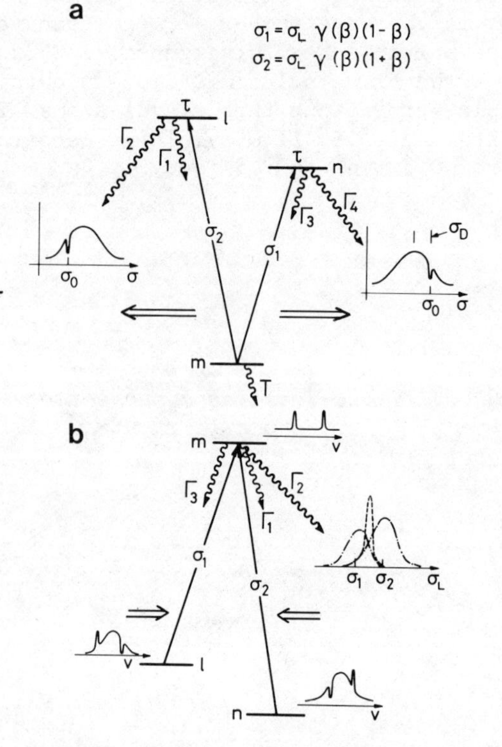

Fig.3: (a) The V - configuration with the relevant parameters as lifetimes τ, transit time T, and transition rates Γ_i and (b) the inverted V - configuration also showing velocity distributions leading to an increased emission at σ_0, given by Eq.(3) due to hole/peak formation and stimulated processes.

ted Raman processes. Again the increase in emission is Doppler free to first order and only limited by the homogeneous linewidth $\gamma_{\ell m} + \gamma_{nm}$, valid for both the incoherent hole/peak formation and the coherent stimulated Raman process.

In the cascade configuration shown in Fig.4, coherent processes dominate. On exact resonance, again given by Eqs.(2) and (3), a resonantly enhanced two-photon absorption $n \rightarrow \ell$ takes place. This absorption is Doppler free to first order and is, because of its coherent nature, not dependent on the first-order broadening due to the transverse velocity distribution. Thus a resolution, only limited by the homogeneous linewidth $\gamma_{\ell n}$ is easily obtained.

Three-Level Atoms : Formal

All these phenomena are described in the density matrix formalism by the equation of motion

$$\frac{d}{dt}\hat{\rho} = -\frac{i}{\hbar}[\kappa_0 - \vec{\mu}\cdot\vec{E}, \hat{\rho}] + R \; , \tag{4}$$

where κ_0 is the atomic Hamiltonian, $\vec{\mu}$ the dipole moment $-e\vec{r}$, $\vec{E}$ the laser field, and R incoherent relaxation terms. The field/atom interaction is conveniently given by the Rabi frequency

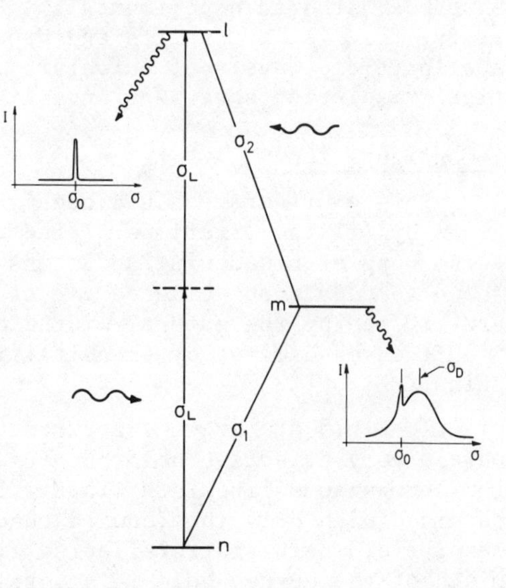

Fig.4: The cascade configuration used in resonantly enhanced two-photon absorption in a fast accelerated atom beam.

$\Omega_i = \vec{\mu}_i \vec{E}_i / 2\hbar$ and the homogeneous linewidth γ is given by

$$\gamma = \frac{1}{\tau} + \frac{2}{T}\,, \tag{5}$$

where $\frac{1}{\tau} = \Sigma\Gamma_i$. T is the longitudinal transit time and Γ_i decay rates (see Fig.3).

As an example, using the rotating wave approximation, the component ρ_{mm} of Eq.(4) becomes, in the V configuration

$$\frac{d}{dt}\rho_{mm} = (n_m^{(0)} - \rho_{mm})/T + \Gamma_1 \rho_{\ell\ell} + \Gamma_3 \rho_{nn}$$

$$+ i\Omega_1(\rho_{mn} - \rho_{nm}) + i\Omega_2(\rho_{m\ell} - \rho_{\ell m})\,, \tag{6}$$

where $n_m^{(0)}$ is the initial population in level m. The decay rates Γ_i are introduced to take into account the effects of optical pumping, whereas the redistribution of population among the various degenerate Zeeman levels is not included in Eq.(4),(6). In all cases studied, the branching ratios favours decay out of the three-level system, thus reducing considerably the effective saturation parameter[25]. Equation (4) can be solved under steady state conditions and integrated over the velocity distribution of the fast accelerated atoms/ /ions as well as taking into account the fast beam kinematics.

RESULTS

The three different three-level configurations discussed above have been studied experimentally. Problems concerning spectral stability, that is line shifts and broadenings as well as transit time and 'particle wavefront curvature' limitations on the obtainable optical resolution have been investigated.

V Configuration

This configuration has been studied in two fundamental limits given by (i) the lifetimes of the excited levels and (ii) the transverse Doppler broadening and transit times involved. In Table I are shown some data on selected V configurations, where the NeI cases are limited by the natural widths of the excited levels and the ^{238}UII cases limited by transit time and transverse Doppler broadening.

In Fig.5 are shown experimental data, showing that indeed we obtain very effective hole burning and a resolution only limited by the homogeneous linewidth. In Fig.5b is shown the Doppler profile obtained with only the forward running laser field present and the laser field being retroreflected along the fast beam in Fig.5a. The width of the burned hole is severely laser power broadened despite low laser powers, a fact related to optical pumping out of the V configuration.

Table I

Atom	Transition		Lifetime(ns)	σ_0(cm^{-1})	E^R (keV)
^{20}NeI	$3s[\frac{3}{2}]^0_2$	$3p'[\frac{3}{2}]_2$	15	16906.401	263.715
		$3p'[\frac{3}{2}]_1$	16		
^{20}NeI	$3s[\frac{3}{2}]^0_2$	$3p'[\frac{1}{2}]_1$	15	16773.413	61.763
		$3p'[\frac{3}{2}]_2$	15		
^{238}UII	$4420_{11/2}$	$21691_{11/2}$	1000	17280.269	34.380
		$21710_{13/2}$	330		
^{238}UII	$289_{11/2}$	$17392_{9/2}$	1370	17124.230	168.007
		$17434_{11/2}$	3800		

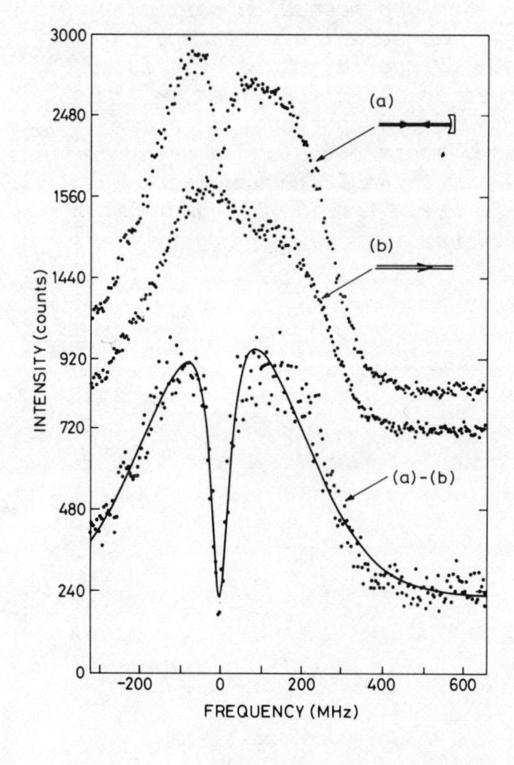

Fig.5: The 16906.401 cm^{-1} resonance in ^{20}NeI found at 263.715 kV.

In Fig.6 are shown resonances obtained at different particle energies, demonstrating that the observed hole burning is indeed Doppler free to first order. The broad Doppler profile is dependent on the total Doppler shift, whereas the hole burning is only shifted by the second-order Doppler effect. This is shown in Fig.7, where the energy detuning σ_D (see Fig.3a) is plotted versus the beam energy. The total Doppler shift completely accounts for the data, showing that ac-stark shifts are negligible.

These points can be further elaborated by solving Eq.(4) in steady state. After velocity integration a homogeneous linewidth of

$$\gamma_{eff} = \frac{\gamma_\ell}{2}\sqrt{1 + \frac{\Omega_2^2}{\Omega_{20}^2}} + \frac{\gamma_n}{2}\sqrt{1 + \frac{\Omega_1^2}{\Omega_{10}^2}} \tag{7}$$

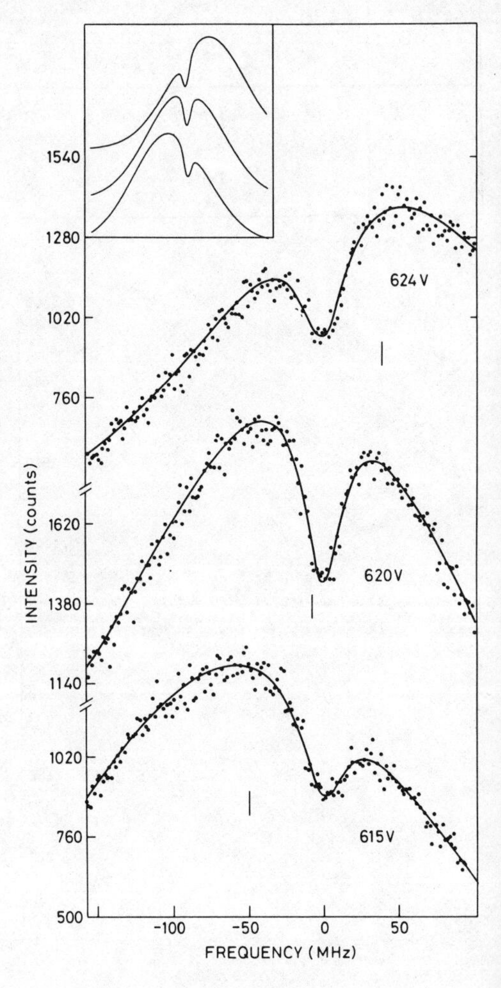

Fig.6: Hole burnings in ^{20}NeI and its dependence on particle energy. The actual beam energy is $(61142 + \delta V)$ volts, where δV is the post acceleration given in the figure. In the inset is shown a steady state calculation using Eq.(4), with Rabi frequencies Ω_1 = 12 MHz, Ω_2 = 25 MHz, and T = 500 ns.

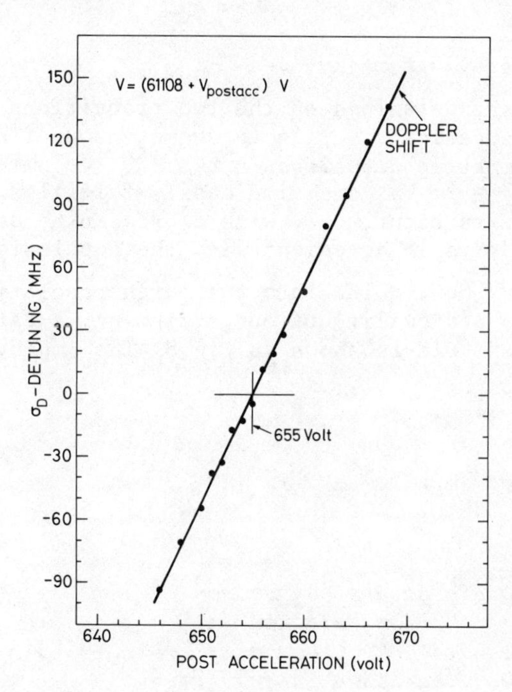

Fig.7: Energy detuning σ_D versus post acceleration.

is found[25], where Ω_{i0} are the reduced Rabi-saturation parameters given by

$$\Omega_{10}^2 = \frac{\gamma_n(1/\tau + 1/T)}{4(2 + \Gamma_4 T)}$$

$$\Omega_{20}^2 = \frac{\gamma_\ell(1/\tau + 1/T)}{4(2 + \Gamma_2 T)} \tag{8}$$

where τ is the lifetime of the upper levels ℓ, n in the V configuration and Γ_2, Γ_4 the transition rates out. Thus a large reduction in saturation intensity is obtained for large branching ratios Γ_2/Γ_1 and Γ_4/Γ_3. In the cases studied, branching ratios in the order of 5 - 10 are thus effectively reducing the laser intensities needed, and makes possible an expansion of the laser field to ensure plane wave fronts and long interaction times.

Equation (4) also predicts an ac-stark shift of[26]:

$$\delta = -\left[\frac{\Delta\Omega_1^2}{\Delta^2 + \gamma^2} + \frac{\Delta'\Omega_2^2}{\Delta'^2 + \gamma^2}\right] \tag{9}$$

where $\Delta = \omega - kv - \omega\frac{v^2}{2c^2} - \omega_{nm}$, $\Delta' = \omega + kv - \omega\frac{v^2}{2c^2} - \omega_{\ell m}$ are the laser detunings on the two transitions and Ω_i the Rabi frequencies, respectively. γ is the homogeneous linewidth, given by Eq.(5). For symmetric detunings $\Delta + \Delta' = 0$ we immediately find $\delta \equiv 0$ for $\Omega_1 = \Omega_2$, a condition easily fulfilled in the experiment. Thus no shifts occur, only with a broadening determined by Eqs.(7) and (8). This is in agreement with the conclusions of Fig.7.

Now, going into the 'high resolution' limit, set by the geometry and quality of our equipment, a saturated absorption, obtained in ^{238}UII is shown in Fig.8. The improved accelerator has been used,

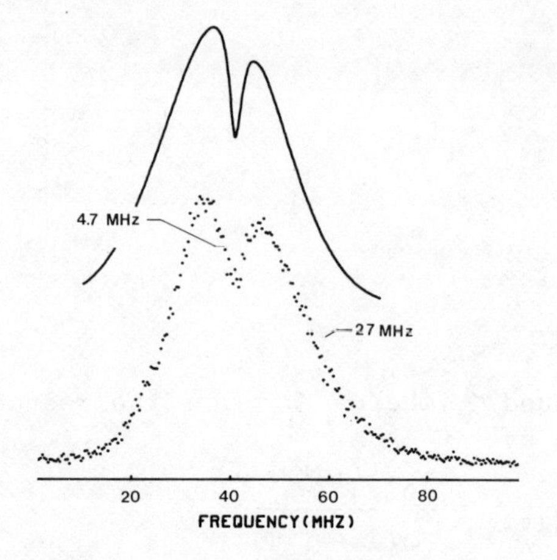

Fig.8: Saturated absorption obtained in the 34.380 kV resonant V configuration in ^{238}UII. Also shown is a steady-state calculation, discussed in the text.

yielding a Doppler width of ≃27 MHz, at a beam energy of ≃34 kV. Also shown in Fig.8 is a steady-state calculation, performed with effective Rabi-frequencies of Ω(4420-21710) = 0.5 MHz and Ω(4420-21691) = 2.3 MHz. These Rabi-frequencies are given by

$$\Omega(4420\text{-}21710) = 2.34\cdot 10^{-9}\sqrt{W\cdot gA\cdot\lambda^3}\sqrt{\frac{169}{4} - M^2} \quad \text{MHz}$$

$$\Omega(4420\text{-}21691) = 3.74\cdot 10^{-9}\sqrt{W\cdot gA\cdot\lambda^3}\,|M| \quad \text{"} ,$$

where W is the laser power in W/cm^2, gA the statistical weight of the upper level times the transition probability, λ the wavelength of the transition, and M the Zeeman quantum number. Polarization of the laser field is linear, resulting in $\Delta M = 0$ transitions. The gA values for these two transitions can be estimated from the known lifetimes[5] and the relative intensities[27], yielding A(4420-21710) ≃ $2.7\cdot 10^5$ sec^{-1} and A(4420-21691) ≃ $3.7\cdot 10^5$ sec^{-1} and branching ratios A(4420-21710)/$\sum_i$A(i-21710) ≃ 0.09 and

A(4420-21691)/ $\sum$ (i-21691) $\simeq$ .33. Now, Eq.(4), solved at steady state and velocity integrated, predicts an enhanced emission from the 21710 cm^{-1} level and a decreased emission from the 21691 cm^{-1} level for small M values and the reverse situation, observed experimentally, for large M. Thus we clearly see the effect of optical pumping within the degenerate Zeeman sublevels. The ratio $\Omega(4420\text{-}21691)/\Omega(4420\text{-}21710) \simeq 4$, predicts $|M|$ values in the range 7/2, 9/2, 11/2 for the distribution within the $4420_{J=11/2}$ level.

With the above Rabi frequencies, calculated using the experimentally measured laser power W, a linewidth of 1.7 MHz is calculated. Several other factors contributing to the observed linewidth of $\simeq$4 MHz, are listed in Table II. Here $v_0 = \sqrt{2kT/M}$ is the thermal

Table II

Effect	Dependence, $\delta\sigma$	Size
Transverse Doppler shift	$\sigma \frac{v_0}{c} \delta\theta$	$\simeq$ 2MHz
Zeeman effect-earth field	$\mu \Delta g_J M \cdot H$	$\simeq$ .3 -
Laser bandwidth		$\simeq$ 1 -
Natural linewidth	$\frac{\gamma_\ell}{2} + \frac{\gamma_n}{2}$	.33 -
Power broadening	Eq.(7)	1.7 MHz

velocity in the ion source, T being the ion source temperature, Δg_J the difference in g values in the excited and metastable levels, and $\delta\theta$ an angle representing the mode mismatch between the laser field and the ion beam. In the two V configurations studied in ^{238}UII, with long-lived upper levels, the 34.380 keV and the 168.007 keV resonances, the worst problem relates to the power broadening. With lifetimes corresponding to distances (flight times) ranging from 15 (330 ns) to 150 cm (4 μsec) along the fast beam, lower laser powers result in very few pumping cycles (and low excitation probability) making the hole burning weak. This combined with the small solid angle of 10^{-4} for observation of the fluorescent light, along a 2-cm beam segment, has made it difficult to reduce the Rabi-frequencies below the homogeneous linewidth.

Resonant Inverted V Configuration

A resonant inverted V configuration, shown in Fig.3b can be realized[28] in the 4s4p ^3P - 4s5s $^3S_1$ multiplet in ^{40}CaI at beam energies of 47.329, 196.905, and 437.308 keV, respectively for the various J combinations. The 196.905 keV resonance {4s4p $^3P_2$; 4s4p $^3P_1$} - 4s5s $^3S_1$ have been selected because it is nearly closed, as seen from $(\Gamma_1 + \Gamma_3)/\Gamma_2 \simeq 9$. Two processes take place in this configuration. The hole burning is again prominent, with velocity classes β_a, β_b excited, but due to strong spontaneous coupling, these excited velocity classes decay back to the initial metastable levels, creating peaks. When these peaks collapse with the holes, an increased emission is observed. This emission is only limited by the homogeneous linewidth and is Doppler free to first order. A steady-state solution of Eq.(4), in velocity space, is shown in Fig.9 for

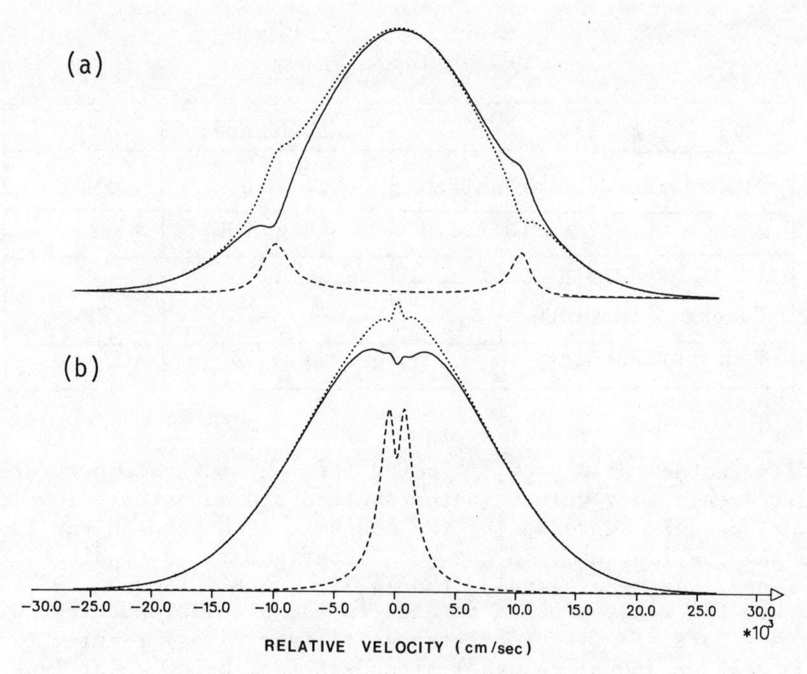

Fig.9: Velocity profiles of the populations of level L (——), M (---), and N (...) in the resonant inverted V configuration for (a) large laser detuning and (b) the laser field being close to resonance σ_0 *, given by Eq.(3).*

two different laser detunings, with zero relative velocity corresponding to exact resonance, given by Eq.(2). For large laser detunings the hole and peak formation is clearly seen. The same features are seen at small laser detunings where, in addition, a sharp feature is observed at zero relative velocity, due to resonant stimulated Raman processes $1 \leftrightarrow n$. This Raman process is Doppler free to

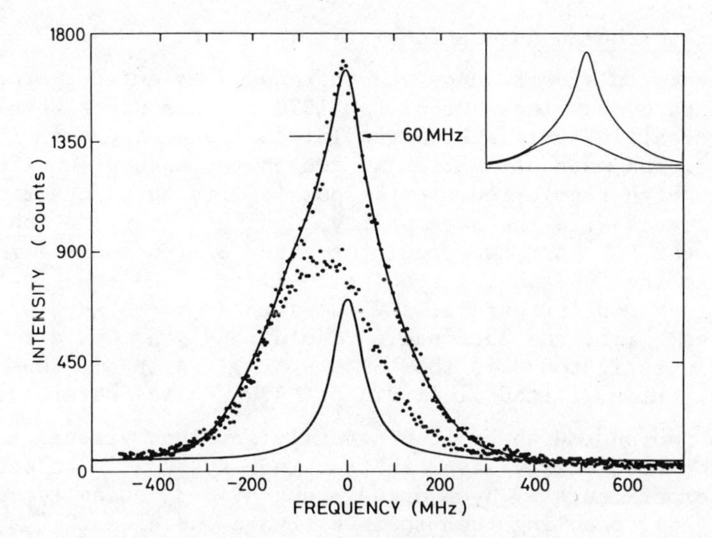

1 Fig. 10: A laser frequency scan through resonance in the inverted V configuration, with an energy detuning of −10 V. Both the signal from only the l-m excitation and the total signal with the return laser field on, are shown, together with the Doppler free contribution from incoherent and coherent processes.

first order and with a power broadened homogeneous linewidth, corresponding to backwards scattering. An experimental result, obtained at a beam energy of 196.895 keV is shown in Fig.10. A sharp Doppler free emission is superimposed onto the two contributions from the $1 \rightarrow m$ and $n \rightarrow m$ transitions, excited by the two counter propagating laser beams, respectively. Solving Eq.(4) in steady state, using the measured Rabi-frequencies of $\Omega(^3P_1 - {}^3S_1) \simeq 60$ MHz and $\Omega(^3P_2 - {}^3S_1) \simeq 80$ MHz and using the experimentally observed Doppler profile of width $\simeq 200$ MHz, the experimental data are readily simulated, as shown in the inset of Fig.10. Also shown in this inset is the direct excitation $1 \rightarrow m$ [$\Omega(^3P_2 - {}^3S_1) \simeq 80$ MHz and $\Omega(^3P_1 - {}^3S_1) = 0$], that is the counter propagating laser field is turned off. The contribution of Resonant Stimulated Raman processes to the signal observed in Fig.10 is to change the populations of the two initial levels and thus only indirectly observable in the fluorescence emitted from the excited level (m). Approximately 10% of the population changes in the initial levels are due to coherent processes (with our Rabi frequencies and branching ratios), the remaining $\simeq 90\%$ being due to incoherent hole burning. The final case to be considered is the resonant cascaded configuration, where coherent processes dominate completely.

Resonant Two-Photon Absorption

Two-photon spectroscopy was introduced as a first-order Doppler-free method by Vasilenko et al. in 1970 and has since been developed into a versatile tool in high-resolution spectroscopy on thermal absorbers[30]. The main attraction in two-photon absorption (TPA), besides the high resolution, is the possibility to induce transitions between levels with the same parity. Hydrogen 1s - 2s[31] and positronium $1\,^3S_1 - 2\,^3S_1$[32] are two prominent cases of fundamental interest. Going into the fundamental resolution limits, the transit times, the second-order Doppler effect, and the second-order laser wave-front curvatures, limit the obtainable resolution[33], as was also experimentally investigated[34] in the $^4S_{3/2} - {}^2P_{3/2}$ TPA in BiI, where a mainly transit time limited linewidth of $\simeq$400 kHz was obtained.

The two-photon absorption normally involves virtual transitions, with associated energy level shifts. This systematic effect has been studied experimentally by Björkholm and Liao[35], using two different laser fields, enabling them to investigate the effect of a resonant intermediate level, also studied theoretically by Salomaa and Stenholm[36]. Naturally occurring three-level systems, with a resonant intermediate level, are rare, but they have been found in the Na_2 molecule[37], where detunings as small as 34 MHz, far smaller than the Doppler width, resulted in a TPA, comparable in size to one photon stepwise excitations.

Using fast accelerated atom/ion beams, the intermediate level can be Doppler shifted completely through resonance. Salomaa and Stenholm[38] proposed this scheme in a theoretical analysis of the resonant enhancement from an intermediate resonant real level. In the following we will discuss the realization of this scheme, both in the low and high field regimes as well as in the time domain.

Rabi frequencies $\Omega_i < \gamma_i$. The resonant cascaded configuration, shown in Fig.4, was first realized in the $3s[\frac{3}{2}]_2 - 3p'[\frac{3}{2}]_2 - 4d'[\frac{5}{2}]_3$ transition array in NeI[20], previously studied in a non-resonant TPA[39]. At a beam energy of $E^R = 119.105$ keV (see Eq.(2)), the intermediate $3p'[\frac{3}{2}]_2$ level allowed the creation of a complete harmonic three-level atom, to be excited at a laser wave number of $\sigma_0 = 16876{\cdot}9196\ \mathrm{cm}^{-1}$, given by Eq.(3). In Fig.11 is shown (lower part) a laser frequency scan through resonance, with the energy of the accelerator set to $E^R = Mc^2[\gamma(\beta) - 1]$. A sharp Doppler free feature, with a homogeneous linewidth of γ_{1n} is observed. In the upper part of Fig.11 is shown the resonant enhancement of the two-photon signal, when changing the particle velocity around β , given by Eq.(2). In this low field regime, no ac-Stark shifts of this TPA are found as well as only modest laser power broadenings. Following Salomaa and Stenholm[36,38], who have solved analytically Eq.(4), with one strong field $\Omega_1(n \rightarrow m)$ and one weak field $\Omega_2(m \rightarrow 1)$, we find the excitation probability of the upper level 1 $(4d'[\frac{5}{2}]_3)$ to be given by the imaginary part of $\rho_{1m} = \tilde{\rho}_{1m} \cdot \exp(-i(kz+\omega t))$, with

Fig.11: In the lower part is shown a laser frequency scan around σ_0, observed in the fluorescence from the final level [4d']. A sharp TPA is clearly observed. Both Rabi frequencies $\Omega_1 \simeq 5$ MHz and $\Omega_2 \simeq 2$ MHz are small compared to the homogeneous widths, resulting in only a small power broadening. In the upper part is shown the resonant enhancement, obtained at σ_0 with the velocity (energy) of the relativistic absorbers being varied around β, given by Eq.(2).

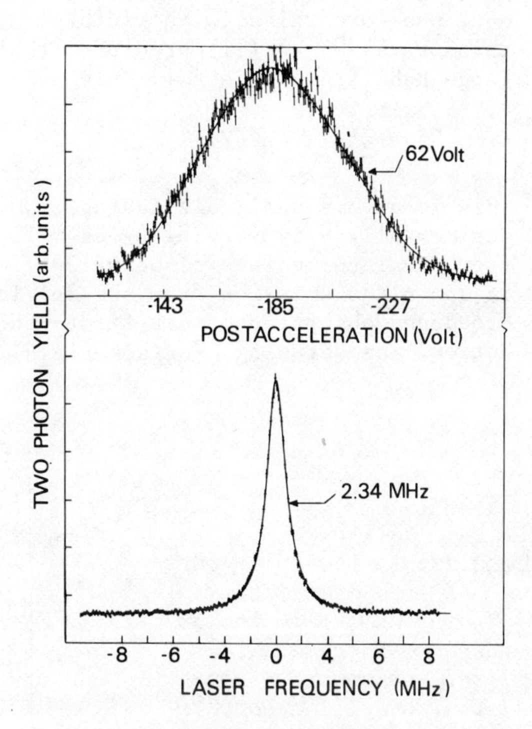

$$\tilde{\rho}_{1m} = \frac{K(\Omega_1^2)}{(\Delta + \omega\beta + \omega\delta\beta)^2 + \Gamma^2} \; \frac{2(\Delta_0 - i\gamma_{1n}) - (\gamma_m/\gamma_{mn})(\Delta + \omega\beta + \omega\delta\beta + i\gamma_{mn})}{[(\Delta_0 - \Delta - \omega\beta - \omega\delta\beta - i\gamma_{1m})(\Delta_0 - i\gamma_{1n}) - \Omega_1^2]} \tag{10}$$

where $\Delta_0 = \omega_{1n} - 2\omega$, $\Delta = \omega_{mn} - \omega$ and $\Gamma^2 = \gamma_{mn}^2 + 2\gamma_{mn}(\gamma_m + \gamma_n)\Omega_1^2/\gamma_m\gamma_n$. β is given by Eq.(2), and $\delta\beta$ is the velocity detuning. K is a function, containing an Ω_1^2 dependence[36]. Assuming the longitudinal velocity spread being much larger than the homogeneous width parameters[36], Salomaa and Stenholm find

$$\mathrm{Im}\tilde{\rho}_{1m} \propto \exp\left(-\frac{\Delta^2}{(ku)^2}\right) \mathrm{Im}\left(\frac{F_-}{\Delta_0 - i\Gamma_-} + \frac{F_+}{\Delta_0 - i\Gamma_+}\right) \tag{11}$$

where u is the velocity spread in the fast beam, assumed to be Gaussian. Also no spontaneous coupling from $1 \to m$ and $m \to n$ is included and each laser field only acts on <u>one</u> transition. This last assumption is rigorously fulfilled in our fast beam, with a total Doppler shift of $\simeq$50 Å. Eq.(11) predicts a resonant enhancement given by $\simeq \Delta^2/\gamma ku$, where Δ is the off-resonant detuning and γ the homogeneous width of the TPA. In the present case we find an enhancement $\simeq 10^9$. Eq.(11) further predicts, in agreement with the experiment, one Lorentzian, with width $\Gamma_- \simeq \gamma_{1n}$ as $F_+ \simeq 0$ with our relaxation rates[36]. Eq.(11) predicts, that no power shifts occur, even for large Rabi frequencies Ω_1. Only a broadening is predicted, given by

$$\Gamma_- = \frac{1}{2}(\gamma_1 + \gamma_n \Gamma/\gamma_{mn}) \quad . \tag{12}$$

This is a very desirable feature, which also has been verified experimentally[20] by varying Ω_1 as well as observing the fluorescence from the intermediate resonant 3p' level. In the upper part of Fig.12 is shown the emission from the 3p' level, showing a direct Doppler broadened absorption and a first-order Doppler free coherent contribution. The detuning σ_D (see Fig.4) again can be plotted against

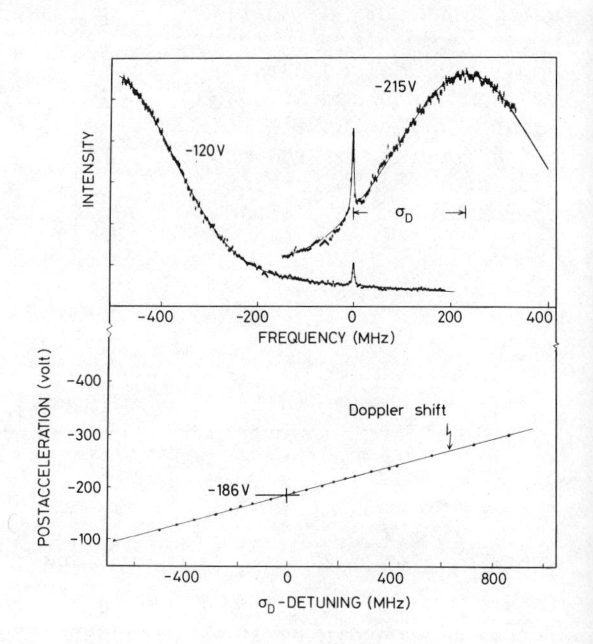

Fig.12: In the upper part is shown the fluorescence from the 3p' resonant intermediate level, showing two contributions (i) a Doppler broadened direct excitation and (ii) a Doppler free coherent part. In the lower part the detuning σ_D is represented against post acceleration.

the particle energy, with no deviations from the total Doppler shift being observed. This confirms the predictions of Eq.(11.12) and also Eq.(9), which is exact to all orders in Ω_1, Ω_2 [26], but does not include the velocity integration. This lack of power shifts is due to the extreme symmetry in the resonant cascaded three-level system, leaving only broadening as an inherent systematic effect to be considered. In addition, the small power broadenings present, the linewidth in Fig.11 is composed of a finite laser bandwidth of $\simeq$ 1 MHz, a residual second-order Doppler broadening of $\simeq$ 1 MHz, a transit time limited width of $\simeq$ .2 MHz and the natural linewidth of 2.7 MHz, corresponding to a lifetime of 60 ns[40].

$\Omega_1 > \gamma_{mn}$ and $\Omega_2 < \gamma_{1n}$. Going to higher fields in one of the interactions an ac-Stark splitting is predicted, both from Eq.(9) and Eq.(10). After velocity integration, this splitting only manifests itself as broadening unless the Rabi frequency Ω_1 is comparable to the Doppler width. Thus a narrow Doppler profile makes this doublet structure more easily observable. Figure 13 shows an experimental result, with the power broadened TPA being split into two resonances.

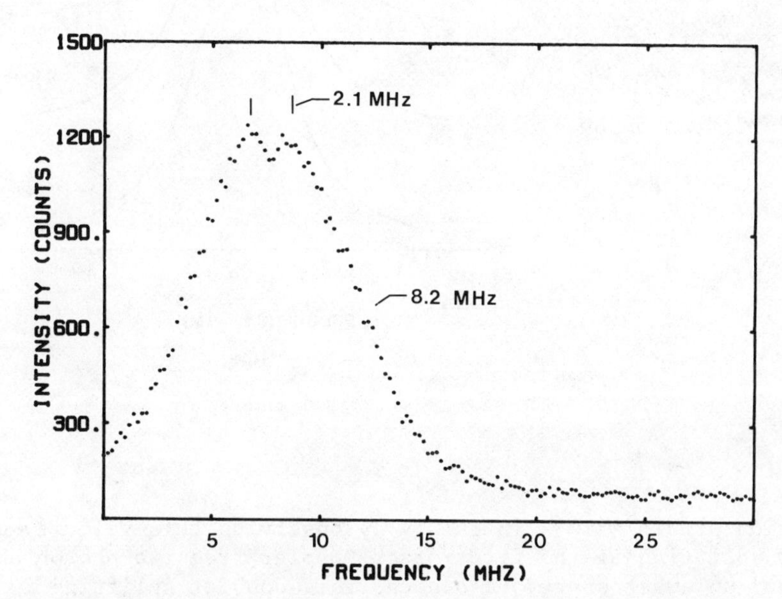

Fig. 13: ac-Stark splitting of TPA at Rabi frequencies $\Omega_1 \simeq$ 100 MHz and $\Omega_2 \simeq$ 2 MHz. Optical pumping among the Zeeman sublevels smear out this splitting, as compared to the calculated spectra.

The observation of this doublet is critically dependent on the exact realization of a resonant three-level atom. As the velocity integration can no longer be performed analytically, we again solve Eq.(4), in steady state, taking into account the proper relaxation rates, branching ratios, and Rabi frequencies and finally performing numerically the velocity integration. The results are shown in Fig.14.

A doublet structure, which is extremely velocity (energy) dependent, is found. It only manifests itself for velocity classes corresponding to the power broadened homogeneous linewidth of the TPA. In the present case this corresponds to an energy variation of ± 1 V around the resonant energy E^R. Thus we have established a method to identify and locate the exact resonant three-level con-

Fig.14: Calculated ac-Stark splitting of TPA at Rabi frequencies $\Omega_1 = 100$ MHz and $\Omega_2 = 2$ MHz as function of particle energy E^R.

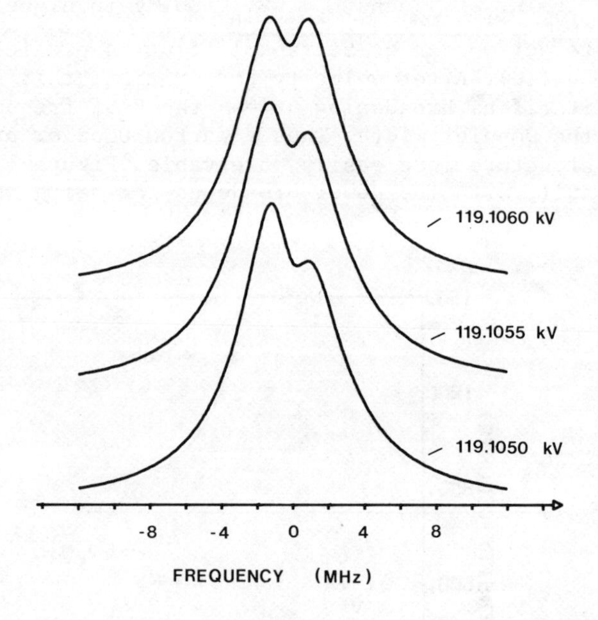

figuration within $4 \cdot 10^{-6}$ in energy by observing this velocity dependent splitting of the first-order Doppler free two-photon absorption. On exact energy resonance, this doublet splitting is completely symmetric around σ_0 , resulting in no shift of the centroid of the absorption, whereas an energy (velocity) detuning results in a net shift of this centroid, given by Eq.(9). These observations are also in agreement with Eq.(11), which predicts a broadening, but no splittings and shifts. They are present in Eq.(10), but disappear in the velocity integration assuming a flat Doppler profile close to resonance. Our experimental conditions, with Rabi frequencies comparable to the Doppler widths make this assumption invalid.

Time domain spectroscopy. By pulse modulating the laser field[5,40], the transient behaviour of the TPA process can be studied, again taking advantage of the clean collision-free environment. Subnatural optical resolution is possible, by performing measurements on 'old atoms'. At present the spectral bandwidth of our laser is not sufficiently good to fully exploit this possibility. But in the time domain, delayed fluorescence[5] can be used to measure excited state relaxation times. In Fig.15 is shown the decay of the $4d'[\frac{5}{2}]_3$ level in TPA, the laser field being modulated by an opto-acoustic wave. The lifetime of this level is found to (60 ± 3) ns, a value used in the two previous sections.

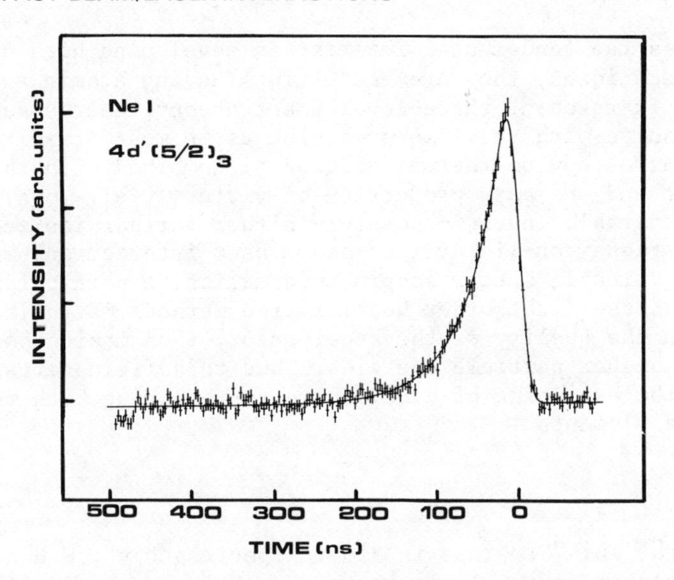

Fig.15: The relaxation of the $4d'[\frac{5}{2}]_3$ *level in NeI, obtained in resonant two-photon absorption in a fast accelerated Ne* atom beam.*

CONCLUSION

The main aspects of fast-beam laser-spectroscopy concern (i) the fundamental limits of optical resolution working with fast relativistic absorbers, the applications in (ii) high-resolution spectroscopy, and (iii) velocity and high-voltage control. The longitudinal transit time limit is in the range 10 - 100 kHz and with our improved accelerator, the second-order Doppler broadening is reduced to ≃ 20 kHz. This can further be reduced by optical cooling of the accelerated particles around a mean velocity β . $(\omega, 2\omega)$ spectroscopy in an inverted V configuration on such a 'cold' beam would then allow super high optical resolution in a forward resonant Raman scattering at frequency ω. Also resonant two-photon absorption in a H(2s) beam will yield a very high optical resolution, with both types of experiments being able to measure the second-order Doppler shift[41)] uniquely to $\simeq 10^{-7}$ without line splitting and independently determine the first-order Doppler shift, when locating the exact velocity resonance, given by the appearance of the Rabi doublet. Going from frequency to time space, the Doppler switching capability is not yet fully explored. In particular the interesting possibilities of obtaining high optical resolution in a Ramsey-type coherent excitation need further attention.

Besides the fundamental interest in developing high optical resolution techniques, they are useful in studying atomic and molecular structure. In resonant three-level spectroscopy, energy separations can be measured with the same precision as in most 'traditional' non-linear spectroscopy on thermal absorbers. Combined with the clean environment and the easy production of virtually all ions, in a variety of metastable levels - possibly either further ionized or neutralized - these non-linear fast-beam laser-interactions will be able to provide valuable spectroscopic information. A particular useful feature of these first-order Doppler free methods is their small dependence on the quality of the accelerator, thus making possible a scaling to higher particle energies. Thus this field certainly has seen only the beginning of a development containing both fundamental and applied elements.

REFERENCES

1. H.J.Andrä, in: "Progress in Atomic Spectroscopy", W.Hanle and H. Kleinpoppen, ed., Plenum Press, New York and London (1979) p.829
2. H.J.Andrä, A.Gaupp, and W.Wittman, Phys.Rev.Lett. 31, 501 (1973)
3. H.J.Andrä, in: Proceedings of the "Sixth International Conference on Fast Ion Beam Spectroscopy", to appear in Nucl.Instrum.Methods (1982)
4. D.J.Pegg and M.L.Gaillard, IEEE Transactions on Nuclear Sciences, NS-28, 1186 (1981)
5. O.Poulsen, T.Andersen, S.M.Bentzen, and U.Nielsen, Phys.Rev.A, 24, 2523 (1981)
6. S.M.Bentzen, U.Nielsen, and O.Poulsen, J.Opt.Soc.Am., 72, Sept. (1982)
7. H.J.Andrä, in: "Beam-Foil Spectroscopy, vol.2, I.A.Sellin and D. J.Pegg, ed., Plenum Press (1976), p.835
8. O.Poulsen and P.S.Ramanujam, Phys.Rev.A 14, 1463 (1976)
9. W.H.Wing, G.A.Ruff, W.E.Lamb,Jr., and J.J.Spezeski, Phys.Rev.Lett. 36, 1488 (1976)
10. S.L.Kaufman, Opt.Comm., 17, 309 (1976)
11. E.-W.Otten, in: Proceedings of the IV International Conference on "Nuclei far from Stability", P.G.Hansen and O.B.Nielsen, ed., CERN 81-09, Geneva (1981) p.3
12. R.A.Holt, S.D.Rosner, T.D.Gaily, and A.G.Adam, Phys.Rev.A 22, 1563 (1980)
13. E.G.Meyers, P.Kuske, H.J.Andrä, I.A.Armour, N.A.Jelley, H.A. Klein, J.D.Silver, and E.Träbert, Phys.Rev.Lett.47, 87 (1981)
14. O.R.Wood II, C.K.N.Patel, D.E.Murnick, E.T.Nelson, M.Leventhal, H.W.Kugel, and Y.Niv, in: "Laser Spectroscopy V", A.R.W.McKellar, T.Oka, and B.P.Stoicheft, ed., Springer Verlag, Berlin, Heidelberg, New York, (1981) p.45
15. J.J.Snyder and J.L.Hall, in: Proceedings of the "Second International Conference on Laser Spectroscopy", S.Haroche et al., eds., Springer Verlag, Berlin, Heidelberg, New York, (1975) p.6

16. M.Dufay, M.Carré,M.L.Gaillard, G.Mennier, H.Winter, and A.Zgainski, Phys.Rev.Lett.37, 1678 (1976)
17. J.Bergquist, S.A.Lee, and J.L.Hall, Phys.Rev.Lett. 38, 159 (1977)
18. G.Borghs, P.de Bisschop, J.-M.Van den Cruyce, M.Van Hove, and R.E. Silverans, Phys.Rev.Lett.,6, 1074 (1981)
19. O.Poulsen, P.Nielsen, U.Nielsen, P.S.Ramannujam, and N.I.Winstrup, Phys.Rev.A, submitted (1982)
20. O.Poulsen and N.I.Winstrup, Phys.Rev.Lett.47, 1522 (1981)
21. J.L.Hall and S.A.Lee, Appl.Phys.Lett.29, 367 (1976)
22. S.Gerstenkorn and P.Luc, Opt.Comm., 36, 322 (1981)
23. T.W.Hänsch, Spectroscopia non Lineare, in: "Proceedings of the International School of Physics Enrico Fermi, vol.42, Academic Press (1976) p.17
24. V.S.Letokhov and C.P.Chebotayev, Non Linear Laser Spectroscopy, Springer Series in Optical Sciences, vol.4, D.L.MacAdam, ed., Springer Verlag, Berlin, Heivelberg, New York (1977), and V.P.Chebotayev and V.S.Letokhov, Prog.Quant.Electr.,4, 111 (1975)
25. P. G. Pappas, M.M.Burns, D.D.Hinshelwood, M.S.Feld, and D.E.Murnick, Phys.Rev.A 21, 1955 (1980)
26. R.G.Brewer and E.L.Hahn, Phys.Rev.A, 11, 1641 (1975)
27. D.W.Steinhaus, L.R.Radziemski, Jr., R.D.Cowan, J.Blaise, G. Guelavili, Z.Ben Osman, and J.Vergés, Los Alamos Scientific Laboratory Report, No.LA-4501
28. O.Poulsen, Nucl.Instrum.Methods, accepted (1982)
29. L.S.Vasilenko, V.P.Chebotayev, and A.V.Shishayev, JEPT.Lett.12, 113 (1970)
30. E.Giacobino and B.Gagnac, in: Progress in Optics XVII", E.Wolf, ed., North Holland (1980) p.85
31. C.Wieman and T.W.Hänsch, in: "Laser Spectroscopy III", J.L.Hall and J.L.Carlsten, ed., Springer Series in Optical Sciences, vol.7, Berlin, Heidelberg, New York (1977) p.39
32. S.Chu and A.P.Mills, Jr., Phys.Rev.Lett.48, 1333 (1982)
33. C.Bordé, C.R.Acad.Sc.Paris, 282B,341 (1976)
34. J.L.Hall, O.Poulsen, S.A.Lee, and J.C.Bergquist, J.Opt.Soc.Am., 68, 697 (1978) and to be published
35. J.E.Björkholm and P.F.Liao, Phys.Rev.Lett.33, 128 (1974)
36. R.Salomaa and S.Stenholm, J.Phys.B 8, 1795 (1975);9,1221 (1976)
37. H.-R.Xia, G.-Y.Yan, and A.L.Schawlow, Opt.Comm.39, 153 (1981)
38. R.Salomaa and S.Stenholm, Opt.Comm.16, 292 (1976)
39. F.Biraben, E.Giacobino, and G.Grynberg, Phys.Rev.A 12, 2444 (1975)
40. O.Poulsen, T.Andersen, S.M.Bentzen, and I.Koleva, in: Proceedings of the "Sixth International Conference on Fast Ion Beam Spectroscopy, to appear in Nucl.Instrum.Methods (1982)
41. J.L.Hall, in: "Atomic Physics 7", D.Kleppner and F.M.Pipkin, ed., Plenum Press, New York and London (1981) p.267

ISOTOPIC SHIFTS

H.H. Stroke[†]

Department of Physics, New York University
4 Washington Place, New York, N.Y. 10003, U.S.A.

INTRODUCTION

Investigations at the interface of atomic and nuclear physics have been fruitful for a considerable time. It is now history how high resolution atomic spectroscopy has contributed to the development of the nuclear shell model[1] through the systematic measurement of nuclear spins and magnetic moments, and also how it pointed out its failures by the observation of large quadrupole moments of nuclei. Had it stopped here, I would not discuss the subject further. In fact, the situation developed quite differently, and much of it has been made possible by important experimental advances, which led to Nobel awards to several of our colleagues at this conference! Thus the meaning and complexity of high resolution spectroscopy have changed significantly from what they were in the 1930's, when purely optical spectrographic diffraction and interference methods were used. The new tools were atomic beam magnetic resonance, optical double resonance, optical pumping - coupled with nuclear magnetic resonance, level crossing spectroscopy, and, most recently, laser spectroscopy, and a variety of combinations of these. The earlier techniques have been described, for example, in the works of Hans Kopfermann[2] and Francis Bitter.[3] Accounts of a number of experiments done with the laser techniques are presented at this conference.

Two particular features of these experiments are crucial to the advances that have been made in our gaining knowledge of nuclear

---

[†]Visiting scientist, Max Planck Institut für Quantenoptik, D8046 Garching, F.R. Germany, summer 1982.

structure through the interaction of the atomic electron with the nucleus: sensitivity and precision. Bitter recognized that systematic measurements of long chains of isotopes would be the most revealing in such studies: The great sensitivity of the optical techniques would allow the spectroscopy of radioactive isotopes and isomers. The high precision could give additional insight into the nucleus by revealing effects of the spatial distribution of the nuclear charge and magnetism.

Generally, these effects are too small, relative to the precision with which we understand the spectrum of an atom, to allow the comparison of a hypothetical point nucleus with the real one with its finite extent. We must therefore content ourselves with isotopic variations, hence the title Isotopic Shifts. For the distribution of the nuclear charge, these are simply the commonly known isotope shifts. For the distribution of nuclear magnetization, the isotopic variations give rise to the so-called hyperfine structure (hfs) anomalies - or Bohr-Weisskopf effect.[4] I should like to point out at the outset that in both of these instances similar structure effects in the spectra of mu-mesic atoms do not lead to small corrections, but may account in some cases for nearly half of the corresponding interaction with the nucleus.[5] The departure from point-like structure can thus be measured in single isotopes. Similarly, we will see that only a moment of the distribution of nuclear charge and magnetization is measurable by the atomic electron-nuclear interaction. In contradistinction, high energy electron scattering experiments can measure the form factor.[6] Thus, while we atomic physicists may wish to make greater claims of what we can contribute to nuclear structure studies, we should recognize the limitations. On the other hand, the remarkable sensitivity of the techniques of atomic spectroscopy commonly allow by now spectroscopic measurements with as few as a total of 10^9 atoms or less. The detection sensitivity (which is not the same) has reached the level of single atoms, as exemplified by the photon burst method of George Greenlees and coworkers[7] and the resonance ionization method of Hurst, et al.[8] Thus one can work with isotopic samples many orders of magnitude smaller than required for the preparation of targets in more conventional nuclear experimentation. This means that one may study by atomic techniques nuclear structure properties reaching far off the line of nuclear stability, in regions inaccessible to these conventional methods, and of other very rare isotopes. In our early work, started with Bitter[3], we were thus able to study optically the first extensive series of mercury isotopes ranging over twelve mass numbers and including several nuclear isomeric states.[9] This was extended another dozen mass numbers in mercury (and since to many other elements) by the work, on-line with particle accelerators and mass separator, by Ernst Otten and H.-J. Kluge and their collaborators at Mainz and CERN.[10] Fig. 1 shows, as an illustration of results, the variations of the radial nuclear charge distribution in mercury as obtained from these experi-

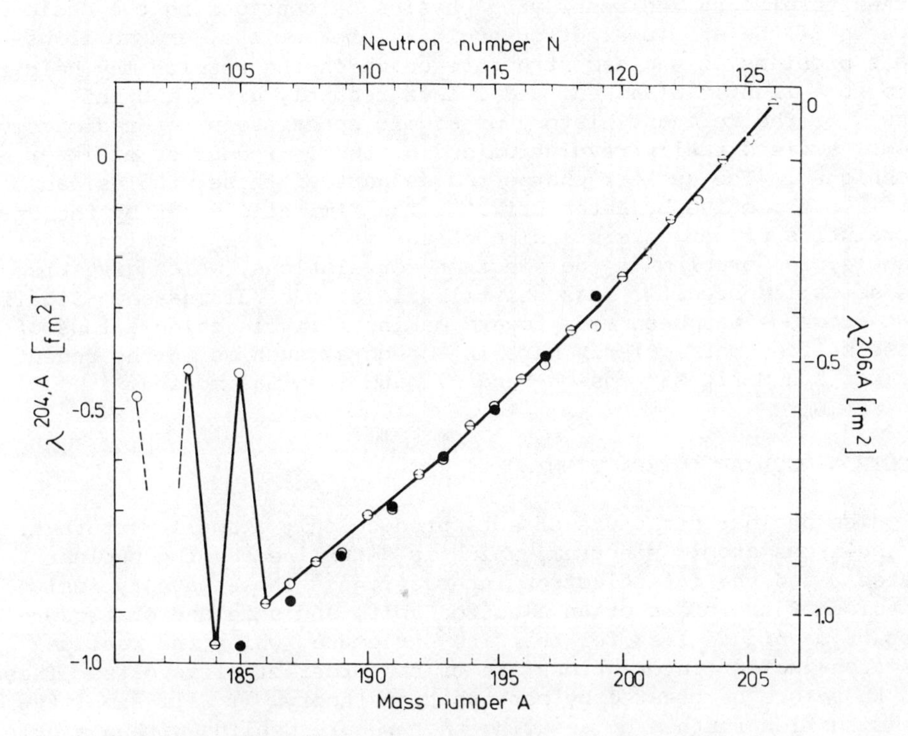

Fig. 1. Changes from ^{204}Hg of the mean values of r^2 of the nuclear charge distribution (given by the isotope shifts). Open circles - ground states, solid circles - nuclear isomers. (Courtesy of E.-W. Otten and H.-J. Kluge.)

ments. Since this initial work, a number of systematic studies were undertaken for chains of isotopes of the alkalis - sodium[11], potassium[12], rubidium[13], cesium[14], and francium[15], as well as other period II elements - calcium[16] (which is of particular interest in nuclear shell theory, as the $f_{7/2}$ shell is filled in going from the magic neutron number N = 20 to 28), barium[17], and cadmium.[18] There have also been studies in noble gases[19], to which we return further on, and rare earth elements.[20] Heavy elements will be discussed separately. It is not possible for me here to do justice to all who have made contributions in this field and I ask for their indulgence. I give references to a number of reviews which, I trust, will lead to many of these works (Refs. 10a-d, 21-25).

My intent is not to describe the experimental techniques, but rather to discuss the underlying physics by considering the basic aspects of the atomic electron-nuclear interactions, and to illustrate problems of nuclear structure on which the results may help to shed some light. Olav Redi and I have recently given a brief review[25] of the relevant history of atomic spectroscopy of radioactive atoms, a field really revolutionized by the introduction of laser techniques. The nuclear charge radius part of these studies has been put in perspective by Peter Brix.[26] The limitations set by the uncertainties of our understanding of the atomic wave functions (for example, the problem of the electron correlations, which give rise to the so-called specific mass shifts) will also be discussed. Significant progress has been made toward making reliable calculations of these shifts, particularly here in Göteborg, such as in the recent work of Ann-Marie Mårtensson-Pendrill and Sten Salomonson.[27]

ELECTRON-NUCLEAR INTERACTION

The nuclear structure effects produce only a small perturbation on the gross atomic spectrum, which is determined by the nuclear charge Z and the interelectron interactions. These have typical energies of the order of an electron volt, and some one or two orders of magnitude less for the fine structure, while the residual electron-nuclear interactions are of the order of microvolts. These can therefore be treated by perturbation theory. I will simplify the situation further by assuming that we are dealing with a single unperturbed level of the atomic electron. (Interesting consequences of level mixing are discussed by Eckart Matthias[28] in the following paper.) The residual electron nuclear interactions can then be expressed compactly by the Hamiltonian

$$\underline{H} = \sum_k \underline{T}_e(k) \cdot \underline{T}_n(k) \quad , \tag{1}$$

where k is the multipole order, and $\underline{T}$ are operators in the electron (e) and nuclear (n) spaces, respectively. The parity of the operators are even for the electric (E) ones, odd for the magnetic (M), so that we have k = 0, 2, 4 for E, k = 1, 3 for M. No higher order electron-nuclear interactions have been measured than a magnetic octupole[29] and an electric hexadecapole[21b], both of which are extremely small. The multipole interactions (k > 0) can be measured for point-like nuclei. Finite nuclear size effects can be detected only if an electron has a non-zero probability to be found at the nucleus. For a single electron this means an $s_{1/2}$ orbit (which would also be the small component of a relativistic $p_{1/2}$ electron). By selection rules, the evaluation of (1) with electron wave functions that correspond to an angular momentum J = 1/2 restricts the values of k to 0 and 1. We can thus study nuclear structure effects in the electric monopole and magnetic dipole interactions.

Electrostatic Interaction

The interaction energies evaluated for (1) are negligible for $k > 2$, and have been detected only in the highest resolution rf resonance experiments in long lived atomic states (e.g. ground and metastable states), where the natural lifetime does not set a limit on the precision. For the spherically symmetric penetrating $J = 1/2$ electrons, the electric quadrupole (E2) interaction is zero. It is clear, however, that changing incompressibly a nuclear charge distribution from a sphere of radius R_o to a quadrupolar distribution, $R = R_o(1 + \beta Y_o^2)$, which has roughly the same volume (Y is a spherical harmonic), will affect its radial moments, in particular $< r^2 >$. This is of importance in isotope shifts. Spectroscopically, one may measure in states other than $J = 1/2$ an electric quadrupole moment Q. For deformed nuclei Q represents the averaging of the deformation in a nuclear rotational state K (K is the projection of the spin I on a body fixed symmetry axis; for a rotational nucleus K = I). If one denotes the intrinsic quadrupole moment along this symmetry axis by Q_o, one finds[30]

$$Q = \frac{3K^2 - I(I + 1)}{(I + 1)(2I + 3)} Q_o \ , \tag{2}$$

where[31]

$$Q_o \simeq 3(5\pi)^{-1/2} R_o^2 Z <\beta> . \tag{3}$$

For even-even nuclei (I = O), Q = O even for $Q_o \neq 0$. However one may obtain Q_o from E2 transition probabilities in nuclei. With the relation[30] between the E2 transition probability T(E2) at frequency ω and its reduced value B(E2)

$$T(E2) = \frac{12\pi}{(5!!)^2} \frac{1}{\hbar} (\frac{\omega}{c})^5 B(E2) \tag{4}$$

one obtains for the transition to the ground state[31]

$$B(E2) = (\frac{3ZR_o^2}{4\pi})^2 <\beta^2> \tag{5}$$

or

$$BE(2) = \frac{5}{16\pi} Q_o^2 \ , \tag{6}$$

if one takes the limiting value $|<\beta>| = <\beta^2>^{1/2}$. There is model dependence in obtaining this relation and care must be used in its application.[10a]

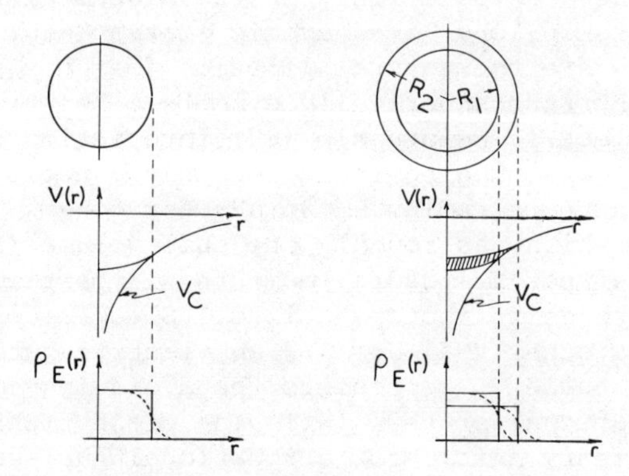

Fig. 2. Potentials V(r) for a point (Coulomb, V_c) and extended nuclear charge distributions. At present, the difference in energy that results from V_c - V(r) cannot be obtained accurately in atoms. The energy shift for two isotopes, which reflects the difference (cross-hatched) between $V_1(r)$ and $V_2(r)$ is measurable.

For the purpose of calculation here we make the simplifying assumption that the averaged spherically symmetric nuclear charge distribution $\rho(r)$, which the s electron probes, is uniform, with radius R (Fig. 2). A more realistic distribution (Fermi shape) is indicated by the dotted lines in the figure. We can calculate the isotope energy shift that corresponds to Fig. 2, with the assumption that the electron wave function is constant over the nucleus, and equal to $\psi_e(0)$. Thus for a transition we have a shift

$$\begin{aligned} \Delta E &\simeq \int_0^\infty \{V_1(r) - V_2(r)\}\, \Delta\rho_e \, d\tau \\ &= \frac{2\pi}{3}\, \Delta\psi^2(0)\; Ze^2\; \{\langle r_1^2\rangle - \langle r_2^2\rangle\}\ , \end{aligned} \tag{7}$$

where $\Delta\rho_e = \Delta\psi^2(0)$, e is the electronic charge. Hence isotope shifts measure the variations of $<r^2>$ of the nuclear charge distribution. The isotopic variations in $<r^2>$ can be expressed as[30]

$$\delta<r^2> = \delta<r^2>_0 \{1 + \frac{5}{4\pi} \delta<\beta^2>\} , \tag{8}$$

where $<r^2>_0$ represents the value for $\beta = 0$, i.e. without deformation. We shall see that with calculated values of $<r^2>_0$ one is thus enabled to extract $\delta<\beta^2>$ from isotope shifts.

Actually, isotope shifts in electronic atoms also measure rapidly decreasing contributions of higher radial moments of the nuclear charge distribution.[32] The shift can then be written in terms of the parameter

$$\lambda(A_1, A_2) \equiv \frac{\Delta E(A_1, A_2)}{C_1} = \sum_{k=1} \frac{C_k}{C_1} <r^{2k}> , \tag{9}$$

where the ratios of the parameters C_k/C_1, calculated by Seltzer[33], are essentially independent of the atomic transition. For muonic atoms yet a different moment is measured, $<e^{-\alpha r} r^k>$, with α a nearly linear function of Z, and k depending on Z and the transition.[34] One can determine the equivalent radius (Barrett radius) of a uniformly charged sphere with the same moment as the actual charge distribution. This allows, in principle, a comparison between electronic and muonic atom data. In practice, muonic shifts may involve different nuclear wave functions (reflecting possible nuclear excitations), and the atomic shifts have to be corrected for mass effects.

Transformation to Center of Mass Coordinate System

When the Hamiltonian for the electron-nuclear interaction is transformed to a coordinate system in which the center of mass is at rest, the kinetic energy operator becomes

$$\underline{T} = \frac{1}{2} \left(\frac{1}{m} + \frac{1}{M}\right) \sum_i p_i^2 + \frac{1}{M} \sum_{i>j} \underline{p}_i \cdot \underline{p}_j , \tag{10}$$

where m and M are respectively the electron and nuclear masses (the parenthesis in (10) is the reciprocal of the reduced mass of the electron, μ), $\underline{p}$ is the electron momentum, and the indices i and j range over all electrons. When one observes an isotope shift it thus consists of the finite nuclear size effect (7) as well as of results of variations in M. These produce 1) a simple scaling of the transition energies[21a] by a factor μ/m (the "normal" or "Bohr" mass shift), and 2) a shift due to the second term on the right of

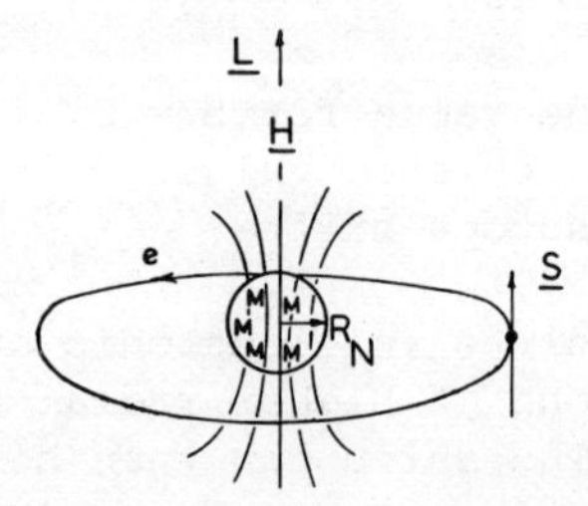

Fig. 3. Interaction of the electron magnetic field (H) due to spin (S) and orbital (L) motion with the extended nuclear magnetization (M).

(10), an electron correlation effect (the "specific" mass shift). The approaches to isolating the latter have been reviewed by Bauche and Champeau.[35] Results of Hartree Fock calculations were found to be only moderately successful, and accurate calculations remain an open problem, although we shall see that recent progress gives encouraging results.[27]

Structure Effects in the Magnetic Interaction

The selection rules allow us to probe only the M1 interaction with the penetrating s electron.[36] I discussed this at some length at the first of these conferences. Here I only recall the underlying physics with reference to Fig. 3. The magnetic dipole interaction energy is

$$E = - \underline{\mu}_N \cdot \underline{H}_e \, , \tag{11}$$

where $\underline{\mu}_N$ is the nuclear magnetic moment, and $\underline{H}_e$ the magnetic field of the electron at the nucleus. It can be evaluated from the expectation value of the Hamiltonian $a\underline{I}\cdot\underline{J}$, where a is the magnetic dipole interaction constant. The finite size effects of the nucleus are twofold: 1) As $\underline{H}_e$ is seen in Fig. 3 to be slightly inhomogeneous over the nuclear region, the energy E in (11) will depend on how the nuclear magnetization ($\underline{M}$) is distributed. Thus the value of a for a point nucleus is modified by a factor $(1 + \varepsilon)$, where ε is the hfs anomaly.[4] 2) Because the nuclear charge is distributed over R_N, compared to the point-like nucleus this affects the electron wave function and in turn $\underline{H}_e$. This gives rise to another correction factor to a(point), $(1 + \delta)$, where δ is known as the Breit-Rosenthal-Crawford-Schawlow correction, and which Rosenberg and I recalculated with a more realistic diffuse nuclear charge distribution.[37] Putting this together, the hfs interaction constant can be written

$$a = (16\pi/3)\, \mu_0^2\, g_I\, \psi^2(0)\, (1 + \varepsilon)(1 + \delta)\, F. \tag{12}$$

F is a relativistic correction, μ_0 the Bohr magneton, $g_I = \mu_N/I$, $\psi^2(0)$ the electron probability at the nucleus. Frequently δ is much smaller than ε, and we neglect it here in our further discussion. In ordinary atoms, ε is of the order of 0.01 (varying up and down by a factor of 10). As for the isotope shifts, we again look at differential effects. We assume that for two isotopes we can equate a_1/a_2 for a point like nucleus to $g_I(1)/g_I(2)$, where the g_I are measured by NMR in a homogeneous magnetic field (hence independent of the distribution of nuclear magnetization). We thus have for the experimental ratio

$$\frac{a_1}{a_2} \simeq \frac{g_1(1 + \varepsilon_1)}{g_2(1 + \varepsilon_2)} \simeq \frac{g_1}{g_2}(1 + \varepsilon_1 - \varepsilon_2) \equiv \frac{g_1}{g_2}(1 + \Delta_{12}). \tag{13}$$

Δ_{12} is the measured differential hfs anomaly. The importance of its determination is that it provides a new operator $\{\underline{S} \times \underline{Y}^2\}^{(1)}$, different from the magnetic moment ($\underline{\mu}_N = g_S\underline{S} + g_L\underline{L}$), with which to test nuclear wave functions. The square bracket is a tensor product of order 1 between the spin operator and the spherical harmonic Y^2. It is a measure of the angular asymmetry of the nuclear spin distribution.

We have extended the Bohr-Weisskopf theory[4], based largely on the nuclear single particle model with the inclusion of core polarization[38] in the evaluation of the effect of finite size of the nuclear magnetization. This model was found quite successful in accounting for nuclear magnetic moments.[39] The form of the result is (Eq. 33, Ref. 38) $-\varepsilon = (1/\mu_N)$ f(nuclear, electronic factors). Moskowitz and Lombardi[40] found that in the mercury isotopes f can be taken as about constant and equal to $\mp\alpha$ for $I = \ell \pm 1/2$ (ℓ, nuclear orbital angular momentum). A more detailed analysis[41] of the quantities entering f lent support to the application of this simple relation in estimating ε. However, if the hfs anomalies are to be used to gain knowledge of nuclear wave functions additional to what the nuclear magnetic moments give, then this empirical relation does not make it possible. An example[42], which illustrates further the inadequacy of the relation, and which was in fact the motivation for our development of the hfs anomaly formalism[38] is shown in Table 1.

Table 1. Hfs anomalies in cesium isotopes. All have $I = 7/2$.

A	μ_N nm	g_I	$-\Delta_{12}$ percent
133	2.56422(28)	0.737	
			+ 0.037(9)
135	2.7134(3)	0.780	
			− 0.020(9)
137	2.8219(3)	0.811	

It is clear that it is impossible to obtain a reversal in sign of Δ for these cesium isotopes from a relation $\varepsilon = \text{const.}/\mu_N$ as the magnetic moments increase monotonically. A physical explanation for the reversal given when these data were obtained[42] was a possible break in the variation of the nuclear size at the shell closure with magic number $N = 82$. This now appears to be supported by recent isotope shift measurements in cesium[14a], Fig. 4. The core polarization (configuration mixing) model did allow us [38] to account for this sign reversal.

It may be worth pointing out that though the nuclear configuration mixing model does generally quite well in accounting for nuclear magnetic moments, there are some striking failures, as in the case of ^{209}Bi, which is doubly magic plus one proton. One can account for the remaining discrepancy remarkably well by pion exchange effects.[39a, 43] The effect of this contribution to the hfs anomaly has yet to be investigated.

ISOTOPE SHIFTS

We have seen that isotope shifts have two origins, non-zero extent of the nuclear charge distribution and non-infinite nuclear mass. They produce isotope shifts in opposite directions: An electron is less tightly bound in the potential of nucleus 2 (Fig. 2) than in 1 ($R_2 > R_1$); the binding energy for a more massive nucleus (A_2) is greater than for the lighter one (A_1). It has thus been usual to assume that in very light atoms we have nearly exclusively

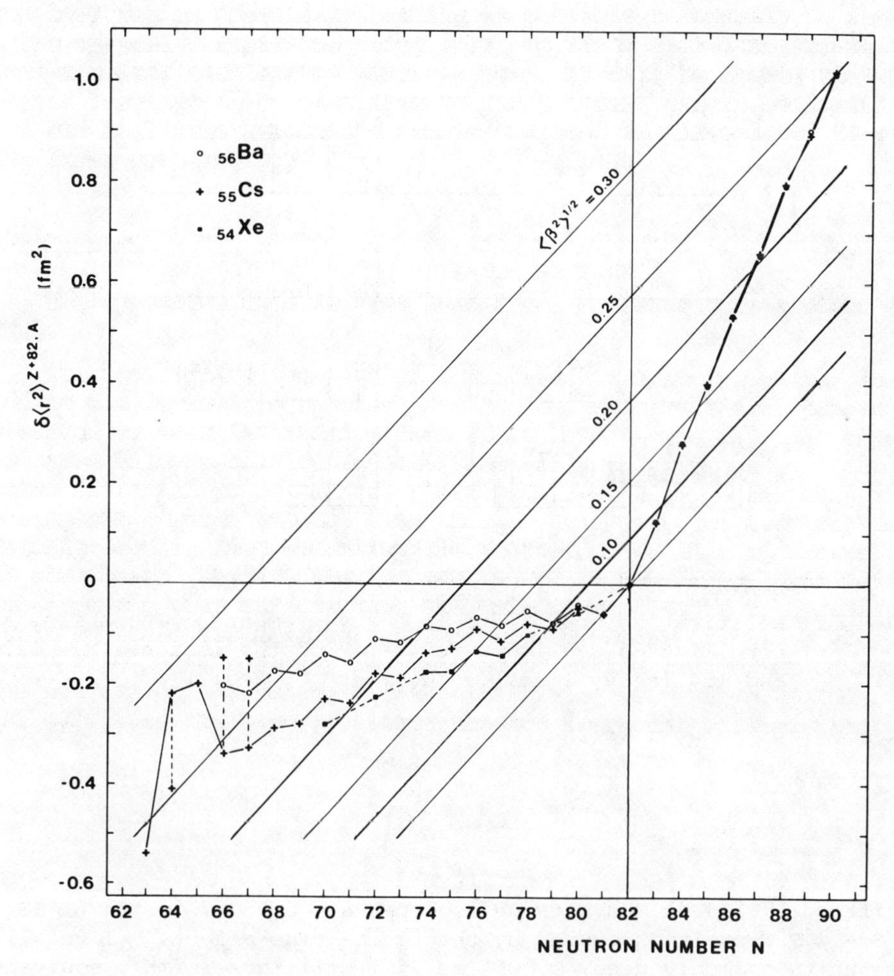

Fig. 4. Variations of $\langle r^2 \rangle$ relative to the value for N = 82 for cesium isotopes. Also shown are those for xenon and barium. (Courtesy R. Neugart, Ref. 10e.)

mass-dependent shifts, while in very heavy nuclei, we have largely the finite size effect. In the intermediate range the two effects nearly cancel each other and the shifts are so small that until the advent of laser spectroscopy their measurement presented a great challenge. But now the picture is changing. While mass-dependent shifts were observed already some time ago[44] in an element as heavy as plutonium, a recent measurement by Neumann[45] put into evidence a

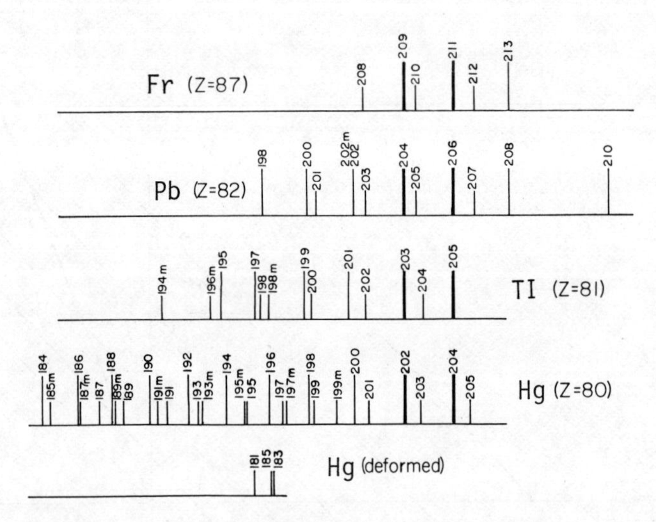

Fig. 5. Relative isotope shifts in heavy elements. The shifts are normalized and aligned for $\delta\langle r^2\rangle$ of the isotones N = 122 and 124 (heavy lines). (Courtesy O. Redi.)

shift of 17.1(9.4) MHz between isotopes A = 6 and 7 in the Li II $2^3S - 2^3P$ transition caused by the finite size effect. (I recall the units commonly used, 0.001 cm^{-1} $\equiv$ 1 milliKayser (mK), equivalent to $\simeq$ 30 MHz, to appreciate this measurement.) The result reflects reasonably well the estimated $\delta\langle r^2\rangle$ for these two isotopes.

Our own approach, which has been directed toward looking at differential nuclear effects in heavy nuclei, relies on the measurement of relative isotope shifts, which give essentially

$$\mathrm{RIS} = \frac{\langle r_i^2\rangle - \langle r_j^2\rangle}{\langle r_k^2\rangle - \langle r_l^2\rangle}, \tag{14}$$

where i...k denote various isotopes. In this manner, at least to our present precision, electronic factors, and in particular the mass dependent shifts, do not play a significant limiting rôle in studying nuclear features. In Fig. 5 we represent results obtained from the data on mercury[9,10], thallium[46], lead[24,47], and francium.[15]

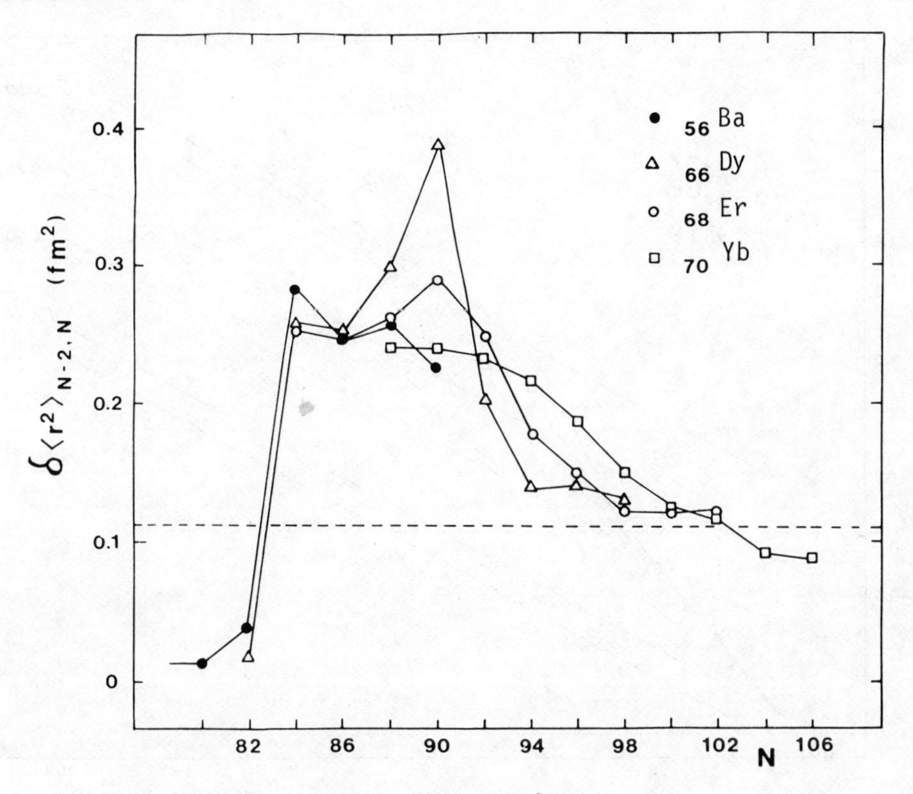

Fig. 6. Two-neutron differences in $\langle r^2 \rangle$ in the region above N = 82. (Courtesy R. Neugart, Ref. 20a.)

Complementary data for bismuth[48] and gold[49] are now being obtained. This representation displays a number of interesting features - preponderant regularities in the even-N isotope shifts, odd-even staggering, isomer shifts, the sudden onset of deformation for neutron deficient mercury isotopes (shown separately in Fig. 5) found by the Mainz group.[10] Large jumps in $\delta\langle r^2 \rangle$ between odd-N and neighboring even-N nuclei, as for ^{181}Hg, ^{183}Hg, and ^{185}Hg, have also been observed in nuclear experiments through variations of the moment of inertia in neutron deficient platinum isotopes in this neutron number region.[50] The jump in deformation, as observed in mercury, has now been studied more systematically by Neugart[20a] and coworkers, with recent results shown in Figs. 6, 7. A dramatic jump in $\langle r^2 \rangle$ occurs for $_{66}$Dy between N = 88 and 90. For ytterbium there is a smooth passage into the strongly deformed region. An analysis leads to a picture in which closed shell proton configurations (e.g. magic numbers) tend to stabilize the spherical shape against the in-

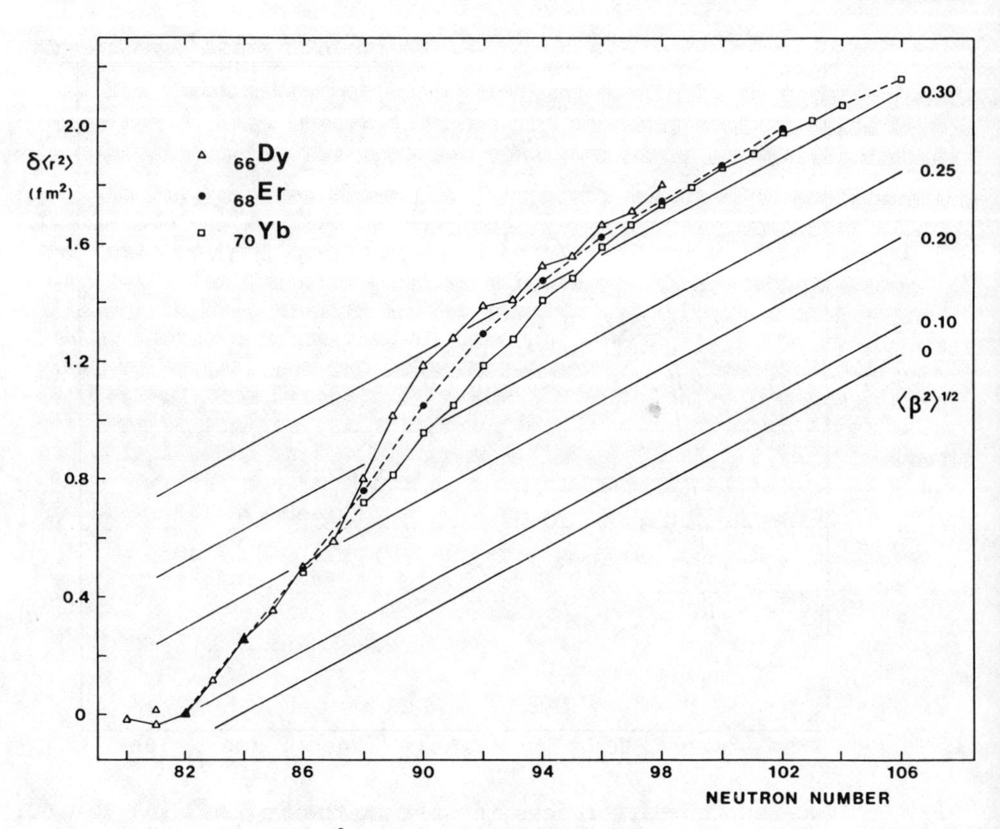

Fig. 7. Change in $<r^2>$ from its value at N = 82. Lines at constant deformation $<\beta^2>$ are predictions of the droplet model (Ref. 51). (Courtesy R. Neugart, Ref. 20a.)

creasing nuclear deformation, up to a point when a jump occurs, as in mercury. The fact that one observes such a jump in dysprosium appears to be explained by the proton subshell closure at Z = 64.

Fig. 5 also permits relative isotonic shift comparisons. Recently, many more such isotonic comparisons have been made possible (Fig. 8), as well as an isobaric one (Fig. 9). We expect that with adequate precision the differences in the isotonic shifts, particularly between even and odd Z elements, can provide new insights into nuclear interactions.

In our studies of the RIS (Eq. 14), our goal is to probe such nuclear interactions. Thus, while the spherical variations of the nuclear charge radius are given quite well phenomenologically in a parametrized form by the droplet model[51], a picture in which the additional nucleon producing the isotope shift polarizes a nucleus may be more revealing. For the odd-even staggering we have thus

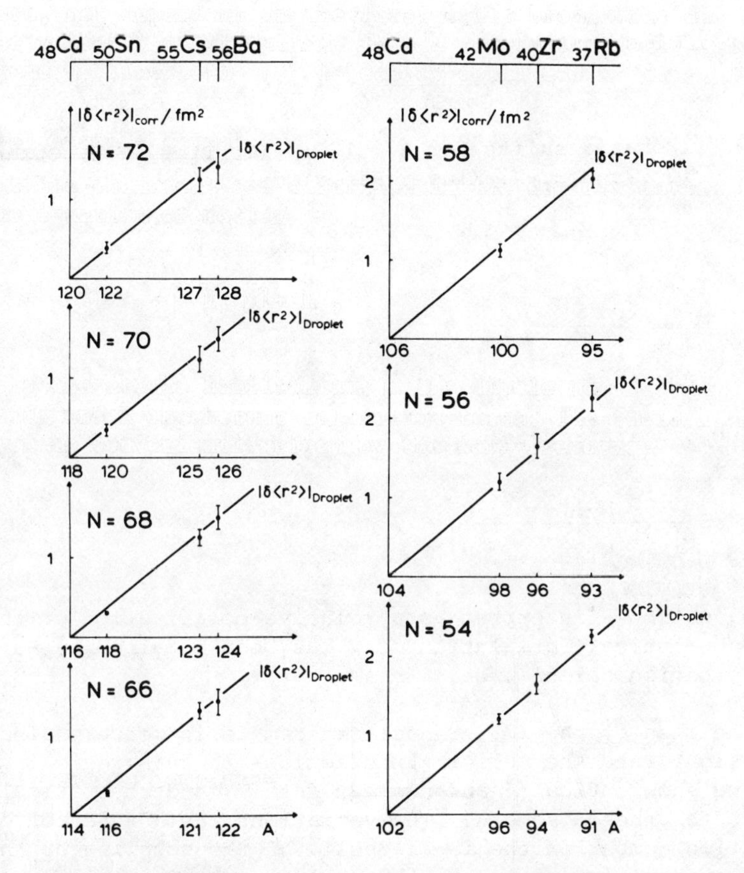

Fig. 8. Isotonic shifts in the region $54 \lesssim N \lesssim 72$. (Courtesy E.W. Otten, Ref. 10c, and H.-J. Kluge, Ref. 10d.)

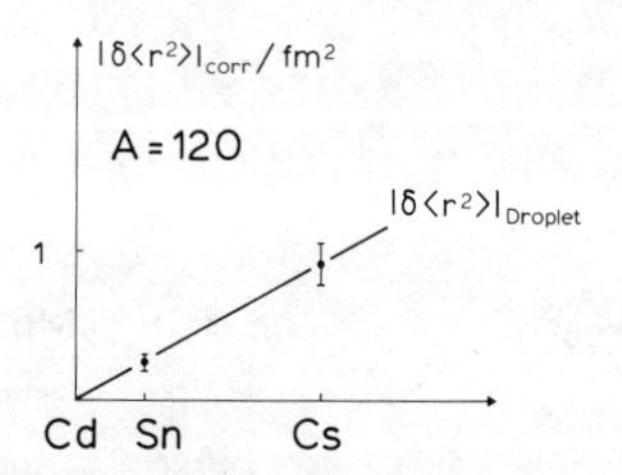

Fig. 9. Isobaric shifts for A = 120. (Courtesy E.W. Otten, Ref. 10c, and H.-J. Kluge, Ref. 10d.)

introduced[52] a parameter

$$\gamma \equiv \frac{\langle r^2\rangle_{N+1} - \langle r^2\rangle_N}{\frac{1}{2}\left[\langle r^2\rangle_{N+2} - \langle r^2\rangle_N\right]} \quad , \tag{15}$$

N even, which we interpret as a relative polarization of the nucleus N. A theoretical formulation of γ has been obtained by Barrett[53] in a perturbation calculation.

The case of the $1i_{13/2}$ nuclear isomers in mercury also gave a new insight into the nuclear interactions in this region. If Figs. 10-12 we show optical[54] and nuclear data[55], and theoretical predictions[56] for these isomers. The variations, as a function of the occupation number in the $i_{13/2}$ shell, of the coupling of the large particle angular momentum to the nuclear rotation axis provides a model that can account quantitatively for their magnetic dipole and electric quadrupole moments[56,10d], and qualitatively for the isomer shifts.[54] In conjunction with the latter, it is interesting to note (Figs. 10, 11) that a consistent discontinuity appears both in γ and E for N = 119.

The droplet model, on the other hand, allows systematic analyses of data. Thus in Fig. 13 one can compare its predictions to the values of $\delta\langle r^2\rangle$ as one approaches the magic numbers of neutrons N = 50 and 82. In both cases the radii shrink more rapidly as one gets

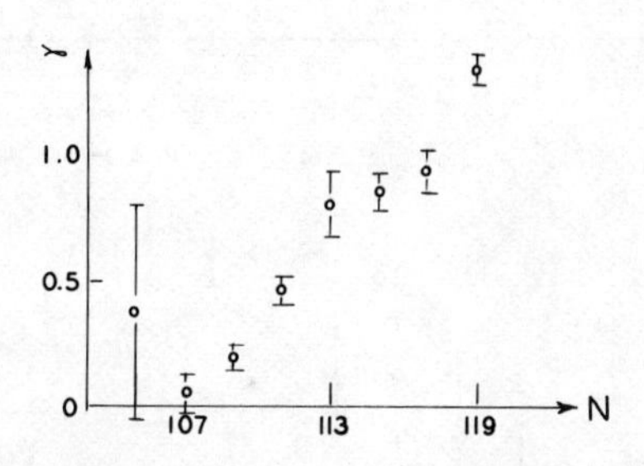

Fig. 10. Odd-even staggering for $i_{13/2}$ nuclear isomers in mercury. (Ref. 54.)

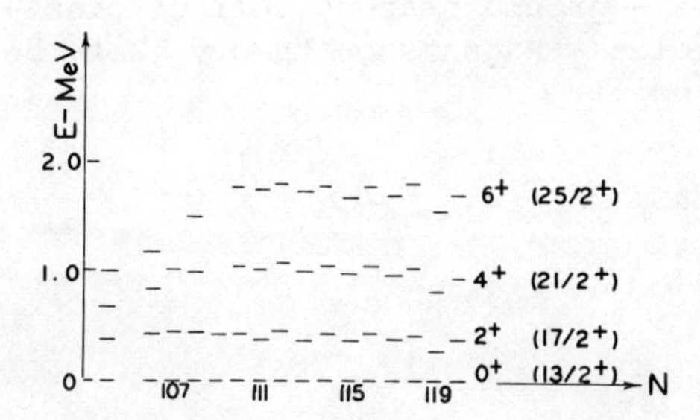

Fig. 11. Mercury nuclear energy levels. For the isomers, the $i_{13/2}$ levels were set at zero energy. (Ref. 55.)

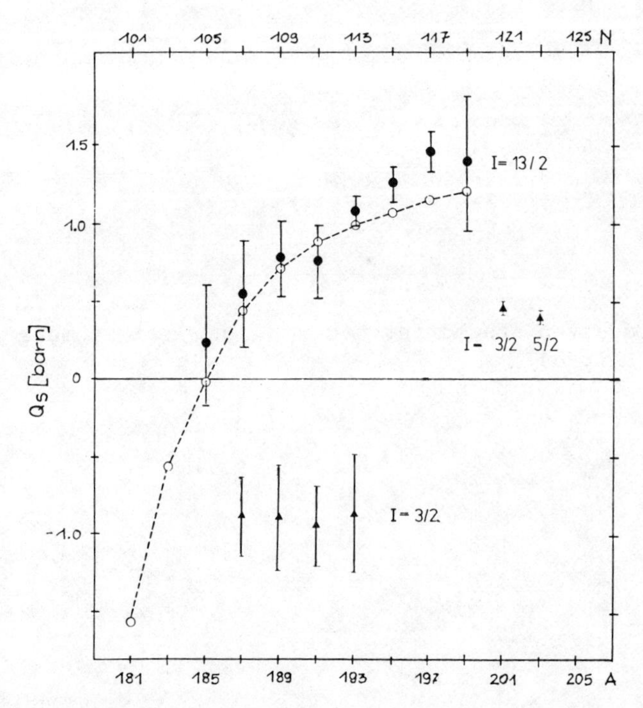

Fig. 12. Spectroscopic quadrupole moments of mercury isotopes: triangles - ground states, full circles - $i_{13/2}$ isomers, open circles - Ragnarsson theory (Ref. 56). (Courtesy H.-J. Kluge.)

to the shell closure - a phenomenon that one may have tended to expect qualitatively. Also, as seen in Fig. 14, there appears to be a break in the variation of $\delta\langle r^2\rangle$ for N = 64, with the likely closure of the $g_{7/2}$ neutron shell. We have seen above that a proton subshell closure at Z = 64 would allow an explanation of the trend of the rare earth isotope shifts. Comparison of the predictions of the

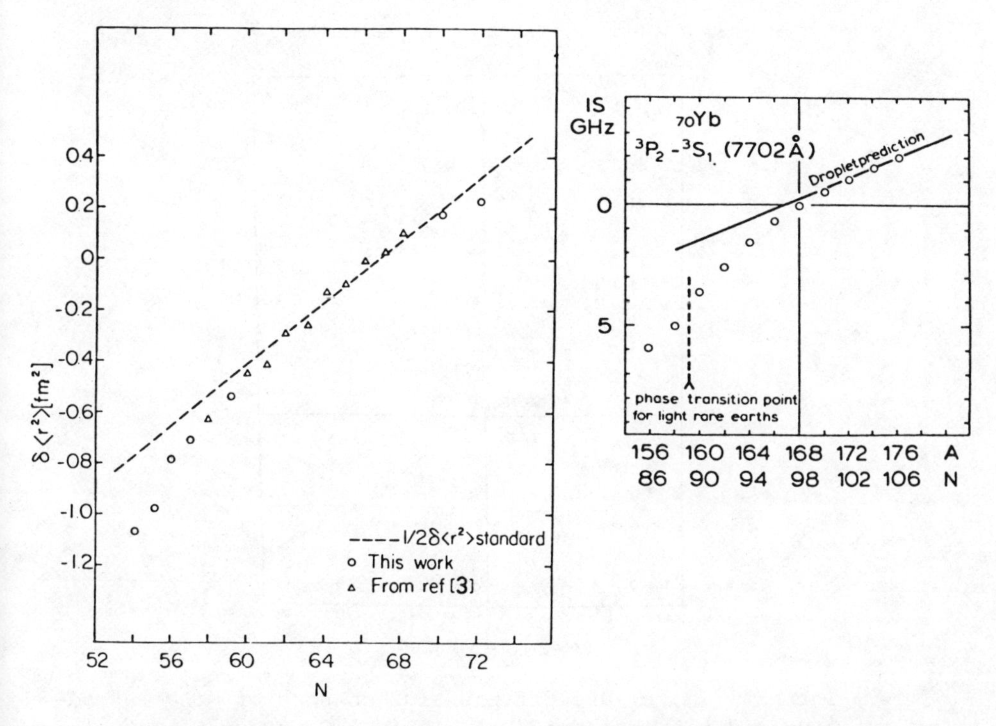

Fig. 13. Variations of $\delta\langle r^2\rangle$ for cadmium (Ref. 18a) and ytterbium isotope shifts (Neugart et al., quoted in Ref. 10c). Ref. 3 in the cadmium figure refers to our Ref. 32 and G.H. Fuller, Nuclear spins and moments, J. Phys. & Chem. Data (USA) 5: 835(1976). (Courtesy H.-J. Kluge.)

droplet model $\delta\langle r^2\rangle_0$ with measurements of $\delta\langle r^2\rangle$ allows the extraction of deformation parameters with the use of Eq. 8, and the assumption that ß is small at a magic neutron number. This is also illustrated in Fig. 7.

An important isotope shift study is that of the series of calcium isotopes where it is found that the doubly magic nuclei ^{40}Ca and ^{48}Ca have the same $\langle r^2\rangle$ (Fig. 15). A detailed discussion of this observation in terms of the variation of the nuclear skin thickness (the region in which the charge density decreases from 90 to 10 percent of its central value) is given by Rebel and Schatz.[24] The values of $\langle r^2\rangle$ of these isotopes, calculated by summing individual nucleon contributions, are in good agreement with these results.[57] Variations in skin thickness have also been suggested to explain other isotope shifts, such as in cadmium[18b] and krypton.[58]

At the extreme of capabilities of the optical experiments to date is the measurement of the shift of the 1 msec spontaneous fission isomer ^{240m}Am. It is known from nuclear physis studies that

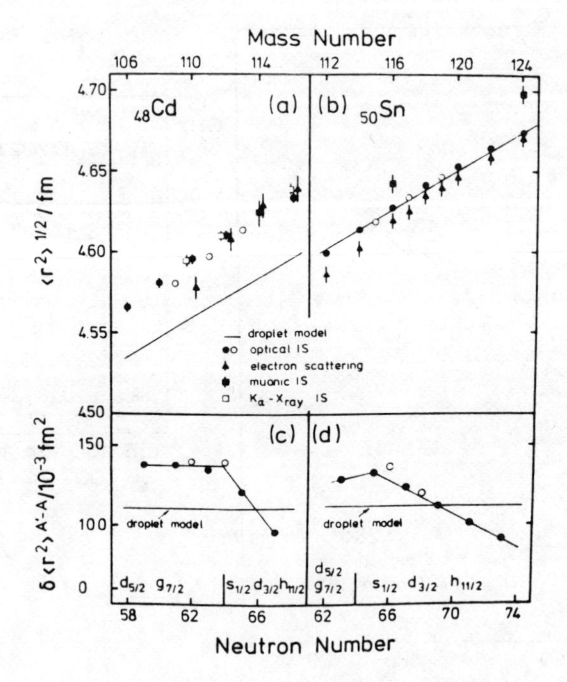

Fig. 14. Integral (a, b) and differential changes of $\langle r^2 \rangle$ in cadmium and tin isotopes (Ref. 18b). (Courtesy E. Matthias.)

the fission process proceeds by a shape isomeric state. The largest shift ever measured by this experiment (^{240m}Am - ^{241}Am $\simeq$ 6.32cm^{-1}) is to be compared to the already large isotope shift $\simeq$ 0.24 cm^{-1} between ^{243}Am and ^{241}Am. This leads to a ^{240}Am radius isomer shift $\delta\langle r^2 \rangle$ = 4.6 fm^2, and, with some model assumption, to a quadrupole moment Q_o = 30.4 b! Thus, with somewhat sophisticated experiments, fairly short lived states in nuclei are becoming accessible to optical studies.

Relation to other Measurements and Theory

The most widely used comparisons of optical isotope shifts have been with results from muonic atom X-rays and electron scattering. These have been reviewed recently by Shera.[60] The analysis of several muonic X-ray transitions allows the extraction of various radial moment parameters, as well as of the Barrett radius defined earlier. The experiments are limited to stable isotopes, as targets of the order of 0.1 g or more are required for such muonic X-ray studies. Absolute calibrations of optical measurements for these stable isotopes can thus be made, which in turn allow more precise extraction of nuclear data for the radioisotopes. This has been done recently for barium.[60,61] The muonic experiments have also allowed the de-

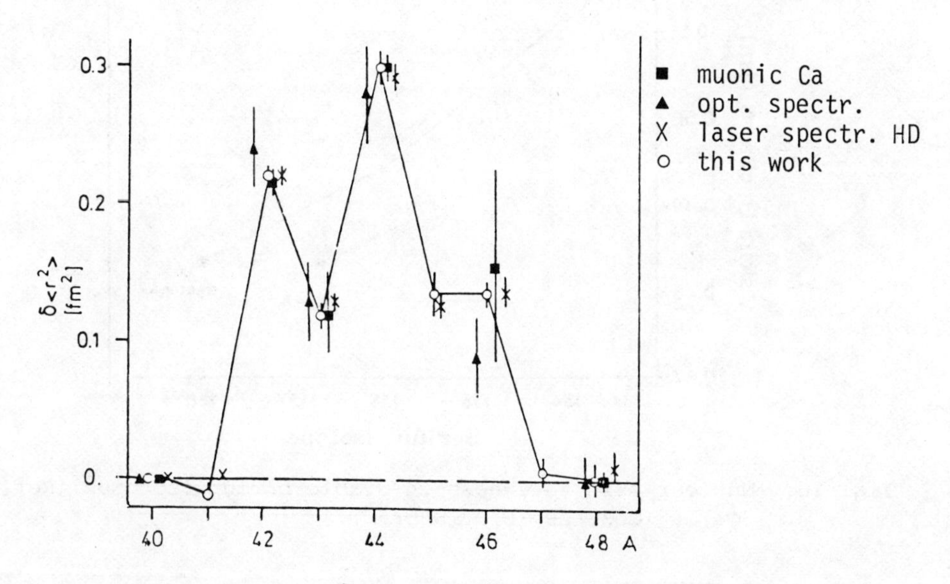

Fig. 15. Variations in $\langle r^2 \rangle$ of the calcium isotopes (from Refs. 16a and 24). "This work" is Ref. 24. (Courtesy H. Rebel and G. Schatz.)

termination of the skin thickness (Fig. 16) in the stable barium isotopes. An important result of a combination of electron scattering and muonic X-ray studies has been the possibility of estimating the size of the higher radial moments ($k > 1$) in Eq. (9). The result is shown in Fig. 17.

The agreement between optical and muonic and electron scattering measurements tends to be quite satisfactory. We have said nothing so far about isotope shift measurements by electron K X-ray measurements. These shifts, very small relative to the line widths, require great precision to yield reliable data. Although a number of experiments have been done, the additional information has been sparse and at times in disaccord with the optical data. Recent high precision experiments in lead[62] appear to be changing this situation and the results may supplement isotope shift data obtained by other techniques. In Table 2 we show the RIS (Eq. 14) obtained in a number of different optical and K X-ray transitions. It is seen that

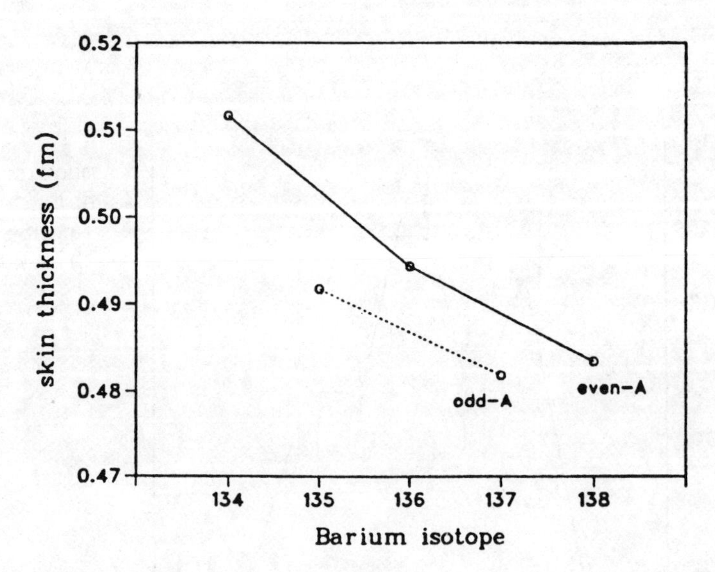

Fig. 16. Nuclear skin thickness in stable barium isotopes (Ref. 60a). (Courtesy E.B. Shera.)

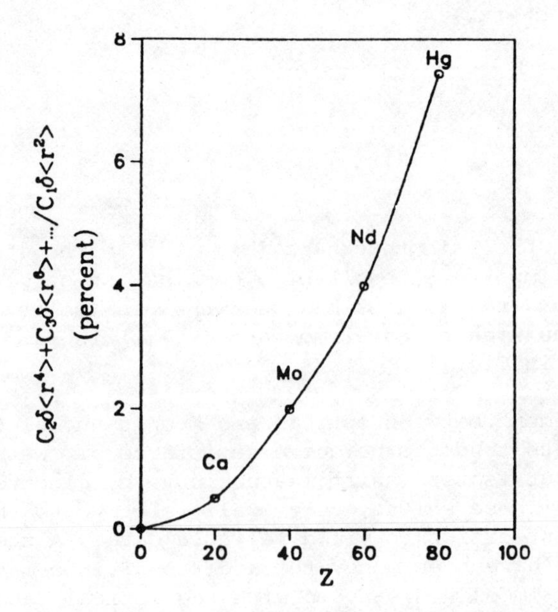

Fig. 17. Estimated ratio of higher order radial moment contributions to $\langle r^2 \rangle$ (Eq. 9) (Ref. 60a). (Courtesy E.B. Shera.)

Table 2. Isotope shifts in lead normalized to $^{208}Pb - ^{206}Pb$

Atomic or ionic transition		Isotope pair 206 - 207	204 - 206	Ref.
Optical				
$6p^2\ ^2P_0$	$-6p7s\ ^3P^0_1$	0.374(11)	0.91(4)	a
$6p^2\ ^3P_0$	$-6p7p\ ^3P_0$	0.375(10)	0.90(1) laser-2γ	b
$6p^2\ ^3P_2$	$-6p7s\ ^3P^0_1$	0.376	0.889	c
$6p^2\ ^1D_2$	$-6p6d\ ^3D^0_2$	0.370	0.90	"
$6p^2\ ^1S_0$	$-6p8s\ ^3P^0_1$	0.372	0.887	"
$6s^26d\ ^2D_{5/2}$	$-6s^25f\ ^2F^0_{7/2}$	0.377	0.88	"
$6s6p^2\ ^2D_{5/2}$	$-6s^27f\ ^2F^0_{7/2}$	0.370	0.89	"
$6s6p^2\ ^4P_{5/2}$	$-6s^25f\ ^2F^0_{7/2}$	0.382	0.89	"
X - ray				
K_α		0.27(11)	1.02(23)	d
K_α		0.34(6)	0.82(10)	e

a) Refs. 47 b, c
b) Ref. 47 a
c) Ref. 64
d) Ref. 62

there is excellent agreement among the optical results in both atomic and ionic transitions, and that the recent X-ray measurements represent a substantial advance over the earlier values.

In addition to the droplet model[51] and the perturbation approaches[53] to account for nuclear radii and their variations, there have been a number of Hartree-Fock calculations.[65] A relatively new entry into these calculations is the interacting boson model (IBA).[66] This model incorporates features of both the shell and collective models of the nucleus, and involves the interaction between paired protons (p) and neutrons (n), each favoring zero spin without the p-n interaction, but spin 2 with the interaction. The energies of the pair excitations show in general very good agreement with the observed nuclear spectra. The IBA model has been used to calculate the shifts in the barium isotopes.[60a] The agreement with experiment is excellent, which is not quite the case with the droplet model. Agreement in tungsten isotopes is somewhat less satisfactory.[67]

Before leaving nuclear structure properties, I should like to recall the observation, first made by Gerstenkorn, of the relation between RIS and relative binding energy differences per nucleon, E. This Gerstenkorn relation[68] is (see note added in proof)

$$\frac{\langle r_1^2\rangle - \langle r_2^2\rangle}{\langle r_2^2\rangle - \langle r_3^2\rangle} = - \frac{E_1 - E_2}{E_2 - E_3} . \tag{16}$$

It appears to hold particularly around magic numbers; even isotope shift inversions are correctly predicted. However no formal theoretical foundation is given. A more extensive survey has been made and some connection to the shell model given.[69] An analysis of barium and xenon data by Matthias and coworkers[70] has led to a modification of Eq. 16 to account for effects of shell closure.

Atomic Effects

While there has been an age old caveat about using optical spectroscopic data for the extraction of absolute values of nuclear parameters from atomic hyperfine structure and isotope shift, there has been considerable progress in overcoming the limitations, in particular the specific mass shift. Sophisticated theoretical methods have been developed, many by the school of our conference chairman, Ingvar Lindgren[71], to allow realistic calculations. The progress is not only theoretical, but was also made possible by the new experimental methods of laser spectroscopy. I want to illustrate this with the results of Niemax and Pendrill[72] for potassium (Fig. 18). For both the 4S - nD and 4S - nS transitions the isotope shifts become equal for high enough principal quantum numbers, and the slope becomes equal to that given by the Bohr (reduced mass) shift only. This equality and value can be expected for optical electrons that are far away from the nucleus and from the remaining electrons. One can then determine the residual (finite size, or "field", plus specific mass) shift of the lower individual level. (Such measurements were reported earlier for rubidium by Stoicheff and Weinberger.[73]) It is then of course relatively easier to calculate the specific mass shift for the level than for the transition, and then to extract the field shift. Such a calculation has been made by Mårtensson-Pendrill and Salomonson[27] with the use of

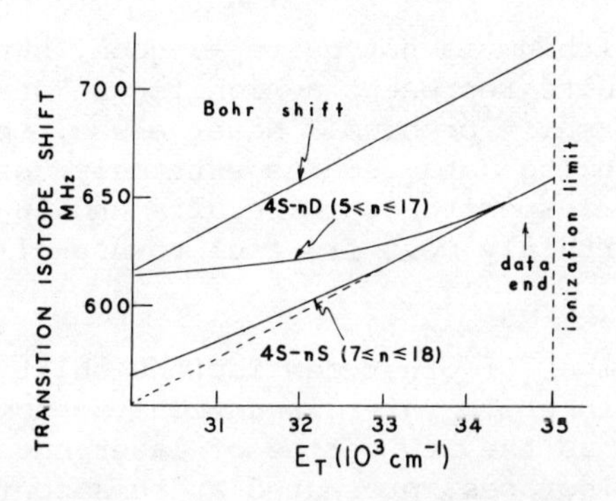

Fig. 18. Potassium isotope shifts in transitions to Rydberg states as a function of energy, or principal quantum number n (Ref. 72a). (Courtesy K. Niemax and L.R. Pendrill.)

many-body perturbation theory. Their work is presented in a poster at this conference. In Table 3 we compare their results to the experimental values of several specific mass shifts in potassium.

Table 3. Specific mass shifts (in MHz) between ^{39}K and ^{41}K (Experiment Ref. 72a, theory Ref. 27)

Level	4S	4P	5D
Theory	-53	-30	-22
Experiment	-61(2)	-28(2)	-53(8)

The agreement in lithium is not quite as good, but still very impressive. The results represent a significant step forward with this specific mass shift problem. Nonetheless, the use of muonic and electron scattering data remains extremely useful in the interpretation of optical spectroscopic results and in supplementing them, and it is certainly most fruitful to use all available information together!

There is a number of other new isotope shift investigations. Some are more pertinent to this discussion of atomic nuclear structure studies, such as the new series of laser spectroscopic measurements of nine tin isotopes, presented at this conference.[74] Others are perhaps more of interest to atomic physics. For example, a systematic study has been initiated of the J dependence of the field isotope shift.[75] On the other hand, recent anomalies in the isotope shifts in samarium have been explained by mixing of closely lying levels by the isotope shift operators, in particular the one describing the field effect.[76]

I have tried to present a picture of the richness of this rapidly developing field of isotope shift studies. Spectroscopists have long used the nuclear field shift to help in the classification of spectra by identifying the presence of penetrating electrons in a configuration. Here we saw how these penetrating electrons serve to study nuclear structure.

ACKNOWLEDGMENTS

I wish to thank the National Science Foundation for support of our work under Grant PHY-8204402. My colleague, Professor Olav Redi, has been intimately associated with all our research program. The collaboration with Professor R.A. Naumann and the Princeton University cyclotron group has been essential in our radioactive atom work. Valuable discussions with Professor H.-J. Kluge and his input relating to this paper were much appreciated. I am grateful to Professor Herbert Walther for his hospitality and stimulating discussions. The friendly reception by the Max Planck Institut für Quantenoptik and the staff at Universität München are also acknowledged.

REFERENCES

1. M.G. Mayer and J.H.D. Jensen, Elementary Theory of Nuclear Shell Structure (Wiley, New York, 1955).
2. H. Kopfermann, Nuclear Moments (Academic, New York, 1958).
3. F. Bitter, Selected Papers and Commentaries, edited by T. Erber and C.M. Fowler (MIT, Cambridge, 1969); F. Bitter, Magnetic

Resonance in Radiating or Absorbing Atoms, Appl. Opt. 1: 1 (1962).

4. A. Bohr and V.F. Weisskopf, The Influence of Nuclear Structure on the Hyperfine Structure of Heavy Elements, Phys. Rev. 77: 94 (1950).
5. J. Hüfner, F. Scheck, C.S. Wu, Muonic Atoms, in Muon Physics 1 (Academic, New York, 1977).
6. H.D. Wohlfahrt, O. Schwentker, G. Fricke, H.G. Andresen, E.B. Shera, Phys. Rev. C22: 264 (1980).
7. G.W. Greenlees, D.L. Clark, S.L. Kaufman, D.A. Lewis, J.F. Tonn, J.H. Broadhurst, High resolution spectroscopy with minute samples, Opt. Commun. 23: 236 (1977); D.A. Lewis, J.F. Tonn, S.L. Kaufman, G.W. Greenlees, Photon-burst method in high-resolution laser spectroscopy, Phys. Rev. A19: 1580 (1979).
8. G.S. Hurst, M.H. Nayfeh, J.P. Young, A demonstration of one-atom detection, Appl. Phys. Lett. 30: 229 (1977); One-atom detection using resonance ionization spectroscopy, Phys. Rev. A15: 2283 (1977).
9. W.J. Tomlinson III, H.H. Stroke, Nuclear Moments and Isotope and Isomer Shifts of Neutron-Deficient Mercury Isotopes 195, 195^m, 194, 193, 193^m and 192, Nucl. Phys. 60: 614 (1964); O. Redi, H.H. Stroke, Isotope Shift of 47-day ^{203}Hg, J. Opt. Soc. Am. 65: 1 (1975).
10. J. Bonn, G. Huber, H.-J. Kluge, E.W. Otten, Spins, Moments and Charge Radii in the Isotopic Series ^{181}Hg - ^{191}Hg, Z. Phys. A276: 203 (1976); P. Dabkiewicz, C. Duke, H. Fischer, T. Kühl, H.-J. Kluge, H. Kremmling, E.-W. Otten, H. Schüssler, Nuclear Shape Staggering in Very Neutron Deficient Hg Isotopes Detected by Laser Spectroscopy, J. Phys. Soc. Japan 44 Suppl.: 503 (1978); (a) E.W. Otten, Laser Spectroscopy for Probing Nuclear Structure off Stability, in Future Directions in Studies of Nuclei far from Stability, Nashville 1979, edited by J.H. Hamilton et al. (North Holland, Amsterdam, 1980) (hereafter referred to as Nashville, 1979); (b) Laser Techniques in Nuclear Physics, Nucl. Phys. A354: 471c (1981); (c) Some Recent Developments in Laser Spectroscopy of Unstable Atoms, in 4th International Conference on Nuclei far From Stability (CERN, Geneva, 1981) Report 81-09 (hereafter referred to as CERN, 1981); (d) H.-J. Kluge, Laser Spectroscopy of Radioactive Isotopes in Resonance Cells, presented at the Conference on Lasers in Nuclear Physics, Oak Ridge, Tennessee, 21-23 April 1982 (to be published by Harwood Academic) (hereafter referred to as Oak Ridge, 1982); (e) R. Neugart, Laser Spectroscopy on Mass-Separated Radioactive Beams, Nucl. Instr. & Meth. 186: 165 (1981).
11. G. Huber, F. Touchard, S. Büttgenbach, C. Thibault, R. Klapisch, H.T. Duong, S. Liberman, J. Pinard, J.-L. Vialle, P. Juncar, P. Jacquinot, Spins, magnetic moments, and isotope shifts of $^{21-31}$Na by high resolution laser spectroscopy of the

atomic D_1 line, Phys. Rev. C18: 2342 (1978); K. Pescht, H. Gerhardt, E. Matthias, Isotope Shift and HFS of D_1 Lines in Na-22 and 23 Measured by Saturation Spectroscopy, Z. Phys. A281: 199 (1977); F. Touchard, J.M. Serre, S. Büttgenbach, P. Guimbal, R. Klapisch, M. de Saint Simon, C. Thibault, H.T. Duong, P. Juncar, S. Liberman, J. Pinard, J.-L. Vialle, Phys. Rev. C25: 2756 (1982); R. Klapisch, Nuclear Properties Studied by Atomic Physics, in Atomic Physics 7, edited by D. Kleppner and F.M. Pipkin (Plenum, New York, 1981), p. 231.

12. H.T. Duong, P. Juncar, S. Liberman, J. Pinard, J.L. Vialle, S. Büttgenbach, P. Guimbal, M. de Saint Simon, J.M. Serre, C. Thibault, F. Touchard, R. Klapisch, Ground state hyperfine structures of ^{43}K and ^{44}K measured by atomic beam magnetic resonance coupled with laser optical pumping, J. Phys. (France) 43: 509 (1982); F. Touchard, et al., Isotope shifts and hyperfine structure of $^{38-47}K$ by laser spectroscopy, Phys. Lett. 108B: 169 (1982).

13. C. Thibault, F. Touchard, S. Büttgenbach, R. Klapisch, M. de Saint Simon, H.T. Duong, P. Jacquinot, P. Juncar, S. Liberman, P. Pillet, J. Pinard, J.L. Vialle, A. Pesnelle, G. Huber, Hyperfine structure and isotope shift of the D_2 line of $^{76-98}Rb$ and some of their isomers, Phys. Rev. C23: 2720 (1981); C. Ekström, S. Ingelman, G. Wannberg, M. Skarestad, Nuclear Spins and Magnetic Moments of some Neutron-Deficient Rubidium Isotopes, Nucl. Phys. A311: 269 (1978); W. Klempt, J. Bonn, R. Neugart, Nuclear Moments and Charge Radii of Neutron-Rich Rb Isotopes by Fast-Beam Laser Spectroscopy, Phys. Lett. 82B: 47 (1979).

14. C. Thibault, F. Touchard, S. Büttgenbach, R. Klapisch, H.T. Duong, P. Jacquinot, P. Juncar, S. Liberman, P. Pillet, J. Pinard, J.L. Vialle, A. Pesnelle, G. Huber, Hyperfine structure and isotope shift of the D_2 line of $^{118-145}Cs$ and some of their isomers, Nucl. Phys. A367: 1 (1981); B. Schinzler, W. Klempt, S.L. Kaufman, H. Lochmann, G. Moruzzi, R. Neugart, E.-W. Otten, J. Bonn, L. von Reisky, K.P.C. Spath, J. Steinacher, D. Weskott, Collinear Laser Spectroscopy of Neutron-Rich Cs Isotopes at an On-Line Mass Separator, Phys. Lett. 79B: 209 (1978); C. Ekström, L. Robertsson, G. Wannberg, J. Heinemeier, Nuclear Spins and Magnetic Moments of some Neutron-Rich Rubidium and Cesium Isotopes, Physica Scripta 19: 516 (1979); H. Gerhardt, E. Matthias, F. Schneider, A. Timmermann, Isotope Shifts and Hyperfine Structure of the 6s - 7p Transitions in the Cesium Isotopes 133, 135, and 137, Z. Phys. A288: 327 (1978); S. Büttgenbach, Hyperfeinstruktur-Untersuchungen an Atomen mit offener 4d- und 5d-Schale un an radioaktiven Cäsium-Isotopen, Habilitationsschrift, Rheinische Friedrich-Wilhelms Univeritát Bonn (1980).

15. S. Liberman, J. Pinard, H.T. Duong, P. Juncar, P. Pillet, J.L. Vialle, P. Jacquinot, F. Touchard, S. Büttgenbach, C. Thibault,

M. de Saint Simon, R. Klapisch, A. Pesnelle, G. Huber, Laser optical spectroscopy on francium D_2 resonance line, Phys. Rev. A22: 2732 (1980); C. Ekström, S. Ingelman, G. Wannberg, M. Skarestad, Nuclear Ground State Spins of the Francium Isotopes $^{208-213,220-222}$Fr, Physica Scripta 18: 51 (1978).

16. P. Grundevik, M. Gustavsson, I. Lindgren, G. Olsson, L. Robertsson, A. Rosén, S. Svanberg, Precision Method for Hyperfine Structure Studies in Low-Abundance Isotopes: The Quadrupole Moment of ^{43}Ca, Phys. Rev. Lett. 42: 1528 (1979); H.-W. Brandt, K. Heilig, H. Knöckel, A. Steudel, Isotope Shift in the Ca I Resonance Line and Changes in Mean-Square Nuclear Charge Radii of the Stable Ca Isotopes, Z. Phys. A288: 241 (1978); (a) E. Bergmann, P. Bopp, Ch. Dorsch, J. Kowalski, F. Träger, G. zu Putlitz, Nuclear Charge Distribution of Eight Ca - Nuclei by Laser Spectroscopy, Z. Phys. A294: 319 (1980); M. Arnold, E. Bergmann, P. Bopp, C. Dorsch, J. Kowalski, T. Stehlin, F. Träger, G. zu Putlitz, Hyperfine structure and nuclear moments of odd calcium isotopes by laser and radiofrequency spectroscopy, Hyperfine Inter. 9: 159 (1981); C.W.P. Palmer, P.E.G. Baird, J.L. Nicol, D.N. Stacey, G.K. Woodgate, Isotope shift in calcium by two-photon spectroscopy, J. Phys. B15: 993 (1982); see also H. Rebel and G. Schatz, Ref. 24.

17. P.E.G. Baird, R.J. Brambley, K. Burnett, D.N. Stacey, D.M. Warrington, G.K. Woodgate, Optical isotope shifts and hyperfine structure in λ 553.5 nm of barium, Proc. R. Soc. Lond. A365: 567 (1979); K. Bekk, A. Andl, S. Göring, A. Hanser, G. Nowicki, H. Rebel, G. Schatz, Laser-spectroscopic Studies of Collective Properties of Neutron Deficient Ba Nuclei, Z. Phys. A291: 219 (1979); R. Neugart, F. Buchinger, W. Klempt, A.C. Mueller, E.W. Otten, C, Ekström, J. Heinemeier, ISOLDE, Fast-Beam Laser Spectroscopy of Neutron-Rich Barium Isotopes, Hyperfine Inter. 9: 151 (1981); R.E. Silverans, G. Borghs, J.-M. Van den Cruyce, Fast ion beam laser spectroscopy on radioactive ^{140}Ba, Hyperfine Inter. 9: 193 (1981), also with G. Dumont, Collinear Fast Beam-Laser Spectroscopy on 5d $^2D_{3/2,5/2}$ Barium Ions, Z. Phys. A295: 311 (1980); W. Fischer, M. Hartmann, H. Hühnermann, H. Vogg, Isotopieverschiebung der Bariumnuklide ^{140}Ba, ^{138}Ba, ^{136}Ba und ^{134}Ba in der Ba II Resonanzlinie 6p $^2P_{1/2}$ - 6s $^2S_{1/2}$, Z. Phys. 267: 209 (1974); K. Wendt, Kollineare Laserspektroskopie an Bariumionen zur Bestimmung der Hyperfeinstruktur und Isotopieverschiebung, Diplomarbeit, Johannes Gutenberg-Universität, Mainz (1981); see also H. Rebel, G. Schatz, Ref. 24.

18. (a) F. Buchinger, P. Dabkiewicz, H.-J. Kluge, A.C. Mueller, E.-W. Otten, ISOLDE, The Isotope Shift of the Radioactive Cd-Isotopes ($102 \leqslant A \leqslant 120$) Determined by On-Line Laser Spectroscopy, Hyperfine Inter. 9: 165 (1981); (b) R. Wenz, A. Timmermann, E. Matthias, Subshell Effect in Mean Square Charge Radii of Stable Even Cadmium Isotopes, Z. Phys. A303: 87 (1981).

19. H. Gerhardt, F. Jeschonnek, W. Makat, E. Matthias, H. Rinneberg, F. Schneider, A. Timmermann, R. Wenz, P.J. West, Nuclear Charge Radii and Nuclear Moments of Kr and Xe Isotopes by High Resolution Laser Spectroscopy, Hyperfine Inter. 9: 175 (1981).
20. H. Brand, B. Seibert, A. Steudel, Laser-Atomic-Beam Spectroscopy in Sm: Isotope Shifts and Changes in Mean-Square Nuclear Charge Radii, Z. Phys. A296: 281 (1980); H. Brand, V. Pfeufer, A. Steudel, Laser-Atomic-Beam Spectroscopy of $4f^7 5d6s$-$4f^7 5d6p$ Transitions in Eu I, Z. Phys. A302: 291 (1981); D.L. Clark, M.E. Cage, G.W. Greenlees, The hyperfine structure and hyperfine anomaly of ^{161}Dy and ^{163}Dy, Phys. Lett. 62A: 439 (1977); D.L. Clark, G.W. Greenlees, Measurements of Changes of Charge Radii and Hyperfine Interactions of the Dy Isotopes, private communication 1982, preprint; R.-J. Champeau, J.-J. Michel, H. Walther, Spectroscopic determination of the nuclear moments of ^{169}Yb; relative isotope shift between the isotopes ^{168}Yb, ^{169}Yb and ^{170}Yb, J. Phys. B7: L262 (1974); D.L. Clark, M.E. Cage, D.A. Lewis, G.W. Greenlees, Optical isotopic shifts and hyperfine splittings for Yb, Phys. Rev. A20: 239 (1979); P. Grundevik, M. Gustavsson, A. Rosén, S. Rydberg, Analysis of the Isotope Shifts and Hyperfine Structure in the 3988 Å ($6s6p\ ^1P_1 - 6s^2\ ^1S_o$) Yb I Line, Z. Phys. A292: 307 (1979); R. Neugart, W. Klemt, C. Ekström, F. Buchinger, A.C. Mueller, K. Wendt, Nuclear Charge Radii and Moments ($\sim$N = 90) from Laser Spectroscopy in Neutron-Deficient Dy, Er, and Yb Isotopes, Oak Ridge, 1982, Abstracts, p. 25; (a) R. Neugart, Collinear Fast-Beam Laser Spectroscopy on Radioactive Isotopes in the Rare-Earth Region, Oak Ridge, 1982, CERN-EP/82-80, 18 June 1982; F. Buchinger, A.C. Mueller, B. Schinzler, K. Wendt, C. Ekström, W. Klempt, R. Neugart, Fast-beam laser spectroscopy on metastable atoms applied to neutron-deficient ytterbium isotopes, Nucl. Instr. & Meth. (1982, to be published).
21. (a) L. Wilets, Isotope Shifts, in Handbuch der Physik XXXVIII/1, edited by S. Flügge (Springer, Heidelberg, 1958), pp. 95, 96; D.N. Stacey, Isotope Shifts and Nuclear Charge Distributions, Rep. Progr. Phys. 29, Part I: 171 (1966); (b) S. Penselin, Recent Developments and Results of the Atomic-Beam Magnetic-Resonance Method, in Progress in Atomic Spectroscopy, Part A, edited by W. Hanle and H. Kleinpoppen (Plenum, New York, 1978), p. 463.
22. P. Jacquinot and R. Klapisch, Hyperfine spectroscopy of radioactive atoms, Rep. Progr. Phys. 42: 773 (1979).
23. O. Redi, Atomic spectra for nuclear studies: New techniques of hyperfine spectroscopy, Physics Today 34: 25, 26 (1981).
24. H. Rebel and G. Schatz, Laser induced fluorescence spectroscopy in atomic beams of radioactive nuclides, Oak Ridge, 1982.
25. O. Redi and H.H. Stroke, Nuclear structure and the atomic spectrum, Oak Ridge, 1982.
26. P. Brix, Altes und Neues über die Radien der Atomkerne, Max-

Planck-Institut für Kernphysik Heidelberg, Report MPI H - V3: (1982).

27. A.-M. Mårtensson and S. Salomonson, Specific Mass Shifts in Li and K Calculated Using Many-Body Perturbation Theory, J.Phys. B15, 2115 (1982).
28. H. Rinneberg, J. Neukammer, E. Matthias, Isotope shifts of perturbed 6sns $^1S_0$ and $^3S_1$ Rydberg states of odd Ba isotopes, Z. Phys. 306: 11 (1982).
29. V. Jaccarino, J.G. King, R.A. Satten, H.H. Stroke, Hyperfine Structure of ^{127}I. Nuclear Magnetic Octupole Moment, Phys. Rev. 94: 1798 (1954).
30. A. Bohr and B.R. Mottelson, Nuclear Structure II (Benjamin, Reading, MA, 1975), 45, 164,382.
31. J.M. Eisenberg and W. Greiner, Nuclear Models (North Holland, Amsterdam, 1970), 13,68.
32. K. Heilig and A. Steudel, Changes in Mean-Square Nuclear Charge Radii from Optical Isotope Shifts, At. Data & Nucl. Data Tables 14: 613 (1974).
33. E.C. Seltzer, K X-Ray Isotope Shifts, Phys. Rev. 188: 1916 (1969).
34. R. Engfer, H. Schneuwly, J.L. Vuilleumier, H.K. Walter, A. Zehnder, At. Data & Nucl. Data Tables 14: 509 (1974).
35. J. Bauche and R.-J. Champeau, Recent Progress in the Theory of Atomic Isotope Shift, Adv. Atom. Mol. Phys. 12: 39 (1976).
36. H.H. Stroke, Hyperfine Structure, in Atomic Physics 1, edited by B. Bederson, V.W. Cohen, F.M.J. Pichanick (Plenum, New York, 1969), p. 523.
37. H.J. Rosenberg, H.H. Stroke, Effect of a Diffuse Nuclear Charge Distribution on the Hyperfine-Structure Interaction, Phys. Rev. A5: 1992 (1972).
38. H.H. Stroke, R.J. Blin-Stoyle, V. Jaccarino, Configuration Mixing and the Effects of Distributed Nuclear Magnetization on Hyperfine Structure in Odd-A Nuclei, Phys. Rev. 123: 1326 (1961).
39. A. Arima, H. Horie, Configuration Mixing and Magnetic Moments of Nuclei, Progr. Th. Phys. (Kyoto) 11: 509 (1954); (a) R.J. Blin-Stoyle, Theories of Nuclear Moments, Revs. Modern Phys. 28: 75 (1958).
40. P.A. Moskowitz, M. Lombardi, Distribution of Nuclear Magnetization in Mercury Isotopes, Phys. Lett. 46B: 334 (1973).
41. T. Fujita, A. Arima, Magnetic Hyperfine Structure of Muonic and Electronic Atoms, Nucl. Phys. A254: 513 (1975).
42. H.H. Stroke, V. Jaccarino, D.S. Edmonds, Jr., R. Weiss, Magnetic Moments and Hyperfine-Structure Anomalies of ^{133}Cs, ^{134}Cs, ^{135}Cs, and ^{137}Cs, Phys. Rev. 105: 590 (1957).
43. H. Miyazawa, Role of Mesons in the Physics of Nuclear Moments, J. Phys. Soc. Japan Suppl. 34: S10 (1973).
44. F. Tomkins, S. Gerstenkorn, Déplacement isotopique relatif dans le spectre d'arc du plutonium, C.R. Acad. Sci. B265: 1311 (1967).

45. R. Neumann, Laser- und Mikrowellenspektroskopie an heliumartigen Lithium-Ionen, Habilitationsschrift, Ruprecht-Karls Universität, Heidelberg (1982), p. 92.

46. R.J. Hull, H.H. Stroke, Hyperfine-Structure Separations, Isotope Shifts, and Nuclear Magnetic Moments of the Radioactive Isotopes ^{199}Tl, ^{200}Tl, ^{201}Tl, ^{202}Tl, and ^{204}Tl, Phys. Rev. 51: 1203 (1961); D. Goorvitch, S.P. Davis, H. Kleiman, Isotope Shift and Hyperfine Structure of the Neutron-Deficient Thallium Isotopes, Phys. Rev. 188: 1897 (1969).

47. (a) A. Timmermann, High Resolution Two-Photon Spectroscopy of the $6p^2\ ^3P_o$ - $7p\ ^3P_o$ Transition in stable Lead Isotopes, Z. Phys. A296: 93 (1980); (b) R.C. Thompson, A. Hanser, K. Bekk, G. Meisel, D. Frölich, High Resolution Measurements of Stable Isotopes in Lead, Z. Phys. A305: 89 (1982); R. Wiggins, O. Redi, H.H. Stroke, R. Lis, R.A. Naumann, Isotope Shift and Hyperfine Structure of ^{205}Pb, Bull. Am. Phys. Soc. 21: 626 (1976); R. Wiggins, O. Redi, H.H. Stroke, HFS of 3 x 10^7 - Y ^{205}Pb by Level-Crossing Spectroscopy, Abstracts, Fifth International Conference on Atomic Physics, Berkeley, CA, 26-30 July 1976, p. 312; F. Moscatelli, R.L. Wiggins, O. Redi, H.H. Stroke, R.A. Naumann, Nuclear Spin of 3 x 10^7 - y ^{205}Pb, Bull. Am. Phys. Soc. 24: 17 (1979); O. Redi, F. Moscatelli, H.H. Stroke, R.L. Wiggins, R.A. Naumann, Isotope shift between 3 x 10^5 - y ^{202}Pb and ^{208}Pb, Bull. Am. Phys. Soc. 24: 624 (1979); F. Moscatelli, R.L. Wiggins, O. Redi, H.H. Stroke, Isotope Shift and Hyperfine Structure in the Optical Spectrum of 52-Hour ^{203}Pb, Bull. Am. Phys. Soc. 25: 493 (1980); (c) F.A. Moscatelli, O. Redi, P. Schönberger, H.H. Stroke, R.L. Wiggins, Isotope shift in the $6p^2\ ^3P_o$ - $6p7s\ ^3P_1^0$ 283.3-nm line of natural lead, J. Opt. Soc. Am. 72: 918 (1982); see also H. Rebel and G. Schatz, Ref. 24.

48. C.A. Mariño, G.F. Fülöp, W. Groner, P.A. Moskowitz, O. Redi, H. H. Stroke, Nuclear Magnetic Moments of ^{205,207,209}Bi Isotopes - HFS of the 15-Day ^{205}Bi 3067-Å Line, Phys. Rev. Lett. 34: 625 (1975).

49. H.-J. Kluge, H. Kremmling, H.A. Schuessler, J. Streib, K. Wallmeroth, Determination of the isotope shift in the D_1 line between ^{197}Au and ^{195}Au, Abstracts, 14th EGAS Conference, Liège, Belgium, 27-30 July 1982.

50. E. Hagberg, P.G. Hansen, P. Hornshøj, B. Jonson, S. Mattsson, P. Tidemand-Petersson, Staggering of the Moments of Inertia of very Neutron-Deficient Platinum Isotopes, Phys. Lett. 78B: 44 (1978).

51. W.D. Myers, Droplet model isotope shifts and the neutron skin, Phys. Lett. 30B: 451 (1969); Droplet Model of Atomic Nuclei (IFI/Plenum, New York, 1977).

52. W.J. Tomlinson, III, H.H. Stroke, Nuclear Isomer Shift in the Optical Spectrum of ^{195}Hg: Interpretation of the Odd-Even

Staggering Effect in Isotope Shift, Phys. Rev. Lett. 8: 436 (1962).

53. R.C. Barrett, Isotope and Isomer Shifts in Mercury, Nucl. Phys. 88: 128 (1966); Nuclear Charge Distributions, Rep. Progr. Phys. 37: 1 (1974).
54. H.H. Stroke, D. Proetel, H.-J. Kluge, Odd-even staggering in mercury isotope shifts: evidence for Coriolis effects in particle-core coupling, Phys. Lett. 82B: 204 (1979).
55. D. Proetel, D. Benson, Jr., A. Gizon, J. Gizon, M.R. Maier, R. M. Diamond, F.S. Stephens, Decoupled bands in odd-mass mercury isotopes, Nucl. Phys. A226: 237 (1974).
56. I. Ragnarsson, Some applications of the shell correction method, Nashville, 1979, p. 367; see also the systematic study - C. Ekström, Nuclear single-particle structure derived from spins and moments of long isotopic chains, CERN, 1981.
57. B.A. Brown, S.E. Massen, P.E. Hodgson, The charge distributions of the oxygen and calcium isotopes, Phys. Lett. 85B: 167 (1979).
58. H. Gerhardt, E. Matthias, H. Rinneberg, F. Schneider, A. Timmermann, R. Wenz, P.J. West, Changes in Nuclear Mean Square Charge Radii of Stable Krypton Isotopes, Z. Phys. A292: 7 (1979).
59. J.R. Beene, C.E. Bemis, Jr., J.P. Young, S.D. Kramer, Study of the Fission Isomer ^{240m}Am (S.F.) using Laser-Induced Nuclear Polarization, Hyperfine Inter. 9: 143 (1981).
60. (a) E.B. Shera, Electronic isotope shifts, muonic atoms, and electron scattering, Oak Ridge, 1982; H.J. Emrich, G. Fricke, M. Hoehn, K. Käaser, M. Mallot, H. Miska, B. Robert-Tissot, D. Rychel, L. Schaller, L. Schellenberg, H. Schneuwly, B. Shera, H.G. Sieberling, R. Steffen, H.D. Wohlfahrt, Y. Yamazaki, Systematics of radii and nuclear charge distributions from elastic electron scattering, muonic X-ray and optical isotope shift, CERN, 1981.
61. W. Kunold, M. Schneider, L.M. Simons, J. Wueest, Isotope shift measurements of muonic X-rays in 138, 136, 134 barium, (preprint, Schweizerisches Institut für Nuklearforschung, July 1982).
62. G.L. Borchert, O.W.B. Schult, J. Speth, P.G. Hansen, B. Jonson, H.L. Ravn, J.B. McGrory, Differences of the Mean Square Charge Radii of the Stable Lead Isotopes Observed through Electronic K X-Ray Shifts, submitted to Nuovo Cimento (1982).
63. J. Blaise, Etudes des Déplacements Isotopiques dans les Spectres Atomiques des Elements Lourds, Ann. Phys. (Paris) 3: 1019 (1958).
64. P.L. Lee, F. Boehm, X-Ray Isotope Shifts and Variations of Nuclear Charge Radii in Isotopes of Nd, Sm, Dy, Yb, and Pb, Phys. Rev. C8: 819 (1973).
65. See, for example, X. Campi, M. Epherre, Calculation of binding energies and isotope shifts in a long series of rubidium isotopes, Phys. Rev. C22: 2605 (1980).

66. A. Arima, F. Iachello, Interacting Boson Model of Collective States I. The Vibrational Limit, Ann. Phys. (N.Y.) 99: 253 (1976); Interacting Boson Model of Collective Nuclear States II. The Rotational Limit, id. 111: 201 (1978); O. Scholten, F. Iachello, A. Arima, id. III. The Transition from SU(5) to SU(3), id. 115: 325 (1978).
67. P.D. Duval, B.R. Barrett, Interacting boson approximation model of the tungsten isotopes, Phys. Rev. C23: 492 (1981).
68. S. Gerstenkorn, On the Isotope Shifts in the Vicinity of Closed Shell, Comm. At. Mol. Phys. 9: 1 (1979).
69. I. Angeli, M. Csatlós, Fine structure in the mass number dependence of rms charge radii, Nucl. Phys. A288: 480 (1977); I. Angeli, Shell effects in nuclear radii and binding energies, Phys. Lett. 82B: 313 (1979).
70. R. Wenz, E. Matthias, H. Rinneberg, F. Schneider, Further Evidence for a Linear Relationship between Changes in Nuclear Charge Radii and Binding Energies per Nucleon, Z. Phys. A295: 303 (1980).
71. I. Lindgren, J. Morrison, Atomic Many-Body Theory (Springer, Heidelberg, 1982).
72. (a) K. Niemax, L.R. Pendrill, Isotope shifts of individual nS and nD levels of atomic potassium, J. Phys. B13: L461 (1980). Also, C.J. Lorenzen, K. Niemax, Level isotope shifts of ^{6,7}Li, J. Phys. B15: L139 (1982); L.R. Pendrill, K. Niemax, Isotope shifts of energy levels in ^{40}K and ^{39}K, J. Phys. B15: L147 (1982); C.J. Lorenzen, K. Niemax, L.R. Pendrill, Isotope shifts of energy levels in the naturally abundant isotopes of the alkaline earth elements, submitted to Phys. Rev. (1982).
73. B.P. Stoicheff, E. Weinberger, Doppler-free two-photon absorption spectrum of rubidium, Can. J. Phys. 57: 2143 (1979).
74. P.E.G. Baird, S. Blundell, G. Burrows, C.J. Foot, G. Meisel, D.N. Stacey, G.K. Woodgate, Laser spectroscopy of the tin atoms, 8th ICAP contribution, 1982.
75. P. Grundevik, M. Gustavsson, G. Olsson, T. Olsson, Hyperfine-Structure and Isotope-Shift Measurements in the 6s5d ↔ 6p5d Transitions of Ba I in the Far-Red Spectral Region, preprint, 1982; R.E. Silverans, J-dependent Isotope Shifts in Ba Ions, (private communication, 1982).
76. C.W.P. Palmer, D.N. Stacey, Theory of anomalous isotope shifts in samarium, J. Phys. B15: 997 (1982).

NOTE ADDED IN PROOF

A theoretical accounting for the Gerstenkorn relation has been proposed recently (E.F. Hefter and I.A. Mitropolsky, The inverse mean field method and relative nuclear radii, preprint, Institut für Theoretische Physik, Universität Hannover, Appelstrasse 2, 3000 Hannover 1, FRG, April 1983).

HYPERFINE STRUCTURE AND ISOTOPE SHIFTS OF RYDBERG STATES IN ALKALINE EARTH ATOMS

E. Matthias, H. Rinneberg, R. Beigang, A. Timmermann, J. Neukammer and K. Lücke

Institut für Atom- und Festkörperphysik
Freie Universität Berlin
D-1000 Berlin West

INTRODUCTION

Rydberg states of atoms with two valence electrons exhibit three markedly different features compared to alkali systems. First of all, the coupling between the two valence electrons is most crucial for the properties of the Rydberg states and, in general, we have to distinguish between series with predominant singlet or triplet character. Secondly, low-lying states of doubly excited configurations cause strong perturbations of the Rydberg series belonging to one excited valence electron. This can lead to both singlet-triplet mixing and/or configuration mixing more or less local in energy. The third feature involves the interaction between the ms valence electron of the ion and the nucleus. The electrostatic part of it reveals itself through the field shift, while for nuclei with $I \neq 0$ the magnetic interaction causes a hyperfine splitting as well as a higher-order shift.

Much effort has been invested to obtain a complete understanding of the detailed structure of Rydberg states of alkaline-earth atoms. In parallel to the experimental mapping of states went a successful parametrization of the observed spectra by the Multichannel Quantum Defect Theory (MQDT). The paper by Cooke[1] presented at the last ICAP described the basic aspects of this approach. Specifically, the level structure was measured and analysed by MQDT for odd[2] and even parity states of Ca (Armstrong et al.[3]), Sr (Esherick[4]), and Ba (Aymar et al.[5]). Lifetimes[6-9], Stark shifts[6], and g-factor measurements[10,11] were also employed to probe state mixing and the type of coupling between the two valence electrons.

It is important to notice, however, that none of these quantities is sensitive to the relative phase of the admixed wavefunctions.

It was pointed out independently by Liao et al.[12] and Barbier and Champeau[13] that the hyperfine splitting of excited states of atoms with two valence electrons is another quantity very sensitive to state mixing. In fact, in ^{3}He the hyperfine interaction of the 1s ion core electron was found to be responsible for singlet-triplet mixing in 2 ^{3}P and 3 ^{3}D states[12]. By measuring $4f^{14}6snd$ Rydberg states of Yb I for $24 \leq n \leq 53$ Barbier and Champeau[13] observed a pronounced hyperfine structure for the odd isotopes Yb-171 and 173 which converges for large n to the hyperfine pattern of the ion. For even isotopes (I=0), on the other hand, they did not record any significant change of the isotope shift as a function of n.

Since then several groups seem to have become interested in this subject, and presently we are aware of hyperfine structure investigations of Rydberg states in Ba and Sr by the groups at Lund[14], Amsterdam[15], and Göteborg[16], further in He by the Stanford group[17], and of isotope shift measurements in alkaline-earth atoms by the group at Kiel[18]. We have measured both hyperfine splittings and isotope shifts of msns and msnd Rydberg states in Ca, Sr, and Ba. In the following some selected results will be presented and discussed in terms of state mixing.

The Rydberg states were excited by Doppler-free two-photon techniques. Coherent two-photon excitation was used predominantly for Ca and Sr and cascade excitation via the 6s6p $^1P_1$ resonance state in Ba. All measurements were carried out using frequency-stabilized cw dye lasers of about 1 MHz bandwidth. The absolute accuracy of the data is approximately 10 MHz. Coherent two-photon excitation was detected by means of a thermionic diode, while two-step excitation was monitored by the change in absorption of the $6s^2\ ^1S_0$ - 6s6p $^1P_1$ resonance transition when tuning through the upper transitions[19]. Space limitations prevent us here from going into the experimental details and the reader is referred to the original papers.

ISOTOPE SHIFTS FOR I=0

The finite charge density of s-electrons at the nucleus gives rise to an electrostatic interaction with the nuclear charge distribution, generating a field shift contribution which, in lowest order of r_N, is of the form[20]

$$\Delta E_{FS}^{AA'} = \frac{2\pi}{3} Z \Delta\rho_{el} \Delta\langle r_N^2\rangle^{AA'} \quad (1)$$

Here, $\Delta\rho_{el}$ is the change of the average relativistic electronic charge density across the nucleus for the optical transition and $\Delta\langle r_N^2\rangle^{AA'}$ is the change of the nuclear mean square charge radius when comparing isotopes with mass number A and A'. In case the nuclear spin I is zero, this electric monopole term constitutes the only contribution of the electromagnetic interaction between the electrons and the nucleus. Apart from a small relativistic share of $p_{1/2}$ electrons, the field shift originates from s-electrons and can consequently be used to detect any change in the total s-electron density throughout a Rydberg series. In alkaline-earth atoms this concerns the ms valence electron of the ion core. The electron density at the origin of ns Rydberg electrons decreases proportional to $(n^*)^{-3}$, where n^* is the effective quantum number, and will henceforth be neglected.

When going up in principal quantum number with one valence electron we basically expect two different effects to occur: (1) At low n values the ion core will contract due to reduced shielding when the outer electron is promoted to higher shells, causing an increase in total electron density at the nucleus with increasing n. (2) Once the region of screening effects is left and the ion core has settled to its final spatial extension, admixtures of doubly excited configurations like ndn'l cause a change of $\Delta\rho_{el}$, since in those states the ms valence electron density is missing. However, core polarization will partly counteract this effect and can be taken into account using appropriate experimental data[21].

To demonstrate this behavior and to exploit the field shift for probing local admixtures of doubly excited states, the isotope shifts of the 6sns $^1S_0$ and 6snd $^{1,3}D_2$ Rydberg series in Ba were studied[21]. The excitation scheme $6s^2\ ^1S_0 \xrightarrow{553.7\ nm} 6s6p\ ^1P_1 \xrightarrow{423-470\ nm} 6sns\ ^1S_0$ or 6snd $^{1,3}D_2$ was used, and in Fig.1 typical isotope shift spectra are shown for the even Ba isotopes. A comparison between the upper and lower spectra immediately reveals the drastic influence of the 5d7d $^3P_0$ and 5d7d $^3F_2$ perturbers, which strongly affect the 6s18s $^1S_0$ and 6s14d $^1D_2$ Rydberg states, respectively.

The change in electronic charge density between the Ba I ground state and the Rydberg states can be quantitatively evaluated as follows. The measured isotope shifts are composed of the shifts for the first and the second transition, the former being weighted by a factor ν_2/ν_1[22]. After subtracting the contribution of the first transition from the experimental data the shifts for the 6s6p $^1P_1 \rightarrow$ 6sns $^1S_0$ or 6snd $^{1,3}D_2$ transitions were obtained. Those were then combined in a King plot[22,23] with the isotope shifts of the 6s $^2S_{1/2} \rightarrow$ 6p $^2P_{1/2}$ resonance transition in Ba^+ to yield the ratios of the changes in electron densities $\Delta\rho_{el}(6snl-6s6p)/\Delta\rho_{el}(6p-6s)$. With the known[21] values $\Delta\rho_{el}(6p-6s)$ and $\Delta\rho_{el}(6s6p-6s^2)$ the change of the

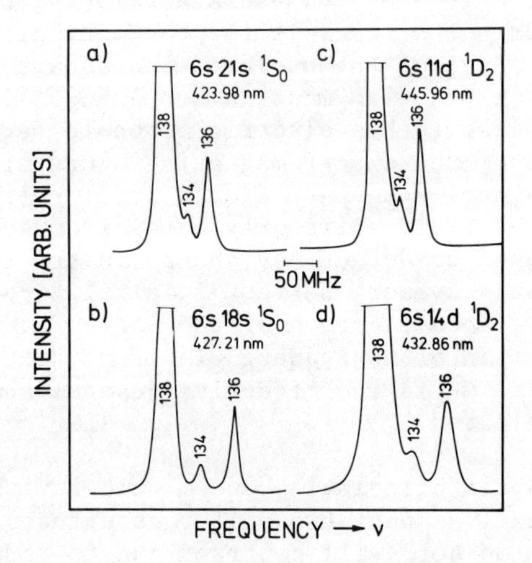

Fig. 1. Isotope shifts of the even Ba isotopes in the two-step excitation of the 6sns $^1S_0$ (a and b) and 6snd $^1D_2$ (c and d) Rydberg states. The upper two spectra belong to unperturbed, the lower ones to perturbed members of the corresponding series.

charge densities of the Rydberg states versus the ground state were derived. The results are displyed in Fig. 2.

In accordance with the expectation mentioned above three conspicuous features in Fig. 2 cath the eye. First of all, $\Delta\rho_{el}$ (6snl $-6s^2$) is negative due to the missing charge density of the Rydberg electron. Second, screening effects occur for $n \leq 10$; for larger n the total s-electron density is constant within the experimental uncertainty and equals the one of the ion. Thirdly, 5d7d $^3P_0$ and $^3F_2$ perturbers drastically change $\Delta\rho_{el}$ very local at n=14 and 18 for the 6snd $^1D_2$ and 6sns $^1S_0$ series, respectively. In fact, in both cases $\Delta\rho_{el}$ is decreased approximately to the value for the $6s^2$-6sp transition. From the deviation of $\Delta\rho_{el}$ compared to the saturation value for unperturbed states the amount of 5d7d admixture can be concluded. As a result, we find a 33 % admixture of 5d7d $^3P_0$ into the 6s18s $^1S_0$ and a 36 %[24] admixture of 5d7d $^3F_2$ into the 6s14d $^1D_2$ Rydberg states. It should be remembered, that $\Delta\rho_{el}$ only yields the amount of state mixing and carries no phase information.

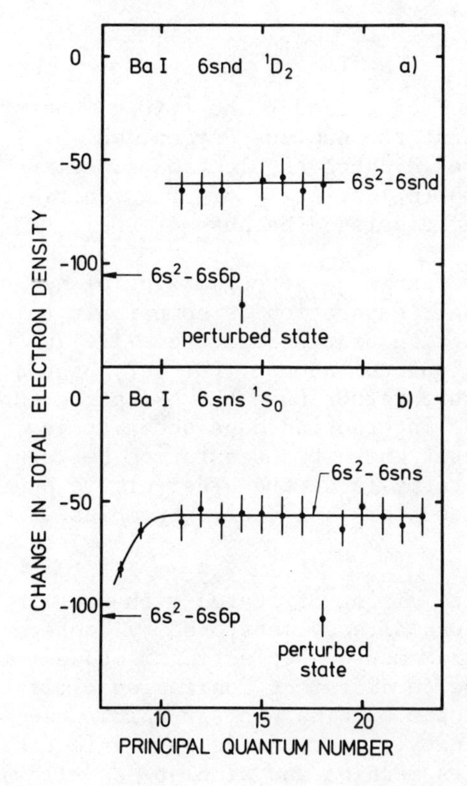

Fig. 2. Change in total charge density, $\Delta\rho_{el}(6snl-6s^2)$ in atomic units, for several states of the 6snd $^1D_2$ and 6sns $^1S_0$ Rydberg series in Ba I. The values for 6s8s and 6s9s were calculated from isotope shift data of Jitschin and Meisel[25].

SINGLET-TRIPLET MIXING INDUCED BY HYPERFINE INTERACTION

Whenever the nuclei possess a spin $I \neq 0$ we encounter a strong magnetic hyperfine interaction between the nuclear dipole moment and the ms valence electron of the ion core. This Fermi-contact interaction is of the form[26]

$$W_{hfs} = a_s \vec{I} \cdot \vec{s} \ , \tag{2}$$

where $a_s = \frac{16\pi}{3} \mu_B \mu_N g_I |\Psi(0)|^2_{ms}$.

Here, g_I is the nuclear g-factor and $|\Psi(0)|^2_{ms}$ describes the density of the ms electron at the nucleus. Again, we shall neglect contributions of the excited Rydberg ns-electron since its Fermi contact term decreases with $(n^*)^{-3}$. Also, orbital contributions of nl electrons can safely be dismissed because of their $\langle r^{-3} \rangle$ dependence.

In general, we expect a strong mixing of neighboring Rydberg states whenever their separation is comparable to or smaller than the hyperfine splitting of the ion core[12,13]. In this case only F will remain a good quantum number. For very high n (roughly n>100) this will be the rule rather than the exception. However, state mixing by hyperfine interaction does occur at lower n for equal F states in those cases where the separation between singlet-triplet or fine structure terms is of the order of the hyperfine coupling constant. In this section instructive examples of each will be discussed.

Singlet-triplet mixing of states with equal F, caused by hyperfine interaction, was already mentioned by Kopfermann[26] in §6 of his book on Nuclear Moments. Neglecting isotope shifts, hyperfine components converge to different ionization limits compared to the ones for even isotopes with no nuclear spin. When the excited electron moves to infinity, the relative spin orientation of the two electrons looses its meaning and singlet as well as triplet hyperfine components merge into the hyperfine doublet of the ms $^2S_{1/2}$ ion ground state. Barbier and Champeau[13], for example, observed such convergence of the hyperfine structure of the $4f^{14}6snd$ states for ^{173}Yb (I=5/2) and ^{171}Yb (I=1/2) up to the principal quantum number n=53. However, this effect is even more conspicuous in the msns $^1S_0$ Rydberg series, because in this case there is no hyperfine splitting and the total quantum number is F=I. This was done in Ca[27], Sr[27-30], and Ba[27,31], and the examples Sr and Ba will be presented in the following.

The 5sns $^1S_0$ series in Sr was excited by coherent two-photon transitions[28] in the range $10 \leq n \leq 70$ with corresponding energies between 43512 and 45907 cm^{-1}. A typical excitation spectrum is shown in Fig. 3. The dramatic shift of the line belonging to the odd isotope Sr-87 with respect to the even ones is evident. This behavior demonstrates that there exists a magnetic contribution to the isotope shift of atoms with nuclear spin $I \neq 0$, induced by the hyperfine interaction. Such effect was noticed as early as 1932 by Schüler and Jones[32] and Casimir[33] and was called second-order hyperfine correction[34]. However, when observing this shift in a Rydberg series up to high n, second order perturbation theory does no longer suffice.

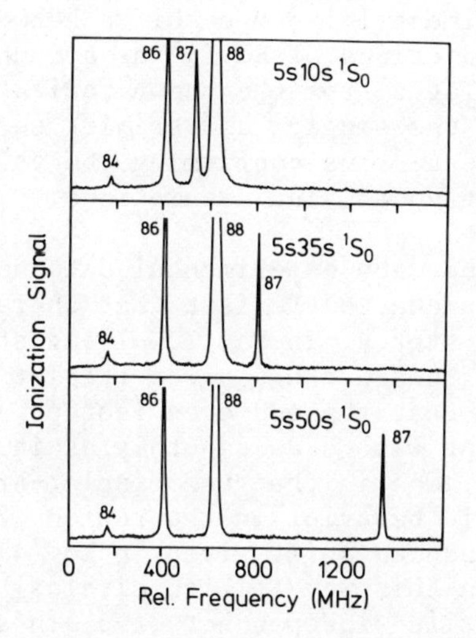

Fig. 3. Shift of the $5s^2$-5sns $^1S_0$ absorption line of the odd isotope Sr-87 for three selected n values.

The hyperfine coupling term in the Hamiltonian connects the F=I components of the $^1S_0$ and $^3S_1$ states, leading to off-diagonal terms of the form[34]

$$<{}^1S_0,F|a_s\vec{I}\cdot\vec{s}|{}^3S_1,F> = \frac{a_s}{2}\sqrt{I(I+1)} \quad . \tag{3}$$

Solving the secular equation results in the following energies of the hyperfine components of the singlet and triplet levels[27,29]:

$$\begin{aligned} E_{F=I+1} &= -\frac{\Delta E_{ST}}{2} + \frac{a_s}{2} I \\ E_{F=I-1} &= -\frac{\Delta E_{ST}}{2} - \frac{a_s}{2}(I+1) \\ E^{\pm}_{F=I} &= -\frac{a_s}{4} \pm \frac{1}{2}\left[a_s^2(I+1/2)^2 + a_s\,\Delta E_{ST} + (\Delta E_{ST})^2\right]^{1/2} \end{aligned} \tag{4}$$

Here, $\Delta E_{ST} = E(^1S_0)-E(^3S_1)$ is the singlet-triplet separation. The E^+ and E^- solutions correspond to the singlet and triplet solutions, respectively, for $a_s << \Delta E_{ST}$. The decrease of ΔE_{ST} with increasing n causes the pronounced n-dependence of the F=I hyperfine components. In Fig. 4 the experimental isotope shifts between Sr-87 and 86 are compared to the predictions of Eq.(4) as a function of n. Values for ΔE_{ST} were calculated from the known ionization energy and the quantum defects for the singlet and triplet series[4]. One can notice quantitative agreement, thus confirming the validity of this simple treatment and of the underlying assumptions.

The agreement between experimental data and the theoretical predictions in Sr is due to the fact that there are no additional admixtures of other states. In Ba, low-lying states of the 5d7d configuration do influence the singlet-triplet mixing[31], and from Fig. 5a one can see that the odd-even isotope shifts (open circles) are no longer varying with n as smoothly as in Sr. Here we encounter two effects. One is that the singlet-triplet separation ΔE_{ST} shows an erratic behavior in the region $17 \leq n \leq 21$. Subtracting the hyperfine-induced shift given in Eq.(4) corrects for it and the data adjusted in this way (filled circles) lie, with one exception, on a straight line independent of n, in accordance with the results of Fig. 2. The second effect is that the data for n=18 do not follow this trend. As was discussed in the previous section, the 6s18s $^1S_0$ state is not pure and carries 33 % admixture of the 5d7d $^3P_0$ state. The hyperfine interaction now mixes the states $|^1S_0, F=I\rangle$, $|^3S_1, F=I\rangle$, and $|^3P_0, F=I\rangle$. This can be treated completely

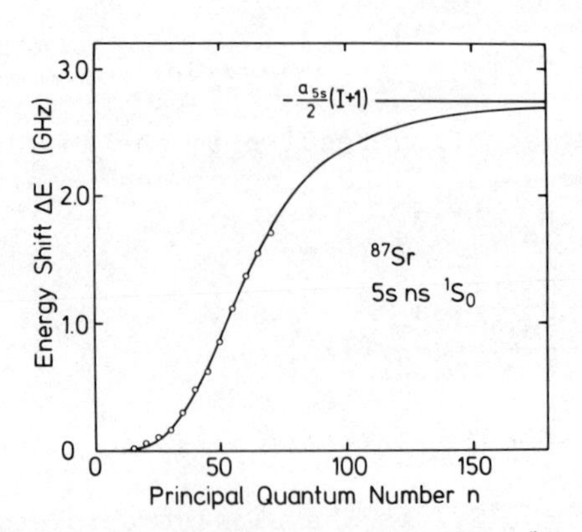

Fig. 4. Hyperfine-induced shift of the 5sns $^1S_0$ levels between the isotopes Sr-87 and 86. The solid line represents the theoretical prediction.

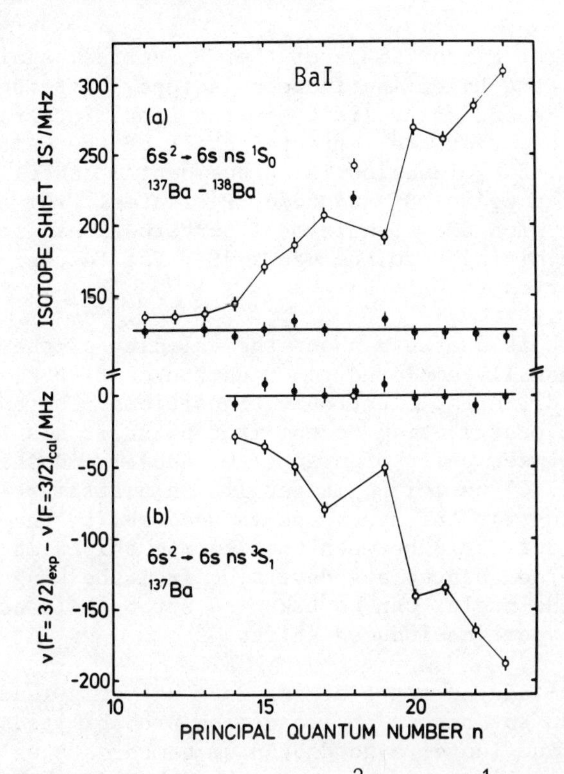

Fig. 5. (a) Isotope shifts of the $6s^2 \to 6sns\ ^1S_0$ transitions between Ba-137 and 138, corrected for the normal mass shifts. (b) Deviation of the F=3/2 component from the position predicted by the Landé interval rule. Open and filled circles represent experimental data and those corrected for the hyperfine-induced shifts, respectively.

analogous to what was said above for the two-state mixing and one finds relations for the energies of the three states involved[31]. These depend again on a_{6s} and ΔE_{ST} but in addition on the amount of admixture and the separation between the $6s18s\ ^1S_0$ state and the $5d7d\ ^3P_0$ perturber.

The repulsion between singlet and triplet F=I states is indicated by comparison between Fig. 5a and b. As can be seen from Eqs.(4), the singlet-triplet mixing causes a shift of the F=3/2 component from the position predicted by the Landé interval rule. Using the F=1/2 to 5/2 splitting these positions were calculated and their deviation from the actually measured ones is plotted in Fig. 5b as a function of n. The data for the $^3S_1$ series approximately represent the mirror image of the $^1S_0$ series. Again, after correcting for the hyperfine-induced isotope shift the data become independent of n and the validity of the Landé interval rule is restored. The fact that the corrected shift of the 6s18s $^3S_1$ state is zero finds its explanation in an accidental cancellation of the influence of the 6s18s $^1S_0$ and 5d7d $^3P_0$ states. From the data in Fig. 5 one can find additional local perturbations, for example the 5d7d $^1P_1$ state mixing into the 6s20s $^3S_1$. For further details the reader is referred to Ref. 31.

One remark is due here about the validity of the King plot procedure[23], usually employed to check the consistency of isotope shifts in two different electronic transitions. The hyperfine-induced magnetic contribution to the isotope shift has the consequence, that shifts between pairs of even (I=0) nuclei and shifts between even and odd ($I \neq 0$) nuclei no longer lie on a straight line. However, after correcting for the hyperfine-induced shift, the validity of the linear relationship between the isotope shifts in two transitions is restored. Hence, any deviation from the King plot of shifts that involve odd nuclei can be taken as strong evidence for the existence of hyperfine-induced shifts.

Hyperfine-induced state mixing does not only lead to level shifts but also to changes in transition probabilities. Due to the fact that J is no longer a good quantum number states can be excited which normally could not be reached by two-photon excitation. For example, states of the 5sns $^3S_1$ series in Sr with $50 \leq n \leq 85$ were populated by Doppler-free two-photon transitions from the $5s^2\ ^1S_0$ ground state[30], because of the added transition strength to the admixed 5sns $^1S_0$ states.

Both effects, level shifts and "forbidden" transitions have been studied[35] in case of the hyperfine spectrum of the 5s19d "$^1D_2$" state of Sr-87. Near n=19, there is a fortuitous crossing of the 5snd $^1D_2$ and 5snd $^3D_3$ series. The hyperfine splitting of the 5s ion core electron is at n=19 comparable to the separation of the two series and consequently all states of equal F are strongly mixed. This results in the curious behavior of the hyperfine spectrum as shown in Fig. 6. The spectra in the upper and lower part of the figure show, in addition to the isotope-shifted lines for Sr-84, 86, and 88, the normal hyperfine splitting of the 5snd $^1D_2$ states of Sr-87, corresponding to a nuclear spin I=9/2 (see also next section). The middle part of the picture displays the strongly perturbed spectrum for n=19. The centers of gravity of the 5s19d states belonging to the $^3D_3$ and $^1D_2$ series have approached each other to about 1.8 GHz. Both states now share the transition strength for two-photon transitions from the ground state and J is no longer a good quantum number. Note that this is only true for the odd isotope Sr-87, while for the even isotopes the 5s19d $^1D_2$ and $^3D_3$

states are strictly orthogonal. Due to the selection rule for two-photon excitation, $\Delta F=0, \pm 1, \pm 2$, only the states $5/2 \leq F \leq 13/2$ are populated. Each F-state has its counterpart in the structure belonging to the other series, resulting in a strong mutual repulsion. Consequently, the spectrum belonging to the $^1D_2$ series is pushed to the right and, in addition, compressed because the 5/2 states repell each other more strongly than the 13/2 ones. A simple calculation for each pair of F-levels, similar to the one mentioned above (Eqs.(4)), did reproduce the observed spectra for 5s19d to an accuracy of ± 10 MHz, assuming a $^1D_2$-$^3D_3$ separation of 1.78 GHz[35].

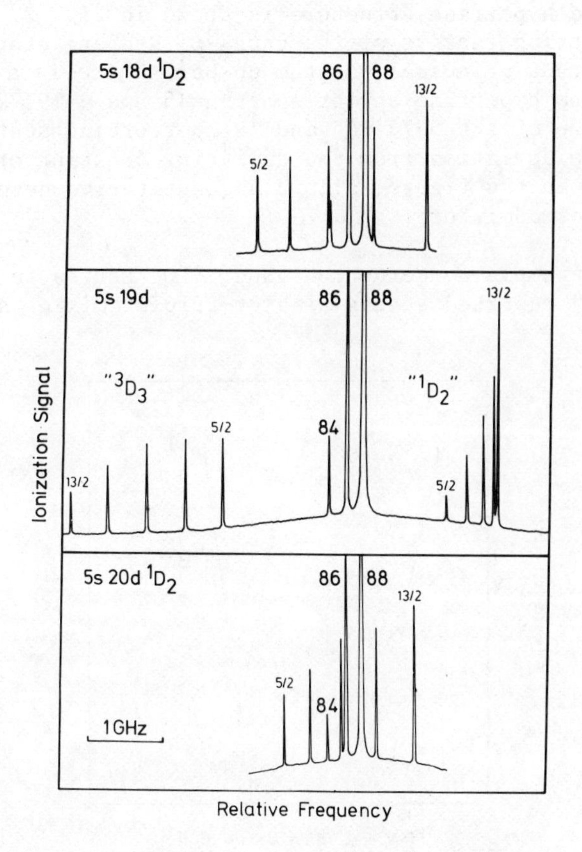

Fig. 6. Hyperfine structure of the 5snd $^1D_2$ Rydberg states of Sr-87 (I=9/2) between n=18 and 20. At n=19 the hyperfine interaction strongly mixes F-states belonging to the $^1D_2$ and $^3D_3$ series.

HYPERFINE STRUCTURE AS A PROBE FOR STATE MIXING

In contrast to the previous section we will now present a few cases where the hyperfine splitting merely serves as an aid for observing state mixing, which is now induced by configuration or spin-orbit interaction. Compared to the field shift discussed above the magnetic hyperfine structure is a much more sensitive probe. In addition, it provides the relative phase of the admixed wave functions[36] which cannot be derived easily from other experimental information. The reason for the sensitivity of the hyperfine structure against admixtures is that for these two-electron systems even highly excited Rydberg states remain predominantly LS-coupled, since both the spin-orbit interaction and the singlet-triplet separation decrease with $(n^*)^{-3}$. This coupling is crucially influenced by the admixture of other configurations. A striking example[37] that even small admixtures of doubly excited configurations alter significantly the observed hyperfine structure is shown in Fig. 7. Here, the hyperfine splitting factors of the 6sns $^3S_1$ Rydberg states of Ba are plotted versus principal quantum number. There is a significant reduction of the hyperfine splitting at n=15 and n=19, 20, caused by the admixture of the 5d7d $^3S_1$ and $^1P_1$ perturbing states, respectively. The deviation from the splitting constant of the free ion, indicated in the figure, allows a quantitative determination of the amount of admixture.

The next example concerns the 5snd $^{1,3}D_2$ series in Sr, for which Esherick[4] reported strong singlet-triplet mixing around n=16

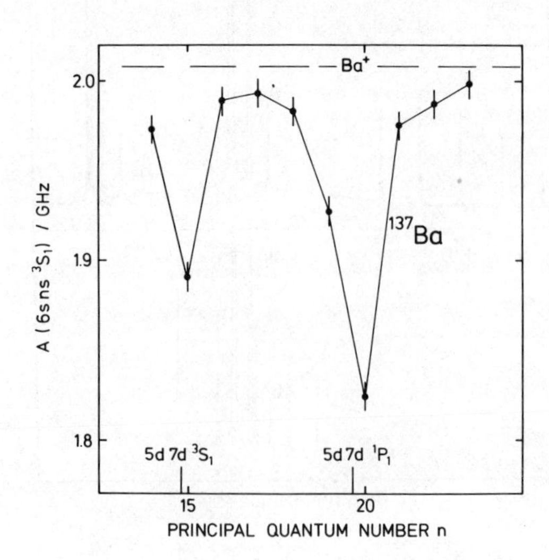

Fig. 7. Hyperfine splitting factor A of the 6sns $^3S_1$ Rydberg states in Ba-137. The splitting factor of the free ion is indicated by the dashed line.

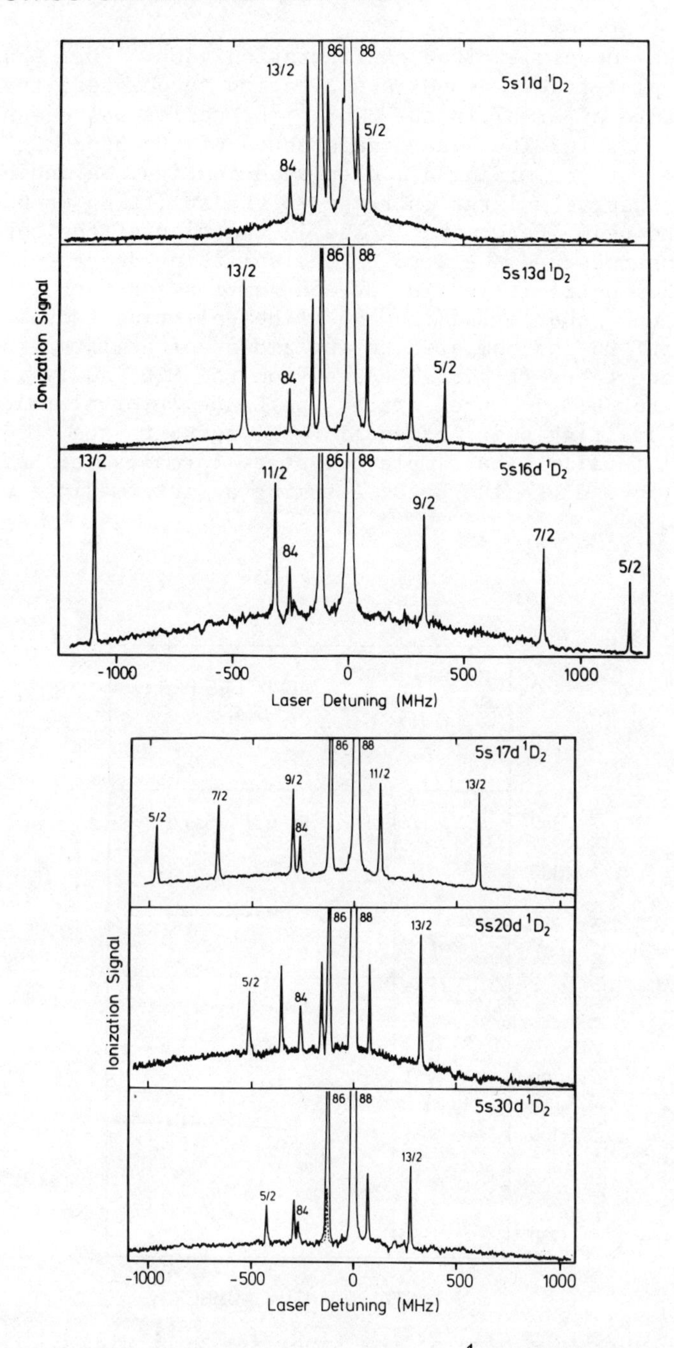

Fig. 8. Hyperfine structure of the 5snd $^1D_2$ series of Sr-87 (I=9/2) for selected n-values, demonstrating the singlet-triplet mixing around n=16.

caused by the doubly excited configuration 4d6s $^{1,3}D_2$. Again employing Doppler-free two-photon excitation techniques, the hyperfine structure of Sr-87 in the 5snd $^{1,3}D_2$ series was measured[38] between $9 \leq n \leq 70$. The hyperfine spectra of the 5snd $^1D_2$ states in the range of strong mixing are displayed in Figs. 8a and b. In Fig. 8a one can observe a large increase in the splitting when going from n=11 to n=16. Between n=16 and 17 the sign of the hyperfine structure changes,and the size of the splitting decreases again (see Fig. 8b) until it levels out and remains constant for $n \geq 30$. This resonance behavior of the hyperfine splitting factor is summarized in Fig. 9b and compared to the amount of admixture of singlet into triplet states (Fig. 9a) taken from the MQDT analysis of Esherick[4]. For all measured states the Landé interval rule was fulfilled. Treating pairs of equal F belonging to the $^1D_2$ and $^3D_2$ series individually, the simple two-state formalism of Ref.34 is sufficient to analyse the data. Assuming an intermediate coupling

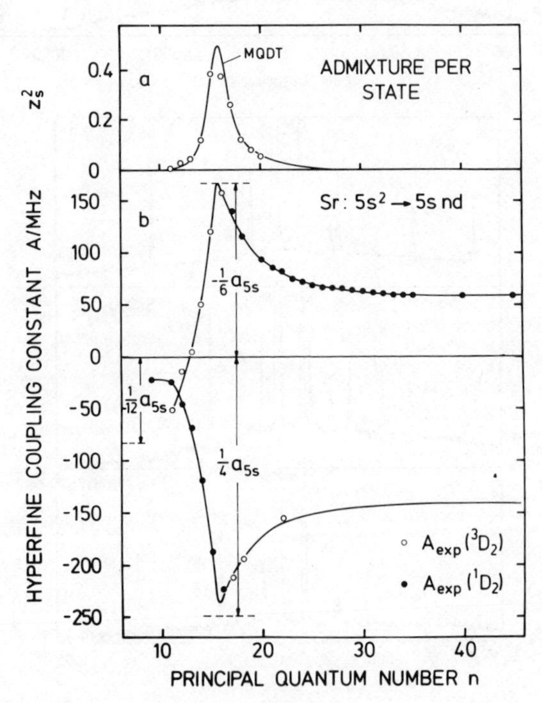

Fig. 9. Dependence on n of the hyperfine splitting of the 5snd $^{1,3}D_2$ states of Sr-87 (b). The solid line is a fit to the data, resulting in an admixture indicated by open circles in part (a) of the figure and compared to the prediction of MQDT[4].

scheme and neglecting any direct contribution of the excited nd electron, the data can be fitted (solid line in Fig. 9b) by a varying singlet-triplet admixture as indicated by the open circles in Fig. 9a. The hyperfine splitting of the 5s16d $^{1,3}D_2$ states represent very closely the value expected for pure $(jj)^2$ coupling, corresponding to about equal amounts of singlet and triplet contributions. However, for $25 < n < 50$ only about 2 % admixture of 5snd $^3D_2$ into the 5snd $^1D_2$ series are needed to explain the observed hyperfine structure.

In a systematic investigation of the hyperfine structure of the 6snd $^1D_2$ Rydberg series of Ba-135 and 137 (I=3/2) another dispersion-like resonance in singlet-triplet mixing, caused by configuration interaction, was observed. It was analysed, for the first time, to yield the relative phase of the admixed wave function. The resonance occurs in the vicinity of the 5d7d $^1D_2$ perturber located between the Rydberg states 6s26d and 6s27d $^1D_2$. To give an impression about the quality of the two-step excitation spectra in Ba a typical one for n=15 is shown in Fig. 10. Besides the hyperfine splitting of the final state, the hyperfine structure of the intermediate 6s6p $^1P_1$ state appears in the spectrum, reduced by a factor $(\nu_2-\nu_1)/\nu_1{}^{22}$. The signals for the odd isotopes 135 and 137 are labelled F_i-F_f according to the respective hyperfine components F_i and F_f of the intermediate and final states involved in the two-step excitation. Since the hyperfine structure of the intermediate

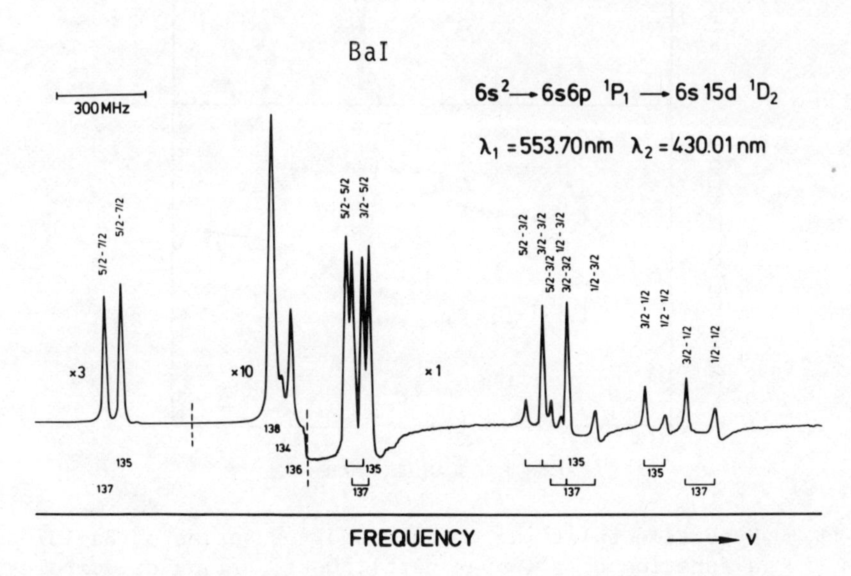

Fig. 10. Typical two-step excitation spectrum of the 6snd $^1D_2$ series in Ba for n=15. The labelling of the lines is explained in the text.

state is well known, this additional splitting can be corrected for quantitatively. In the upper part of Fig. 11 the variation with n of the hyperfine components E_f of the 6snd $^1D_2$ Rydberg states of Ba-137 is shown. The frequencies are measured with respect to Ba-138. It should be noted, that the isotope shifts of both transitions enter into the data displayed in the upper part of Fig. 11.

For the following discussion of the singlet-triplet mixing we express the wave function of the 6snd $^1D_2$ states as

$$|6snd\ ^1D_2> = \Lambda(^1D_2)\ \overline{|6snd\ ^1D_2>} + \Omega(^1D_2)\ \overline{|6snd\ ^3D_2>} + \varepsilon(^1D_2)\ \overline{|5d7d\ J=2>} \tag{5}$$

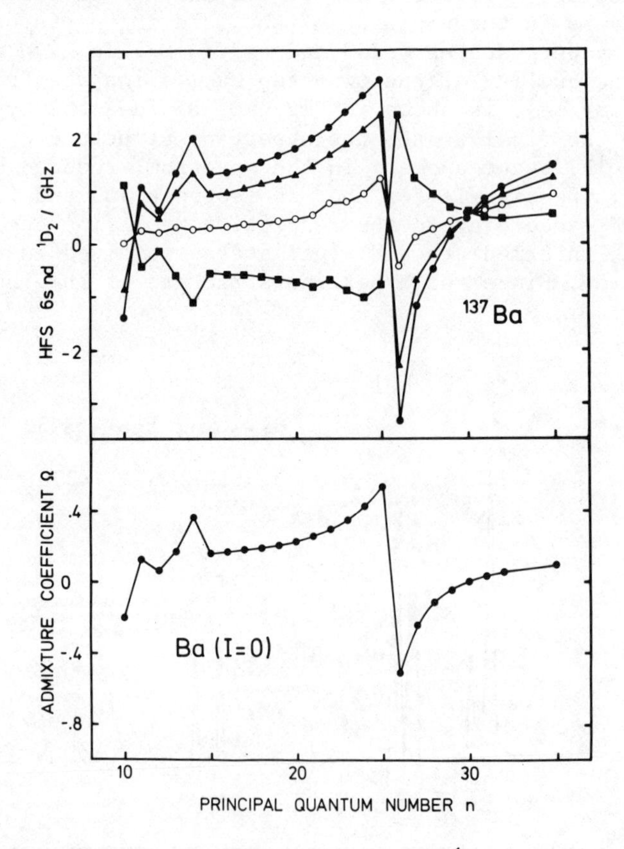

Fig. 11. Hyperfine splitting of the 6snd $^1D_2$ series of Ba-137 as a function of n (upper part). Dots, triangles, circles, and squares correspond to the F_f=1/2, 3/2, 5/2, and 7/2 components, respectively. The lower part of the figure shows the admixture coefficient derived from the data

with $|\varepsilon| = \sqrt{1-\Lambda^2-\Omega^2}$. The basis vectors, indicated by a horizontal bar, correspond to the exactly LS-coupled intermediate basis $|\bar{\alpha}\rangle$, conventionally used in MQDT analysis[24]. Using the wave function given in Eq.(5) the hyperfine splitting factor can be expressed in first order as

$$A(^1D_2) = \frac{a_{6s}}{12}\{\Omega^2 - 2\sqrt{6}\,\Omega\,\Lambda\} \tag{6}$$

Neglecting core polarization of a pure 5d7d configuration, the last term in Eq.(5) does not contribute to the observed splitting factor directly. By choosing arbitrarily the absolute phase of the wave function in such a way, that the largest admixture coefficient Λ enters with a positive sign, the singlet-triplet mixing parameter $\Omega(^1D_2)$ was derived. Its variation with n is shown in the lower part of Fig. 11. Starting from $\Omega(^1D_2) = 0.2$ at n=15 the $|^1D_2\rangle$ state vector rotates toward the jj-coupled basis state $|6snd_{5/2}\rangle$. Between n=25 and 26 a rapid drop in $\Omega(^1D_2)$ to negative values occurs. The state at n=26 corresponds to an approximately jj-coupled $|6snd_{3/2}\rangle$ vector. With increasing n the state vector approaches again its original composition, represented by $\Omega(^1D_2) \sim 0.2$. Whereas the configuration interaction of the 5d7d $^1D_2$ perturber with the 6snd $^1D_2$ Rydberg series causes the resonant variation of $\Omega(^1D_2)$, spin-orbit interaction leads to an almost constant singlet-triplet mixing, represented by the residual value $\Omega(^1D_2) \sim 0.2$. This was inferred from a three-channel quantum defect analysis of the 6snd $^{1,3}D_2$ series of Ba, carried out in the vicinity of the 5d7d $^1D_2$ perturber, using as channels the basis vectors appearing in Eq.(5). In this analysis the hyperfine structure was used as input data in addition to term values[39]. Good agreement between experimental and calculated singlet-triplet mixing parameters $\Omega(^1D_2)$ was achieved. The inclusion of the hyperfine structure data into the MQDT analysis allows the relative phases of the $U_{i\alpha}$ scattering matrix to be determined. A previous nine-channel quantum defect analysis[5], using term values only, did not reproduce the singlet-triplet mixing of the 6snd $^1D_2$ and $^3D_2$ series correctly[36].

Besides the total perturber fraction the three-channel QDT predicts the amplitudes of the $|\overline{6snd\ ^1D_2}\rangle$ and $|\overline{6snd\ ^3D_2}\rangle$ basis vectors for the perturber itself. These amplitudes were found to be at variance with results obtained from lifetime and Stark shift measurements. The hyperfine structure of the 5d7d $^1D_2$ state, shown in Fig. 12. is in excellent agreement with the three-channel QDT[39]. We conclude that the analysis of Stark shift and lifetime measurements have to be improved to yield the correct mixing amplitudes.

The resonance in singlet-triplet mixing observed for the 6snd $^1D_2$ series does occur for the 6snd $^3D_2$ series as well. However, the hyperfine structure observed for $10 \leq n \leq 28$ is strongly

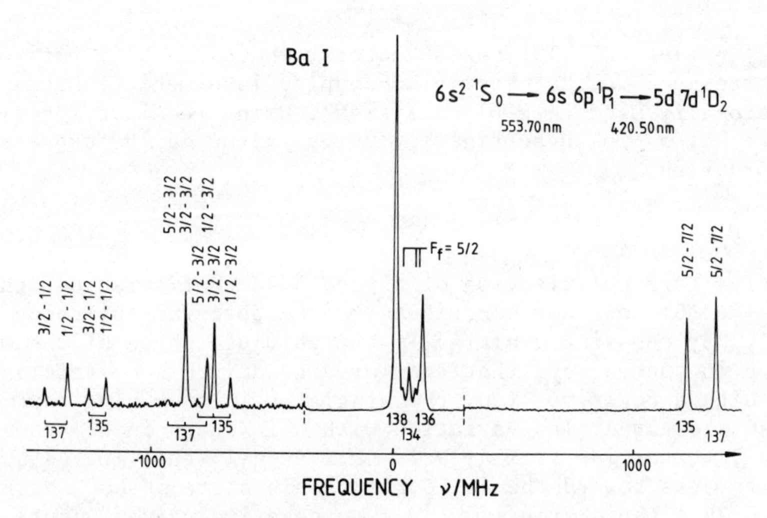

Fig. 12. Hyperfine structure in the two-step excitation spectrum of the 5d7d $^1D_2$ perturber state.

perturbed by close lying $^3D_1$ fine structure components for $19 \leq n \leq 25$. Although the resonant behavior of the singlet-triplet mixing is therefore no longer as conspicuous as for the 6s6d $^1D_2$ Rydberg series, a quantitative analysis allows to derive mixing amplitudes $\Lambda(^3D_2)$ and $\Omega(^3D_2)$, being in good agreement with those for the singlet series.

We will conclude this chapter by presenting an example of singlet-triplet mixing in the 4snd $^1D_2$ series of Ca. It also stands for the sensitivity which can be achieved by two-photon excitation in combination with thermionic diode detection. The only isotope with a nuclear spin is Ca-43 (I=7/2), the abundance of which is only 0.135 % in a natural sample. Fig. 13 shows the signal-to-noise ratio that was obtained for the hyperfine spectrum of the 4s12d $^1D_2$ Rydberg states. The adjacent lines belong to the even isotopes Ca-42 and 44 which have a natural abundance of 0.65 % and 2.09 %, respectively.

The observed dependence of the hyperfine structure on n is illustrated in Fig. 14a for the 4snd $^1D_2$ series up to n=25. The main feature of the figure is the sign change of the hyperfine splitting factor between n=8 and 9 and, again, between n=14 and 15. It corresponds directly to the sign changes of the singlet-triplet separation, $\{E(^1D_2)-E(^3D_2)\}$ (Fig. 14b), thus giving qualitative

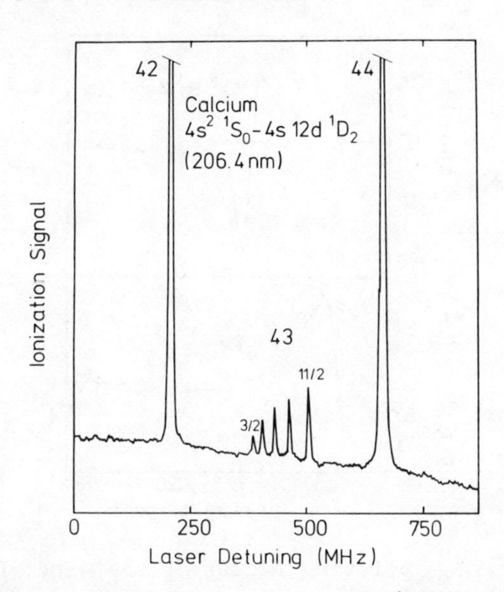

Fig. 13. Hyperfine spectrum of the 4s12d $^1D_2$ state in Ca-43 observed by two-photon excitation in combination with a thermionic diode.

evidence for singlet-triplet mixing. We notice, on the other hand, that the splitting factor A in Fig. 14a is rather small compared to the hyperfine coupling constant a_{4s} = -825 MHz which means that only a small triplet admixture is needed to account for the data. In fact, Armstrong et al.[3] in their MQDT analysis did not find it necessary to include the admixture of triplets into the 4snd $^1D_2$ series.

Presently, the data shown in Fig. 14a have not been finally analysed. However, we will briefly mention the main points that should be considered. The observed sign changes in Fig. 14 are caused by singlet-triplet mixing. The small size of the hyperfine splitting, however, makes it susceptible to the influence of the perturbing states 3d5s $^1D_2$, located between n = 8 and 9, and $3d^2$ $^1D_2$, close to n = 15. Hence a simple analysis in the intermediate coupling scheme[34], using singlet-triplet mixing coefficients derived from fine structure data, does not suffice for the region below n=15. Including the admixtures of the perturbers into the 4snd $^1D_2$ series by means of the mixing coefficients given in Ref.3 does considerably improve the agreement between calculated and measured hyper-

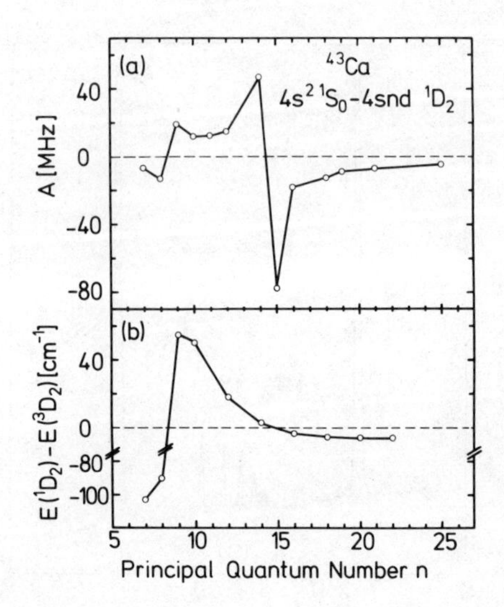

Fig. 14. Hyperfine splitting factor of the 4snd $^1D_2$ series in Ca-43 (part a) and separation of 4snd $^1D_2$ and $^3D_2$ states (part b) as a function of n.

fine splittings. Still it cannot reproduce the observed splitting at n=9 and we suspect that also the 3d5s $^3D_2$ must be taken into account to reach quantitative agreement for all n[40]. A more complete analysis along these lines is under way.

This work was supported by the Deutsche Forschungsgemeinschaft, Sonderforschungsbereich 161.

REFERENCES

1. W. E. Cooke, "Two Electron Rydberg States", p. 167 in Atomic Physics 7, D. Kleppner and F. M. Pipkin ed., (Plenum Press, N.Y. 1981)
2. J. A. Armstrong, J.J. Wynne, and P. Esherick, J. Opt. Soc. Am. 69, 211 (1979)
3. J. A. Armstrong, P. Esherick, and J.J. Wynne, Phys. Rev. A15, 180 (1977)
4. P. Esherick, Phys. Rev. A15, 1920 (1977)
5. M. Aymar, P. Camus, M. Dieulin, and C. Morillon, Phys. Rev. A18, 2173 (1978); M. Aymar and O. Robaux, J. Phys. B: At. Mol. Phys. 12, 531 (1979)

6. T. F. Gallagher, W. Sandner, and K. A. Safinya, Phys. Rev. A23, 2969 (1981)
7. K. Bhatia, P. Grafström, C. Levinson, H. Lundberg, L. Nilsson, and S. Svanberg, Z. Physik A303, 1 (1981)
8. M. Aymar, R.-J. Champeau, C. Delsart,and J.-C. Keller, J. Phys. B: At. Mol. Phys. 14, 4489 (1981)
9. M. Aymar, P. Grafström, C. Levinson, H. Lundberg, and S. Svanberg, J. Phys. B: At. Mol. Phys. 15, 877 (1982); and P. Grafström, Jiang Zhan-Kui, G. Jönsson, C. Levinson, H. Lundberg and S. Svanberg, Phys. Rev. A, in press.
10. J. J. Wynne, J. A. Armstrong, and P. Esherick, Phys. Rev. Letters 39, 1520 (1977)
11. P. Grafström, C. Levinson, H. Lundberg, S. Svanberg, P. Grundevik, L. Nilsson, and M. Aymar, Z. Physik 308, 95 (1982).
12. P. F. Liao, R. R. Freeman, R. Panock, and L. M. Humphrey, Opt. Comm. 34, 195 (1980)
13. L. Barbier and R.-J. Champeau, J. Physique 41, 947 (1980)
14. P. Grafström, Jiang Zhan-Kui, G. Jönsson, S. Kröll, C. Levinson, H. Lundberg, and S. Svanberg, Z. Physik, in press
15. E. R. Eliel and W. Hogervorst, contribution to this conference
16. P. Grundevik, H. Lundberg, L. Nilsson, and G. Olsson, Z. Physik A306, in press; and contribution to this conference
17. T. W. Hänsch and H. Gerhardt, private communication
18. C.-J. Lorenzen, K. Niemax, and L.R. Pendrill, contribution to this conference
19. J. Neukammer and H. Rinneberg, J. Phys. B: At. Mol. Phys. Sept. 1982
20. I. Lindgren, private communication
21. J. Neukammer, E. Matthias, and H. Rinneberg, Phys. Rev. A25, 2426 (1982)
22. C. Delsart and J.-C. Keller, Opt. Comm. 15, 91 (1975)
23. For the King plot procedure see, e.g., K. Heilig and A. Steudel, At. Data Nucl. Data Tables 14, 613 (1974)
24. J. Neukammer and H. Rinneberg, J. Phys. B: At. Mol. Phys., in press
25. W. Jitschin and G. Meisel, Z. Physik A295, 37 (1980)
26. H. Kopfermann, "Nuclear Moments", Academic Press Inc., New York (1958)
27. R. Beigang and A. Timmermann, Phys. Rev. A25, 1496 (1982)
28. R. Beigang, E. Matthias, and A. Timmermann, Z. Physik A301, 93 (1981)
29. H. Rinneberg, Z. Physik A302, 363 (1981)
30. R. Beigang and A. Timmermann, Phys. Rev. A26, in press
31. H. Rinneberg, J. Neukammer, and E. Matthias, Z. Physik A306, 11 (1982)
32. H. Schüler and E.G. Jones, Z. Physik 77, 801 (1932)
33. H. Casimir, Z. Physik 77, 811 (1932)
34. A. Lurio, M. Mandel, and R. Novick, Phys. Rev. 126, 1758 (1962)

35. R. Beigang, D. Schmidt, and A. Timmermann, J. Phys. B: At. Mol. Phys. 15, L201 (1982)
36. H. Rinneberg and J. Neukammer, Phys. Rev. Letters 49, 124 (1982)
37. J. Neukammer and H. Rinneberg, J. Phys. B.: At. Mol. Phys. 15, L425 (1982).
38. R. Beigang, E. Matthias, and A. Timmermann, Phys. Rev. Letters 47, 326 (1981) and (E) Phys. Rev. Letters 48, 290 (1982)
39. H. Rinneberg and J. Neukammer, to be published
40. K. Lücke, Diplomarbeit 1982, Freie Universität Berlin

CONCLUDING REMARKS

Arthur L. Schawlow

Department of Physics
Stanford University
Stanford, California 94305

In his opening remarks, Professor Siegbahn raised the question "What is atomic physics?" For that question, one can only give the answer that is given when someone asks about what is physics. Atomic physics is what atomic physicists do, and it is very well described by the kind of reports presented here. What a wonderfully rich and diverse field it is! This is all the more remarkable considering the length of time that it has been reported as dead, first by the nuclear physicists and then by the particle physicists who also included nuclear physics among the dead fields. Atomic physics goes on, and keeps producing surprises.

We have some new tools. About fifteen years ago, Henry Crosswhite came to my office and told me "There is a great new use for lasers." "You're joking," I replied. "There can't possibly be any use for lasers. They're useless by definition." "Yes there is," he said. "It's called holography. Now all they have to do is find a use for holography." Well, there are uses for holography, including the making of excellent diffraction gratings for spectroscopy. Lasers also have been improved so that they are even useful for atomic physics, especially since they are now tunable over a wide range of wavelengths.

There are other new and powerful tools, such as computers, both for controlling experimental apparatus and logging data, and for carrying out complex theoretical calculations. Accelerators can provide multiply charged and highly excited ions, or even produce rare isotopic species by nuclear reactions. Ion traps can hold a few charged atoms for study, and they can make up for some of the deficiencies of lasers. For instance, we thought years ago about observing the two-photon Doppler-free spectrum of positronium. All we needed was a laser at 4860 Angstroms, capable of giving a few hundred kilowatts of continuous-wave power. But no such laser exists! Now with the ion trap, as Mills and Chu have shown, the ions can be gathered up as they are emitted randomly, and held until the pulsed laser is ready to fire. Alternatively, the spectrum can be observed if the positrons can be obtained in bursts from a pulsed laser, as in the Yale-Livermore-Stanford collaboration.

These positron investigations and related experiments on the various atoms involving muons are really opening up some exciting new fields. It will be possible to study in detail some purely leptonic systems, free from the complications of nuclear structure. Will we find anything simple? It should be, since we don't know anything that is theoretically any simpler. But as the experiments improve, the quantum electrodynamics people have to grind out more and more terms in their computer formulation of the theory. This is an entirely different sort of complexity from that of, say, a heavy atom or a molecule, which is a complicated structure of simple constituents. For the leptonic atoms, we can't help wondering if there isn't something simpler, and whether we are really looking at things in the right way. Do we have all the depth of understanding that we can hope for? If the pre-Copernican astronomers had access to twentieth century computers, it might have been possible to use many more epicycles, and thereby give a very precise account of planetary motions. Would we want to abandon a theory like quantum electrodynamics because the complexity of calculations seems excessive? To be fair, quantum electrodynamics does give a systematic procedure to identify the needed terms, numerous as they are. Clearly it will not be superseded until either some of its predictions come into conflict with experimental data, or until someone finds a simpler theory of comparable accuracy.

Dirac, in his recent Lindau lecture, objected to quantum electrodynamics on other grounds. He described it as just a prescription for calculating, rather than a real theory like Heisenberg's matrix quantum mechanics. He believes that a better theory might somehow get away from using a classical Hamiltonian in quantum mechanics. Some other kind of Hamiltonian might be found, possibly incorporating new degrees of freedom, which would inherently get rid of the infinities that now have to be subtracted rather arbitrarily in quantum electrodynamics. That is a hint as to where we might progress. At any rate this problem is going to be of continued interest because the powerful new tools will permit increasingly refined measurements. We will be able to measure some lines of simple atoms thousands of times more accurately than we can now, and the theoretical calculations seem destined to grow ever more complicated, unless some new approach is found.

Of course, one of the ways physics progresses is by looking for simple models of the phenomena, even though they may only account for the behavior under a limited range of circumstances. In a real sense, waves and particles are metaphors for both matter and light. George Gamow once told about how, in the 1920s, the Soviet theoretical physicist Frenkel wrote a popular article to inform the Russian public about the advances in atomic physics. In it, he used the then popular scientific joke that "On Mondays, Wednesdays and Fridays we talk about electrons as particles. On Tuesdays, Thursdays and Saturdays, we talk about them as waves." This separation of complementary aspects was not understood by some officials, and was condemned. Frenkel might have been in serious trouble but for the intervention of Joffe, but he had to go for a time into exile in a cold place, namely Minnesota. According to Gamow, that is how it happened that Frenkel's books on quantum mechanics were written at the University of Minnesota.

Physical metaphors, such as the dual concepts of particles and waves in dealing with the light and atoms, are more than just conveniences, but rather are practical necessities. These dual descriptions serve us very well. Certainly, in the case of light, when dealing with lasers it is very convenient to not think of photons, but rather to treat the light inside the laser as a classical electromagnetic wave. If you do use a photon description, the concept of phase is not at all natural, and is difficult to take into account. Willis Lamb once was so frustrated by misguided attempts to use photon concepts in discussing lasers that he said that you should require a license to use the word photon. He also said that he should give the licenses, because he would give very few of them. Nearly always in laser action light quanta are so numerous that their discreteness can be ignored. In other matters, such as the photoelectric effect, we often need consider only the quantization of the light energy.

But the experiments on Bell's inequalities are making it difficult for us to continue using some of our familiar physical metaphors in the old ways. We are used to thinking that light waves are produced at an atom with definite polarizations and are subsequently detected by remote detectors. However, the experiments show that if anything is propagated, it seems to convey more polarization information than a transverse wave. Conceivably, one might find a new metaphor for the light, more complex than the wave picture. But perhaps there is no escape from considering the emission of light waves and measuring of their polarizations as one single event, even though that surely makes life more complicated. As Julian Schwinger pointed out some years ago, while the universe can only be comprehended as one interacting whole, the only route we know to that understanding is by studying the parts. Clearly quantum mechanics imposes some severe limitations on the ways that we can reasonably divide systems into parts for separate study.

I have heard that Niels Bohr once said something like "If quantum mechanics doesn't frighten you a little, you haven't understood it." In the light of these experiments, it is difficult to preserve any kind of a realistic picture. As an experimentalist, I like to think that there is something there that we call an atom, and that we can make good measurements on it if we are careful not to disturb it too much. But the experiments on polarization of correlated photons don't bear out these expectations. The measurement process seems to require a fully quantum mechanical treatment of the atom and measuring apparatus together.

Let me turn now to the subject of precision measurements. My esteemed colleague at Stanford, Felix Bloch, has made, over the years, some interesting and profound remarks. He once told me that "If you want to make a more precise measurement, it is not enough to be careful. You have to be clever." At this conference, we have heard some wonderfully ingenious techniques, such as the work on quantized Hall effect, the measurements on muonic atoms and other spectroscopic methods. There are some quite amazing advances. We have heard that the velocity of light, c, has been defined officially, because the old krypton length standard was not as good as could be realized by frequency measurements extending up to the optical region. Then the meter is just the distance traveled by light in 1/c seconds. While we can no longer measure the velocity of light, we can still measure the old prototype meter bars to see if their length changes with time. Or we can check gravitationally determined astronomical distances, such as the distance to the moon, to see whether they exhibit any secular variations. Thus we can still do the same sorts of experiments, but we don't call them measuring the velocity of light.

There have been tremendous advances in sensitivity, to a large extent made possible by the combination of lasers and computers. They enable us to study very small numbers of atoms or molecules. Indeed, in the positronium experiments, the average number of atoms present in the apparatus is much less than one. The experiments have become possible because the small numbers of atoms have been made available in bunches. We can now also observe states in atoms or molecules with very small populations, or extremely rare isotopes produced in small numbers by accelerators or even by neutrino interactions.

We are continuing to learn new techniques of high-resolution laser spectroscopy, each with its own special area of usefulness. Saturation, polarization, polinex and intermodulated fluorescence methods use counterpropagating laser beams to eliminate Doppler broadening by selecting those molecules with small velocity components along a particular direction. For hydrogen atoms, thermal motions can be reduced by cooling them to low temperatures without condensation, in a suitably coated container. Traps can confine ions so that they remain in one region for a long time, thereby minimizing both Doppler and transit-time broadening. Remarkably, it is also possible to eliminate Doppler broadening by accelerating ions to a high velocity with a constant electric field.

I am not at all qualified to discuss the theoretical papers at this conference. Nevertheless, I have been much impressed by the great progress. It is the task of physics not only to explain the positronium and hydrogen atoms, but to explain the universe as it exists. This certainly includes all of the atoms and their various ionized species. If physicists had never ventured beyond hydrogen, we would still lack such an important concept as exchange. It is impressive that the calculations of ionic inner shell energy levels are accurate enough that they have to include not only the electron self energy but also the nuclear size effects. We are getting increasing understanding of what happens in collisions. We have new experimental tools to study collisions, in the combinations of lasers with atomic beams, where we benefit also from continued improvements in vacuum technology. The collisional calculations are indeed complex, but great ingenuity is leading to impressive advances.

There has been some remarkable progress in clearing up old problems. It is now possible to understand several cases where reasonable theories had predicted unbound states, while experiment shows them to be bound.

One might say that we come to meetings like this one, to find out what has been going on, and to find out everything that is known about the various branches of our science. But researchers know that to discover new things, we never have to know everything that is known about a subject. All we have to do is recognize one thing that is not known. The speakers have been very good in telling us where their part of the field has reached, and what they think is not yet known. Of course, inevitably members of the audience who come with different backgrounds, will recognize other gaps in existing knowledge. They will see things that they can and want to do to fill in those gaps. So we return to our laboratories refreshed and inspired, to see what we can do with what we have learned.

INDEX